REINFORCED CONCRETE
Second Edition

REINFORCED CONCRETE DESIGN

Second Edition

S.U. Pillai
Principal
Regional Engineering College
Calicut, India
[Formerly Visiting Professor of
 Civil Engineering
Royal Military College of Canada]

D.W. Kirk
Professor of Civil Engineering
Dean, Canadian Force Military College
Royal Military College of Canada

McGraw-Hill Ryerson Limited

Toronto Montreal New York Auckland Bogotá Cairo Hamburg Lisbon London Madrid
Mexico Milan New Delhi Panama Paris San Juan São Paulo Singapore Sydney Tokyo

REINFORCED CONCRETE DESIGN 2/E

ISBN 0-07-549469-8

67890D9876543

Printed and bound in Canada

Canadian Cataloguing in Publication Data

Pillai, S.U.
Reinforced concrete design

2nd ed.
First ed. (1983) published under title: Reinforced concrete design in Canada.
Includes bibliographical references and index.
ISBN 0-07-549469-8

1. Reinforced concrete.   2. Reinforced concrete construction — Canada.   I. Kirk, D.W. (Donald Wayne), date-
II. Title.   III. Title: Reinforced concrete design in Canada.

TA683.2.P54 1988   624.1'8341   C87-094179-8

COVER PHOTO CREDIT

McGraw-Hill Ryerson Ltd. would like to thank the following companies and individuals for their cooperation in obtaining the cover photograph:

Stadium Corporation of Ontario Limited

The Robbie/Adjeleian/NORR Consortium

Ellis-Don Limited, Toronto

Special thanks is expressed to C. Michael Allen, P.Eng. for his assistance in obtaining the photograph of the SkyDome.

REINFORCED CONCRETE DESIGN
Second Edition

CONTENTS

PREFACE

Consequent to the adoption of the SI system of weights and measures in Canada, the Canadian Codes and Standards have been converted to the SI units. The SI versions of the major codes related to the Design of Reinforced Concrete Structures, namely, the Canadian Standards Association Standards CAN3-A23.1, CAN3-A23.2, and CAN3-A23.3, were published in 1977. Until that time, Canadian universities used American textbooks, since both the Code provisions and the system of units used in both countries were essentially the same. With the introduction of SI units in Canada, there was a definite need for a textbook on Reinforced Concrete Design written with reference to the Canadian Codes, and in SI units. It is in this context that the first edition of this book was conceived and prepared in 1983.

The second edition of this text has been prepared primarily to reflect the changes that have been incorporated in the relevant specification, CSA Standard A23.3-M84 — Design of Concrete Structures for Buildings. The 1984 edition of CSA A23.3 has revised the safety provision by adopting the load and resistance factors specified in the National Building Code of Canada and used by the corresponding specifications for Steel Structures for Building — CSA S16.1-M84. This fundamental change has necessitated, apart from the associated changes in the theoretical discussions, re-working of all the examples in the text. The revised Standard A23.3-M84 has also introduced an alternative method of design for shear and torsion. This is presented in the text in the new Chapter 8.

Presently, the *Concrete Design Handbook*, published by the Canadian Portland Cement Association, is extensively used in Canadian universities. This handbook contains numerous design aids, all of which cannot be provided in a general textbook, and is a desirable supplement to any textbook on Reinforced Concrete Design.

The present book has eighteen chapters and deals with all aspects of Reinforced Concrete Design generally taught at the Baccalaureate level. Prestressed Concrete Design, however, has not been included, as the authors felt that, it being a major subject in itself, a one-chapter introduction on it may be too restrictive.

Appendix A contains helpful suggestions for maintaining systematic bookkeeping, and, for convenience, all design aids contained in the textbook proper are listed in Appendix B.

The authors thank the Canadian Portland Cement Association, the Canadian Standards Association, and the Reinforcing Steel Institute of Ontario for giving permission to reproduce some of their published materials.

S.U. Pillai
D.W. Kirk

CHAPTER 1

INTRODUCTION AND MATERIAL PROPERTIES

1.1 INTRODUCTION

The following sections provide a brief introduction to the materials in, and the properties of, concrete. For a detailed study of concrete technology, reference should be made to any of the standard works on the subject (Refs. 1.1, 1.2, 1.3).

When cement, aggregates (sand, gravel, crushed rock) and water are mixed together in the proper proportions and allowed to harden, a solid rocklike mass is obtained which is called *concrete*. The cement and water in the paste react chemically (*hydration*) and the paste hardens binding the aggregates. The process of hydration continues as long as the concrete remains saturated, the temperature remains favourable and any unhydrated cement is present. To aid hydration, the freshly placed concrete is cured by keeping it continually moist and under controlled temperature until the desired quality is attained.

The gain in strength of concrete is very rapid in the first few days, after which the rate of gain decreases steadily as shown in Fig. 1.1. Concrete is generally designed

Fig. 1.1 *Variation of compressive strength with age (Ref. 1.3)*

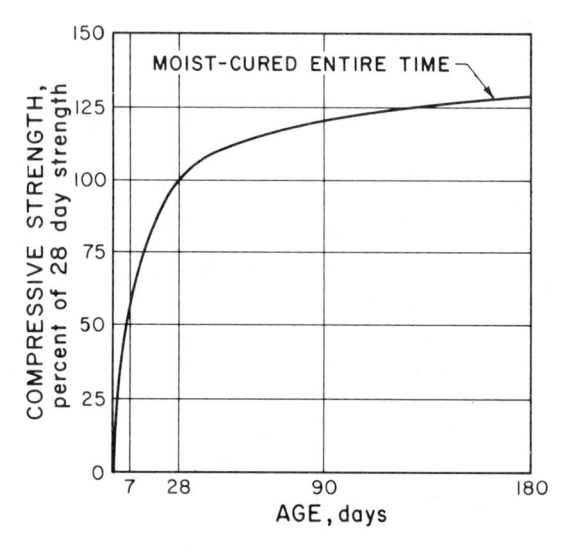

on the basis of its strength at 28 days, by which time it has acquired most of its strength. Normal concrete under proper curing acquires about 75 percent of its design strength (28 day strength) by the end of two weeks. This strength is usually adequate to carry the dead loads and construction loads so that the formwork can be removed about this time. Considerable variations in the properties of concrete can be achieved by variations in mix proportions, type of cement used, type of aggregates, methods of curing, and by the use of special *admixtures*. Plain concrete has a high compressive strength but a very low tensile strength. Hence when areas of a concrete structural member are subjected to tensile stresses (either due to direct tensile forces or indirectly due to flexural tension, diagonal tension, temperature changes or shrinkage and volume changes) it is *reinforced* with steel bars embedded in these locations while casting the member. The resulting composite is termed *reinforced concrete*.

1.2 CEMENT AND WATER-CEMENT RATIO

Hydraulic cements, most notably *portland cements*, which set and harden by chemically reacting with water are used to make structural concrete. CSA Standard A5-M83 provides for five types of portland cements as follows:

Type 10, Normal
Type 20, Moderate
Type 30, High-early-strength
Type 40, Low-heat-of-hydration
Type 50, Sulphate-resisting

Type 10 Normal cement is used in general concrete construction where the special properties of the other types are not required. Other types of special cements such as air-entraining portland cement, white and coloured cements are also available.

The ratio, by mass, of water to cement in the mix is termed the *water–cement ratio* (w/c), and is the major factor which controls the strength and many other properties of the concrete. The minimum w/c ratio required in a concrete mix to ensure complete hydration of the cement is about 0.35 to 0.40 (even though the purely chemical requirement is somewhat lower). However, higher w/c ratios are usually used in order to improve the *workability*; that is, to make the mixture more plastic and workable so that it can be placed in the formwork and compacted without difficulty. A reduction in the water content will result in increased compressive strength, bond strength, and resistance to weathering, and lower porosity, creep, absorption, and volume changes in the hardened concrete.

1.3 AGGREGATES

The aggregates make up 60 to 80 percent of the total volume of concrete. Aggregates are usually divided into *fine aggregate*, consisting of sand with particle sizes up to 10 mm and *coarse aggregate* composed of gravel, crushed rock or other similar materials. Coarse aggregate has particle sizes in excess of 1.25 mm, but usually the bulk of it has sizes in excess of 5 mm. Commonly used aggregates such as sand, gravel and crushed stone produce *normal density concrete* with a mass density of about 2150 to 2500 (average 2320) kg/m³. For design computations, normal density concrete, reinforced with steel, is commonly assumed to have a mass density of 2400 kg/m³. Both the size and grading of the aggregates are important considerations in proportioning the mix. These aggregates should meet the requirements of CSA Standard A23.1-M77 — *Concrete Materials and Methods of Concrete Construction.*

Occasionally, structural low density aggregates are used to produce low density concrete for special applications. Such aggregates consist of expanded shale, clay, slate, and slag and the resulting concrete has a density of about 1350 to 1850 kg/m³ and compressive strengths comparable to those of normal density concrete. Requirements to be met by structural low density aggregates are given in ASTM Standard C330 — *Lightweight Aggregate for Structural Concrete.* Low density concrete having a 28 day compressive strength in excess of 15 MPa and an air dry density not exceeding 1850 kg/m³ is termed *Structural Low Density Concrete,* and low density concrete having a 28 day compressive strength in excess of 15 MPa and an air dry density between 1850 and 2150 kg/m³ is termed *Structural Semi-low Density Concrete.* Similarly, high density aggregates such as limonite, barite, ilmenite, magnetite, haematite, iron (shots and punchings) and steel slugs are used occasionally to produce high density concrete for special applications such as radiation shielding and counterweighting.

1.4 DESIGN AND CONTROL OF CONCRETE MIXES

For given materials, the quality of the hardened concrete is most significantly influenced by the w/c ratio and the proportion of the ingredients. Figure 1.2 shows typical variation of compressive strength with w/c ratio for normal concrete. Maximum permissible w/c ratios for different types of structures and exposure conditions are recommended in CSA Standard A23.1-M77, section 14.

The mix is designed on the basis of the desired mean compressive strength, workability (slump), availability of materials, and economy. CSA Standard A23.1-M77, section 14, recommends ranges of slump to be used for different types of construction. The concrete mix proportion is usually expressed as the ratio, by volume, or more accurately by mass, of cement to fine aggregate to coarse aggregate, in that order (for example, 1:2:4) together with a w/c ratio.

Concrete is normally specified by the required *compressive strength at 28 days,* f'_c, as determined by tests on standard specimens 150 mm diameter by 300 mm long. A concrete is considered satisfactory (CSA Standard A23.1-M77) if the averages of all sets of three consecutive strength tests equal or exceed the specified strength, with no individual test more than 3.5 MPa below the specified strength. Thus the specified *compressive strength of concrete,* f'_c, is the strength that will be equalled or exceeded by the average of all sets of three consecutive strength tests, with no individual test result falling more than 3.5 MPa below f'_c. Results of standard strength tests on cylinders cast from concrete of identical proportions show inevitable scatter,

Fig. 1.2 *Typical variation of 28 day compressive strength with water–cement ratio*

the amount of scatter depending on the degree of control. To account for these variations in strength test results, it is necessary to proportion the concrete mix for a *mean strength* higher than the specified strength, f_c'. The minimum magnitude of this overallowance is determined from a statistical analysis of test results. Methods and procedures for proportioning of concrete mixes are not included in this book. Several such methods are given in detail in Refs. 1.1–1.6.

In addition to cement, aggregates, and water, other materials termed collectively as *admixtures* are sometimes added to the mix in order to impart special properties to the concrete. For instance, admixtures are often used to entrain air, to improve workability, and to accelerate or retard setting and rate of strength gain. Air-entraining admixtures, which entrain microscopic air bubbles into the concrete, are widely used particularly in cold environments as they improve the resistance of concrete to freezing and thawing and to surface scaling caused by deicing agents. CSA Standards A266.1, A266.2 and A266.6 give specifications for admixtures. Guidelines for the use of admixtures are contained in CSA Standard A266.4 and the ACI Committee 212 Report (Ref. 1.7).

Proper mixing, placing, compacting, and curing are all essential for the production of quality concrete. A very concise and pertinent presentation of all the major aspects relevant to the manufacture of good quality concrete, including examples of mix design, with reference to the requirements of Canadian Standards, is given in Ref. 1.3.

1.5 COMPRESSIVE STRENGTH AND STRESS–STRAIN CURVE

The compressive strength of concrete is one of its most important and useful properties. Structural concrete is primarily used to withstand compressive stresses. Many important properties which influence the strength and deformation of concrete structures, such as tensile strength, shear strength, and elastic modulus can be approximately correlated to the compressive strength and thereby are normally determined using the compressive strength. However, it should be noted that all such correlations are not precise but only approximate or empirical in nature and are generally applicable to normal concrete with nominal compressive strength in the range of 20 to 45 MPa.

The determination of compressive strength is very simple and is generally done by the standard compression test (CSA Test Method A23.2-9C) on 150 mm by 300 mm cylinders at 28 days. The standard cylinder has a length-to-diameter ratio of two. For cylinders with this ratio appreciably less than two, ASTM Standard C42 provides the correction factors given in Table 1.1. The size of the cylinders also influences the strength test results. A variation of compression test results with the cylinder diameter expressed as a percentage of the strength of the standard 150 mm by 300 mm cylinder is shown in Fig. 1.3 (Refs. 1.8, 1.9).

Table 1.1

Ratio of length of cylinder to diameter	Strength correction factor
2.00	1.00
1.75	0.98
1.50	0.96
1.25	0.93
1.00	0.87

In usual structural applications the concrete strength specified is in the range of 20 to 35 MPa for reinforced concrete and 35 to 45 MPa for prestressed concrete. Typical stress–strain curves for normal concrete in this range, obtained from standard compression tests, are shown in Fig. 1.4. The initial part of the curve, up to a stress level equal to about one-half of the maximum, is nearly linear. As the stress reaches about 70 to 90 percent of the maximum, minute internal cracks parallel to the direction of the applied load are initiated in the mortar throughout the concrete mass (Ref. 1.10). The lateral tensile strain (due to the Poisson's ratio effect) at this loading may exceed the failure strain of concrete in tension and may be the cause of this cracking. As a result of the associated larger lateral extensions, the apparent Poisson's ratio starts to increase sharply (Ref. 1.11). The maximum stress, f'_c, is reached at a strain of approximately 0.002, and beyond this an increase in strain is accompanied by a decrease in stress. This is believed to be the result of extensive

Fig. 1.3 *Effect of diameter of cylinder on compressive strength (Ref. 1.9)*

microcracking in the mortar. With most testing machines the concrete cylinder fails abruptly soon after the maximum load is attained. However, the descending branch of the curve can be obtained by using a suitably stiff testing machine or special techniques (Ref. 1.12). For the usual range of concrete strengths, the strain at failure is in the range of 0.003 to 0.0045. The higher the strength of concrete, the less the failure strain. For design purposes the CSA Standard A23.3-M84 (10.2.3) limits the usable strain at the extreme concrete compression fibre to a maximum of 0.003.

Fig. 1.4 *Typical stress–strain curves for concrete in compression*

Careful tests (Ref. 1.12) have confirmed that the stress–strain relation for the compression zone of a flexural member is nearly identical to that obtained for a standard cylinder test.

A knowledge of the *modulus of elasticity* of concrete is necessary for all computations of deformations as well as for design of sections by the working stress design procedure. The term *Young's modulus of elasticity* has relevance only in the linear elastic part of a stress–strain curve. For a material like concrete whose stress–strain curve is nonlinear throughout the range, an elastic modulus may be defined in any of the four ways shown in Fig. 1.5. Amongst these the secant modulus at a stress of about one-half the compressive strength would represent approximately the average modulus of elasticity in compression throughout the working stress range. ASTM Standard C469 describes a test for determining the static chord modulus between a lower point A (Fig. 1.5) corresponding to a strain of 50×10^{-6} and an upper point D corresponding to a stress equal to 40 percent of the concrete strength. The lower point is set far enough from the origin to be free of the influence of any initial errors in the setup.

Fig. 1.5 *Various definitions of static modulus of elasticity*

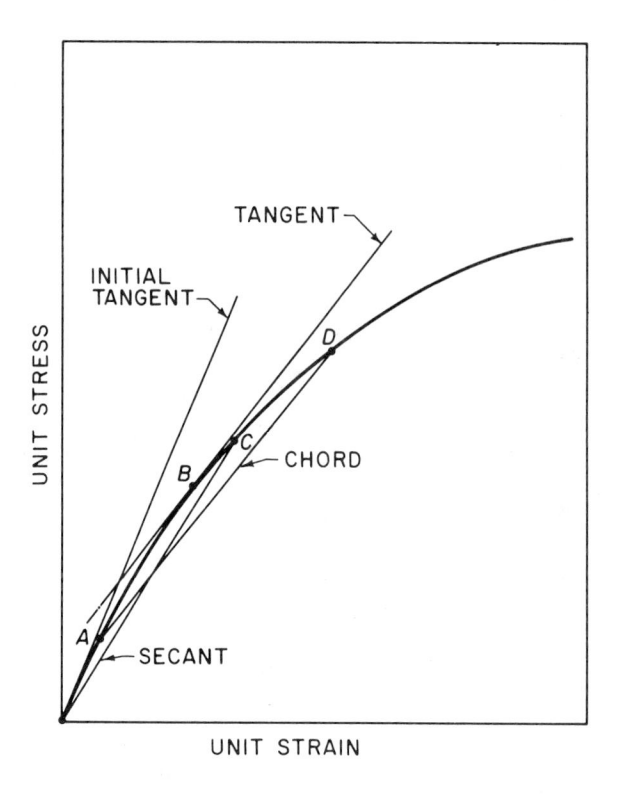

CSA Standard A23.3-M84 (8.5.1) gives an approximate expression for the modulus of elasticity of concrete, in terms of its strength and mass density as:

$$E_c = \gamma_c^{1.5}\, 0.043\, \sqrt{f'_c} \qquad\qquad (1.1)$$

where E_c is the modulus of elasticity in MPa, γ_c is the mass density of concrete in kg/m^3, and f'_c is the specified compressive strength in MPa. This empirical equation was determined from test results on concrete with γ_c in the range of 1500 to 2500 kg/m^3 and represents the secant modulus at a stress of approximately 0.5 f'_c. For normal density concrete (γ_c = 2400 kg/m^3), E_c may be taken as 5000 $\sqrt{f'_c}$.

Poisson's ratio, which is the ratio of the transverse strain to the axial strain under uniform axial stress, is generally in the range of 0.15 to 0.20 for concrete and is usually taken as 0.20 for design. Observed variations of axial, transverse and volumetric strains in a concrete prism tested to failure in uniaxial compression are shown in Fig. 1.6 (Ref. 1.11). At a stress equal to about 80 percent of the compressive strength, there is a point of inflection in the volumetric strain curve. As the stress is increased beyond this point the rate of volume reduction decreases and, soon after, the volume begins to increase. It is believed that this inflection point coincides with the initiation of major microcracking in the concrete leading to large lateral extensions. Poisson's ratio appears to be essentially constant for stresses below the inflection point. At higher stresses, the apparent Poisson's ratio begins to increase sharply.

Fig. 1.6 *Strains in concrete prism under uniaxial compression (Ref. 1.11)*

1.6 BEHAVIOUR UNDER TENSION

Concrete is not normally designed to resist direct tension. However, tensile stresses do develop in concrete members as a result of flexure, shrinkage, and temperature changes. Principal tensile stresses may also result from multiaxial states of stress. Cracking in concrete is a result of tensile failures. A knowledge of the tensile strength of concrete is necessary for the control of cracking.

As has been pointed out, concrete is very weak in tension, the direct tensile strength usually being only about 7 to 15 percent of the compressive strength (Ref. 1.8). A direct tension test on a concrete specimen where a purely axial tensile force must be applied, free of any misalignments and secondary stresses at the grips, is difficult to perform. Hence the tensile strength of concrete is generally determined by indirect tension tests, most commonly by *flexure tests* or *cylinder splitting tests*.

In the flexure test most commonly employed (CSA Test Method A23.2-8C), a standard, plain concrete, simple beam of square or rectangular cross-section is loaded to failure with third-point loading. Assuming a linear stress distribution across the cross-section, the theoretical maximum tensile stress reached in the extreme fibre, termed the modulus of rupture, f_r, is computed with the flexure formula as

$$f_r = M/Z \qquad\qquad (1.2)$$

where M is the maximum bending moment and Z is the section modulus. The actual stress distribution is not linear and the modulus of rupture so computed is greater than the direct tensile strength, usually by about 60 to 100 percent (Ref. 1.8). However, the modulus of rupture, f_r, will be the appropriate limiting tensile strength to be used in the computation of the cracking moment, M_{cr}, of a beam using the flexure formula $M = fZ$ as the same assumptions are involved in calculations for both f_r and M_{cr}.

In the splitting test (CSA Test Method A23.2-13C), a standard cylinder of the same type as used for the compression test is loaded in compression on its side along a diametral plane until failure by splitting along the loaded plane (Fig. 1.7). In an elastic homogeneous cylinder this loading produces a nearly uniform tensile stress across the loaded plane as shown in Fig. 1.7c. The splitting tensile strength, f_{ct}, may be computed as:

$$f_{ct} = \frac{2P}{\pi dL} \qquad\qquad (1.3)$$

where P is the maximum applied load, d is the diameter and L the length of the cylinder. The splitting test is the easiest to perform and gives more uniform results than the other tension tests. Wright (Ref. 1.13) has shown that for normal density concrete, the splitting strength is about 1.5 times the direct tensile strength and about two-thirds of the modulus of rupture. The splitting tensile strength, f_{ct}, for

Fig. 1.7 *Cylinder splitting test for tensile strength*

normal density concrete varies between 0.5 to 0.58 times $\sqrt{f_c'}$ and as an average is taken as:

$$f_{ct} = 0.56 \sqrt{f_c'} \text{ MPa} \tag{1.4}$$

The failure strain of concrete in uniaxial tension is in the range of 0.0001 to 0.0002 (average 0.00015) and the stress–strain curve in tension can be approximated as a straight line up to failure. The modulus of elasticity in tension does not appear to differ from its value in compression. Since the tensile strength of concrete is very low and often ignored in design, the tensile stress–strain relation is of little practical value.

The tensile strength of concrete is found to be approximately proportional to $\sqrt{f_c'}$. On this basis, tensile strength is often taken as a constant times $\sqrt{f_c'}$ for design purposes. Thus CSA Standard A23.3-M84 takes the modulus of rupture, f_r, of concrete as:

$$f_r = 0.6 \lambda \sqrt{f_c'} \text{ MPa} \tag{1.5}$$

where f_c' is in MPa and λ is a factor to account for the density of concrete. Recommended values for factor λ are:

   $\lambda = 1.00$ for normal density concrete;
   $\lambda = 0.85$ for structural semi-low density concrete; and
   $\lambda = 0.75$ for structural low density concrete.

1.7 SHEAR STRENGTH

The strength of concrete in pure shear is of little practical relevance in design. The case of pure shear rarely exists in actual structures. Further, a state of pure shear is accompanied by principal tensile stresses of equal magnitude on a diagonal plane and since the tensile strength of concrete is far less than its shear strength, failure invariably occurs in tension. This, incidentally, makes it difficult to experimentally determine the resistance of concrete to pure shearing stresses. A reliable assessment of the shear strength of concrete can be obtained only from tests under combined stresses. On the basis of such studies, the strength of concrete in pure shear has been reported as approximately 20 percent of the compressive strength (Ref. 1.9). In normal design practice, the shear stress in concrete has to be limited to values well below the shear strength in order to control the associated principal tensile (diagonal tension) stresses and cracking.

1.8 COMBINED STRESSES

Structural members are usually subjected to various combinations of axial force, bending moments, transverse shear forces and twisting moments. Concrete is therefore subjected to combinations of normal (tensile or compressive) and shearing stresses. Such a general three-dimensional state of stress acting on an element can be transformed into an equivalent set of three normal stresses (principal stresses) acting in three orthogonal directions. When one of these three principal stresses is zero, the state of stress is termed *biaxial*. The failure strength of materials under combined stresses is normally defined by appropriate failure criteria. As yet, there is no universally accepted criterion for failure of concrete.

Concrete subjected to a biaxial state of stress has been studied extensively due to its relative simplicity in comparison with the triaxial case and because of its common occurrence in flexural members, plates and thin shells. Figure 1.8 shows the general shape of the biaxial strength envelopes for concrete, obtained experimentally (Refs. 1.11, 1.14), along with proposed approximations. In general, the strength of concrete in biaxial compression is greater than in uniaxial compression by up to 27 percent. The biaxial tensile strength is nearly equal to its uniaxial tensile strength. However, in the region of combined compression and tension, the compressive strength decreases nearly linearly with an increase in the accompanying tensile stress. The failure strains under biaxial states of stress depend on the nature of the stress states — tensile or compressive. Failure itself occurs generally by tensile splitting along a plane at right angles to the maximum tensile strain. For biaxial compression, as in the case of uniaxial compression, the volumetric strain begins to increase as the peak stress is approached. This is again due to the progressive increase of microcracking in the mortar. Observed failure modes suggest that tensile strains are of vital importance in the failure criteria and failure mechanism of concrete for both uniaxial and biaxial states of stress (Ref. 1.14).

Fig. 1.8 *Failure stress envelope — biaxial stress*

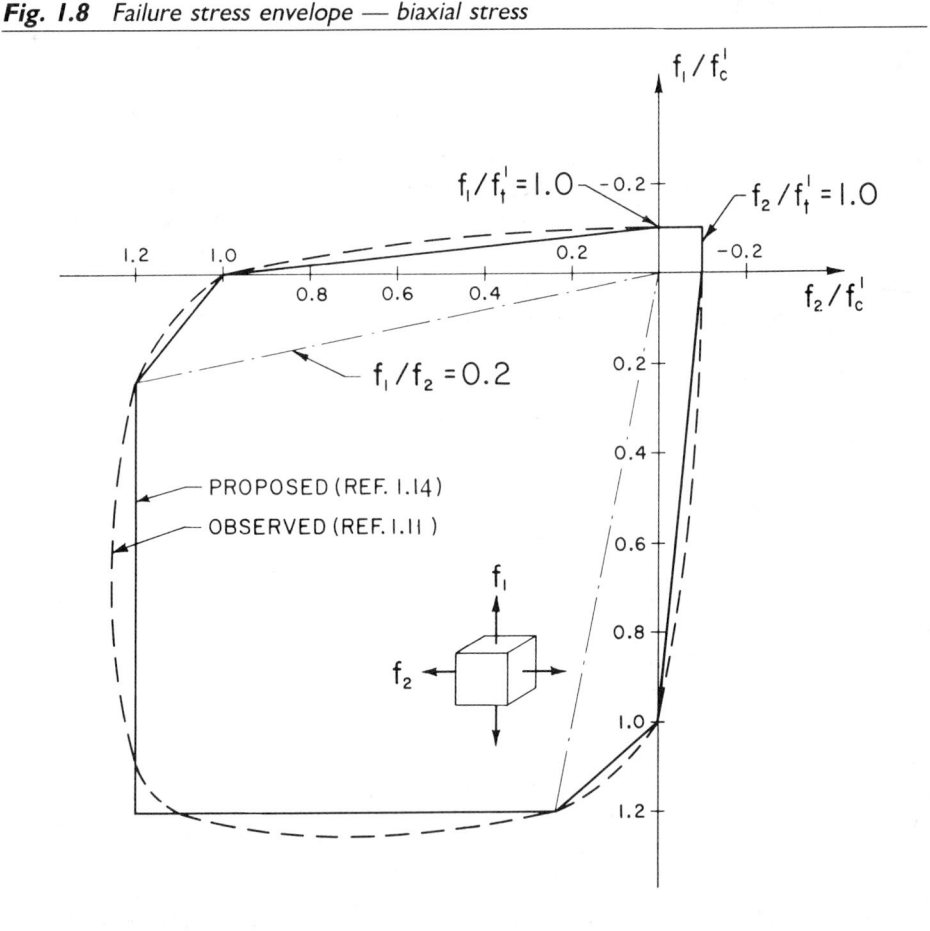

Another commonly used but approximate failure criterion for concrete is the Mohr's failure envelope (Fig. 1.9). Here the stresses at failure for any given test under uniaxial stress or combined stresses are represented by the corresponding Mohr's circle. Figure 1.9 presents the results of a direct tension test, a standard compression test, and triaxial compression test of which only the smallest and largest principal stresses are considered. A state of pure shear can be created by applying torsion to a thin-walled hollow cylinder. However, this creates direct tensile and compressive stresses of equal magnitude on diagonal planes and failure is inevitably in tension along the diagonal plane. The result of such a test is also shown in Figure 1.9. Since the circles correspond to conditions at failure, an envelope drawn to them is termed a *failure envelope*. Points on the failure envelope define limiting states of stress which can exist in the material prior to failure. For any combination of stress for which the Mohr's circle is tangent to the failure envelope, failure will occur. The intercept of the envelope on the vertical axis, v_c', gives the apparent strength of concrete in pure shear. The strength criteria presented in Figs. 1.8 and 1.9 are not in complete mutual agreement. The former is obtained by curve fitting of biaxial

Fig. 1.9 *Mohr's failure envelope for concrete*

test data, while the latter is an extension of a standard failure criterion for concrete. For biaxial compression, Mohr's theory gives more conservative results due to the neglect of the confining influence of the intermediate principal stress. However, for the cases of biaxial compression–tension and tension–tension, the influence of the intermediate principal stress on the failure strength is negligible.

1.9 TRIAXIAL STRENGTH

Strength and behaviour of concrete under triaxial stress systems has received considerable attention lately, and a large volume of experimental data is currently available (Ref. 1.15). Based on this, several strength criteria for concrete subjected

Fig. 1.10 *Schematic representation of the ultimate strength surface (Ref. 1.16)*

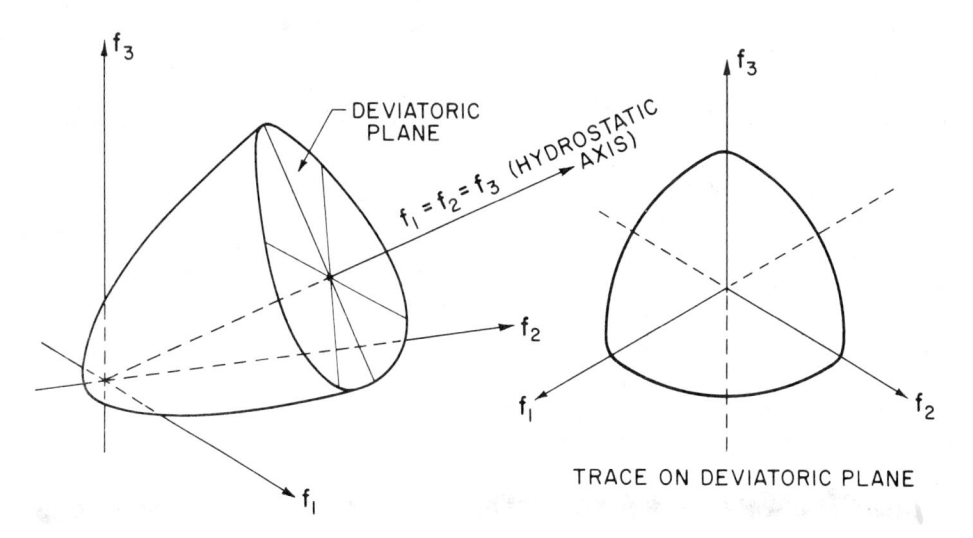

to short-term multiaxial stresses have been proposed. A generalized strength criterion for triaxial states of stress, proposed by Kotsovos (Ref. 1.16) is shown as an ultimate strength surface in Fig. 1.10. This criterion can be expressed mathematically in terms of the magnitude of the non-dimensionalized (with respect to f'_c) octahedral normal stress component and the magnitude and direction of the octahedral shear stress component. The orthogonal axes in Fig. 1.10 represent the three principal stresses. This ultimate strength surface was found to provide a close fit for extensive biaxial and triaxial strength test results. The trace of the surface on a coordinate plane, which gives the strength envelope for the biaxial case, is in very close agreement with the experimentally obtained envelope in Fig. 1.8. This criterion shows promise as a unified strength criterion common for uniaxial, biaxial and triaxial stress states.

Tests on concrete under triaxial compression have shown that both the compressive and the shear strength increase with increase in confining pressures.

1.10 STRESS–STRAIN RELATIONS FOR MULTIAXIAL STRESS STATES

For the use of nonlinear computer-based methods in the analysis of concrete structures subjected to multiaxial stress states, an essential input is a realistic stress–strain relation. Although a variety of models characterizing the stress–strain relations and failure criteria for concrete under multiaxial stress states have been proposed in recent years (Refs. 1.17, 1.18, 1.19), none of them has received general acceptance. A critical evaluation of some of these proposals is given in Ref. 1.19.

1.11 CREEP

Concrete under sustained loads shows an increase in strain with time (Fig. 1.11). The time-dependent part of strain resulting from stress is termed *creep*. In Fig. 1.11, ϵ_{inst} is the instantaneous elastic strain on the application of stress. If the stress is maintained at this level the strain will continue to increase with time along the solid curve although at a progressively decreasing rate. At any given time, the increase in strain above the initial elastic strain is termed the creep strain, ϵ_{cp}. Measurements indicate that small increases in creep strain take place after as long as 30 years (Ref. 1.20, Fig. 1.12). The 30 year creep may be on the average about 1.36 times the one year creep. The ultimate creep strain may be as much as 1.5 to 3 times the instantaneous elastic strain, ϵ_{inst}. If the sustained load is removed the strain follows the curve shown by the broken lines in Fig. 1.11. There is an instantaneous recovery of strain by an amount equal to the elastic strain due to the load removed at this age. This is followed by a gradual decrease in strain which is termed creep recovery. The exact mechanism of creep in concrete is as yet not fully understood, but creep is generally attributed in varying degrees to internal movement of adsorbed water, viscous flow or sliding between the gel particles, moisture loss and the growth in microcracks.

There are several independent and interacting factors related to the material prop-

Fig. 1.11 *Typical strain–time curve for concrete in uniaxial compression*

TIME SINCE APPLICATION OF STRESS

Fig. 1.12 *Variation of creep with time (Ref. 1.20)*

TIME AFTER LOADING

erties and composition, curing and environmental conditions, and loading conditions that influence the magnitude of creep strains. The major factors related to the composition of the mix that influence creep are the type and amount of cement, w/c ratio, aggregate type and content, and admixtures. Creep increases with increases in w/c ratio, cement content, and air entrainment. Since creep is associated primarily with the cement paste portion of the concrete, an increase in aggregate content reduces the creep. Physical properties of the aggregates such as the modulus of elasticity and porosity also influence the creep.

The ambient relative humidity and temperature are significant factors related to the environment that influence creep. A decrease in relative humidity as well as an increase in temperature increases the creep strain. Both these factors increase the loss of moisture. For unsealed specimens exposed to a drying atmosphere there is also a size effect associated with the moisture gradient in the specimen. The effect of size on creep can be expressed in terms of the ratio of volume to surface area or in terms of an average thickness of the specimen. An increase in this factor results in a decrease in creep.

Within the range of working stresses, creep is proportional to the applied stress. Further, creep strain in general is inversely proportional to the strength of concrete at the time of application of stress. Thus creep can be expressed as a linear function of the stress/strength ratio. Creep is also dependent on the age of the concrete at the time of loading. At higher stresses creep increases with applied stress at an increasing rate, and at a stress in the region of 0.8 to 0.9 f_c' creep leads to failure in the course of time. Cyclic loading produces higher creep strain than a static load of equal intensity acting for the same total time.

Measurement of creep is usually made on specimens loaded in uniaxial compression (ASTM Standard C512-76). Creep under multiaxial compression is less than that associated with uniaxial stress, and may be as low as 50 percent of the latter for a hydrostatic state of stress (Ref. 1.21).

Except under very high stress (0.75 to 0.9 f_c'), creep does not affect the strength of concrete. However, it often influences the stress distribution in members and increases the strains and deflections. Thus in columns the creep of concrete under stress results in a gradual transfer of load from concrete to the reinforcement so that the stress distribution will be different from that indicated by elastic theory. When the stress in the concrete is the result of an imposed deformation as with the settlement of a support in a continuous structure, the effect of creep is to reduce the induced stresses. In the same manner creep leads to a loss of prestress in prestressed concrete. In reinforced concrete, however, the major concern for creep is with regard to the increased deformations.

A review of different equations proposed for predicting creep along with comparisons with experimental data and numerical examples is presented in Ref. 1.22. In one widely recommended empirical method (Refs. 1.22, 1.23) a creep coefficient, C_t (defined as the ratio of creep strain ϵ_{cp} to the initial elastic strain ϵ_{inst}), is computed for "standard conditions" using Eq. 1.6. For departures from standard conditions, correction factors are applied to the ultimate (in time) creep coefficient, C_u, in Eq. 1.6. Standard conditions mean a relative humidity of not more than 40 percent, a loading age of seven days for moist-cured concrete and one to three days for steam-cured concrete, an average thickness of 150 mm (6 in.), a slump of 102 mm (4 in.) or less, and an air content of 7 percent or less.

For standard conditions

$$C_t = \frac{t^{0.6}}{10 + t^{0.6}}C_u \tag{1.6}$$

where $C_t = \epsilon_{cp}/\epsilon_{inst}$ and t is the time after application of load, in days. The ultimate creep coefficient, C_u, varies in the range of 1.30 to 4.15 and has an average value of 2.35. Since the value of C_t predicted using Eq. 1.6 depends primarily on the value taken for C_u, it is preferable that C_u for local conditions be established based on tests or local experience, and the average value of 2.35 used only in the absence of such data. Recommended correction factors for variations from "standard conditions" are presented in Ref. 1.22. The two major correction factors (CF) are given below:

1) for age of loading later than seven days for moist cured and one to three days for steam cured concrete,

$$(CF)_{AL} = 1.25\, t_a^{-0.118} \text{ for moist cured}$$
$$(CF)_{AL} = 1.13\, t_a^{-0.095} \text{ for steam cured}$$

(1.7)

where t_a is the age of concrete when load is applied, in days

2) for relative humidity greater than 40 percent,

$$(CF)_H = 1.27 - 0.0067H$$

(1.8)

where H is the ambient relative humidity in percent

1.12 SHRINKAGE AND TEMPERATURE EFFECTS

Slight changes in the volume of concrete occur during and after hardening (Ref. 1.2). The decrease in volume other than that due to externally applied forces and temperature changes, and resulting mostly from moisture loss during drying is broadly termed *shrinkage*. Unlike creep, shrinkage is not related to an applied stress and is reversible to a greater extent. Shrinkage and creep are not independent phenomena; however, for convenience it is normal practice to treat their effects as separate, independent and additive. All the factors related to constituent material properties, composition of mix, curing and environmental conditions, member size and age that affect creep also affect shrinkage. Typical variations of shrinkage with time are shown in Fig. 1.13.

Major concerns for shrinkage in concrete structures are related to its potential for inducing cracking and deformations. When shrinkage is restrained, as it often is in concrete structures, tensile stresses develop which, if excessive, may lead to cracking. Similarly, a differential shrinkage due to a moisture or thermal gradient or due to a differential restraint to shrinkage (as for example, the reinforcement being placed nearer to one side of a beam) will result in internal stresses, curvature and deflections. Shrinkage, like creep, also leads to a loss of prestress in prestressed concrete.

Since the primary cause of shrinkage is moisture loss from the cement paste phase of the concrete, it can be minimized by keeping the unit water content in the mix as low as possible and the total aggregate content as high as possible.

Shrinkage is usually expressed as a linear strain (mm/mm). ASTM Standard C157-75 provides a test procedure for the measurement of length change of cement mor-

Fig. 1.13 *Typical variation of shrinkage with time*

tar and concrete. Shrinkage measurements in the range of 200×10^{-6} to 1200×10^{-6} mm/mm have been reported (Ref. 1.2) for a small plain concrete specimen stored for six months at 50 percent relative humidity and 21°C. Shrinkage is less in reinforced concrete due to the restraint offered by the reinforcing and is ordinarily in the range of 200×10^{-6} to 300×10^{-6} mm/mm.

Because of the several interacting variables involved, prediction of shrinkage, as for creep, can only be approximated. An empirical method recommended by ACI Committee 209 (Refs. 1.22, 1.23) uses an approach similar to that for creep discussed in the previous section. The shrinkage strain at time t, $(\epsilon_{sh})_t$, for "standard conditions" (relative humidity 40 percent or less, average thickness of part under consideration 152 mm or less, and slump 100 mm or less) is given by:

1) for shrinkage after seven days, for moist curing,

$$(\epsilon_{sh})_t = \frac{t}{35 + t} (\epsilon_{sh})_u \tag{1.9}$$

2) for shrinkage after age one to three days, for steam curing,

$$(\epsilon_{sh})_t = \frac{t}{55 + t} (\epsilon_{sh})_u \tag{1.10}$$

where t is the time in days from age seven days for moist curing and from age one to three days for steam curing; and $(\epsilon_{sh})_u$ is the ultimate shrinkage strain.

Ref. 1.22 suggests an average value for $(\epsilon_{sh})_u$ of 780×10^{-6} mm/mm for both moist and steam cured concrete for use in the absence of more precise data. Correction factors to be applied to the ultimate shrinkage, to account for variations in humidity, thickness of member, slump, cement content, percentage of fine aggregate, and air content are given in Ref. 1.22. Except for the first which is given below, the influence of these correction factors is only marginal and can be neglected under normal conditions.

For relative humidity, *H*, greater than 40 percent:

$$\text{Shrinkage correction factor} = 1.40 - 0.01\,H \text{ for } 40\% < H < 80\%$$
$$= 3.00 - 0.03\,H \text{ for } 80\% < H < 100\% \quad (1.11)$$

Concrete expands with a temperature rise and contracts as the temperature drops. An average value of the coefficient of thermal expansion is 10×10^{-6} mm/mm/°C. This is close to the value for steel (which is about 12×10^{-6} mm/mm/°C) so that there is little likelihood of any differential thermal expansion and associated relative movements between the steel and surrounding concrete. Thermal contraction has effects similar to shrinkage.

1.13 STEEL REINFORCEMENT

1.13.1 GENERAL

As already explained, concrete is strong in resisting compressive stresses, but very weak in tension. For this reason it has to be "reinforced" with a material strong in tension if tensile stresses are to be resisted. Usually this is done by embedding steel in the form of bars, wires or welded wire fabric, at the required locations. Because steel is much stronger than concrete even in compression, it can also be introduced into concrete to carry compressive stresses if it is desired to reduce the dimensions of the concrete section.

To be effective, the reinforcement and the surrounding concrete must deform under load as an integral unit without any relative movement or slip between them. In order to assist the bond between the steel bars and concrete, and mechanically inhibit longitudinal movement of the bar relative to the concrete around it, reinforcing bars are provided with lugs or protrusions (called "deformations") on their surface. Plain bars are normally permitted only as spiral reinforcing in columns and, in sizes smaller than 10 mm diameter, as stirrups and ties.

Requirements for concrete reinforcing steel are given in CSA Standard A23.1-M77. The primary requirements are a minimum specified yield strength, f_y (specified by "Grade"), minimum ultimate tensile strength, f_u, minimum percentage elongation and cold bending capability. The mass density of reinforcing steel is normally taken as 7850 kg/m^3 and the coefficient of thermal expansion as 12×10^{-6} per degree Celsius.

1.13.2 TYPES, SIZES, AND GRADES

The two types of reinforcing steel generally in use are hot-rolled round bars and cold-drawn wires. The former is the most widely used type in reinforced concrete. Specifications for deformed and plain bars used for concrete reinforcement are given in CSA Standard G30.12. A summary of the standard sizes and dimensions of deformed bars and their number designations is given in Table 1.2. Reinforcing bars are classified into three grades based on minimum specified yield strength: 300, 350, and 400 (MPa). Grade 400 bars (with $f_y = 400$ MPa) are the most frequently used type of reinforcement.

Table 1.2 Metric Deformed Bar Size and Designation Numbers —
CSA G30.12 and G30.16

| Bar Designation No.[1] | Nominal Mass kg/m | Nominal Dimensions[2] | | |
		Diameter mm	Cross-Sectional Area mm^2	Perimeter mm
10	.785	11.3	100	35.5
15	1.570	16.0	200	50.1
20	2.355	19.5	300	61.3
25	3.925	25.2	500	79.2
30	5.495	29.9	700	93.9
35	7.850	35.7	1000	112.2
45	11.775	43.7	1500	137.3
55	19.625	56.4	2500	177.2

1. Bar designation numbers approximate the nominal diameter of the bar in millimetres.
2. The nominal dimensions of a deformed bar are equivalent to those of a plain round bar having the same mass per metre as the deformed bar.

There are situations where it is necessary to ensure that the designed strength is not exceeded by much, and it may be undesirable to have a yield strength much higher than the specified value. For instance, in seismic design, a ductile flexural failure with considerable yielding is desired. For use in such cases an upper limit on the yield strength and a minimum ratio of actual tensile strength to actual yield strength may be specified. Steel conforming to CSA G30.16 and ASTM A706 generally meets these superior ductility requirements as well as chemical composition restrictions for improved weldability.

All deformed bars are identified by a distinguishing set of marks rolled on the bar surface. The marking system, shown in Fig. 1.14, is given in the following order: a letter or symbol designating the producer mill, the bar designation number, the type of steel, and the number 400 or a continuous line through five or more spaces for 400 grade steel only.

Cold-drawn wires used as reinforcement may be smooth (CSA G30.3) or deformed (CSA G30.14). Smooth wire is denoted by the letter W followed by a number indicating the cross-sectional area in mm^2. Similarly, deformed wire is designated by the letter D followed by the number representing the area as for the smooth wire. In both cases for metric size wires a letter M is prefixed to the above designations (example: MW25.8, MD22.6). Cold-drawn wires are used mostly (in slabs and similar situations) in the form of welded wire fabric, which consist of a series of wires arranged in orthogonal patterns (forming square or rectangular openings) and welded together at all intersections. These are denoted by the letters WWF followed by spacings of longitudinal and transverse wires in mm and the designations of the longitudinal and transverse wires, in that order (example: WWF 152 × 152 — MD25.8 × MD25.8). Welded wire fabric must conform to CSA G30.5 if made of

Fig. 1.14 *Marking systems for reinforcing bars*

smooth wires and to CSA G30.15 if made of deformed wires. A summary listing of all the standard reinforcing steels, giving the type of steel, specification number, bar sizes, grades and major tensile strength properties is presented in Table 1.3.

1.13.3 STRESS–STRAIN CURVES

Typical stress–strain curves for reinforcing steel are shown in Fig. 1.15. For all steels there is an initial linear elastic portion with constant slope which gives the modulus of elasticity, E_s. For hot rolled bars in the lower grades (300, 350 MPa) the elastic part is followed first by a yield plateau where the strain increases at almost constant stress (yielding) and then a strain hardening range in which the stress once again

Fig. 1.15 *Typical stress–strain curves for reinforcing steels*

Table 1.3 Standard Reinforcing Steel Data

Type of Reinforcement	Type of Steel and CSA Specification	Size No. Inclusive	Grade	Minimum Yield Strength f_y, MPa	Minimum Tensile Strength f_u, MPa	Minimum Percent elongation for 200 mm Gauge Length[2]
Bars						
Deformed and plain[1]	Billet G30.12	10–55	300	300	450	10–12
		10–55	350	350	550	7–9
		10–55	400	400	600	7–9
Deformed weldable	low alloy G30.16	10–55	400	Min. 400 Max. 540	Min. 550 but ≥ 1.25 (test f_y)	12–13
Cold-drawn wires[3]						
Smooth	G30.3	MW12.9–MW87.3		485	550	
Deformed	G30.14	MD12.9–MD87.3		517	585	
Welded wire fabric[3]						
Smooth	G30.5	(steel conforming to G30.3)		{450* / 385‡}	{515* / 485‡}	
Deformed	G30.15	(steel conforming to G30.14)		485	550	

(1) Plain bars in sizes from 50 mm² (8 mm diameter) to 500 mm² (25 mm diameter).

(2) Specified minimum elongation depends on the bar size.

(3) At present time the metric wire sizes and dimensions have been "soft" converted from imperial units.

* For size MW7.7 and larger

‡ For size smaller than MW7.7

increases with further increase in strain (although at a decreasing rate) until the peak stress (tensile strength, f_u) is reached. Finally, there is a descending portion of the curve with the nominal stress decreasing until fracture occurs. In general, the higher the strength, the shorter the length of the yield plateau. High-strength high-carbon steels and cold-drawn wires may not exhibit a distinct yield plateau and the elastic phase may be followed by a strain hardening phase.

The structurally significant properties of reinforcing steel are its yield strength, f_y, and modulus of elasticity, E_s. The yield strength for reinforcing steel is taken as the stress corresponding to a specified strain as indicated in Fig. 1.16. The modulus of elasticity varies very little and is generally taken as 200 GPa for all reinforcing steels.

For design calculations, CAN3-A23.3-M84 (9.4) does not permit the use of a value of f_y in excess of 550 MPa except for prestressing steel. Further, for compression reinforcement having a specified f_y in excess of 400 MPa, the value of f_y assumed in design calculations shall be taken as 400 MPa or as the stress corresponding to a strain of 0.35 percent. When Grade 300 and 400 reinforcement conforming to CSA Standard G30.12 is used to resist earthquake-induced flexural and axial forces, the actual yield stress based on mill tests shall not exceed the specified yield stress by more than 125 MPa and the actual ultimate tensile strength shall not be less than 1.25 times the actual tensile yield strength (CAN3-A23.3-M84 (21.2.5.1)).

Fig. 1.16 *Initial portion of stress–strain curves for reinforcing steels*

1.14 CODES AND SPECIFICATIONS

This book is written primarily with reference to the National Building Code of Canada 1985 (NBC1985, Ref. 1.24). NBC requires that the design of plain, reinforced and prestressed concrete conform to Canadian Standards Association (CSA) Standard CAN3-A23.3-M84, *Design of Concrete Structures for Buildings* (Ref. 1.25). CSA has several other standards applicable to concrete construction and dealing with areas such as materials, construction methods and bridge design. The important standards which control the design of concrete structures in Canada are given below:

1. National Building Code of Canada
2. CSA Standard CAN3-A5-M83   *Portland Cements*
3. CSA Standard CAN3-A23.1-M-77   *Concrete Materials and Methods of Concrete Construction*
4. CSA Standard CAN3-A23.2-M77   *Methods of Test for Concrete*
5. CSA Standard CAN3-A23.3-M84   *Design of Concrete Structures for Buildings*
6. CSA Standard CAN3-A23.4-M78   *Precast Concrete — Materials and Construction*
7. CSA Standard CAN3-A266.1-M78   *Air-Entraining Admixtures for Concrete*
8. CSA Standard CAN3-A266.2-M78   *Chemical Admixtures for Concrete*
9. CSA Standard CAN3-A266.4-M78   *Guidelines for the Use of Admixtures in Concrete*
10. CSA Standard CAN3-A266.6-M85   *Superplasticizing Admixtures for Concrete*
11. CSA Standard G30.12-M77   *Billet Steel Bars for Concrete Reinforcement*
12. CSA Standard G30.16-M77   *Weldable Low Alloy Steel Deformed Bars for Concrete Reinforcement*
13. CSA Standard CAN3-S6-M78   *Design of Highway Bridges*

The design of reinforced concrete buildings in the United States is mostly governed by the *Building Code Requirements for Reinforced Concrete* (ACI 318-83) of the American Concrete Institute (Ref. 1.26). The requirements in the ACI Code are for the most part identical to those of CSA Standard CAN3-A23.3. To avoid lengthy repetitions in subsequent chapters, the simple reference to Code will mean CSA CAN3-A23.3-M84.

1.15 INTERNATIONAL SYSTEM OF UNITS

The International System of Units (SI Units) is used throughout this book. With this system, all loads must be converted from their mass (kilogram) into force (Newton = $kg \cdot m/s^2$) by multiplying with the acceleration due to gravity (g = 9.80665 $\simeq$ 9.81 m/s^2). For a description of SI units, reference may be made to the Canadian Metric Practice Guide (Ref. 1.27).

REFERENCES

1.1   Troxel, G.E., Davis, H.E., and Kelly, J.W., *Composition and Properties of Concrete*, 2nd ed., McGraw-Hill, New York, 1968, 529 pp.

1.2   Neville, A.M., *Properties of Concrete*, 2nd (Metric) ed., Pitman Publishing Co., London, 1973.

1.3   *Design and Control of Concrete Mixtures*, Canadian Metric edition,

Engineering Bulletin, 4th ed., Canadian Portland Cement Association, Ottawa, Ontario, 1984.

1.4 ACI Standard 211.1-81, *Standard Practice for Selecting Proportions for Normal, Heavyweight, and Mass Concrete*, American Concrete Institute, Detroit, Michigan, 1981, 32 pp.

1.5 ACI Standard 211.2-81, *Standard Practice for Selecting Proportions for Structural Lightweight Concrete*, American Concrete Institute, Detroit, Michigan, 1981, 18 pp.

1.6 ACI Standard 211.3-75, *Standard Practice for Selecting Proportions for No-Slump Concrete*, American Concrete Institute, Detroit, Michigan, Revised, 1980.

1.7 ACI Committee 212, Report, (ACI 212.2R-81) *Guide for Use of Admixtures in Concrete*, American Concrete Institute, Detroit, Michigan, 1981, 13 pp.

1.8 Price, W.H., *Factors Influencing Concrete Strength*, Proc. ACI, Vol. 47, 1951, p. 417.

1.9 Kesler, C.E., *Hardened Concrete-Strength*, Tests and Properties of Concrete, ASTM Special Technical Publication No. 169-A. Am. Soc. for Testing and Materials, 1966, pp. 144–159.

1.10 Hsu, T.T.C., et al., *Microcracking of Plain Concrete and the Shape of the Stress–Strain Curve*, J. ACI, Vol. 60, Feb. 1963, pp. 209–223.

1.11 Kupfer, H., Hilsdorf, H.K. and Rüsch, H., *Behaviour of Concrete Under Biaxial Stresses*, J. ACI, Vol. 66, Aug. 1969, pp. 656–666.

1.12 Hognestad, E., Hanson, N.W., and McHenry D., *Concrete Stress Distribution in Ultimate Strength Design*, J. ACI, Vol. 52, Dec. 1955, pp. 455–479.

1.13 Wright, P.J.F., *Comments on an Indirect Tensile Test on Concrete Cylinders*, Magazine of Concrete Research, No. 20, 1955, p. 87.

1.14 Tasuji, M.E., Slate, F.O., and Nilson, A.H., *Stress–Strain Response and Fracture of Concrete in Biaxial Loading*, J. ACI, Vol. 75, Jul. 1978, pp. 306–312.

1.15 Gerstle, K.H., et al., *Strength of Concrete under Multi-axial Stress States*, Douglas McHenry International Symposium on Concrete and Concrete Structures (Mexico City, Oct. 1976). ACI Publication SP-55, American Concrete Institute, Detroit, 1978, pp. 103–131.

1.16 Kotsovos, M.D., *A Mathematical Description of the Strength Properties of Concrete under Generalized Stress*, Magazine of Concrete Research, Vol. 31, No. 108, Sept. 1979, pp. 151–158.

1.17 Kotsovos, M.D., and Newman, J.B., *A Mathematical Description of the Deformational Behaviour of Concrete under Complex Loading*, Magazine of Concrete Research, Vol. 31, No. 107, June 1979, pp. 77–90.

1.18 Gerstle, K.H., *Simple Formulation of Biaxial Concrete Behaviour*, J. ACI, Vol. 78, Jan.–Feb. 1981, pp. 62–68.

1.19 Chen, W.F., and Ting, E.C., *Constitutive Models for Concrete Structures*, Proc. ASCE, Vol. 106, No. EM1, Feb. 1980, pp. 1–19.

1.20 Troxell, G.E., Raphael, J.M., and Davis, R.E., *Long Time Creep and*

Shrinkage Tests of Plain and Reinforced Concrete, Proc. ASTM, Vol. 58, 1958, pp. 1101–1120.

1.21 McDonald, J.E., *Creep of Concrete under Various Temperature, Moisture, and Loading Conditions*, Douglas McHenry International Symposium on Concrete and Concrete Structures, ACI Publication SP-55, American Concrete Institute, Detroit, 1978, pp. 31–53.

1.22 Branson, D.E., *Deformation of Concrete Structures*, McGraw-Hill, Inc., New York, 1977, 546 pp.

1.23 ACI Committee 209, *Predictions of Creep, Shrinkage and Temperature Effects in Concrete Structures*, Designing for the Effects of Creep, Shrinkage and Temperature in Concrete Structures, ACI Publication SP-27, American Concrete Institute, Detroit, 1971, pp. 51–93.

1.24 *National Building Code of Canada*, National Research Council of Canada, Ottawa, 1985.

1.25 CSA Standard CAN3-A23.3-M84 — *Design of Concrete Structures for Buildings*, Canadian Standards Association, Rexdale, Ontario, 1984.

1.26 *Building Code Requirements for Reinforced Concrete*, ACI 318-83, American Concrete Institute, Detroit, Michigan, 1983.

1.27 CSA Standard Z234.1-79, *Canadian Metric Practice Guide*, Canadian Standards Association, Rexdale, Ontario, 1979.

CHAPTER 2

REINFORCED CONCRETE BUILDINGS

2.1 INTRODUCTION

Reinforced concrete is extensively used for the construction of various types of structures such as buildings, bridges, retaining walls, water tanks, grandstands, etc. The basic principles involved in the *design of sections* for all these structures are essentially the same. In general, the most common type of reinforced concrete struc-

Fig. 2.1 *Floor plan of a typical office building*

ture is the building and hence it will be considered in some detail. There are many types of reinforced concrete systems which are used in building construction, such as precast, lift slab, post-tensioned, composite, and modular systems. The most common construction, however, is cast-in-place. The two main categories of building structures are office buildings and residential type structures such as apartments and hotels.

Office buildings (Fig. 2.1) are characterized by service cores and flexible floor spaces. Service cores contain elevators, stairwells, mechanical and electrical supply ducts, equipment rooms, janitors' closets, and washrooms. Floor slabs may span the complete distance from the exterior walls to the core or they may be supported on intermediate columns. The distance to the core is generally no greater than 12 m because that represents an acceptable maximum distance from a natural light source. For clear spans, structural economy also tends to limit this distance.

More expensive structural schemes with column-free spaces between the exterior walls and the core are sometimes desired by office tenants who wish to provide flexibility in floor layouts by using the open office concept. Other tenants who require individual offices can generally accept less expensive structural layouts with intermediate columns on 7.5 to 9 m grids which can be conveniently concealed by partitions.

Fig. 2.2 *Floor plan of a typical hotel building*

6 BAYS @ 8.5 m

22 m

Apartment buildings or hotels (Fig. 2.2) can accommodate most architectural layouts with smaller structural spans of 6 to 7.5 m. Structural shear walls are commonly used as partitions because of their acoustic properties. Floor to floor heights are less for apartment buildings because there is generally no requirement to provide dropped ceilings to conceal mechanical and electrical services, and the soffits of flat plate floor systems can be used directly as ceilings for the floor level below.

2.2 LOADS

Loads which must be considered in the design of structures are (Ref. 2.1):

D — *dead loads;*

L — *live load* due to intended use and occupancy;

Q — *live load* due to wind or earthquake, whichever produces the most unfavourable effect;

T — loads due to contraction or expansion caused by temperature changes, shrinkage, moisture changes, creep, movement due to differential settlement or combinations thereof.

Where a structure or member is likely to be subjected to loads other than those listed above, such loads shall also be taken into account in the design. Where it is known from experience, or it can be shown theoretically, that the neglect of some or all of the effects due to loads, T, does not adversely affect the strength and serviceability of the structure, they need not be considered in the design calculations. For design purposes it is often convenient to group loads into gravity loads and lateral (horizontal) loads.

Dead loads consist of the weight of the structure itself and all other permanent installations such as partitions, ceiling and floor finishes, piping and lighting fixtures, permanent equipment, and the forces due to prestressing. The magnitude and distribution of these loads are fixed in relation to time and can be estimated with reasonable accuracy from known or assumed dimensions. When dimensions are assumed, their accuracy must be verified and they must be revised if necessary as

Table 2.1   Average Mass-Densities of Building Materials

Material	kg/m³
Brick masonry	1600–2400
Concrete, normal density	2400
Concrete, light weight	1440–1920
Granite	2250–2650
Steel	7850
Aluminum	2640
Water	1000
Snow, fresh	100
Snow, old	200–400
Snow, wet	640–800
Sand	1600–1920
Glass	2560
Asphalt	1280–1600
Mortar	1600
Timber, redwood	415
douglar fir	513
pine	560–640
oak	865

the design progresses. Average mass densities of the more commonly used engineering materials are given in Table 2.1. For ceilings, fixtures, irregularly spaced partitions, etc., it may be convenient to reduce their dead load to a uniformly distributed load (kN/m²) over the entire floor area. For ordinary reinforced concrete buildings, the average gross dead load considering the whole building may be in the range of 5 to 7 kN/m² of floor area. Such approximate figures may be helpful for the preliminary choices of foundation type and structural layout. In computing the load effects on the structure, dead loads are considered to act at all times, except that during the construction period these loads may be somewhat different from their values expected on the finished structure.

Live loads are the loads on the structure resulting from its use and occupancy (including vertical loads due to cranes); snow, ice and rain; earth and hydrostatic pressure; and horizontal components of static or inertia forces. They are more or less of a temporary duration and can vary considerably in magnitude and distribution. It is possible to have very heavy loads concentrated on small areas as with a book shelf or a safe box. However, the average load over larger areas may be much smaller. Furthermore, the worst load effect on a structure or member may not in all cases be when the entire structure is fully loaded. Hence the conditions of live load being absent and of live load acting only on part of the structure (pattern loading) must also be considered for design.

Minimum specified live loads on floors and roofs in buildings for various uses and occupancies are given in subsection 4.1.6 of the National Building Code of Canada (NBC) 1985 (Ref. 2.1). This consists of a uniformly distributed load (kN/m²) applied over the entire area or on any portion of the area (Table 2.2), or a concentrated load applied over an area of 750 mm by 750 mm (Table 2.3), whichever produces the most critical effects. When a member supports a large tributary area of floor and/or roof (e.g., the columns in the lower storeys of a multistorey building), it is highly unlikely that this entire area would be loaded uniformly with the full specified live load. Hence for large tributary areas a reduction in the specified live load intensity is allowed.

For roofs, the live loads due to the accumulation of snow, ice and rain may be more critical than the specified load based on use and occupancy. Specified snow load is taken as the product of the ground snow load specified in Chapter 1, "Climatic Information for Building Design in Canada" of Ref. 2.2 for various locations in Canada and the snow load factors detailed in subsection 4.1.7 of Ref. 2.1. Live loads on structures also include, where applicable, the lateral pressure due to liquid and lateral earth pressure.

Live loads, even if moving or dynamic in type, are generally treated as static loads. However, when their dynamic, vibratory and impact effects are significant, as with heavy moving equipment such as cranes, elevators, etc., these effects are accounted for by increasing the live load by an impact factor. Specified impact factors for building structures are given in subsection 4.1.10 of Ref. 2.1. Specified live loads on bridge structures due to traffic and their impact effects are given in Ref. 2.3.

Wind loads Wind pressure on the surface of a structure results from the differential pressures caused by the obstruction to the free flow of the wind. Such loads,

**Table 2.2 Minimum Specified Distributed Live Loads for Floors and Roofs
(Abstracted from National Building Code of Canada — 1985,
Subsection 4.1.6.)***

Use of Area of Floor or Roof	Minimum Specified Load, kN/m^2
Assembly areas with fixed seats that have backs over at least 80 percent of the assembly area and used for churches, courtrooms, lecture halls, theatres and classrooms with or without fixed seats	2.4
Assembly areas other than those listed above	4.8
Balconies, exterior and interior	4.8
Corridors, lobbies and aisles over 1200 mm wide (except upper floor corridors of residential areas of apartments, hotels and motels)	4.8
Equipment areas and service rooms	3.6
Exits and fire escapes	4.8
Factories	6.0

Garages for
    Passenger cars
    Unloaded buses and light trucks
    Loaded buses and trucks and all other trucking spaces
	2.4
	6.0
	12.0

Kitchens (other than residential)	4.8

Libraries
    Stack rooms
    Reading and study rooms
	7.2
	2.9

Office areas in office buildings and other buildings (not including record storage and computer rooms) located in
    Basement and first floor
    Floors above first floor
	4.8
	2.4

Operating rooms and laboratories	3.6
Recreation areas that cannot be used for assembly purposes	3.6
Residential areas Sleeping and living quarters in apartments, hotels, motels, boarding schools and colleges.	1.9
Retail and wholesale areas	4.8
Roofs	1.0

Table 2.2   continued

Use of Area of Floor or Roof	Minimum Specified Load, kN/m²
Sidewalks and driveways over areaways and basements	12.0
Storage areas	4.8
Toilet areas	2.4
Warehouses	4.8

*Abstracted from Ref. 2.1

Table 2.3   Minimum Specified Concentrated Live Loads for
Floors and Roofs (Ref. 2.1)

Area of Floor or Roof	Minimum Specified Load, kN
Roof surfaces	1.3
Floors of classrooms	4.5
Floors of offices, manufacturing buildings, hospital wards and stages	9.0
Floors and areas used by passenger cars	11.0
Floors and areas used by vehicles not exceeding 3600 kg gross weight	18.0
Floors and areas used by vehicles exceeding 3600 kg but not exceeding 9000 kg gross weight	36.0
Floors and areas used by vehicles exceeding 9000 kg gross weight	54.0
Driveways and sidewalks over areaways and basements	54.0

therefore, depend on the wind velocity; the orientation, area and shape of the surface; and the density of the air which itself depends on the atmospheric pressure and air temperature. Within the height range of usual structures, the wind speed increases with height, and is influenced by the roughness of the terrain. Wind speed also fluctuates constantly and the peaks in the wind speed represent gusts. Because of their limited spatial extent, wind gusts are unlikely to act simultaneously over the full extent of the structure.

Although the determination of wind forces on a structure is a dynamic problem, for buildings of moderate proportions which are not susceptible to wind induced vibrations, it has been the usual practice to treat such forces as equivalent static loads expressed in kN/m² of exposed surface area. Wind loads are derived from pressures resulting from a mean wind velocity and amplified for the effect of gusts.

A reference wind speed, $\bar{v}$, based on mean wind speeds taken at a standard elevation of 10 m above ground in open exposure is used to determine a reference wind pressure (dynamic pressure of moving air), q, as:

$$q = \tfrac{1}{2}\rho\bar{v}^2 = C\bar{v}^2 \qquad (2.1)$$

where ρ is the mass density of air. For Canadian conditions, with q in kN/m² and $\bar{v}$ in km/h, a value of 50×10^{-6} is recommended for C. Values of q, with probabilities of exceedance in any one year of 1/10, 1/30 and 1/100, for various selected locations in Canada are given in Chapter 1 of Ref. 2.2. Values specified for q seldom exceed 1.5 kN/m². Ref. 2.1, subsection 4.1.8, recommends that the specified external static pressure (or suction), p, due to wind acting normal to the surface be calculated as:

$$p = q\, C_e\, C_g\, C_p \qquad (2.2)$$

In Eq. 2.2, C_e is an exposure factor which allows for the variation of the wind speed with height and the effects of the variations in the surrounding terrain (shielding) and ranges between 0.9 and 2.0; C_g is a gust effect factor, with recommended values of 2.0 for the building as a whole and main structural members and 2.5 for small elements including cladding; and C_p is the external pressure coefficient and allows for the effects of shape, wind direction and the profile of wind velocity. These are generally determined from wind tunnel experiments. Recommended values for C_p are given in the Commentary on Wind Loads in Chapter 4 of Ref. 2.2. Equations similar to 2.2 above are also recommended for specified internal pressure (or suction).

The net specified wind pressure on the surface of a building is the algebraic difference between the external and internal pressures. Similarly the algebraic difference between the external wind pressures on the windward and leeward sides of the building is the specified net wind pressure for the design of the whole structure.

Exposed elements such as cladding, walls, and roof must be designed to carry the net load due to the wind distributed over the exposed area. However, for the design of the structural frame, it is normal practice to assume the wind load as concentrated at the frame joints. The concentrated wind load at a joint is taken as the total wind force acting on the exterior surface tributary to that joint. The area tributary to a joint is the portion of the exterior surface bounded by the horizontal centrelines between the joint and the floors above and below, and the vertical centrelines between the joint and the frames on either side (see Fig. 2.3b).

For design of cladding, as well as for design of structural members for deflection and vibration, the reference velocity pressure, q, is based on a probability of being exceeded in any one year of 1/10. For design of structural members for strength, this probability is reduced to 1/30, except for post-disaster buildings for which the probability is kept as low as 1/100. Post-disaster buildings are those which are expected to be functional in the event of a disaster, such as hospitals, fire stations, power stations, etc.

An example of wind load calculations by the static approach is given in Example 2.1. Buildings which are likely to be susceptible to wind load vibrations such as tall and slender buildings must be designed for wind load effects either by experimental methods such as wind tunnel tests or by a dynamic approach to the action of wind gusts. Details of such a dynamic approach are given in the Commentary on Wind Loads in Chapter 4 of Ref. 2.2.

Earthquake loads The sudden horizontal translation of the foundation of a structure due to earthquake ground motion leads to horizontal inertia forces distributed throughout the mass of the structure. Seismic loads on structures depend on the peak horizontal ground acceleration, the peak horizontal velocity and the characteristics of the structure and foundation. For simplicity in analysis, the loading due to earthquake motion is generally specified in Ref. 2.1, subsection 4.1.9, as equivalent static loads acting horizontally. Specified earthquake loads are based on the requirements that the structure be able to resist moderate earthquakes without significant damage and to resist major earthquakes without collapse.

Based on a statistical analysis of earthquake data, seismic zoning maps for Canada have been prepared giving contours of peak horizontal ground accelerations (in units of g, the acceleration due to gravity) and contours of peak horizontal ground velocities in m/s, both having a probability of exceedance of 10 percent in 50 years (Ch.4, Ref. 2.2). The peak ground motion paremeters that have a probability of exceedance of 10 percent in 50 years (that is, 0.0021 per annum) are the basis for design specifications in the NBC 1985. Based on the ranges of the ratio, a, of peak horizontal ground acceleration to the acceleration due to gravity (taken nominally as 10 m/s^2) and of the ratio, v, of peak horizontal ground velocity to a velocity of 1 m/s, seven seismic zones are defined and for each zone a, v, Z_a and Z_v are specified:

where a = zonal acceleration ratio, from 0.00 to 0.40;

v = zonal velocity ratio, from 0.00 to 0.40;

Z_a = acceleration-related seismic zone, from 0 to 6; and

Z_v = velocity-related seismic zone, from 0 to 6.

Values for v, Z_a and Z_v for selected locations across Canada are tabulated in Chapter 1 of Ref. 2.2. The specified minimum lateral seismic force at the base of the structure, V, is given as:

$$V = v\,SKIFW \tag{2.3}$$

where, v is the zonal velocity ratio, equal to the specified zonal horizontal ground velocity expressed as a ratio of 1 m/s; S is a seismic response factor for the structure which reflects the dependence of seismic force on the fundamental period of the structure as well as the contributions of the higher modes for tall buildings; K is a numerical coefficient which reflects the influence of material and type of construction on ductility and energy-absorptive capacity of the structure; I is an importance factor of the structure; F is a foundation factor; and W is the weight of the structure. A portion of the total V (up to 15 percent) is assumed to be concentrated at the top of the structure and the remainder distributed along the height of the building (including the top level) in proportion to the product of the weight at each level and its height above the base.

As an alternative to this, the distribution of the total lateral seismic force, V, over the height of a building may also be determined by a dynamic analysis, and guidelines for such a procedure are presented in Chapter 4 of Ref. 2.2.

Other loads Temperature variations as well as shrinkage and creep of materials cause dimensional changes in members which, if restrained by other parts of the structure, give rise to internal strains and stresses. Similarly, differential settlement of the foundation of continuous structures also produces internal forces. Structures must be designed to withstand such loads. These forces may, however, be minimized by making provisions in the structure (expansion joints, construction joints, discontinuity, etc.) to accommodate differential movements.

2.3 REINFORCED CONCRETE STRUCTURAL SYSTEMS

2.3.1 LOAD TRANSFER SCHEMES

Reinforced concrete buildings are three-dimensional structures; however, they are generally constructed as an assemblage of more or less two-dimensional or planar subsystems lying primarily in the horizontal and vertical planes (e.g., floors, roofs, walls, plane frames, etc. See Fig. 2.3). As an integrated system, the structure must carry and transfer to the foundation and the ground below all gravity loads on it as well as all horizontal loads and associated overturning moments resulting from wind or earthquake or asymmetry. The structural system can, for the sake of convenience, be separated into gravity load resisting schemes and lateral load resisting schemes, although these two are mutually interacting and complementary.

Gravity load resisting schemes in building structures consist of a horizontal subsystem (floors and roofs) and a vertical subsystem (columns, walls, shafts, transfer girders, hangers, etc.).

2.3.2 FLOOR SYSTEMS

The floor systems (horizontal subsystems) in buildings pick up the gravity loads acting on them and transfer them to the vertical elements (columns, walls, etc.). In this process the floor system is primarily subjected to flexure and transverse shear and the vertical elements are subjected to axial force (usually compression), bending and shear (Fig. 2.3a). The floor also functions as a horizontal diaphragm connecting together and stiffening all the vertical elements. When these connections are rigid, "frame action" is also developed which greatly improves the efficiency of the system. In addition, the floor system, acting as diaphragms, picks up and distributes the horizontal loads generated in its tributary area (from wind or earthquake loadings) to the vertical elements of the main lateral load-resisting system (Fig. 2.3b). The diaphragms further help to maintain the overall cross-sectional geometry of the structure.

In cast-in-place reinforced concrete construction, the floor system usually consists of one of the following: (1) slabs supported on bearing walls, (2) flat plates, (3) flat slabs, (4) joist floor and, (5) beam and slab systems.

Wall supported slabs A bearing wall and slab system is used mainly in low-rise residential type buildings. Slabs are generally 100 mm to 200 mm thick with spans from 3 m to 7.5 m. When the slab is supported only on opposite sides (Fig. 2.4a)

Fig. 2.3 *Load transfer systems*

a) VERTICAL LOAD TRANSFER

b) LATERAL LOAD TRANSFER

the slab bends in one direction only and the slab is called a one-way slab. The entire load is transmitted one way to the two supports. A very similar situation exists in a long rectangular slab (length greater than about twice the width) supported on all four sides (Fig. 2.4b). The bending of the slab is predominantly along the short span and most of the load is transmitted across the short span on to the two longer

Fig. 2.4 *One- and two-way slab action*

supports. When the slab dimensions approach a square (Fig. 2.4c) the bending action along both spans becomes relatively significant. In this case, load is transmitted in both directions and the slab is termed a two-way slab. In two-way slabs where corners are held down, there will be a zone of negative bending in the corner areas. In Fig. 2.4a, b, and c, the behaviour of the slab will be essentially unaltered if the walls are replaced by stiff beams (Fig. 2.4d) which deflect very little. When the slab is fixed to or continuous over a support (Fig. 2.4e), negative moments are developed in the slab over such supports.

Flat plate A flat plate floor is a reinforced concrete slab of uniform thickness supported directly on columns (Fig. 2.5). Edge (spandrel) beams are commonly

Fig. 2.5 *Flat plate*

added to stiffen exterior side frames which have heavy cladding loads. Flat plate floors are from 125 to 250 mm thick and are used with spans of 4.5 to 6 m. Flat plate construction is suitable for such buildings as apartments and hotels where floor loads are small, long spans are not required, and where plate soffits can be used as ceilings. As the spans and/or loads increase, the shear and flexural stresses in the slab at the columns become very high. For moderate spans and loads, extra reinforcement may be used in this region, in the form of embedded steel beams or closely spaced bars. However, for longer spans and heavier loads, the effective depth of the slab at the support regions has to be increased as is done in flat slab construction.

Flat slabs These are plates which are stiffened by drop panels and/or column capitals (Fig. 2.6). Longer spans and heavier floor loads are possible with flat slab systems because of increased shear and negative moment capacity. Flat slabs are suitable for office construction when drop ceilings are provided and are used for spans up to about 10 m.

Fig. 2.6 *Flat slab*

Concrete joist floors Also called ribbed slab, this consists of a thin slab (usually 50 to 100 mm thick) cast integrally with relatively narrow, closely spaced stems (ribs) arranged in a one or two-way pattern (Figs. 2.7 and 2.8). Ribs will have a thickness of 100 mm or more and a depth usually less than 3.5 times the minimum width. Two-way patterns are referred to as waffle slabs. Waffle slabs are built in 0.9 to 1.5 m modules, span from 9.0 to 12 m, and are about 300 to 600 mm deep

Fig. 2.7 *One-way joist floor*

(a) PLAN

(b) SECTION 'AA' (ENLARGED)

overall. One-way joists use pans 0.5 to 0.80 m wide to form the ribs, and have span-depth ratios similar to those of waffle slabs.

Beam and slab This type of construction consists of a slab built monolithically with beams (girders) forming a grid pattern where the slab can be one- or two-way in nature depending on the dimensions of the slab panel (Figs. 2.9a, b, c). Slabs are commonly 100 to 150 mm thick and span 3 to 7.5 m. Beam and slab construction

Fig. 2.8 *Two-way joist floor (waffle slab)*

(a) PLAN

(b) SECTION 'AA' (ENLARGED)

is a light-weight system which adapts easily to floor openings and can be used for a wide range of column spacings.

2.3.3 VERTICAL FRAMING ELEMENTS

The four types of vertical framing elements used in cast-in-place reinforced concrete construction include columns, walls, transfer girders and hangers. Columns may be of any reasonable shape and vary in size depending upon the tributary floor area, number of storeys, and strength of concrete and reinforcing steel. A 300 by 300 mm square column is considered to be the minimum size for common column spacings. Frequently, high strength concrete (up to 65 MPa) is used to minimize column size especially in the lower floors.

Fig. 2.9 *Beam and slab floor systems*

(a) TWO-WAY BEAM AND SLAB

(b) TWO-WAY BEAM AND SLAB   (c) ONE-WAY BEAM AND SLAB

Bearing walls vary in thickness depending upon slab spans, the number of storeys, and the influence of lateral loads on their design. In medium-rise residential structures, walls are commonly 125 to 200 mm thick. Where walls form part of the major lateral load resisting system such as core walls in office buildings, they may be considerably thicker.

Transfer girders are occasionally necessary where the architect or client wishes to increase column spacing in the lower floors for convention halls, lobbies, or parking areas. The perimeter girder shown in Fig. 2.10 is 1.22 m by 3.05 m deep and is used to transfer the exterior loads at ground level to cast-in-place reinforced concrete columns, spaced at 15.2 m centres.

Steel hangers can be used to suspend the floors of a multistorey structure from a central reinforced concrete core. Hanger loads can be transferred to the core by large cantilevered beams, cross-braced trusses, or Vierendeel trusses at the roof or at intermediate floors. Hangers take up very little of the floor space and provide an interesting effect at ground level where columns are eliminated.

Fig. 2.10 *Transfer girder*

Courtesy SWR Engineering Limited

2.3.4 LATERAL LOAD RESISTING SYSTEMS

Lateral load resisting structural systems are usually composed of one or more of the basic units such as frames, shear walls and tubes (exterior wall frames). Frames are composed of columns and slabs or beams, and rely on the rigidity of the joints and the resistance to flexure of the individual elements to resist lateral loads (Fig. 2.3b, and 2.11a). They are used as the sole lateral load resisting system in structures with up to 15 to 20 storeys.

Shear walls are solid deep vertical elements in buildings such as core walls, transverse walls, or facade walls (Fig. 2.11b). Shear wall deformation under lateral load is similar to the bending of a solid vertical cantilever.

Frames are commonly interconnected to shear walls. This system combines the structural advantages of both frames and walls (Fig. 2.11c). In the lower storeys, the wall restrains frame deformations and in the upper storeys the frame helps to

Fig. 2.11 *Lateral load resisting schemes*

| (a) RIGID FRAME ACTION | (b) SHEAR WALL | (c) INTERCONNECTED FRAME AND SHEAR WALL | (d) TUBE ACTION |

reduce the bending deformation of the wall. Interacting frames and walls are considered in structures up to 40 storeys.

Tubular schemes (Fig. 2.11d) are essentially peripheral moment resistant frame and wall systems. The National Life Assurance Company of Canada Building (Fig. 2.12) illustrates the tube concept with closely spaced exterior columns and relatively deep spandrel beams. The floor framing spans 12.2 m from perimeter to core providing column-free flexible office space.

Tubes can also be combined with central shear cores in the tube-in-tube system. In this system, the stiff exterior frame (or tube) and the shear core mutually complement each other. Tube-in-tube schemes are efficient up to 80 storeys.

An extension of the tube system is the rigid or modular tube in which the exterior frame walls are linked by interior shear walls. These walls greatly increase the structural efficiency of the system by reducing the "shear lag" in the "flange column" of the tube. Shear lag in tube structures is the phenomenon in which all columns in the windward and leeward sides of the structure (flange columns) do not contribute equally in resisting lateral loads.

Figure 2.13 illustrates the basic lateral load resisting systems and their practical range of application for buildings. In summary, overall structural integrity is obtained by mutually interacting horizontal and vertical subsystems which jointly resist both gravity loads and lateral loads.

Fig. 2.12 *National Life Assurance Company of Canada Building, Toronto*
(Architects: Neish Owen Rowland & Roy, Toronto
Structural Engineers: SWR Engineering Limited, Toronto)

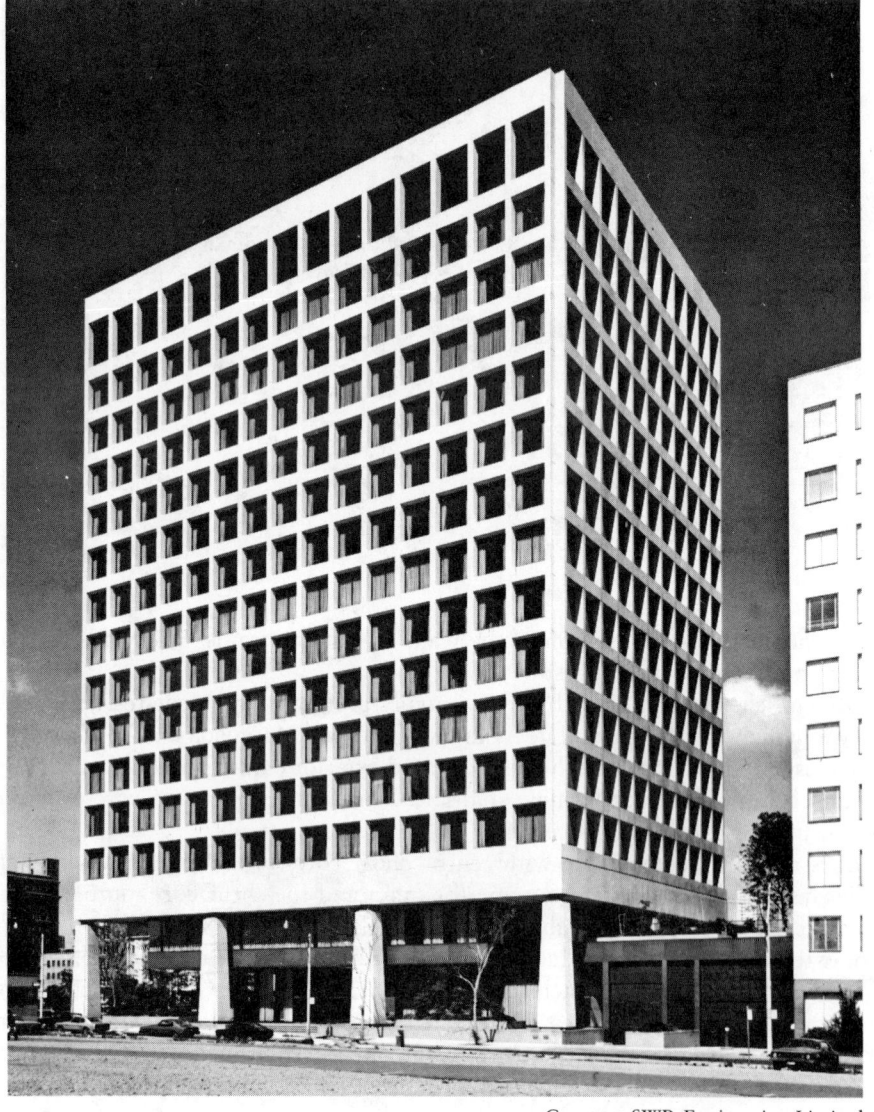

Courtesy SWR Engineering Limited

2.4 STRUCTURAL ANALYSIS AND DESIGN

A structure is a three-dimensional system; however, it is generally constructed as
an assemblage of two-dimensional or planar subsystems lying primarily in the hor-

Fig. 2.13 *Range of application of various framing systems*

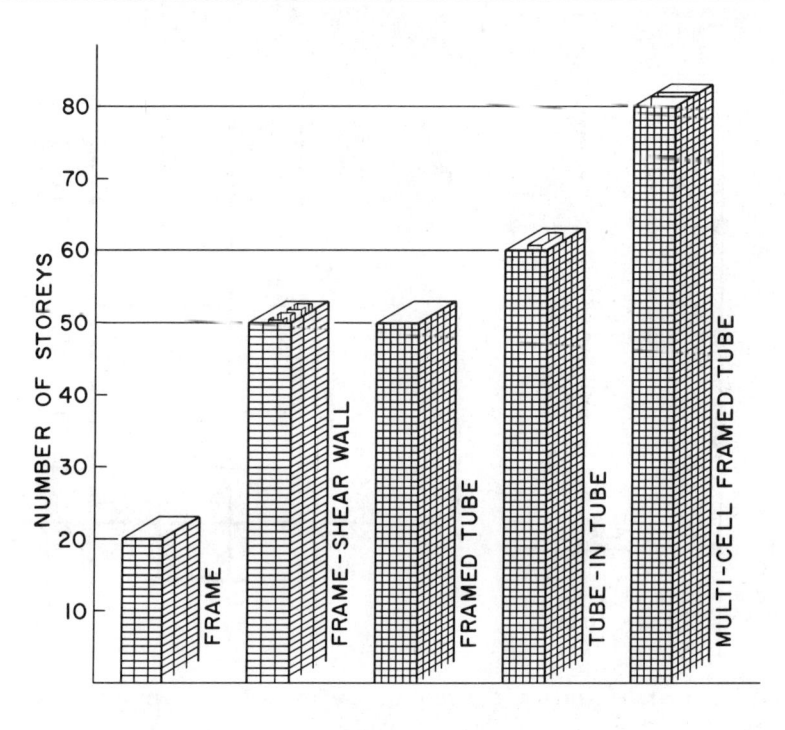

izontal and vertical planes, such as floors, roofs, walls and plane frames. The planar subsystems themselves may be an assemblage of individual elements such as slabs, beams and columns. The purpose of the analysis for load effects is to obtain all the forces and moments acting on these elements and the object of design is to provide adequate member sizes, reinforcement and connection details to withstand these forces and moments satisfactorily. Thus the basic design process involves the proportioning of individual members such as slabs, beams, columns, walls, etc., knowing the forces acting thereon, and the detailing of their connections. The methods for the design of these basic elements are described in subsequent chapters. Methods of analysis of structures for member forces are detailed in books on structural analysis (Refs. 2.4 and 2.5). However, Codes often permit approximations and these are discussed in Chapter 8.

EXAMPLE 2.1

The framing details for an 8-storey office building to be located in Toronto, Ontario, are shown in Figs. 2.14a,b. Compute the lateral wind loads to be used in the design of an interior frame in the short direction.

Fig. 2.14 Example 2.1

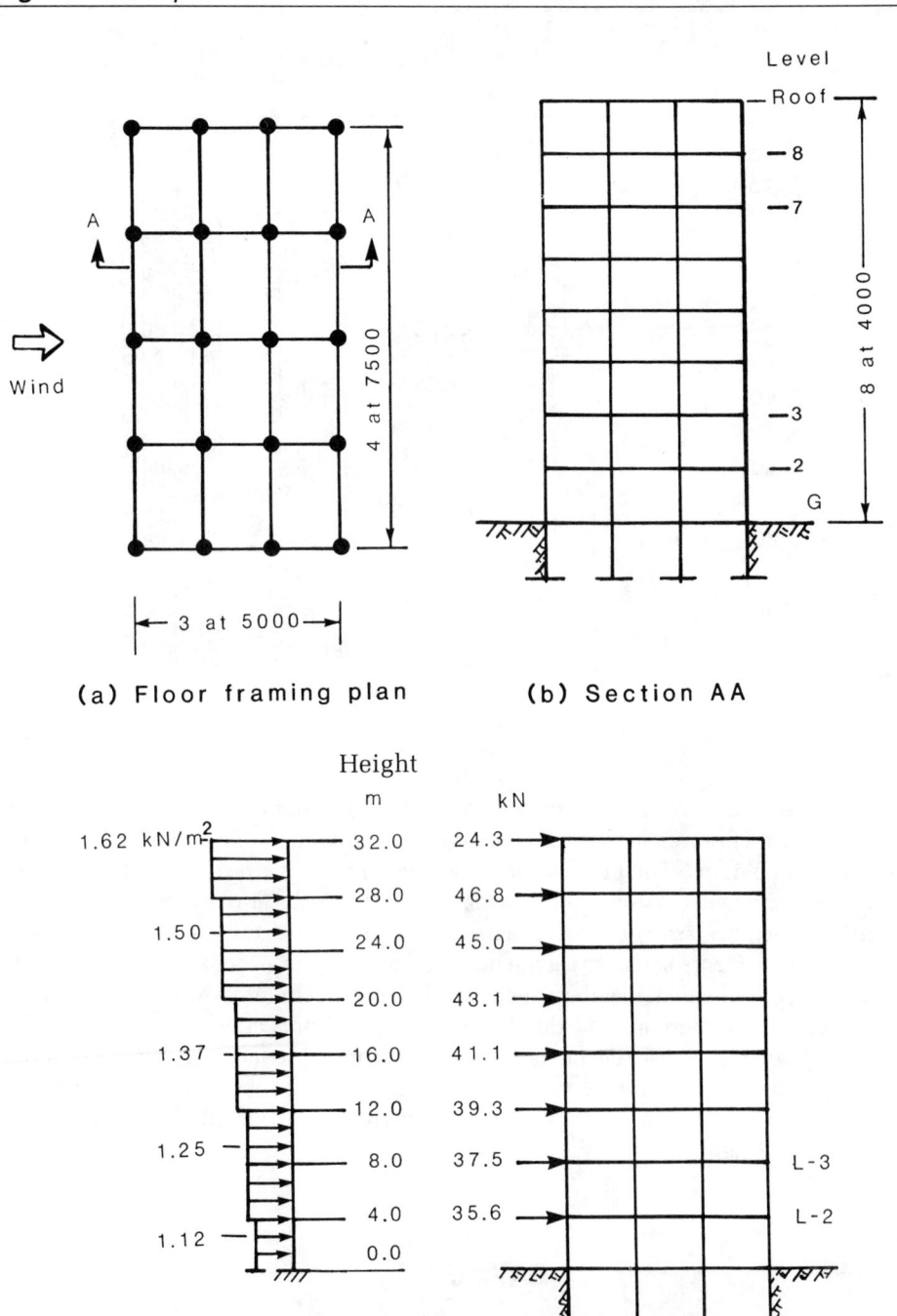

(a) Floor framing plan

(b) Section AA

(c) Net wind load

(d) Lateral wind load on interior frame

SOLUTION

The reference velocity pressure for a 30-year wind (to be used in the design of structural members) for Toronto (Ref. 2.2) is:

$$q = 0.48 \text{ kN/m}^2$$

The exposure factor, C_e, for various heights specified in Ref. 2.1 is given in column 2 of Table 2.4 below. The gust effect factor, $C_g = 2.0$. Referring to Fig. B-10 in Chapter 4 of Ref. 2.2 (for height > width), external pressure coefficient, C_p, has values of 0.8 for the windward face and -0.5 (suction) for the leeward face.

For net wind load, $C_p = 0.8 - (-0.5) = 1.3$
The total net wind pressure is,

$$p = q \, C_e \, C_g \, C_p$$
$$= 0.48 \, C_e(2)(1.3) = 1.248 \, C_e \text{ kN/m}^2$$

The variations of p with height are given in column 3 of Table 2.4. For convenience in computation, it will be conservatively assumed that the wind pressure intensity changes at the floor level as indicated in column 4 of Table 2.4 and in Fig. 2.14c. The tributary surface area for wind load per joint for wind along the short direction is $7.5 \times 4 = 30 \text{ m}^2$ (except for the joint at roof level, for which area = $7.5 \times 2 = 15 \text{ m}^2$).

The wind loads are:
at level 2, $1.12 \times 15 + 1.25 \times 15 = 35.6$ kN
at level 3, $1.25 \times 30 = 37.5$ kN, and so on.
The lateral wind loads acting on the frame are shown in Fig. 2.14d.

Table 2.4 Example 2.1

Height m	C_e	$p = 1.248 \, C_e$ kN/m²	Use for Height, m
0–6	0.9	1.12	0–4
6–12	1.0	1.25	4–12
12–20	1.1	1.37	12–20
20–30	1.2	1.50	20–28
30–44	1.3	1.62	28–32

REFERENCES

2.1 *National Building Code of Canada 1985*, Associate Committee on the National Building Code, National Research Council of Canada, Ottawa, 1985, 454 pp.

2.2 *The Supplement to the National Building Code of Canada 1985*, Associate Committee on the National Building Code, National Research Council of Canada, Ottawa, 1985, 278 pp.

2.3   CSA Standard CAN3-S6-M78 *Design of Highway Bridges*, Canadian Standards Association, Rexdale, Ontario, Canada, 1978, 272 pp.

2.4   Weaver, W., and Gere, J.M., *Matrix Analysis of Framed Structures*, 2nd Ed., Van Nostrand Company, New York, 1980, 492 pp.

2.5   Wang, C.K., *Statically Indeterminate Structures*, McGraw-Hill, New York, 1953.

CHAPTER 3

STRUCTURAL SAFETY

3.1 INTRODUCTION

Primary considerations in structural design are safety, serviceability, and economy. *Safety* requires that the structure function without damage under the normal expected loads (specified loads), and further that under abnormal but probable overloads including earthquake or extreme winds the likelihood of collapse (exceeding the load-carrying capacity, overturning, sliding, fracture and fatigue) be minimal. *Serviceability* requires that under the expected loads the structure performs satisfactorily with regard to its intended use, without discomfort to the user due to excessive deflection, permanent deformations, cracking, dynamic effects (vibration, acceleration), corrosion or any other similar ill-effects. Further, the overall structural system must have sufficient *structural integrity* to minimize the likelihood of a progressive type of collapse initiated by a local failure caused by an abnormal event or severe overload. Safety and serviceability can be improved by increasing the design margins of safety but this increases the cost of the structure. In considering overall economy, any such increased cost associated with increased safety margins must be balanced against the potential losses that could result from any damage.

3.2 SAFETY CONSIDERATIONS

The major quantities which are considered in the design calculations, namely the loads, dimensions, and material properties, are subject to varying degrees of uncertainty and randomness. Further, there are idealizations and simplifying assumptions used in the theories of structural analysis and design. There are also several other variable and often unforeseen factors that influence the prediction of ultimate strength and performance of a structure, such as construction methods, workmanship and quality control, probable service life of the structure, possible future change of use, frequency of loadings, etc. Thus the problem facing the designer is to design economically on the basis of "prediction through imperfect mathematical theories of the performance of structural systems constructed by fallible humans from material with variable properties when these systems are subjected to an unpredictable natural environment" (Ref. 3.1). It is obvious that any realistic and rational quantitative representation of safety must be based on statistical and probabilistic analysis.

Fig. 3.1   *Frequency distribution of load effects S and resistance R*

Considering all the uncertainties and randomness, the load effects, S, on a structure and the resistance (or strength), R, of the structure may be represented by two probability distribution curves as shown schematically in Fig. 3.1. It is assumed here that R and S are independent, which is approximately true for the case of static loading. If $R > S$ the structure is safe and if $R < S$ the structure will fail. It is apparent from Fig. 3.1 that there is always a probability, however small, that failure may occur. For a load effect S_1, the probability that the resistance $R < S_1$ is given by:

$$\int_0^{S_1} g(R)\, dR$$

The probability that the load effect is S_1 is given by $f(S_1)$ and accounting for all possible values of S_1 from 0 to α, the probability of failure (that is, $R < S$) is obtained as:

$$P_F = \int_0^\alpha f(S) \left[\int_0^s [\, g(R)\, dR \right] dS \tag{3.1}$$

Similarly, the probability of unserviceability such as exceeding a deflection limit may be obtained (Fig. 3.2) as:

$$P_u = \int_{\Delta_{all}}^\alpha h(\Delta)\, d\Delta \tag{3.2}$$

where $h(\Delta)$ is the frequency distribution for the serviceability parameter (deflection, crack width, etc.) and Δ_{all} is the limiting allowable value. A rational solution to the design problem would be to design the structure so as to limit its probability of failure (failure means collapse or unserviceability) to an "acceptable" low level. However, such a procedure is extremely difficult, if not impossible, for standard structures for several reasons (Ref. 3.1). Since losses associated with failures are influenced by economic, social and moral considerations which are difficult to quan-

Fig. 3.2 *Probability of serviceability*

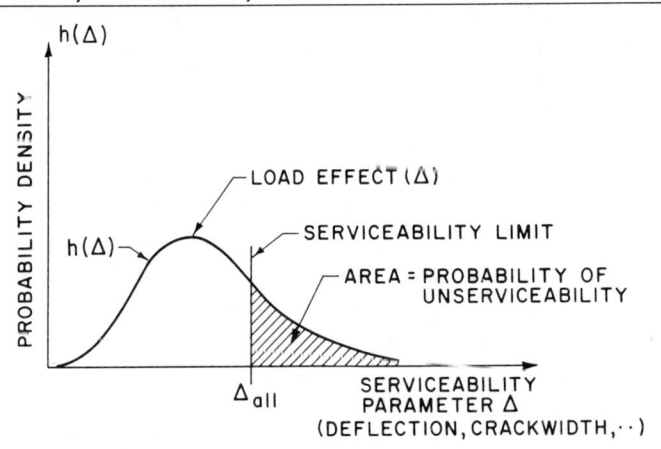

tify, generally acceptable failure probabilities are hard to define. Further, frequency distributions such as those shown in Figs. 3.1 and 3.2 are known only for a few simple cases. In addition much of the uncertainty associated with prediction of structural performance is of a non-statistical nature (e.g., relationship between laboratory strength and field strength, calculation and construction errors, relation between assumed and actual load distributions) (Ref. 3.3).

3.3 LIMIT STATES DESIGN

Because of the difficulties of a completely probabilistic design, simpler yet probabilistically sound design procedures have been developed (Refs. 3.1–3.5). One such procedure which has received increasing attention in recent years is the Limit States Design (LSD), also called Load and Resistance Factor Design (LRFD) (Refs. 3.1, 3.4, 3.5).

The concept of Limit States Design is simple. The probability of failure of a structure can be reduced by underestimating its resistance, R, and/or overestimating the load effects, S, and ensuring that $R \geq S$. Refining this concept on a probability basis, the LSD criteria (see Fig. 3.1) take the general form (expressed in simplified terms):

$$\phi R_n \geq \alpha S_n \tag{3.3}$$

where R_n is the *nominal resistance* computed on the basis of nominal material properties and dimensions; ϕ is the *resistance factor* (also known as capacity reduction factor or performance factor), always less than unity, which reflects the uncertainties in determining R_n; S_n is the *nominal load effect* based on specified loads; and α is a *load factor*, usually greater than unity, which reflects potential overloads and uncertainties associated with the determination of S_n. As already mentioned the primary requirements of a structure are that it be serviceable and safe from collapse. R_n, the resistance, is computed as the value corresponding to a condition of a failure or *Limit State* being just reached and the relation in Eq. 3.3 ensures that this limit state is avoided. *Limit states* define the various structural conditions representing collapse and unserviceability that are to be avoided. Those states concerning safety are called *ultimate limit states* (e.g., strength, buckling, overturning, sliding, fracture, fatigue, etc.), and those states which restrict the intended use and occupancy are called *serviceability limit states* (e.g., deflection, cracking, spalling, vibration, etc.). R_n is a generalized force (moment, axial force, shear force, etc.) for ultimate limit states and is an allowable limit of structural response (such as allowable deflection, crack width, stress, vibration, etc.) for serviceability limit states. A relation of the type in Eq. 3.3 must be satisfied for each limit state and for each set of load combinations applicable for the member or structure being designed.

Equation 3.3 is a simplified representation of Limit States Design criteria. In practice, the right hand side of this equation will be partitioned into the separate load effects (dead load, live load, wind load, etc.) with corresponding individual load factors so that:

$$\alpha S_n = \alpha_D S_{Dn} + \alpha_L S_{Ln} + \alpha_W S_{Wn} + \text{----} \tag{3.4}$$

where S_{Dn}, S_{Ln}, S_{Wn} are nominal load effects due to dead, live and wind loads, respectively, and α_D, α_L, α_W are corresponding load factors.

Individual load factors for each load type can better reflect the varying degrees of uncertainties associated with different types of load. For example, dead load, being predictable with better precision, would have a lower load factor than that for live loads. A load factor may also have different values depending on whether the effect of the load is to increase or decrease the total load effect. Thus in considering failure by overturning where the dead load helps prevent failure, the probability of failure is higher should the actual dead load be less than anticipated. In this case the load factor for dead load will have to be less than unity. Equation 3.3 can be made more realistic by introducing additional factors or by adjusting the factors ϕ and α to reflect the lower probability of several loads acting concurrently at their full intensities (load combination factor), to account for the seriousness of failure of the member or structure under consideration (importance factor), to allow for the type of failure (sudden as opposed to gradual failure with warning), and to allow for uncertainties associated with structural analysis.

Acceptable values for all these factors (collectively termed partial safety factors) can be determined by statistical evaluation of data on strength, loads and other aspects related to the factor and then calibrating the failure probability to those associated with current accepted practice (see Section 3.4).

Equation 3.3 may be rearranged as:

$$\frac{\phi}{\alpha} R_n \geqslant S_n \tag{3.5}$$

$$R_n \geqslant \frac{\alpha}{\phi} S_n \tag{3.6}$$

These two forms are comparable with the allowable stress design procedure (Eq. 3.7) wherein a Factor of Safety is applied to the material strength (underestimating the resistance, R) and the plastic design procedure (Eq. 3.8) where all the working loads are multiplied by Load Factors (overestimating load effect, S), respectively.

$$\left(\frac{1}{F.S.}\right) R_n \geq S_n \tag{3.7}$$

$$R_n \geq (L.F.) S_n \tag{3.8}$$

Since the use of a single Factor of Safety as in Eq. 3.7 does not provide uniform reliability and hence economy, structural codes have been gradually moving towards a method involving several partial safety factors (ACI, CSA, NBC). The separate consideration of loads, materials and performance by means of statistically determined partial safety factors in Eqs. 3.3 and 3.4 makes the design more responsive to the differences between types of loads, kinds of materials, types of structural behaviour, and degrees of seriousness of different modes of failure (failure here means unsatisfactory performance).

3.4 PARTIAL SAFETY FACTORS

A basis for the development of partial safety factors is presented in Ref. 3.4. Referring to Fig. 3.1, failure is characterized by $R < S$, or alternatively the probability of failure, P_F, is given by:

$$P_F = P\left[\ln\frac{R}{S} < 0\right] \tag{3.9}$$

Figure 3.3 shows the frequency distribution for $\ln R/S$, and the probability of failure is given by the shaded area to the left of the origin. This area is a function of the number, β, of standard deviations, $\sigma_{\ln R/S}$, between the origin and the median, $(\ln R/S)_m$. The number β is a relative measure of the degree of safety and is defined as the *safety index*. The higher the value of β, the less the probability of failure and the higher the factor of safety. By adjusting β to be constant for all structural elements, a uniform safety could be achieved. If the actual distribution curves such as shown in Fig. 3.3 were known, the value of β could be evaluated and related to a desired probability of failure. In the absence of such data, acceptable values for β are established by a process of *calibrating* the probabilistic failure criteria to existing design standards (such as the allowable stress design method). On this basis, values of 3 to 4 have been used for β.

Reference 3.4 uses such a *calibration* technique to determine the required partial safety factors. Values for the resistance factor, ϕ, and the load factor, α, for different load combinations are assumed and then β is determined for standard structural elements and for a domain of relevant design parameters (such as dead load/live load ratios, dead load/wind load ratios). The variation of β for the same elements and over the same domains is also determined for the traditional design/specification

Fig. 3.3 *Definition of safety index*

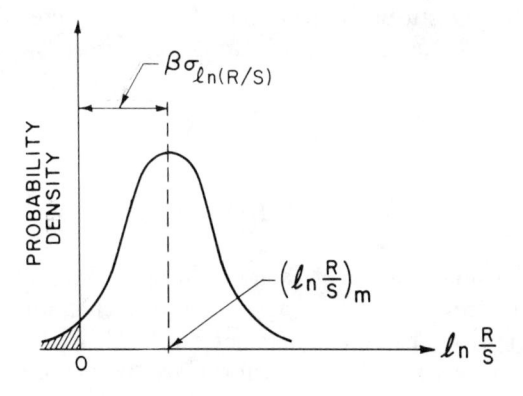

which on the basis of past experience represents a measure of acceptable safety factors. The values of the ϕ and α factors are than adjusted until the value of β is reasonably constant over the whole domain and is comparable with that of the traditional design practice over at least part of the domain (Fig. 3.4). This process ensures a safety index which is more consistent than, and comparable to, those inherent in the existing design practice.

Fig. 3.4 *Schematic representation of calibration of limit states design*

DOMAIN OF DESIGN PARAMETERS

3.5 LIMIT STATES DESIGN BY CANADIAN CODES

CSA Standard CAN3-A23.3-M84, *Design of Concrete Structures for Buildings*, is based on NBC 85 (Ref. 3.6) and these codes present a unified limit states design philosophy, with common *safety* and *serviceability* criteria for all materials and methods of construction. In this method, the design is aimed at avoiding the attainment of *Limit States* which define the onset of various types of collapse and unserviceability. Both ultimate limit states which concern safety (e.g., attainment of load carrying capacity, fracture, overturning, etc.) and serviceability limit states (e.g., limiting deflection, cracking, vibration, etc.) are considered in the design. The probability of attainment of any of the limit states (that is, the probability of failure) is kept to an acceptably low level with the use of load factors, α, load combinations factors, ψ, and importance factors, γ, applied to *specified loads*, and resistance factors, ϕ, applied to specified material properties or to the resistance of a member, connection or structure which for the limit state under consideration takes into account the variability of dimensions and material properties, workmanship, type of failure and uncertainty in the prediction of resistance. The resistance of a cross-section, member, connection or structure, including application of the appropriate resistance factors is termed the *factored resistance*. A structure and its components are designed such that:

$$\text{factored resistance} \geq \text{the effect of factored loads} \qquad (3.10)$$

Here the *effect of factored loads* means the structural effect (such as bending moments, shear force, axial force, torsion) due to the factored load combinations obtained by multiplying the *specified loads* by appropriate load factors, α, a load combination factor, ψ, and an importance factor, γ. This factored load combination is taken as:

$$\alpha_D D + \gamma\psi[\alpha_L L + \alpha_Q Q + \alpha_T T] \qquad (3.11)$$

and correspondingly Eq. 3.10 can be expressed as:

$$R_r \geq \text{effect of } [\alpha_D D + \gamma\psi(\alpha_L L + \alpha_Q Q + \alpha_T T)] \qquad (3.11a)$$

where R_r = the factored resistance, including the application of resistance factors ϕ

and D, L, Q, T are specified loads, with

> D = dead loads
> L = live loads due to intended use and occupancy (including vertical loads due to cranes); snow, ice and rain; earth and hydrostatic pressure; horizontal component of static or inertia forces
> Q = wind or earthquake whichever produces the more unfavourable effect
> T = loads due to the cumulative effects of temperature, creep, shrinkage, and differential settlement
>
> α = load factors which take into account variability of the load and load patterns, and analysis of their effects
> ψ = load combination factor which takes into account the reduced probability of simultaneous occurrence of several loads at their factored values
> γ = importance factor to take into account the consequences of collapse
> ϕ = resistance factor applied to specified material properties or to the resistance of a member to account for the variability of material

properties and dimensions, workmanship, type of failure (brittle or ductile) and uncertainty in the prediction of resistance

For serviceability limit states, the design criterion in Eq. 3.10 can be expressed in the form of Eq. 3.12, in which the left hand side, S_{lim}, represents the limiting value of a serviceability parameter such as deflection, crack width, etc., and the right hand side is the same parameter computed for the combinations of specified loads shown (i.e., with load factors, $\alpha = 1.0$).

$$S_{lim} \geq \text{effect of } [D + \psi(L + Q + T)] \tag{3.12}$$

No single structural theory for analysis and design (e.g., elastic theory, plastic theory, ultimate strength theory, etc.) may be universally applicable to all limit states. The appropriate theory to be used is generally indicated in the relevant standard (e.g., CSA A23.3-M84 for concrete). In general, elastic theory is applicable to serviceability limit states and fatigue; and one of the ultimate strength theories (allowing for inelastic material behaviour) is applicable to the ultimate limit states. The factors applied to the specified loads (γ, ψ, α) are common to all materials and types of construction and are specified in Section 4.1 of NBC 85 (Ref. 3.6). The resistance factors, ϕ, which are applied to the specified material properties or to the member resistance are specified in the respective structural material standard (e.g., CSA A23.3-M84 for concrete building structures, CSA S16.1 for steel building structures, etc.). The specified loads and specified material properties used to compute member resistance are defined statistically on the basis of probability of occurrence (e.g., a return period of 10 to 100 years for wind, snow and earthquake loads; and a 10 percent maximum probability of underrun for material properties (Ref. 3.7)). For live loads due to use and occupancy, for which statistical data is not available, nominal values are specified which represent upper limits of loads that could be expected normally. Specified loads and load combinations to be used in design of buildings are given in Section 4.1 of NBC (Ref. 3.6). Similar data for highway bridges are given in Ref. 3.8.

3.6 LOAD AND RESISTANCE FACTORS FOR DESIGN OF REINFORCED CONCRETE BUILDINGS

CSA Standard CAN3-A23.3-M77 contained a set of load factors which were different from those specified in the National Building Code of Canada. Furthermore, the *factored resistances* used to be computed by applying a factor, ϕ, termed *capacity reduction factor* to the nominal resistance of the cross section of the member computed using specified material strengths.

The 1984 revision of CSA Standard CAN3-A23.3-M84 has adopted the load factors specified in the National Building Code of Canada and has further adopted the method of applying *resistance factors* to the material strengths (although in a few cases the resistance factor is applied to the member resistance) as previously explained in Section 3.5.

The various factors to be used with Eq. 3.10–3.12, as specified in CAN3-A23.3-M84 are as follows:

(i) Load factor, α, to be used in Eq. 3.11:

    (a) $\alpha_D = 1.25$, except that when considering overturning, uplift or stress reversal where the effect of dead load is to resist these, α_D is taken as 0.85;

    (b) $\alpha_L = 1.50$;

    (c) $\alpha_Q = 1.50$; and

    (d) $\alpha_T = 1.25$.

(ii) Load combination factor, ψ:

    (a) $\psi = 1.0$, when only one of the loads L, Q and T is included in Eq. 3.11 or 3.12;

    (b) $\psi = 0.70$, when two of the loads L, Q and T are included in Eq. 3.11 or 3.12; and

    (c) $\psi = 0.60$, when all three of the loads L, Q and T are included in Eq. 3.11 or 3.12.

Design is to be made for the most unfavourable effect considering L, Q or T acting alone or in combinations with corresponding ψ values.

(iii) Importance factor, γ, to be used in Eq. 3.11 is not less than 1.0 for all buildings, except that for buildings where it can be shown that collapse is not likely to cause injury or other serious consequences, it is not less than 0.8.

(iv) Material resistance factors, ϕ:

    (a) $\phi_c = 0.60$ for concrete strengths in checking ultimate limit states;

    (b) $\phi_p = 0.90$ for prestressing tendons;

    (c) $\phi_s = 0.85$ for reinforcing bars; and

    (d) $\phi_a = 0.90$ for structural steel

Thus in calculation of resistance at ultimate limit state, the material strengths to be used are $\phi_c f'_c$ and $\phi_s f_s$, where f'_c is the specified concrete strength and $f_s = E_s \epsilon_s \leq f_y$ is the stress in steel as specified. The specified material strength and the factored material strength are shown in the stress–strain relationship in Fig. 3.5.

Fig. 3.5 *Factored material strength*

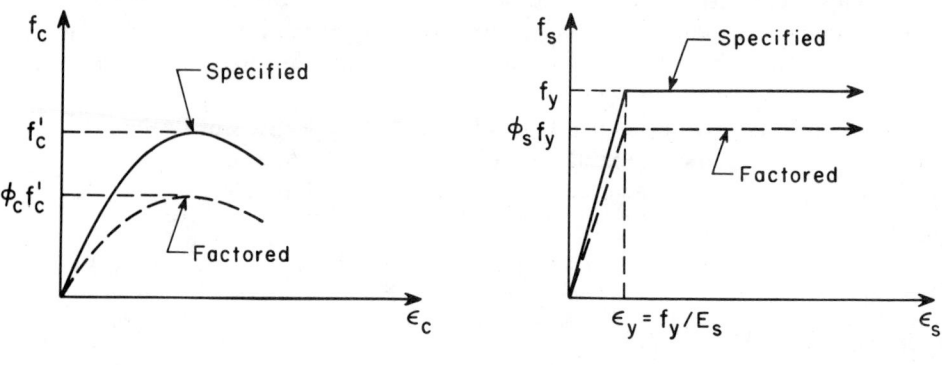

(a) Concrete, $\phi_c = 0.60$ (b) Reinforcing steel, $\phi_s = 0.85$

In addition to the material resistance factors given above, the Code also specifies member resistance factors of $\phi_m = 0.65$ for use in certain slender column calculations and $\phi_m = 0.70$ for certain brittle members in earthquake-resistant design.

3.7 NOTATION

The load effects such as moment, shear force, etc., at vertical sections on members as determined from the analysis of the structure under factored load combinations will be identified by a subscript f (e.g., factored moment, M_f, factored axial load, P_f, factored shear force, V_f, etc.). These factored load effects may also be designated as the *design moment*, *design shear*, etc., as these represent the minimum strengths for which the section/member must be designed. The subscript r will be used for the factored resistance, which is the resistance of a cross-section or member computed theoretically using factored material strengths (or, where applicable, a member resistance factor) and the design dimensions of the member. The design criteria for ultimate limit state require that:

$$M_r \geqslant M_f, \quad V_r \geqslant V_f, \quad P_r \geqslant P_f$$

Hence the design basis is:

$$M_r = M_f, \quad V_r = V_f, \quad P_r = P_f \tag{3.13}$$

3.8 TOLERANCES

Overall cross-sectional dimensions of reinforced concrete members such as slabs, beams, columns, and walls are normally specified in multiples of 10 mm, except that 5 mm increments may be used for thin slabs. Dimensional tolerances are specified as the maximum allowable variations (plus or minus) from the specified dimensions. Allowable variations for cross-sectional dimensions of beams, girders, and columns and for thickness of walls and slabs specified in CSA A23.3-M84 are given in Table 3.1. Allowable tolerances for placing reinforcement are presented in Table 3.2.

Table 3.1 Dimensional Tolerances for Cross-Sections of Beams, Girders, and Columns and for Thickness of Walls and Slabs Other Than Slabs on Grade (CSA A23.3-M84)

Dimension	Allowable variation
Less than 0.3 m	± 8 mm
Greater than 0.3 m but below 1 m	± 12 mm
Greater than 1 m	± 20 mm

Table 3.2   Tolerances for Placing Reinforcement

	Application	Tolerance
a)	For clear concrete protection of reinforcement	± 8 mm
b)	When depth of flexural member, thickness of a wall or smallest dimension of a column is:	
	(i) 200 mm or less	± 8 mm
	(ii) larger than 200 mm but less than 600 mm	± 12 mm
	(iii) 600 mm or larger	± 20 mm
c)	Lateral spacing of bars	± 30 mm
d)	Longitudinal location of bends and bar ends	± 50 mm
e)	As item (d) at discontinuous ends of members	± 20 mm

Note: Regardless of these tolerances, the concrete cover shall in no case be reduced by more than one third of the specified cover.

REFERENCES

3.1   Cornell, C.A., *A Probability-Based Structural Code*, Journal of ACI, Vol 66 (12), Dec. 1969, pp. 974–985.

3.2   Allen, D.E., *Limit State Design — A Unified Procedure for the Design of Structures*, Eng. Journal, Vol 53 (2), Feb. 1970, pp. 18–29.

3.3   Ang, A.H.S., and Amin, M., *Safety Factors and Probability in Structural Design*, ASCE Journal of Struct. Div., Vol 95 (7), July 1969, pp. 1389–1405.

3.4   Allen, D.E., *Limit States Design — A Probabilistic Study*, Can. Journal of Civ. Eng., Vol 2, 1975, pp. 36–49.

3.5   Ravindra, M.K., and Galambos, T.V., *Load and Resistance Factor Design for Steel*, ASCE Journal of Struct. Div., Vol 104, No. ST9, Sept. 1978, pp. 1337–1353.

3.6   *National Building Code of Canada 1985*, National Research Council of Canada, Ottawa, 1985, 454 pp.

3.6   *Supplement to National Building Code of Canada 1985*, National Research Council of Canada, Ottawa, 1985, 278 pp.

3.8   CSA Standard CAN3-S6-M78, *Design of Highway Bridges*, Canadian Standards Association, Rexdale, Ontario, Canada, 1978, 272 pp.

CHAPTER 4

BEHAVIOUR IN FLEXURE

4.1 THEORY OF FLEXURE FOR HOMOGENEOUS MATERIALS

Sections of straight prismatic members loaded transversely through the shear centre are subjected to a bending moment, M, and a transverse shear force V (Fig. 4.1). For equilibrium, the resultants of the internal stresses in the section must balance the moment and shear force due to external loads. For convenience, the influences of the bending moment and shearing forces are considered separately. The behaviour in shear will be considered in Chapter 6. Considering only the applied moment, M, for static equilibrium, the resulting internal normal stresses across the cross-section must reduce to a pure couple (Fig. 4.1c).

The fundamental assumption in flexural theory is that plane cross-sections taken normal to the beam axis before loading remain plane even after the beam is subjected to bending. The assumption is generally valid for beams of usual proportions (Refs. 4.1, 4.2). For initially straight members, this assumption means that the normal strains at points in the beam section are proportional to the distance of such points from the neutral axis (Figs. 4.2a, b). The flexural stress at any point such as A (Fig. 4.2b) in the cross-section depends on the strain, ϵ_A, at that level and is given by f_A,

Fig. 4.1 *Flexure of transversely loaded beam*

Fig. 4.2 *Homogeneous section under flexure*

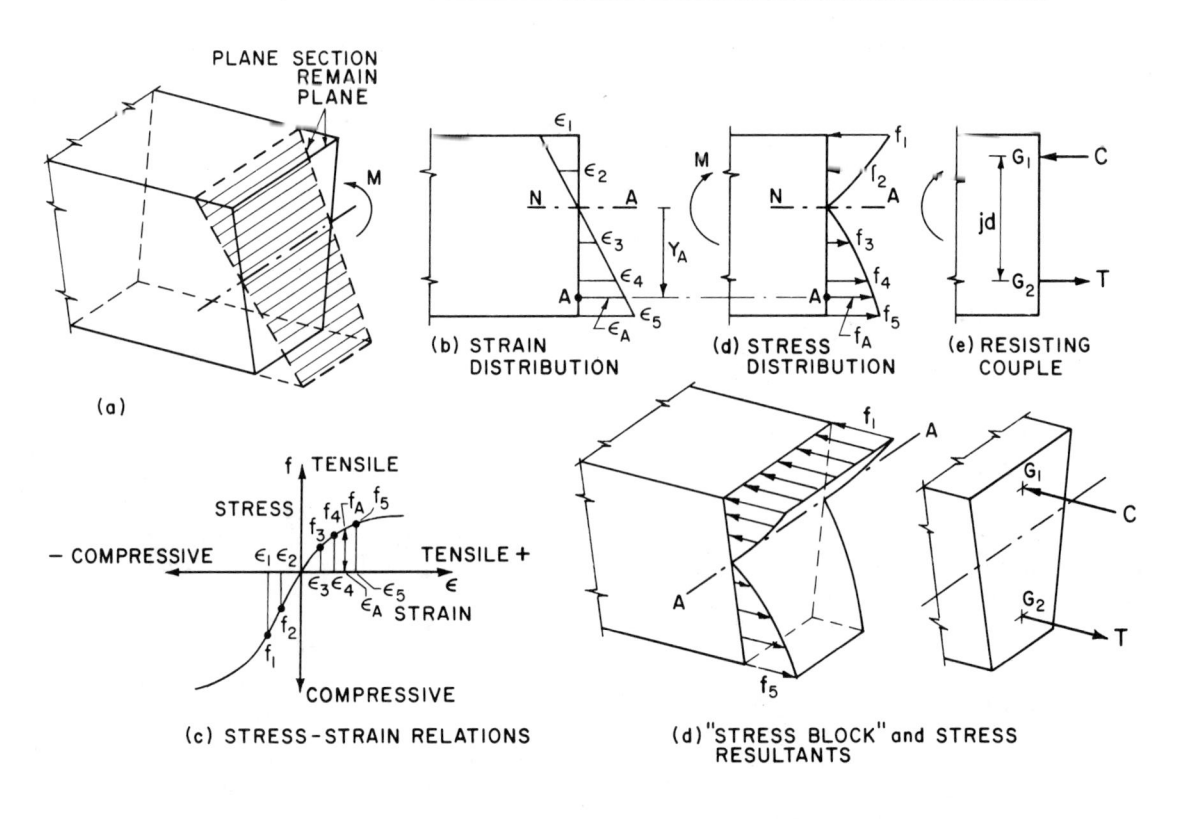

(b) STRAIN DISTRIBUTION

(d) STRESS DISTRIBUTION

(e) RESISTING COUPLE

(a)

(c) STRESS-STRAIN RELATIONS

(d) "STRESS BLOCK" and STRESS RESULTANTS

the stress corresponding to the strain ϵ_A, as determined from the stress–strain relation for the material (Fig. 4.2c). The neutral axis location is such that the equilibrium condition $C = T$ is satisfied, and the moment of the internal resisting couple is given by $M = (C$ or $T) \, jd$ where jd is the lever arm of the couple. G_1 and G_2 represent the centres of compression and tension (locations through which C and T act), respectively.

If the material of the beam obeys Hooke's Law (that is, it is linearly elastic), the stress will be proportional to strain ($f = E\epsilon$ where E is the Young's modulus of elasticity), and the stress distribution in the section will also be linear (Fig. 4.3). (It may be noted that the material does not have to be *elastic* to satisfy these equations and it is only necessary that stress be proportional to strain. However, traditionally a linear stress–strain relation has been associated with elastic behaviour and generally the latter is implied to mean the former as well.) In this case (assuming that E has the same value in both tension and compression) it is easily shown (Ref. 4.3) that the neutral axis passes through the centroid of the section and the stress, f, at any point in the section at a distance y from the neutral axis, is given by Eq. 4.1, or alternatively the resisting moment in terms of the extreme fibre stress is given by Eq. 4.2.

$$f = \frac{My}{I} \tag{4.1}$$

$$M = \frac{f_{max} I}{y_{max}} \tag{4.2}$$

where f_{max} is the stress in the extreme fibre
y_{max} is the distance of the extreme fibre from the neutral axis
I is the moment of inertia of the section
M is the bending moment at the section

Fig. 4.3 *Linear elastic stress distribution in flexure*

(a) BEAM   (b) SECTION   (c) STRAIN   (d) STRESS DISTRIBUTION   (e) STRESS-STRAIN RELATIONSHIP

4.2 ELASTIC BEHAVIOUR OF COMPOSITE SECTION — CONCEPT OF TRANSFORMED SECTION

The concept of a transformed section is useful in reinforced concrete when elastic stress distribution is considered. In a composite beam made of two (or more) different elastic materials (Fig. 4.4a) in which the two parts are bonded together to act as an integral unit (that is, there is no slip between the two parts), the assumption that initially plane cross-sections remain plane while subject to bending is valid. Hence the strain variation in the section will be linear (Fig. 4.4b) and flexural stress at any point in the section is given by the product of the strain at that point and the elastic modulus of the material as shown in Fig. 4.4c. At a depth y below the neutral axis where the strain is ϵ_y, the stress in material 1 is $f_{1y} = E_1 \epsilon_y$ and the stress in material 2 is $f_{2y} = E_2\epsilon_y = nf_{1y}$, where $n = E_2/E_1$ is termed the modular ratio. If the section is to be transformed into an equivalent homogeneous section made up of one material, say material 1, the transformation of material 2 must be such that the forces across the section are unaltered. The total force in the element δA (Fig. 4.4a, c) of material 2 is

$$dF_2 = b_2 dy f_{2y} = b_2 dy(nf_{1y}) = (nb_2)dy\, f_{1y}$$

Fig. 4.4 *Transformed section*

(a) ACTUAL SECTION   (b) STRAIN ϵ

(c) ACTUAL STRESSES

(d) TRANSFORMED SECTION
ALL OF MATERIAL 1

(e) STRESSES IN
TRANSFORMED
SECTION

Thus $b_2 dy$ of material 2 is equivalent to $nb_2 dy$ of material 1. Hence material 2 may be transformed to material 1 by changing its dimension parallel to the neutral axis (here b_2) by a factor equal to the modular ratio $n = E_2/E_1$. In the transformed homogeneous section (all of material 1), both strains and stresses vary linearly and hence the neutral axis will pass through its centroid. The stress distribution computed for the transformed section ($f_T = My/I_T$) gives the actual stresses in the real section for the material of which the transformed section is made (here $f_1 = f_T$). For the second material, the actual stresses at any point will be n times the stress at that level computed for the transformed section ($f_2 = n f_T$).

EXAMPLE 4.1

A reinforced concrete beam of rectangular section has the cross-sectional dimensions shown in Fig. 4.5a. The concrete, which is of normal density, has a compressive strength of 30 MPa and a modulus of rupture of 3.3 MPa. The yield strength of the steel is 400 MPa. Compute (*a*) the stresses due to an applied bending moment of 45 kN·m and (*b*) the bending moment at which cracking of concrete will be initiated.

Fig. 4.5 Example 4.1

(a) Beam properties   **(b) Transformed section**   **(c) Stress**

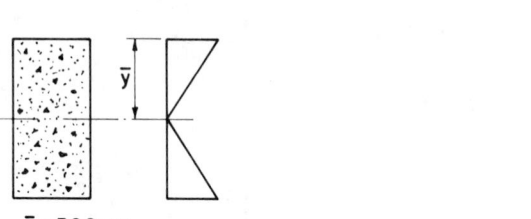

$\bar{y} = 300\,mm$
$I_g = 5.4 \times 10^9 mm^4$

(d) Gross concrete section

SOLUTION

a) Initially it will be assumed that at the applied moment the concrete has not yet cracked. Prior to cracking the full section is effective and since both tensile and compressive stresses will be very low the stress–strain relation for concrete in both tension and compression may be assumed to be linear (see Section 1.5) with the same value for elastic modulus E in both compression and tension. It is further assumed that there is perfect bond between the reinforcing bars and surrounding concrete.

From Eq. 1.1, for concrete, $E_c = 5000 \sqrt{30} = 27\,400$ MPa and

for steel $\quad E_s = 200$ GPa

Reinforced concrete sections are usually transformed into equivalent concrete sections.

The modular ratio, $n = \dfrac{E_s}{E_c} = \dfrac{200\,000}{27\,400} = 7.3$

The transformed section is shown in Fig. 4.5b. Note that the transformed section consists of the *gross* concrete section plus n times the area of steel A_s minus the concrete area displaced by the embedded bars, A_s. The depth to neutral axis (centroid) $\bar{y}$ and the moment of inertia I_T of the transformed section may be computed as $\bar{y} = 315$ mm and $I_T = 6.02 \times 10^9$ mm⁴. The compressive stress in the concrete at the top fibres of the section is:

$$f_{cc} = \frac{M y_c}{I_T} = \frac{45 \times 10^6 \times 315}{6.02 \times 10^9} = 2.35 \text{ MPa}$$

and the tensile stress at the bottom fibres of concrete is:

$$f_{ct} = \frac{M y_t}{I_T} = \frac{45 \times 10^6 \times 285}{6.02 \times 10^9} = 2.13 \text{ MPa} < f_r$$

The stress in steel is given by n times the stress in concrete at the steel level, and is:

$$f_s = n \left(\frac{f_{ct}}{y_t} y_s \right) = 7.3 \frac{2.13 \times 215}{285} = 11.7 \text{ MPa}$$

Since the tensile stress in concrete is below the modulus of rupture, the assumption that the section is uncracked is correct. Both concrete and steel stresses fall within the initial linear ranges of the stress–strain relations and hence the assumption of linear elastic behaviour is justified.

b) As the applied moment is increased, the stresses also increase and cracking of concrete on the tension side will begin as the maximum tensile stress, f_{ct}, reaches the tensile strength in flexure, i.e., the modulus of rupture, f_r. Equating the maximum tensile stress to f_r, the corresponding moment, M_{cr}, is given by

$$f_{cr} = \frac{M_{cr} y_t}{I_T}$$

$$M_{cr} = f_{cr} \frac{I_T}{y_t} = 3.3 \times \frac{6.02 \times 10^9}{285} \text{N·mm} = 69.7 \text{ kN·m}$$

Notes: In the precracking stage, the influence of the reinforcement is usually negligible. In the above example, if the reinforcing steel area was neglected and the section considered as having a gross concrete section 300 mm by 600 mm, (Fig. 4.5d), $\bar{y} = 300$ mm and the moment of inertia $I_g = 5.4 \times 10^9$ mm⁴, the cracking moment is:

$$M_{cr} = \frac{3.3 \times 5.4 \times 10^9}{300} \times 10^{-6} = 59.4 \text{ kN·m}$$

which is in error by 14.7 percent on the conservative side. For computing M_{cr} in deflection calculations, CSA A23.3-M84(9.5.2.3) recommends the use of the gross concrete section properties instead of the more accurate full transformed section properties.

4.3 THE MODULAR RATIO AND TRANSFORMED STEEL AREA

In reinforced concrete, the modular ratio, $n = E_s/E_c$, is usually taken as the nearest whole number, but not less than 6. CSA Standard CAN3-S6-M78-*Design of Highway Bridges* (Ref. 4.4) recommends that the value of n for low density concrete be taken as the same as for normal density concrete of the same strength, except for calculation of deflections.

The tensile reinforcement area, A_s, can be transformed to concrete by multiplying by the modular ratio, n, since the concrete in tension is usually neglected in strength computations. For compression reinforcement area, A_s', for elastic behaviour, the effective transformed area is $(n - 1) A_s'$, which allows for the concrete in compression displaced by the steel area A_s'.

The creep of concrete and the nonlinearity of the stress–strain relation for concrete at higher stresses lead to larger compressive strains in the compression steel and surrounding concrete than the elastic theory would indicate. To account for this, CSA S6-M78 (8.5) recommends the use of an effective modular ratio of $2E_s/E_c$ for transforming any compression reinforcement for stress computations in flexural members. The associated steel stress will then be $2nf_c$ where f_c is the compressive stress in the concrete at the steel level.

4.4 FLEXURE OF REINFORCED CONCRETE BEAMS

In the discussion that follows, explaining the general behaviour of a reinforced concrete section under flexure, the material resistance factors, ϕ_c and ϕ_s, are not taken into account. However, in design in accordance with the Code, the ϕ-values will have to be incorporated as will be illustrated later on.

The computations in Example 4.1 show that concrete on the tension side of a beam begins to crack at very low stresses. A plain concrete beam will fail at a moment equal to the cracking moment M_{cr} due to the tensile failure of the concrete while the maximum compressive stress in the concrete is still only a small fraction of the compressive strength. In order to utilize the higher compressive strength of concrete, the beam is reinforced with steel bars placed on its tension side. Thus reinforced concrete is a nonhomogeneous material consisting of a judicious combination of concrete and reinforcing steel. It is assumed that there is perfect bond between the reinforcing bars and surrounding concrete (Chapter 9). Hence the strain variation in the cross-section is linear (proportional to the distance from the neutral axis), with the strain in the reinforcement being the same as that in the surrounding concrete.

Now consider a beam with gradually increasing load, Fig. 4.6. In the initial phase when the applied moment is less than the cracking moment, M_{cr}, the maximum tensile stress in concrete is less than the tensile strength, the concrete is uncracked, the entire section is effective, and the strain and stress distribution is as shown in Fig. 4.6c (see Example 4.1). As the applied moment is increased beyond the cracking moment, flexural tensile cracks develop in the bottom (tensile) fibres of the beam and propagate gradually towards the neutral axis (Fig. 4.6d). Since the cracked portion of the concrete is now ineffective in resisting tensile stresses, the effective concrete section is reduced by that much and the neutral axis also shifts upwards. Cracks cannot be eliminated altogether from reinforced concrete flexural members under the normal range of applied loads. However, by proper design (Chapter 11) they can be controlled so that there will be several well-distributed fine hairline cracks rather than a few wide cracks. Minute hairline cracks do not seriously damage either the external appearance of the member or the corrosion protection of the steel bars and hence are acceptable in normal situations.

Because of the tensile cracking of concrete at very low stresses, it is generally assumed in flexural computations that concrete has no tensile resistance. This is

Fig. 4.6 *Behaviour of reinforced concrete beam under increasing moment*

(a) Beam and loading

(b) Stress–strain relation for concrete

(i) Beam uncracked (ii) Effective section (iii) Transformed section (iv) Strain (v) Stress

(c) Uncracked phase

(i) Beam cracked (ii) Effective section (iii) Transformed section (iv) Strain (v) Linear stress distribution

(d) Cracked section, straight-line (elastic) stress distribution

Fig. 4.6 *Continued*

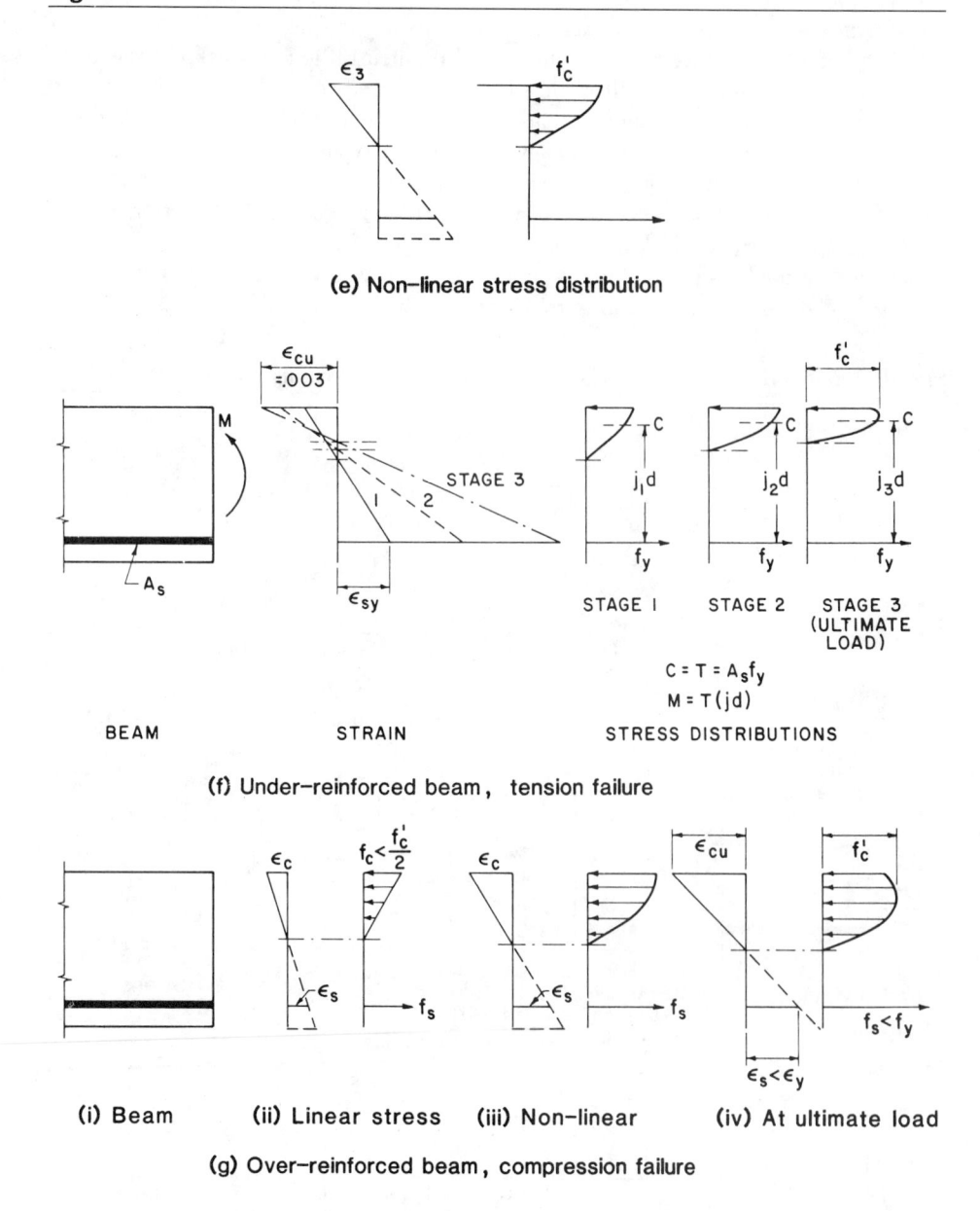

(e) Non-linear stress distribution

BEAM STRAIN STRESS DISTRIBUTIONS

$C = T = A_s f_y$

$M = T(jd)$

(f) Under-reinforced beam, tension failure

(i) Beam **(ii) Linear stress** **(iii) Non-linear** **(iv) At ultimate load**

(g) Over-reinforced beam, compression failure

equivalent to considering the concrete on the tension side of the neutral axis to be virtually nonexistent in the computations for flexural strength. The effective section of the member is as shown in Fig. 4.6d. (It should be noted, however, that the concrete on the tension side is necessary to hold the reinforcing bars in place, to resist shear and torsion, and to provide protection for the steel against corrosion and fire.) In fact, during the first-time loading, a small part of the concrete on the tension side, and close enough to the neutral axis for the tensile strain to be less than that corresponding to f_r, will still be uncracked and effective. However, the magnitude of the resulting tensile force and the internal moment due to it are negligibly small compared to the forces and moments due to the compressive stress in the concrete and tensile stress in the steel. Further, on a previously loaded member, any prior overloading may have caused the tensile cracks to penetrate high enough into the beam to effectively destroy this little contribution from the tensile strength of concrete. Hence the assumption that concrete resists no flexural tensile stresses is a satisfactory and realistic one, and will be followed in subsequent discussions.

If the applied moment, M, in Fig. 4.6d is such that the maximum concrete stress, f_c, is less than about 0.5 f'_c, the stress will be nearly proportional to the strain (see part OB of curve in Fig. 4.6b, and Section 1.5). Then the distribution of both strain and stress in the section is linear as shown in Fig. 4.6d and is given by the linear elastic theory applied to the composite section (see Section 4.2). The magnitude of stresses in usual structural members under the *specified loads* (service load conditions) is generally below about 0.5 f'_c so that the service load stress distribution is closely approximated by the straight-line distribution shown in Fig. 4.6d. For this reason the straight-line distribution is generally used to check the serviceability criteria (crack control, see CSA A23.3-M84(9.2, 9.5 and 10.6)) for which the applied loads are the actual service loads (specified loads with a load factor of 1). The linear elastic theory is also the basis of design in the *working stress design* method (also known as *allowable stress design* or *service load design*) where members are designed for the effects of the specified loads (service loads or working loads) with the resulting stresses limited to a fraction of the material strength in order to ensure a satisfactory margin of safety. Expressions for the stresses and moment of resistance for reinforced concrete sections using the linear elastic stress distribution and the concept of cracked-transformed sections are derived in Section 4.6.

As the applied loads and moments are further increased, the concrete strains and stresses enter into the nonlinear range BCD in Fig. 4.6b. For example, if the maximum compressive strain reaches a value of ϵ_3 which corresponds to a stress of f'_c, the compressive stress distribution will have the shape of the curve OBC in Fig. 4.6b, as shown in Fig. 4.6e. The behaviour of the beam in the inelastic range prior to ultimate failure in bending is dependent on the amount of reinforcing steel and the two possible cases are discussed below.

The maximum usable strength of reinforcing bars is given by the yield strength f_y. Any increase in steel stress beyond f_y is accompanied with large strains and hence large curvatures, crack widths and deflections which are normally unacceptable. Similarly, concrete is generally assumed to fail in compression upon reaching an ultimate compressive strain, ϵ_{cu}, values for which are recommended in codes (CSA A23.3-M84(10.2.3) recommends $\epsilon_{cu} = 0.003$). A section in which these two limiting conditions, namely tension reinforcement yield strain, ϵ_y, and ultimate concrete strain, ϵ_{cu}, are attained simultaneously is termed a *balanced* section. A section containing reinforcement less than that of a balanced section will be termed an *under-reinforced* section. In under-reinforced sections, as the applied moment is increased, a stage will be reached when the tension reinforcement strain reaches ϵ_y and the stress reaches its yield strength, f_y. At this stage the maximum concrete strain will be less than ϵ_{cu} and the strain and stress distributions will be somewhat as shown by stage 1 in Fig. 4.6f, with $C = T = A_s f_y$ and $M = A_s f_y j_1 d$. A slight increase in load at this stage causes the steel to yield and stretch a large amount. The marked increase in tensile strains causes the neutral axis to shift upwards thus reducing the area under compression. Since total tension, T, remains essentially constant at $A_s f_y$, to maintain equilibrium ($C = T$) with the reduced area in compression, the compressive stresses and strains increase. This situation is represented by stage 2 in Fig. 4.6f. The rise of the neutral axis causes a slight upward shift of the centre of compression, thus increasing the lever arm and moment of resistance ($M = A_s f_y j_2 d$) slightly. The relatively rapid increase in strain on the tension side also manifests in wider and deeper tensile cracks and increased beam deflection. This process continues until the maximum compressive strain reaches its ultimate value and the reduced compressive area fails by crushing of concrete (stage 3, Fig. 4.6f). It is to be noted that there is usually only a slight increase in the moment of resistance between stage 1 and stage 3 resulting from the marginal increase in the lever arm jd (and also due to any slight increase in steel stress above f_y when the steel has no sharply defined yield plateau, Fig. 1.5). However, there is a substantial increase in curvature, deflection, and width and spread of cracking between stage 1 and stage 3. In this type of beam it is the yielding of tensile reinforcement that leads, as a secondary effect, ultimately to the crushing failure of concrete and hence this type of failure is termed a *tension failure*. A tension failure is gradual, with ample prior warning of impending failure in the form of increasing curvatures, deflections and extensive cracking, and therefore is preferred in design practice. A schematic presentation of the variation of moment with curvature for an under-reinforced beam is shown in Fig. 4.7a. The large increase in curvature (rotation per unit length) before failure at relatively constant load indicates a *ductile* type of failure. The strength of such sections can be taken as that at which the steel yields.

In an *over-reinforced* beam, containing reinforcement in excess of that required for a *balanced section*, the maximum compressive strain in the concrete reaches the failure strain, ϵ_{cu}, and the concrete fails in compression before the steel reaches its yield point. This type of failure is termed a *compression (or brittle) failure*. In this case the steel remains elastic right up to failure. Once the compressive strain in concrete enters the nonlinear range, compressive stresses do not increase proportionally with strain whereas the tensile stresses do. Hence in order to keep the tensile and compressive forces equal ($C = T$), the area of concrete in compression has to increase causing a lowering of the neutral axis (Fig. 4.6g). As the concrete fails in compression, the strains in the member and hence the curvature ψ, the deflection and the crack width all remain relatively low. Hence the failure occurs suddenly with little warning, and is often of a brittle and explosive nature. Because of this, over-reinforced flexural members are generally not permitted by the Code (CSA A23.3-M84). A typical moment-curvature relation for an over-reinforced section is shown in Fig. 4.7b. Computation procedures for stresses and strength are described in subsequent Sections.

It should be noted that one other type of failure is possible although extremely rare in normal practice. This is failure by fracture of the reinforcing steel, which can happen with extremely low amounts of reinforcements and under dynamic loading.

Fig. 4.7 Moment — curvature relations

**(a) Under-reinforced beam,
tension (ductile) failure**

**(b) Over-reinforced beam,
compression (brittle)
failure**

As already mentioned, under specified loads, a reinforced concrete flexural member will be in the cracked phase (Fig. 4.6d). Because the concrete is considered ineffective in tension, the *effective depth*, d, of the member is taken as the distance from the extreme compression fibre to the centroid of the tensile reinforcement. Since the moment of resistance of a beam is proportional to the lever arm, jd, the tensile reinforcing bars are placed as close to the extreme tension fibre as is practicable. However, a minimum amount of concrete cover for the bars is essential to protect the bars from damage due to corrosion and fire. The requirements for placement of reinforcement are given in CSA A23.1-M77 (12), and are discussed in more detail in Section 5.2. In order to meet the minimum spacing requirements, reinforcing bars may have to be placed in two or more closely spaced layers. In such situations it is usual to assume, for flexural computations, that all the steel area is concentrated at the centroid of the steel bars at the effective depth, d, and that all the bars sustain the same stress and strain.

4.5 ANALYSIS AND DESIGN

There are two kinds of problems encountered in reinforced concrete practice. In the first kind, termed *analysis* (or review) problems, the complete cross-sectional dimensions including steel area and the material properties of the member are known. It is desired to compute (1) the stresses in the materials under a given load or (2) the allowable loads (moments) that the member can resist. The second type of problem involves *design*. In this type of problem, the specified material properties and the load effects (moments, shears, etc.) are known and it is required to compute the dimensions and reinforcement details of the member necessary to satisfactorily sustain the known load effects.

4.6 SERVICE LOAD STRESSES — STRAIGHT-LINE THEORY

4.6.1 ANALYSIS FOR STRESSES — SECTION CRACKED AND ELASTIC

Sections designed for strength (ultimate limit states) under factored loads must be checked for serviceability (deflection, crack width, etc.) under the specified loads. Under service loads, flexural members are generally in the cracked phase with linear distribution of strains and stresses, as explained in Section 4.4 and Fig. 4.6d, and shown again in Fig. 4.8. The computation of stresses using the straight-line theory (elastic theory) for a cracked section is discussed below.

Figure 4.8a shows a beam of rectangular section, subjected to a specified load moment, M. For this beam, the corresponding transformed section, neglecting the concrete on the tension side of the neutral axis, is shown in Fig. 4.8b. The centroid of the transformed section locates the neutral axis. Expressing the depth to neutral axis as a fraction, k, of the effective depth, d, and equating the moments about the neutral axis of the compression and tension areas:

$$\frac{b(kd)^2}{2} = n A_s(d - kd) \tag{4.3}$$

Solving this equation,
$$k = \frac{\sqrt{2B + 1} - 1}{B} \tag{4.4a}$$

where
$$B = \frac{bd}{nA_s} = \frac{1}{\rho n}$$

and
$$\rho = \frac{A_s}{bd} = \text{the reinforcement ratio}$$

Alternatively,
$$k = \sqrt{2 \rho n + (\rho n)^2} - \rho n \tag{4.4b}$$

The moment of inertia of the section, I_{cr}, is given by:

$$I_{cr} = \frac{b(kd)^3}{3} + nA_s (d - kd)^2 \tag{4.5}$$

Fig. 4.8 *Cracked section, elastic stress distribution*

(a) Beam   (b) Transformed   (c) Stress distribution

section

Knowing the neutral axis location and moment of inertia, the stresses in the concrete (and steel) in this composite section may be computed from the flexure formula $f = My/I$, as explained in Section 4.2 and shown in Example 4.1. The same results could be obtained more simply and directly considering the static equilibrium of internal forces and externally applied moment.

With the distribution of concrete and steel stresses shown in Fig. 4.8c, the internal resultant forces in concrete and steel forming the internal resisting couple are respectively C and T and the lever arm of this couple is jd, where

$$C = b\, kd \frac{f_c}{2}, \quad T = A_s f_s \tag{4.6}$$

$$jd = d - \frac{kd}{3}, \text{ or } j = \left(1 - \frac{k}{3}\right)$$

Equating the external applied moment M and the moment of resistance,

$$M = C\, jd, \text{ or } M = T\, jd$$

$$M = \frac{bkd\, f_c}{2} jd, \text{ or } M = A_s f_s\, jd \tag{4.7}$$

from which

$$f_c = \frac{M}{\left(\frac{1}{2} kj\right) bd^2} \tag{4.8}$$

and

$$f_s = \frac{M}{A_s jd} \tag{4.9}$$

4.6.2 BEAMS OF OTHER SHAPES

Although in the discussion above, a beam of rectangular section was used for the sake of simplicity, the procedure is identical for sections of other shapes such as *Tee, Ell, I,* etc. Frequently beams are reinforced in compression, with steel bars placed near the compression edge of the cross-section. Such beams are termed *doubly reinforced.* The compression steel area is then transformed to concrete by multiplying the area by $n-1$, the -1 allowing for the concrete displaced by steel (see also Section 4.4 for inelastic effects). The stresses can thus be determined using the basic principles described above.

EXAMPLE 4.2

The beam shown in Fig. 4.9a is subjected to a specified load bending moment of 140 kN·m. The concrete strength f'_c is 30 MPa and the yield strength of steel f_y is 400 MPa. Compute the stress in concrete and steel.

Fig. 4.9 *Analysis for stresses, straight line theory, Example 4.2*

(a)   (b)   (c)   (d)

SOLUTION

Since concrete is assumed to take no flexural tension, the section will be treated as cracked. The modular ratio $n = E_s/E_c = 200\,000/(5000 \sqrt{30}) = 7$, and the transformed section is shown in Fig. 4.9b. To compute the depth to the neutral axis, take moments of areas about the neutral axis (Eq. 4.3):

$$\frac{300 \, (kd)^2}{2} = 7 \times 2000 \, (530 - kd)$$

Solving,

$$kd = 181 \text{ mm}, \, k = 0.341, \, j = 1 - \frac{0.341}{3} = 0.886$$

Concrete stress (Eq. 4.8) is:

$$f_c = \frac{140 \times 10^6}{0.5 \times 0.341 \times 0.886 \times 300 \times 530^2} = 11.0 \text{ MPa}$$

Steel stress (Eq. 4.9) is:

$$f_s = \frac{140 \times 10^6}{2000 \times 0.886 \times 530} = 149 \text{ MPa}$$

Note that the same results are obtained by the flexure formula

$$f = \frac{My}{I}$$

The moment of inertia of the cracked transformed section (Eq. 4.5) is calculated as $I_{cr} = 2.30 \times 10^9$ mm⁴. The compressive stress in the top fibre of the concrete is:

$$f_c = \frac{My_c}{I_{cr}} = \frac{140 \times 10^6 \times 181}{2.30 \times 10^9} = 11.0 \text{ MPa}$$

The steel stress is n times the stress in the transformed section at the steel level so that

$$f_s = n\frac{M(d - kd)}{I_{cr}} = \frac{7 \times 140 \times 10^6(530 - 181)}{2.30 \times 10^9}$$

$$= 149 \text{ MPa as before}$$

EXAMPLE 4.3

A T-beam having the cross-sectional dimensions shown in Fig. 4.10a is subjected to a specified load bending moment of 250 kN·m. Compute the steel and concrete stresses. The material properties are $f_c' = 20$ MPa and $f_y = 400$ MPa.

SOLUTION

Considerations about the width of flange that is effective in resisting flexural compression are described in Section 4.9.3. For the present it will be assumed that the full flange width is effective.

The modular ratio is:
$$n = \frac{E_s}{E_c} = 9$$

The reinforcement area is transformed to concrete. The section is considered cracked and the concrete area on the tension side of the neutral axis is ignored. For elastic stress distribution under specified load, the neutral axis passes through the centroid of the effective transformed section.

Fig. 4.10 *Straight line theory, T-beams, Example 4.3*

(a) Section

(b) Transformed section
N.A. in flange

(c) Transformed section
N.A. in web

(d) Stress distribution

The neutral axis may lie in the flange as shown in Fig. 4.10b (that is $kd < h_f$) or it may pass through the web as in Fig. 4.10c, $(kd > h_f)$.

To locate the neutral axis, first assuming that it lies in the flange, equating the moments of areas about neutral axis (Fig. 4.10b) gives the equation:

$$b\frac{(kd)^2}{2} = nA_s (d - kd)$$

$$1000\frac{(kd)^2}{2} = 9 \times 4200 (500 - kd)$$

Solving, $\qquad kd = 160$ mm

Since this is greater than $h_f = 100$ mm, the assumption is not correct and the neutral axis passes through the web. Taking moments of areas for the case in Fig. 4.10c:

$$b_w\frac{(kd)^2}{2} + (b - b_w)h_f\left(kd - \frac{h_f}{2}\right) = nA_s (d - kd) \qquad \text{OK}$$

Solving, $\qquad kd = 169$ mm $> h_f$

The stresses may be found using the stress distribution or by using the flexure formula.

Considering the stress distribution shown in Fig. 4.10d, the moment is given by the moment of the compressive forces about the centroid of steel. Taking the compressive area as the difference between two rectangles $b \times kd$ and $(b - b_w) \times (kd - h_f)$, the moment equation gives:

$$M = b \times kd \times \frac{f_c}{2}\left(d - \frac{kd}{3}\right) - (b - b_w)(kd - h_f)\frac{f_{c1}}{2}\left(d - h_f - \frac{kd - h_f}{3}\right)$$

Substituting for various terms and noting that

$$f_{c1} = \frac{f_c(kd - h_f)}{kd} = f_c\frac{(169 - 100)}{169} = 0.409\ f_c,$$

$$250 \times 10^6 = 1000 \times 169\frac{f_c}{2}\left(500 - \frac{169}{3}\right) - (1000 - 250)(169 - 100)$$

$$\times \frac{0.409}{2}f_c\left(500 - 100 - \frac{69}{3}\right)$$

Solving, $f_c = 7.46$ MPa

From the stress distribution diagram,

$$\frac{f_s/n}{f_c} = \frac{d - kd}{kd}$$

$$f_s = nf_c\frac{d - kd}{kd} = 131 \text{ MPa}$$

Alternatively, to use the flexure formula,

$$f = \frac{My}{I_{cr}}$$

$$I_{cr} = \frac{1000 \times 169^3}{3} - \frac{(1000 - 250) \times 69^3}{3} + 9 \times 4200\,(500 - 169)^2$$

$$= 5.67 \times 10^9 \text{ mm}^4$$

$$f_c = \frac{My_c}{I_{cr}} = \frac{250 \times 10^6 \times 169}{5.67 \times 10^9} = 7.46 \text{ MPa as before}$$

and $\quad f_s = n\dfrac{My_s}{I_{cr}} = 9 \times \dfrac{250 \times 10^6 \times (500 - 169)}{5.67 \times 10^9} = 131 \text{ MPa}$

EXAMPLE 4.4

The cross-section dimensions of a beam with compression reinforcement are shown in Fig. 4.11a. The beam is subjected to a specified load moment of 175 kN·m. Compute the specified load stresses in concrete and steel. The material properties are $f_c' = 30$ MPa and $f_y = 400$ MPa.

Fig. 4.11 *Example 4.4*

(a) Section   (b) Transformed section   (c) Stress distribution   (d) Resisting couple

SOLUTION

Referring to the stress distribution in Fig. 4.11c, the concrete stress at compression reinforcement level, f_{cs}, is:

$$f_{cs} = \frac{f_c}{kd}(kd - d') = \frac{f_c(191 - 65)}{191} = 0.66\, f_c$$

Equating the applied moment to the moment of resisting forces (here taken as the moment of compressive forces C_c in concrete and C_s in steel, about tension reinforcement):

$$M = b \times kd \times \frac{f_c}{2}\left(d - \frac{kd}{3}\right) + (2n - 1) \times A'_s \times f_{cs} \times (d - d')$$

Substituting,

$$175 \times 10^6 = 300 \times 191 \times \frac{f_c}{2}\left(530 - \frac{191}{3}\right) + (2 \times 7 - 1) \times 1000$$
$$\times\, 0.66\, f_c(530 - 65)$$

Solving, $f_c = 10.1$ MPa

$$f_{cs} = 0.66\, f_c = 6.67 \text{ MPa}$$

Compression steel stress $= 2n\, f_{cs} = 93.2$ MPa

Tension steel stress $= f_s = n\, f_c \dfrac{d - kd}{kd} = 7 \times 10.1\, \dfrac{530 - 191}{191}$

$$= 125 \text{ MPa}$$

[Using flexure formula, $I_{cr} = 3.32 \times 10^9$ mm⁴,

$$f_c = \frac{M y_c}{I_{cr}} = 10.1 \text{ MPa as above}]$$

The specified load stresses are computed by the straight line theory applied to the cracked transformed section, shown in Fig. 4.11b. As explained in Section 4.3, in transforming the compression reinforcement (for stress computations here) the effective modular ratio is taken as $2n$. Allowing for the concrete displaced by A_s', the net transformed area for compression reinforcement is $(2n - 1)A_s'$. The modular ratio, $n = E_s/E_c = 200\ 000/(5000\ \sqrt{30}) = 7$.

To locate the neutral axis (centroid) of the transformed section, equating the moment of areas about the neutral axis (see Fig. 4.11b),

$$b\,\frac{(kd)^2}{2} + (2n - 1)\,A_s'\,(kd - d') = nA_s\,(d - kd)$$

Substituting,

$$300\,\frac{(kd)^2}{2} + (14 - 1)\,1000\,(kd - 65) = 7 \times 3000 \times (530 - kd)$$

Solving, $\qquad\qquad\qquad\qquad kd = 191 \text{ mm}$

4.6.3 ALLOWABLE SPECIFIED (SERVICE) LOAD MOMENTS

When it is desired to compute the allowable specified load moment on a beam of known cross-section, such that the steel and concrete stresses are not to exceed specified allowable stresses, the procedure is very similar to that given in Section 4.6.1 and Example 4.2. Referring to Fig. 4.8, the location of the actual neutral axis is given by Eq. 4.4. Rearranging Eqs. 4.8 and 4.9, the allowable moment capacity M_{all} can be expressed in terms of the maximum allowable steel and concrete stresses, $f_{s,all}$ and $f_{c,all}$ as:

$$M_{all} \leqslant A_s\, f_{s,all} jd \tag{4.10}$$

and

$$M_{all} \leqslant \frac{1}{2}f_{c,all}\, k\, j\, bd^2 \tag{4.11}$$

In a given beam section, the allowable stresses in both steel and concrete may not be reached simultaneously. Hence the lower of the two moments computed by Eqs. 4.10 and 4.11 will give the real allowable moment and the corresponding stress (f_s or f_c) will reach the allowable maximum first.

Alternatively, from the known neutral axis location and the stress diagram it is possible to determine first whether the steel or concrete stress controls. For example, referring to Fig. 4.12, if it is assumed that the concrete stress reaches the limit first, the associated steel stress is obtained as:

$$f_s = nf_{c,all}\,\frac{d - kd}{kd} \tag{4.12}$$

Fig. 4.12

$$\frac{f_s}{n} = \frac{f_{c,all}}{kd} \times (d-kd)$$

$f_s < f_{s,all}$, CONCRETE STRESS CONTROLS

$f_s > f_{s,all}$, STEEL STRESS CONTROLS

If f_s so calculated is less than $f_{s,all}$, the concrete stress does in fact control and Eq. 4.11 gives M_{all}. If, on the other hand, f_s from Eq. 4.12 exceeds $f_{s,all}$, it is the steel stress that reaches the limiting value first and hence Eq. 4.10 will give the correct value of M_{all}.

EXAMPLE 4.5

CSA Standard S6-M78 — *Design of Highway Bridges* recommends allowable service load stresses of 0.40 f'_c for concrete in flexural compression and 165 MPa for tensile stresses in Grade 400 reinforcement. Compute the allowable service load moment for the beam in Example 4.2.

SOLUTION

Allowable $f_c = 0.4 \times 30 = 12$ MPa
Allowable $f_s = 165$ MPa
Depth to neutral axis (from Example 4.2), $kd = 181$ mm

Assuming the concrete stress reaches the allowable value of 12 MPa, the steel stress will then be (Fig. 4.13):

Fig. 4.13

$$f_s = 7 \times \frac{12}{181} \times 349 = 162 \text{ MPa}$$

$$< 165 \text{ MPa} = \text{allowable}$$

Concrete stress controls, and

$$M = \frac{1}{2}f_c b \, kd \left(d - \frac{kd}{3} \right)$$

$$= \frac{1}{2} \times 12 \times 300 \times 181 \left(530 - \frac{181}{3} \right) \times 10^{-6} = 153 \text{ kN} \cdot \text{m}$$

4.7 STRENGTH IN FLEXURE

4.7.1 USING ACTUAL STRESS–STRAIN RELATIONS

In Chapter 3 it was explained that in Limit States Design, safety is ensured by designing members to have a factored resistance equal to or greater than the required strength which is the load effects resulting from factored load combinations. The factored resistance, M_r, of a member is computed theoretically using factored material strengths and member dimensions. The evaluation of the *nominal resistance* in flexure, M_n, obtained using the specified material strengths without the ϕ-factors, and of the *factored resistance, M_r*, obtained using the specified material strengths with the ϕ-factors, is described in the following sections.

As described in Section 4.4, the ultimate failure in flexure of a reinforced concrete beam, whether under-reinforced or over-reinforced, takes place by crushing of concrete, with the maximum compressive strain in concrete reaching its failure value, ϵ_{cu} (Fig. 4.6f and g). At failure, the compressive strain in concrete is assumed to vary linearly from zero at the neutral axis to ϵ_{cu} at the extreme fibre. If the material resistance factors, ϕ_c and ϕ_s are not included, the distribution of compressive stress in concrete (frequently referred to as the stress block) will be nearly identical to the stress–strain relation for concrete in compression and will be as shown in Fig. 4.14c. (This assumes that stress–strain relations of concrete for concentric compression is applicable to flexure). The actual shape of the stress–strain relation for concrete is not unique and depends on several factors such as the cylinder strength, density, rate and duration of loading. However, in order to compute the flexural strength corresponding to the stress pattern shown in Fig. 4.14c, it is not essential to know the exact shape of the stress distribution in concrete, and it is sufficient to know only the location and magnitude of the resultant compressive force, C. Considering a beam of rectangular section (Fig. 4.14), the magnitude and location of the resultant compression, C, can be expressed in terms of two nondimensional parameters, namely, α, which represents the ratio of the average compressive stress in the compression zone to the cylinder strength, f_c', and β, which is the ratio of the depth of resultant force, C, to the depth of neutral axis, c (Fig. 4.14c). For the rectangular section where the area under compression has a uniform width, α and β are basic properties of the stress–strain curve. Values of α and β can be obtained mathe-

Fig. 4.14   *Conditions at ultimate load*

(a)

Section

(b)

Strain

(c)

Stress and
forces, not
including
ϕ - factors

(d)

Ref. 4·2

(e)

Stresses and
forces with
ϕ - factors

(f)

Rectangular
stress block

matically if the actual shape of the stress–strain relation is known or more appropriately they may be determined experimentally (Refs. 4.2 and 4.5).

For the rectangular section in Fig. 4.14, knowing the values of ϵ_{cu}, α and β, considerations of equilibrium of forces and compatibility of strains are sufficient to compute the value of the nominal bending strength, M_n. Equating the compressive and tensile forces,

$$\alpha f_c' \, b \, c = A_s f_s \tag{4.13}$$

Moment of the resisting couple $C = T$ is:

$$M_n = Cjd = \alpha f_c' \, bc(d - \beta c) \tag{4.14a}$$

or

$$M_n = Tjd - A_s f_s \, (d - \beta c) \tag{4.14b}$$

Equations 4.13 and 4.14 express equilibrium conditions and must always be satisfied. For a tension failure (under-reinforced beam), the steel yields prior to failure so that at the nominal strength, $f_s = f_y$ and Eq. 4.13 directly gives the value of c as:

$$c = \frac{A_s f_y}{\alpha f_c' b} \tag{4.15}$$

Substituting for c in Eq. 4.14a or b, the ultimate strength is obtained as:

$$M_n = A_s f_y \left(d - \frac{\beta A_s f_y}{\alpha f_c' b} \right) \tag{4.16}$$

Using the notations:

$$\rho = \frac{A_s}{bd} \tag{4.17}$$

and

$$\omega = \frac{A_s f_y}{bd f_c'} \tag{4.18}$$

where ρ is a reinforcement ratio expressing A_s as a fraction of effective concrete area, and ω is termed the tension reinforcement index. Equation 4.16 may also be expressed as:

$$M_n = \rho f_y \left(1 - \frac{\beta \rho f_y}{\alpha f_c'} \right) bd^2 \tag{4.19}$$

$$= \omega \left(1 - \frac{\beta}{\alpha} \omega \right) f_c' \, bd^2 \tag{4.20}$$

Equation 4.20 shows that for a tension failure the only property of the stress block needed to compute the flexural strength is the ratio, β/α.

For a compression failure (over-reinforced beam) the concrete crushes before the steel yields so the steel stress at failure f_s ($<f_y$) has to be computed before Eqs. 4.13 and 4.14 can be solved for M_n. For this, considering the strain distribution in Fig. 4.14b, the steel strain, ϵ_s, and hence stress, f_s, can be computed as:

$$\epsilon_s = \epsilon_{cu} \frac{d - c}{c} \tag{4.21}$$

and

$$f_s = E_s \epsilon_s = E_s \epsilon_{cu} \frac{d - c}{c} \tag{4.22}$$

Substituting Eq. 4.22 for f_s in Eq. 4.13, one obtains:

$$\alpha f_c' \, bc = A_s \, E_s \epsilon_{cu} \frac{d - c}{c} \tag{4.23}$$

Equation 4.23 can be solved for c, which on substitution into Eq. 4.14 yields the flexural strength. In this case it is necessary to know all three parameters, ϵ_{cu}, α and β, related to the stress distribution.

Equations 4.13 to 4.23 can be applied to T-sections also, provided the neutral axis at the nominal strength is within the compression flange. When the area under compression is not a rectangle (as is the case with the neutral axis of a T-beam being below the flange) the entire shape of the stress block is required for an *exact* analysis.

4.7.1(a) PARAMETERS ϵ_{cu}, α and β

Assuming a flexural stress distribution in the compression zone of a rectangular beam as shown in Fig. 4.14d, the parameters ϵ_{cu}, k_1, k_2 and k_3 defining the properties of the stress block have been evaluated experimentally (Ref. 4.2, 4.5). $k_1 k_3$ and k_2 correspond respectively to α and β in Fig. 4.14c. The experimental measurements show considerable scatter in the value of ϵ_{cu}, however, the generally conservative lower bound value of 0.003 recommended by the Code (Ref. 4.6) was found satisfactory. Over the normal range of concrete strengths ($f_c' = 15$ to 60 MPa) the average value of k_3 was found to range between 0.96 and 1.04 for normal density concrete and between 0.98 and 1.01 for low density concrete. This indicates that the peak compressive stress in flexure is in fact very nearly equal to the cylinder strength. These studies confirmed that the shape of the stress block is similar to the stress–strain curve in uniaxial compression. Lower bound values recommended for the constants α and β (Ref. 4.5) are given in Fig. 4.15.

4.7.1(b) FACTORED MOMENT RESISTANCE, M_r

The factored moment resistance, M_r, is computed using the factored material strengths, $\phi_c f_c'$ and $\phi_s f_s$. The concrete stress distribution and the force in the reinforcement for this case are shown in Fig. 4.14e. Again, for a tension failure, when $\epsilon_s \geq \epsilon_y$, equating $C = T$ gives:

$$k\phi_c f_c' \, bc = A_s \phi_s f_y$$

$$c = \frac{A_s \phi_s f_y}{k\phi_c f_c' \, b} \tag{4.15a}$$

The moment of the couple, $C = T$ gives:

$$M_r = A_s \phi_s f_y (d - \bar{k} c)$$

$$= A_s \phi_s f_y \left(d - \frac{\bar{k} \, A_s \phi_s f_y}{k \, \phi_c f'_c \, b} \right) \qquad (4.16a)$$

Note that the concrete stress distribution shown in Fig. 4.14e differs from that shown in Fig. 4.14c only by the constant factor ϕ_c. Hence the constants k and $\bar{k}$ (Fig. 4.14e) are the same as α and β (Fig. 4.14c) respectively. If the values for α and β given in Fig. 4.15 are used, the ratio $k/\bar{k} = \alpha/\beta \simeq 1.70$ and Eq. 4.16a reduces to:

$$M_r = A_s \phi_s f_y \left(d - \frac{A_s \phi_s f_y}{1.7 \, \phi_c f'_c \, b} \right) \qquad (4.16b)$$

For compression failure, $\epsilon_s < \epsilon_y$ and the steel strain is given by Eq. 4.21. Equating C and T gives:

$$k \, \phi_c f'_c \, bc = A_s \phi_s E_s \epsilon_{cu} \frac{d - c}{c} \qquad (4.23a)$$

Equation 4.23a can be solved for c. The factored moment resistance is then obtained as the moment of the couple $C = T$:

$$M_r = A_s \phi_s f_s \, (d - \bar{k} c) \qquad (4.23b)$$

4.1.7(c) BALANCED STRAIN CONDITION

In Section 4.4 it was explained that a beam section in which the tension reinforcement reaches its yield strain, ϵ_y, just as the concrete in compression reaches its ultimate strain, ϵ_{cu}, is termed a balanced section. At balanced strain condition, the ratio of neutral axis depth to effective depth, c/d, is obtained from Eq. 4.21 or Fig. 4.14b as:

$$\left(\frac{c}{d} \right)_b = \frac{\epsilon_{cu}}{\epsilon_{cu} + \epsilon_s} = \frac{\epsilon_{cu}}{\epsilon_{cu} + f_y/E_s}$$

$$= \frac{E_s \epsilon_{cu}}{E_s \epsilon_{cu} + f_y} \qquad (4.24)$$

With the values specified in CSA Standard A23.3-M84 (8.52 and 10.2.3) of $E_s = 200\,000$ MPa and $\epsilon_{cu} = 0.003$, the neutral axis depth ratio for balanced strain conditions is given by:

$$\left(\frac{c}{d}\right)_b = \frac{600}{600 + f_y} \tag{4.25}$$

In order to ensure a ductile failure, the Code limits the maximum reinforcement in flexural members such that the actual neutral axis depth ratio, c/d, calculated using factored material strengths $\phi_c f_c'$ and $\phi_s f_y$ is less than that corresponding to the balanced strain conditions. That is, the reinforcement shall be such that:

$$\frac{c}{d} \leq \frac{600}{600 + f_y} \tag{4.26}$$

4.7.2 STRENGTH USING EQUIVALENT RECTANGULAR STRESS DISTRIBUTION

Referring to Fig. 4.14e, the strength is uniquely determined provided the magnitude and location of the resultant compressive force, C, in concrete is known, no matter what the shape of the stress block. Hence the stress block may be imagined to have any convenient shape so long as the magnitude and location of C is unaltered. The simplest form of such a fictitious stress distribution is a rectangular stress block shown in Fig. 4.14f, where a uniform stress of $\gamma \phi_c f_c'$ acts over a depth $a = \beta_1 c$. In order to have the same location for the resultant force, C, as in Fig. 4.14e,

$$\frac{a}{2} = \frac{\beta_1 c}{2} = \bar{k} c$$

or the factor β_1 giving the depth of stress block in terms of c is

$$\beta_1 = 2\bar{k} \tag{4.27}$$

For the magnitude of C to remain unaltered,

$$C = k\phi_c f_c' bc = \gamma \phi_c f_c' ba = \gamma \phi_c f_c' b \beta_1 c$$
$$\gamma = \frac{k}{\beta_1} = \frac{k}{2\bar{k}} \tag{4.28}$$

Experimental results in Ref. 4.5 show that for concrete strengths in the range of 15–100 MPa, the mean value of $\bar{k}/k$ varies over a narrow range between 0.54 and 0.62 for normal density concrete (and between 0.57 and 0.70 for low density concrete). With the lower bound values for k and $\bar{k}$ recommended for design and presented in Fig. 4.15, the ratio $\bar{k}/k$ has a constant value of 0.588 (or $\gamma = k/2\bar{k} = 0.85$) for all concrete strengths. Hence the uniform stress in the equivalent rectangular stress block can be taken as $0.85 \phi_c f_c'$. The depth of the stress block is given by the factor $a/c = \beta_1 = 2\bar{k}$, where the recommended value of $\bar{k}$ is given in Fig. 4.15. Values of γ and β_1, based on the recommended values k and $\bar{k}$ (Eqs. 4.27 and 4.28), are also plotted in Fig. 4.15.

Fig. 4.15 *Parameters α, β*

4.8 CODE RECOMMENDATIONS

The assumptions forming the basis of flexural computations by CSA A23.3-M84 are given in Clause 10.2. The strains in steel and concrete are assumed to vary linearly with distance from neutral axis, with the maximum compressive strain in concrete limited to 0.003. Steel stress is taken as $\phi_s f_s = \phi_s E_s \epsilon_s \leq \phi_s f_y$. Tensile strength of concrete is neglected. In compression, the stress–strain relation for concrete may be assumed as trapezoidal, parabolic or any other shape provided the strength so predicted is in agreement with test results. As one such acceptable stress distribution, the Code recommends an equivalent rectangular concrete stress distribution with a uniform stress of 0.85 $\phi_c f_c'$ over a depth $a = \beta_1 c$ (Fig. 4.16) where β_1 is taken as 0.85 for concrete strengths f_c' up to and including 30 MPa and, beyond this, it is reduced continuously at a rate of 0.08 for each additional 10 MPa of strength, but with a minimum value for β_1 of 0.65. The uniform stress (0.85 $\phi_c f_c'$) and the depth of stress block ($\beta_1 c$) recommended here are essentially the same as those determined experimentally (Ref. 4.5) and shown in Fig. 4.15.

4.9 MAXIMUM REINFORCEMENT RATIO

CSA Standard A23.3-M84(10.3.3) limits the maximum tension reinforcement in flexural members such that the ratio of neutral axis depth to effective depth, c/d, satisfies the condition:

$$\frac{c}{d} \leq \frac{600}{600 + f_y} \tag{4.29}$$

Fig. 4.16 *Rectangular stress block, CSA A23.3*

(a) CSA Rectangular stress block

(b) At maximum steel ratio

For calculating the neutral axis depth, c, the factored concrete strength, $\phi_c f'_c$, and the factored reinforcement bar force, $A_s \phi_s f_s$, are used. The requirement of Eq. 4.29 ensures a tension (ductile) failure for the beam (see also Section 4.7.1c).

In practice, reinforcement corresponding to the upper limit given by Eq. 4.29 will usually result in gross cross-sectional dimensions for the beam which are relatively small. Such slender beams may not have adequate stiffness to control deflection. Furthermore it is difficult to place such a large amount of reinforcement in small cross-sections. Limitations of deflection, convenience in placement of reinforcement and economy in design generally dictate larger overall beam dimensions with correspondingly lower reinforcement ratios, usually in the range of 30 to 40 percent of the maximum limit given by Eq. 4.29. These lower reinforcement ratios further improve the ductility of the beams.

4.9.1 RECTANGULAR SECTION WITH TENSION REINFORCEMENT ONLY (FIG. 4.16b)

The equivalent rectangular concrete stress distribution recommended in CSA A23.3-M84 is shown in Fig. 4.16b. The right hand side of Eq. 4.29 corresponds to the balanced strain condition with $\epsilon_{cu} = 0.003$ and $\epsilon_s = \epsilon_y$. Hence at the upper limit of c/d given by Eq. 4.29, the tension reinforcement would have yielded. Equating $C = T$,

$$0.85 \; \phi_c f'_c \, b(\beta_1 c) = A_s \phi_s f_y \qquad (4.30)$$

$$c = \frac{A_s \phi_s f_y}{0.85 \, \beta_1 \phi_c f_c' \, b}$$

$$\frac{c}{d} = \frac{A_s \phi_s f_y}{0.85 \, \beta_1 \phi_c f_c' \, bd}$$

Equating this to the limiting value of c/d in Eq. 4.29 yields:

$$\frac{c_{max}}{d} = \left[\frac{\phi_s f_y}{0.85 \, \beta_1 \phi_c f_c'} \frac{A_s}{bd} \right]_{max} = \frac{600}{600 + f_y}$$

Denoting $A_s/(bd)$ by ρ, the reinforcement ratio,

$$\rho_{max} = \bar{\rho} = \left(\frac{0.85 \, \beta_1 \phi_c f_c'}{\phi_s f_y} \right) \left(\frac{600}{600 + f_y} \right) \tag{4.31}$$

Equation 4.31 may also be obtained directly by substituting the maximum value of c from Eq. 4.29 into Eq. 4.30. Equation 4.31 gives the maximum reinforcement ratio, $\bar{\rho}$, permitted by the Code for a beam of rectangular section. As an example, for a specified concrete strength $f_c' = 30$ MPa and a steel yield strength $f_y = 400$ MPa; with $\beta_1 = 0.85$ (see Section 4.8), $\phi_c = 0.6$ and $\phi_s = 0.85$; substitution in Eq. 4.31 gives:

$$\rho_{max} = 0.0230$$
$$= 2.3\%$$

4.9.2 RECTANGULAR BEAM WITH COMPRESSION REINFORCEMENT

There are situations where the overall dimensions of a beam section are restricted so that even with the maximum permissible reinforcement provided, the strength of the section is inadequate to carry the loads. In such cases additional tension and corresponding compression reinforcement is provided to enhance the strength of the section. There are also occasions when reinforcement may have to be provided near both top and bottom faces of the beam such as at sections where there is a reversal of moment, at supports of continuous beams, for beams whose long time deflections are to be controlled, and for members detailed to resist seismic forces in a ductile manner. Such beams are called doubly reinforced beams.

Figure 4.17 shows the dimensions of a rectangular beam with compression reinforcement A_s' located at a depth d' below the compression face. The stress in the compression reinforcement, f_s', depends on the strain, ϵ_s', at its level, which may or may not have reached the yield level. From the strain triangle,

$$\epsilon_s' = \frac{0.003}{c}(c - d') = 0.003 \left(1 - \frac{d'}{c} \right)$$

and
$$f_s' = E_s \epsilon_s' = 600 \left(1 - \frac{d'}{c} \right) \leq f_y'$$

Fig. 4.17 *Doubly reinforced section, maximum reinforcement ratio*

PART 1   PART 2

(a) Section   (b) Strain   (c) Resisting couple

where f'_y is the specified yield strength of compression reinforcement. As the tension reinforcement reaches its maximum limit, the value of c is given by Eq. 4.29 as

$$c_{max} = \frac{600}{600 + f_y}d$$

Substituting this value for c in the equation for f'_s, the compression reinforcement stress corresponding to the maximum permissible amount of tension reinforcement is obtained as:

$$f'_s = 600\left(1 - \frac{600 + f_y}{600}\frac{d'}{d}\right) \leq f'_y \tag{4.32}$$

For convenience, the internal resisting couple may be split into two parts, the first consisting of the compressive force C_c in the concrete of area, bc, and a corresponding tension steel area, A_{s1}, with force, T_1, and the second part consisting of the compressive force, C_s, in the compression reinforcement and an associated tensile force, T_2, on steel area, A_{s2}. This two part approach is shown in Fig. 4.17c. The first part is identical to the rectangular beam with tension reinforcement only and the maximum reinforcement ratio for this part was derived as $\bar{\rho}$ in the previous Section, so that:

$$\frac{A_{s1}}{bd} = \bar{\rho}$$

Equating the tensile and compressive forces for the second part,

$$T_2 = C_s$$
$$A_{s2}\phi_s f_y = A'_s\phi_s f'_s$$

In computing C_s, if allowance is made for the concrete area in compression actually displaced by A'_s but already included in C_c, the effective stress in A'_s may be more accurately taken as $(\phi_s f'_s - 0.85\phi_c f'_c)$ and the relation $T_2 = C_s$ becomes:

$$\phi_s A_{s2} f_y = A'_s(\phi_s f'_s - 0.85\ \phi_c f'_c)$$

$$\frac{A_{s2}}{bd} = \frac{A'_s}{bd} \times \frac{(\phi_s f'_s - 0.85\ \phi_c f'_c)}{\phi_s f_y} = \rho' \frac{(\phi_s f'_s - 0.85\ \phi_c f'_c)}{\phi_s f_y}$$

where $\rho' = A'_s/(bd)$ is the compression reinforcement ratio
The total tension reinforcement ratio for limiting conditions is:

$$\rho_{max} = \frac{A_s}{bd} = \frac{A_{s1} + A_{s2}}{bd}$$

$$\frac{A_{s1}}{bd} = \bar{\rho} + \rho' \frac{(\phi_s f'_s - 0.85\ \phi_c f'_c)}{\phi_s f_y} \qquad (4.33)$$

where $\bar{\rho}$ and f'_s are given by Eqs. 4.31 and 4.32 respectively. Note that a normal level of accuracy does not warrant the allowance for the compression concrete area displaced by A'_s, so that the term $0.85\phi_c f'_c$ in Eq. 4.33 may be neglected. This is particularly so when the compression reinforcement stress, f'_s, is high relative to f'_c. Equation 4.33 can then be simplified as:

$$\rho_{max} = \bar{\rho} + \rho' \frac{f'_s}{f_y} \qquad (4.33a)$$

When the compression reinforcement yields and has the same strength as the tension reinforcement, $\rho_{max} = \bar{\rho} + \rho'$.

As an illustrative example, for a beam of rectangular section with $f'_c = 30$ MPa, $f_y = 400$ MPa, $f'_y = 350$ MPa, $d'/d = 0.15$ and $\rho' = A'_s/bd = 0.012$, the maximum permissible tension reinforcement ratio, $\rho = A_s/bd$, can be worked as follows.

Compression reinforcement stress at the maximum value of c/d is given by Eq. 4.32 as:

$$f'_s = 600 \left(1 - \frac{600 + 400}{600} \times 0.15 \right) = 450 > f'_y$$

so $\qquad f'_s = f'_y = 350$ MPa

ρ_{max} is given by Eq. 4.33. $\bar{\rho}$, the maximum reinforcement ratio for a singly reinforced section for these material properties, was computed in the example at the end of Section 4.9.1 as $\bar{\rho} = 0.023$. Thus Eq. 4.33 yields

$$\rho_{max} = 0.023 + 0.012 \frac{0.85 \times 350 - 0.85 \times 0.6 \times 30}{0.85 \times 400}$$

$$= 0.0330$$

If the simplified Eq. 4.33a is used, ρ_{max} is obtained as 0.0335.

4.9.3 T-BEAMS

Beams having T-shaped sections (Fig. 4.18a) are occasionally used in structures. Further, floor slabs are usually cast monolithically with supporting beams, in which case a portion of the slab adjacent to the beam will effectively add to the latter's strength, resulting in a beam of T-section (Fig. 4.18b). The slab portion of the T-section is termed the flange and the beam portion below it the web or stem.

In a T-beam with a relatively wide flange, the flexural stress in the flange is not uniform over its width, but varies from a maximum over the web to progressively lower values at points further removed from the web. The theory of flexure used in reinforced concrete design assumes a uniform stress across the *width* of the section. Hence for T-shaped sections, it is convenient to define a reduced effective flange width over which the stress distribution is considered to be uniform. CSA Standard A23.3-M84(8.10) defines the effective widths for flanged sections. For symmetrical T-beams, the effective flange width is limited to 0.4 of the span length for simply

Fig. 4.18 *T-beams*

(a) T-beam

(b) T-beam action in beam and slab floor

supported beams and 0.25 of the span length for continuous beams; however, the overhang on either side of the web shall not exceed 12 times the slab thickness nor half of the clear distance to the next beam. For simply supported beams, b = least of: 0.4 span, 24 $h_f + b_w$, and spacing s of beams.

The dimensions of a T-beam section considered fully effective are shown in Fig. 4.19. When the depth of neutral axis, c, falls within the flange thickness (Fig. 4.19a) (that is $c \geq h_f$), the analysis of the T-beam is identical to that of a rectangular beam of width b and depth d, since the difference in concrete area between these two shapes is entirely in the tension zone, an area which is neglected in the strength calculation. There is also the possibility where (see Fig. 4.19b) the neutral axis falls below the flange ($c > h_f$) but the depth of stress block taken as $a = \beta_1 c$ is still within the flange thickness ($a \leq h_f$). Although the average stress 0.85 f'_c and the recommended β_1 values characterizing the rectangular stress block are based on a compression concrete area of uniform width, extensive comparison with test results

Fig. 4.19 *Possible neutral axis locations for T-sections*

(a) $c \leq h_f$, as for rectangular beam, $(b \times d)$

(b) $a < h_f < c$, may be treated as rectangular beam, $(b \times d)$

(c) $a > h_f$, treated as T-beam

has shown (Ref. 4.7) that these values can be used with sufficient accuracy even when the concrete area under compression is not a rectangle. Thus for all cases of $a \leq h_f$, the T-beam is usually analysed as a rectangular beam of width b.

For the case of the neutral axis passing through the web, with $a = \beta_1 c > h_f$, (Fig. 4.19c), the T-beam must be analysed to account for the different width of the compression areas in the flange and in the web. However, the rectangular stress block with uniform stress $= 0.85\, f_c'$ is still a satisfactory assumption (Ref. 4.7). The maximum reinforcement conditon with T-beam action is shown in Fig. 4.20. Once again, the resisting couple may be considered in two parts: a web contribution consisting of the compressive force, C_w, in the concrete in the web portion and an equivalent steel area, A_{sw}, with tensile force, T_w; and a flange contribution consisting of the compression, C_f, in the overhanging portions of the flanges and the corresponding tension, T_f, on steel area, A_{sf}. (See Fig. 4.20.) For the first part, the maximum reinforcement ratio is as for a rectangular section, $b_w \times d$, so that

$$A_{sw} = \bar{\rho}\, b_w\, d$$

Fig. 4.20 *Balanced condition for T-beam*

For the second part, equating $T_f = C_f$

$$A_{sf}\phi_s f_y = 0.85\,\phi_c f_c'(b - b_w)h_f$$

$$A_{sf} = \frac{0.85\,\phi_c f_c'}{\phi_s f_y}\,(b - b_w)h_f$$

Designating
$$\rho_f = \frac{A_{sf}}{b_w d} = \frac{0.85\,\phi_c\,f_c'}{\phi_s f_y}\,\frac{(b - b_w)}{b_w}\,\frac{h_f}{d} \qquad (4.34)$$

the balanced reinforcement ratio in terms of the web area, $b_w d$ is:

$$\rho_{max} = \frac{A_s}{b_w d} = \frac{A_{sw} + A_{sf}}{b_w d} = \bar{\rho} + \rho_f \qquad (4.35)$$

In Eq. 4.35, $\bar{\rho}$ is the maximum reinforcement ratio for a rectangular beam, $b_w \times d$, with tension reinforcement only (Eq. 4.31).

In T-beams of usual proportions, because of the large concrete area in compression, the maximum reinforcement limitation is generally met unless extremely high (and costly) concentrations of reinforcement are provided in the narrower web.

4.10 MINIMUM REINFORCEMENT IN FLEXURAL MEMBERS

CSA A23.3-M84(10.5) specifies minimum values for tension reinforcement in flexural members. For members other than slabs, ρ_{min} is given by Eq. 4.36 unless the area of reinforcement provided at every section is at least one third greater than that required by analysis.

$$\rho_{min} = 1.4/f_y \qquad (4.36)$$

In T-beams and joists with webs in tension, $\rho = A_s/bd$ for this purpose is computed using the width of web b_w for b. The above minimum reinforcement ratio roughly gives the steel area required to have a strength equal to the cracking moment of an identical plain concrete section and is necessary to avoid a sudden failure at $M = M_{cr}$.

The minimum flexural reinforcement specified for slabs of uniform thickness is $0.002\,A_g$ in each direction, where A_g is the gross area of section (A23.3-M84 (7.8)). This minimum reinforcement in slabs is meant to provide some control of the cracking due to shrinkage and temperature effects. For exposure conditions where crack control is essential, reinforcement in excess of $0.002\,A_g$ should be provided. In such cases, the concept of skin reinforcement given in CSA-A23.3-M84 (10.6.7) may be used for guidance. This is illustrated in Fig. 4.21. The area of skin reinforcement is equal to $\rho_{sk}A_{cs}$, where A_{cs} is the area of concrete in strips along the exposed face, shown shaded in Fig. 4.21, and $\rho_{sk} = 0.008$ for *interior exposure* and $\rho_{sk} = 0.01$ for *exterior exposure* condition.

Fig. 4.21 *Skin reinforcement for crack control*

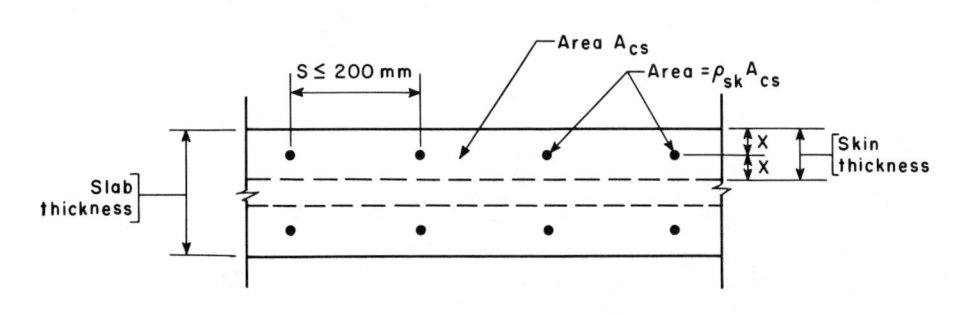

4.11 ANALYSIS FOR STRENGTH BY CSA CODE

4.11.1 RECTANGULAR SECTIONS

The Code permits only under-reinforced beams with the maximum reinforcement ratio limited by Eq. 4.29. The value of ρ_{max} is computed by Eq. 4.31 and this condition is easily checked. If this requirement is satisfied (for over-reinforced beams, see Section 4.11.5), the member fails primarily in tension which leads to a secondary compression failure as explained in Section 4.4, and at the factored resistance, M_r, the steel stress equals $\phi_s f_y$, and the maximum concrete strain is equal to 0.003. Using the rectangular stress block recommended in the Code, the internal stress resultants are shown in Fig. 4.22. Equating C and T,

$$0.85\phi_c f_c' ba = A_s \phi_s f_y$$

$$a = \frac{A_s \phi_s f_y}{0.85\phi_c f_c' b} = \frac{\rho \phi_s f_y}{0.85\phi_c f_c'} d \qquad (4.37)$$

Taking moment of the resisting couple, the factored resistance M_r is:

Fig. 4.22

$$M_r = A_s \phi_s f_y \left(d - \frac{a}{2} \right) \tag{4.38}$$

$$= \rho \phi_s f_y \left(1 - \frac{\rho \phi_s f_y}{1.7 \, \phi_c f_c'} \right) b d^2 \tag{4.38a}$$

$$= K_r b d^2 \tag{4.39}$$

where
$$K_r = \rho \phi_s f_y \left(1 - \frac{\rho \phi_s f_y}{1.7 \, \phi_c f_c'} \right)$$

K_r has the unit of stress and is conveniently expressed in MPa (N/mm^2). With b and d expressed in mm, the design strength is:

$$M_r = K_r b d^2 \, \text{N} \cdot \text{mm}$$
$$= K_r b d^2 \times 10^{-6} \, \text{kN} \cdot \text{m} \tag{4.40}$$

Values of K_r for different combinations of f_c' and f_y and for ρ ranging from ρ_{max} to ρ_{min} are tabulated in Table 4.1 as an aid to analysis. The table also gives the corresponding values of β_1, c_{max}/d, a_{max}/d, ρ_{max} and ρ_{min}.

EXAMPLE 4.6

For the beam in Example 4.2, (Fig. 4.9a) compute the factored resistance by the CSA Code.

SOLUTION

a) First the computation will be made using basic principles rather than the final formula. Referring to Fig. 4.22 for notation, the section properties are: $b = 300$ mm, $d = 530$ mm, $A_s = 2000$ mm^2, $f_c' = 30$ MPa, $f_y = 400$ MPa. For these material properties, $\rho_{max} = 0.0230$ (Eq. 4.31), and $\rho_{min} = 1.4/f_y = 0.0035$. The actual reinforcement ratio $\rho = 2000/(300 \times 530) = 0.0126$, which is within the permissible limits. The stress block, and steel and concrete forces are as shown in Fig. 4.22.

Equating $C = T$,

$$0.85 \, \phi_c f_c' \, b \, a = A_s \phi_s f_y$$

$$0.85 \times 0.6 \times 30 \times 300a = 2000 \times 0.85 \times 400, \quad a = 148 \text{ mm}$$

Factored resistance, $$M_r = A_s \phi_s f_y \left(d - \frac{a}{2} \right)$$

$$= 2000 \times 0.85 \times 400 \left(530 - \frac{148}{2} \right) \times 10^{-6} \, \text{kN} \cdot \text{m}$$

$$= 310 \, \text{kN} \cdot \text{m}$$

Table 4.1 Reinforcement Ratios ρ and Resistance Factors K_r for Rectangular Sections

$$\rho = A_s/(bd); \quad \rho_{max} = 0.85\,\beta_1\,(600/(600 + f_y))\,\phi_c f'_c/(\phi_s f_y); \quad \rho_{min} = 1.4/f_y$$
$$(c_{max}/d) = 600/(600 + f_y); \quad M_r = K_r\,bd^2 \times 10^{-6}; \quad K_r = \rho\phi_s f_y[1 - \rho\phi_s f_y/(1.7\,\phi_c f'_c)];$$
$$\phi_s = 0.85; \quad \phi_c = 0.6; \quad \text{Units of } f'_c, f_y, K_r \text{ in MPa}, M_r \text{ in kN·m}, b \text{ and } d \text{ in mm}, A_s \text{ in mm}^2.$$

f_y (c_{max}/d)	300 0.667					350 0.632					400 0.600				
f'_c	20	25	30	35	40	20	25	30	35	40	20	25	30	35	40
β_1	0.85	0.85	0.85	0.81	0.77	0.85	0.85	0.85	0.81	0.77	0.85	0.85	0.85	0.81	0.77
a_{max}/d	.567	.567	.567	.540	.514	.537	.537	.537	.512	.487	.510	.510	.510	.486	.462
$\rho = \rho_{max}$	.0227	.0283	.0340	.0378	.0411	.0184	.0230	.0276	.0307	.0333	.0153	.0191	.0230	.0255	.0277
$K_r =$	4.14	5.18	6.21	7.04	7.78	4.01	5.01	6.01	6.80	7.51	3.88	4.84	5.81	6.57	7.25
$\rho = 0.95\,\rho_{max}$	.0215	.0269	.0323	.0359	.0390	.0175	.0219	.0262	.0292	.0317	.0145	.0182	.0218	.0242	.0263
$K_r =$	4.01	5.02	6.02	6.81	7.52	3.88	4.84	5.81	6.57	7.25	3.74	4.68	5.62	6.34	6.99
$\rho = 0.9\,\rho_{max}$	.0204	.0255	.0306	.0340	.0370	.0166	.0207	.0248	.0276	.0300	.0138	.0172	.0207	.0230	.0249
$K_r =$	3.88	4.84	5.81	6.57	7.25	3.74	4.67	5.61	6.33	6.97	3.61	4.51	5.41	6.10	6.72
$\rho = 0.85\,\rho_{max}$	.0193	.0241	.0289	.0321	.0349	.0156	.0196	.0235	.0261	.0283	.0130	.0163	.0195	.0217	.0236
$K_r =$	3.73	4.66	5.59	6.31	6.96	3.59	4.49	5.39	6.07	6.69	3.46	4.33	5.19	5.85	6.44
$\rho = 0.8\,\rho_{max}$	.0181	.0227	.0272	.0302	.0329	.0149	.0184	.0221	.0246	.0267	.0122	.0153	.0184	.0204	.0222
$K_r =$	3.58	4.47	5.36	6.05	6.66	3.44	4.30	5.16	5.81	6.39	3.31	4.14	4.97	5.59	6.15
$\rho = 0.75\,\rho_{max}$	.0170	.0213	.0255	.0284	.0308	.0138	.0173	.0207	.0230	.0250	.0115	.0143	.0172	.0191	.0208
$K_r =$	3.41	4.27	5.12	5.77	6.34	3.28	4.10	4.92	5.53	6.08	3.16	3.94	4.73	5.32	5.84
$\rho = 0.7\,\rho_{max}$	.0159	.0198	.0238	.0265	.0287	.0129	.0161	.0193	.0215	.0233	.0107	.0134	.0161	.0179	.0194
$K_r =$	3.24	4.05	4.87	5.47	6.01	3.11	3.89	4.67	5.25	5.76	2.99	3.74	4.49	5.04	5.53

	f_c' 300					350					400				
(c_{max}/d)	0.667					0.632					0.600				
f_c'	20	25	30	35	40	20	25	30	35	40	20	25	30	35	40
β_1	0.85	0.85	0.85	0.81	0.77	0.85	0.85	0.85	0.81	0.77	0.85	0.85	0.85	0.81	0.77
$\rho = 0.65\,\rho_{max}$	.0147	.0184	.0221	.0246	.0267	.0120	.0150	.0179	.0200	.0217	.00995	.0124	.0149	.0166	.0180
$K_r =$	3.07	3.83	4.60	5.17	5.67	2.94	3.67	4.41	4.95	5.43	2.82	3.53	4.23	4.75	5.21
$\rho = 0.6\,\rho_{max}$	.0136	.0170	.0204	.0227	.0246	.0110	.0138	.0166	.0184	.0200	.00918	.0115	.0138	.0153	.0166
$K_r =$	2.88	3.60	4.32	4.85	5.32	2.76	3.45	4.13	4.64	5.08	2.64	3.30	3.97	4.45	4.87
$\rho = 0.55\,\rho_{max}$	.0125	.0156	.0187	.0208	.0226	.0101	.0127	.0152	.0169	.0183	.00842	.0105	.0126	.0140	.0152
$K_r =$	2.68	3.35	4.03	4.51	4.95	2.57	3.21	3.85	4.32	4.73	2.46	3.07	3.69	4.13	4.53
$\rho = 0.5\,\rho_{max}$	.0113	.0142	.0170	.0189	.0205	.00920	.0115	.0138	.0153	.0167	.00765	.00956	.0115	.0128	.0139
$K_r =$	2.48	3.10	3.72	4.17	4.56	2.37	2.96	3.56	3.98	4.36	2.27	2.84	3.40	3.81	4.17
$\rho = 0.45\,\rho_{max}$	.0102	.0128	.0153	.0170	.0185	.00828	.0104	.0124	.0138	.0150	.00689	.00861	.0103	.0115	.0125
$K_r =$	2.27	2.84	3.40	3.81	4.17	2.17	2.71	3.25	3.64	3.98	2.07	2.59	3.11	3.48	3.80
$\rho = 0.4\,\rho_{max}$	.00907	.0113	.0136	.0151	.0164	.00736	.00920	.0110	.0123	.0133	.00612	.00765	.00918	.0102	.0111
$K_r =$	2.05	2.56	3.07	3.44	3.76	1.96	2.44	2.93	3.28	3.58	1.87	2.34	2.80	3.13	3.42
$\rho = 0.35\,\rho_{max}$	.00793	.00992	.0119	.0132	.0144	.00644	.00805	.00966	.0107	.0117	.00536	.00669	.00803	.00893	.00970
$K_r =$	1.82	2.28	2.73	3.05	3.33	1.74	2.17	2.60	2.91	3.18	1.66	2.07	2.49	2.78	3.03
f_c'	20	25	30	35	40	20	25	30	35	40	20	25	30	35	40
β_1	0.85	0.85	0.85	0.81	0.77	0.85	0.85	0.85	0.81	0.77	0.85	0.85	0.85	0.81	0.77
$\rho = 0.3\,\rho_{max}$	.00680	.00850	.0102	.0113	.0123	.00552	.00690	.00828	.00921	.0100	.00459	.00574	.00688	.00765	.00832
$K_r =$	1.59	1.98	2.38	2.66	2.90	1.51	1.89	2.27	2.53	2.76	1.44	1.80	2.16	2.41	2.63
$\rho = 0.25\,\rho_{max}$	.00567	.00708	.00850	.00945	.0103	.00460	.00575	.00690	.00767	.00834	.00383	.00478	.00574	.00638	.00693
$K_r =$	1.34	1.68	2.01	2.25	2.45	1.28	1.60	1.92	2.14	2.33	1.22	1.52	1.83	2.04	2.22
$\rho = 0.2\,\rho_{max}$	—	.00567	.00680	.00756	.00821	—	.00460	.00552	.00614	.00667	—	.00382	.00459	.00510	.00554
$K_r =$	—	1.36	1.64	1.82	1.99	—	1.30	1.55	1.73	1.89	—	1.23	1.48	1.65	1.80
$\rho = 0.15\,\rho_{max}$	—	—	.00510	.00567	.00616	—	—	.00414	.00460	.00500	—	—	—	.00383	.00416
$K_r =$	—	—	1.25	1.39	1.51	—	—	1.18	1.32	1.43	—	—	—	1.25	1.36
$\rho = \rho_{min}$	.00467	.00467	.00467	.00467	.00467	.00400	.00400	.00400	.00400	.00400	.00350	.00350	.00350	.00350	.00350
$K_r =$	1.12	1.13	1.14	1.15	1.16	1.12	1.13	1.14	1.15	1.16	1.12	1.13	1.14	1.15	1.16

b) Using the design aid in Table 4.1
   From table, $\rho_{max} = 0.0230$
   and $\rho_{min} = 0.0035$

   Actual $\rho = 0.0126$, $< \rho_{max}$, and $> \rho_{min}$, hence OK.
   $$\rho = 0.0126 = \frac{0.0126}{0.0230} \times \rho_{max} = 0.548\ \rho_{max}$$

   From Table 4.1, $K_r = 3.40$ for $\rho = 0.5\ \rho_{max}$
                   and $K_r = 3.69$ for $\rho = 0.55\ \rho_{max}$
          Interpolating, for $\rho = 0.548\ \rho_{max}$, $K_r = 3.68$
                   $M_r = K_r\ bd^2 \times 10^{-6}$
                       $= 3.68 \times 300 \times 530^2 \times 10^{-6}$
                       $= 310$ kN·m, as before

4.11.2 SLABS AS RECTANGULAR BEAMS

Slabs of uniform thickness subjected to a moment distributed over its width (Fig. 4.23) may be considered as a wide shallow beam for purposes of analysis and design. In such slabs, reinforcing bars are usually spaced uniformly over the width of the slab and hence, for convenience, computations are generally based on a typical 1 m wide strip considered as a beam (Fig. 4.23c). The area of reinforcement per metre wide strip is given by $A_s = 1000\ A_b/s$, where A_b is the area of one bar and s is the bar spacing in mm. The moment on the 1 m wide design strip is equal to M_f, the moment per metre width on the slab at the design location. The procedure is illustrated in Example 4.7.

----------------------------- **EXAMPLE 4.7** -----------------------------

A one-way slab (Fig. 4.24) having an overall thickness of 150 mm is reinforced with No. 10 deformed bars placed at an effective depth of 125 mm, and at a spacing of 150 mm centre to centre. Use $f'_c = 20$ MPa and $f_y = 300$ MPa. Compute its factored resistance per metre width.

SOLUTION

Area of No. 10 bar $= 100$ mm²

$$A_s \text{ per metre wide strip} = \frac{1000 \times 100}{150} = 667 \text{ mm}^2$$

$$\rho \text{ provided} = \frac{667}{1000 \times 125} = 0.005$$

ρ_{min} for slabs (CSA A23.3-M84(7.8)) $= 0.002$
ρ_{max} (Table 4.1) $= 0.0227$
$\rho_{min} < \rho < \rho_{max}$ OK

Fig. 4.23 *Analysis of slabs*

TYPICAL STRIP
Im WIDE

M_u/m width

(a) Slab

A_b

S | S

(b) Cross section

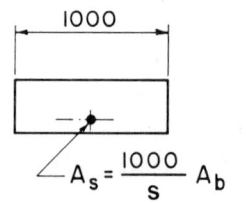

1000

$A_s = \dfrac{1000}{s} A_b$

(c) Design strip

Fig. 4.24 *Example 4.7*

150

125

150

No.10
@150mm

(a) Slab

1000

125

$A_s = 667\,mm^2$

(b) 1 m wide strip

Depth of stress block is obtained from $C = T$

$$0.85\ \phi_c f_c'\ b\ a = A_s \phi_s f_y$$

$$0.85 \times 0.6 \times 20 \times 1000\,a = 667 \times 0.85 \times 300, \quad a = 16.7\ mm$$

$$M_r = \phi_s A_s f_y \left(d - \frac{a}{2} \right)$$

$$= 0.85 \times 667 \times 300 \left(125 - \frac{16.7}{2} \right) \times 10^{-6}$$

$$= 19.8 \text{ kN·m, per metre width of slab}$$

4.11.3 RECTANGULAR BEAM WITH COMPRESSION REINFORCEMENT

It may first be assumed that the compression reinforcement also has yielded. This can be checked and calculations revised if it is found that the assumption is wrong. The factored strength may be considered as consisting of two parts (Fig. 4.25). The compressive force in the compression reinforcement, C_s, and an equal amount of tension, T_1, forms the first part, M_1. The remaining tensile reinforcement and the concrete compression, C_c, contribute the second part of the strength, M_2. Equating the forces, C_s and T_1, the tension reinforcement area, A_{s1}, contributing to M_1 is obtained as:

$$A_{s1}\phi_s f_y = A_s'(\phi_s f_y' - 0.85\ \phi_c f_c')$$

$$A_{s1} = A_s'\left(\frac{\phi_s f_y' - 0.85\phi_c f_c'}{\phi_s f_y} \right) \tag{4.41}$$

where f_y' is the yield strength of the compression reinforcement. Note that the above equation allows for the concrete that is replaced by A_s' by reducing the effective

Fig. 4.25 *Beam with compression reinforcement, all steel yielded*

$$C_s = A_s'\ (\phi_s f_y' - 0.85\ \phi_c f_c')$$
$$T_1 = A_{s1}\phi_s f_y = C_s$$
$$A_{s1} = A_s'\frac{(\phi_s f_y' - 0.85\ \phi_c f_c')}{\phi_s f_y}$$
$$M_1 = C_s(d-d')$$

$$A_{s2} = A_s - A_{s1}$$
$$C_c = 0.85\ \phi_c f_c'\ ba$$
$$T_2 = A_{s2}\phi_s f_y$$
$$a = \frac{A_{s2}\phi_s f_y}{0.85\ \phi_c f_c'\ b}$$
$$M_2 = A_{s2}\phi_s f_y (d - \frac{a}{2})$$

$$c = a/\beta_1$$
$$\epsilon_s' = \frac{0.003\ (c-d')}{c}$$

stress in the compression reinforcement to $\phi_s f'_y - 0.85 \; \phi_c f'_c$. The area, A_{s2}, acting as tension reinforcement for the concrete part is:

$$A_{s2} = A_s - A_{s1} \qquad (4.42)$$

Equating the forces in part 2, $(C_c = T_2)$

$$a = \frac{A_{s2}\phi_s f_y}{0.85 \; \phi_c f'_c b} \qquad (4.43)$$

Since the depth of neutral axis $c = a/\beta_1$, from the strain diagram, the strain in the compression reinforcement can now be computed to confirm whether that steel has in fact yielded.

$$\epsilon'_s = \frac{0.003}{c} (c - d')$$

If $\epsilon'_s > \epsilon'_y = f'_y/E_s$, the compression reinforcement has in fact yielded and the two parts of the factored strength are computed as:

$$M_1 = A'_s(\phi_s f'_y - 0.85 \; \phi_c f'_c)(d - d')$$

$$M_2 = A_{s2}\phi_s f_y \left(d - \frac{a}{2}\right)$$

and $$M_r = (M_1 + M_2) \qquad (4.44)$$

If it is found that $\epsilon'_s < \epsilon'_y$, the compression reinforcement stress $f'_s < f'_y$ and the calculations have to be revised. One approach for this would be to use an iterative procedure, assuming to begin with, say, $f'_s = E_s\epsilon'_s$ and then check and revise following the procedure above. Alternatively an exact solution can be obtained as follows (see Fig. 4.26).

Fig. 4.26 *Compression reinforcement not yielding*

$\epsilon'_s = 0.003 \frac{c-d'}{c}$

$f'_s = E_s \epsilon'_s$

$T = A_s \phi_s f_y$

$C_s = A'_s(\phi_s f'_s - 0.85 \; \phi_c f'_c)$

$C_c = 0.85 \; \phi_c f'_c b \beta_1 c$

Let c be the actual depth of neutral axis, which is as yet unknown. The compression reinforcement stress then is:

$$f_s' = E_s \epsilon_s' = E_s \frac{0.003}{c} (c - d')$$

Equating the total tension and compression across the section:

$$A_s \phi_s f_y = A_s' \left(\phi_s E_s \times 0.003 \frac{c - d'}{c} - 0.85 \phi_c f_c' \right) + 0.85 \phi_c f_c' b \beta_1 c \quad \textbf{(4.45)}$$

The only unknown, c, in this quadratic equation may be determined. Knowing c, the factored resistance may be computed as:

$$M_r = C_s(d - d') + C_c \left(d - \frac{a}{2} \right)$$

$$= \left[A_s' (\phi_s E_s \times 0.003 \frac{(c - d')}{c} - 0.85 \phi_c f_c') (d - d') \right.$$

$$\left. + 0.85 \phi_c f_c' b \beta_1 c \left(d - \frac{\beta_1 c}{2} \right) \right] \quad \textbf{(4.46)}$$

EXAMPLE 4.8

Analyze the beam shown in Fig. 4.27 and compute the factored moment that can be applied. Use $f_c' = 30$ MPa, $f_y' = 350$ MPa for the compression steel and $f_y = 400$ MPa for the tension steel.

SOLUTION

First check the steel ratio against allowable limits. For doubly reinforced beams, the maximum steel ratio may be computed using Eq. 4.33.

Alternatively, starting from basics, the maximum depth of the neutral axis is:

$$c_{max} = \frac{600}{f_y + 600} d = \frac{600}{400 + 600} \times 530 = 318 \text{ mm}$$

For this depth, the compression steel strain is:

$$\epsilon_s' = \frac{0.003 (c - d')}{c} = \frac{0.003 (318 - 55)}{318}$$

Fig. 4.27 *Example 4.8*

$A'_s = 2$ - No. 25 $= 1000\,\text{mm}^2$

$f'_c = 30\,\text{MPa}$, $f_y = 400\,\text{MPa}$, $f'_y = 350\,\text{MPa}$

$A_s = 3$ - No. 35 $= 3000\,\text{mm}^2$

Hence the compression steel will be yielded under the limiting reinforcement condition and $f'_s = 350$ MPa.

Total compression at maximum tension reinforcement is

$$C = C_s + C_c = A'_s \,(\phi_s f'_s - 0.85\,\phi_c f'_c) + 0.85\,\phi_c f'_c\, b\, \beta_1 c$$

$$= 1000(0.85 \times 350 - 0.85 \times 0.6 \times 30) + 0.85$$

$$+ 0.6 \times 30 \times 300 \times 0.85 \times 318$$

$$= 1.52 \times 10^6 \,\text{N}$$

$$A_s = \frac{C}{\phi_s f_y} = \frac{1.52 \times 10^6}{0.85 \times 400} = 4470 \,\text{mm}^2$$

$$\rho_{max} = \frac{A_s}{bd} = \frac{4470}{300 \times 530} = 0.0281$$

$$\rho_{min} = 1.4/f_y = 0.0035$$

$$\text{Actual } \rho = \frac{3000}{300 \times 530} = 0.0189$$

$$\rho_{min} < \rho < \rho_{max} \qquad\qquad\qquad \text{OK}$$

The section is under-reinforced and the tension reinforcement will yield as the factored strength is reached. To compute M_r (refer to Fig. 4.25) assume the compression reinforcement is also yielded (which is later confirmed).

$$A_{s1} = A'_s \frac{(\phi_s f'_y - 0.85\,\phi_c f'_c)}{\phi_s f_y} = \frac{1000\,(0.85 \times 350 - 0.85 \times 0.6 \times 30)}{0.85 \times 400} = 830 \,\text{mm}^2$$

$$A_{s2} = A_s - A_{s1} = 3000 - 830 = 2170 \,\text{mm}^2$$

$$a = \frac{A_{s2}\phi_s f_y}{0.85\,\phi_c f_c b} = \frac{2170 \times 0.85 \times 400}{0.85 \times 0.6 \times 30 \times 300} = 161 \,\text{mm}$$

$$c = a/\beta_1 = 161/0.85 = 189 \,\text{mm}$$

Check f_s':

$$\epsilon_s' = \frac{0.003\,(c - d')}{c} = \frac{0.003\,(189 - 55)}{189}$$

$$= 0.00213 > \epsilon_y' = 0.00175$$

$$f_s' = f_y' = 350\ \text{MPa} \qquad \text{OK}$$

Strength, $M_1 = A_s'(\phi_s f_y' - 0.85\,\phi_c f_c')(d - d')$

$$= 1000(0.85 \times 350 - 0.85 \times 0.6 \times 30)(530 - 55) \times 10^{-6}$$

$$= 134\ \text{kN} \cdot \text{m}$$

$$M_2 = A_{s2}\phi_s f_y\left(d - \frac{a}{2}\right)$$

$$= 2170 \times 0.85 \times 400\left(530 - \frac{161}{2}\right) \times 10^{-6}$$

$$= 332\ \text{kN} \cdot \text{m}$$

$$M_r = M_1 + M_2 = 134 + 332$$

$$= 466\ \text{kN} \cdot \text{m}$$

EXAMPLE 4.9

Compute the factored flexural strength of the beam shown in Fig. 4.28. Take $f_c' = 30$ MPa, and $f_y = 400$ MPa.

Fig. 4.28 *Example 4.9*

SOLUTION

Check the steel ratio first

$$c_{max} = \frac{600}{600 + f_y} d$$

The distance to the centroid of tension and compression reinforcement may be computed from the given dimensions as $d = 316$ mm and $d' = 66$ mm, respectively.

$$c_{max} = \frac{600}{600 + 400} \times 316 = 190 \text{ mm}$$

$$f'_s = E_s \epsilon'_s = 200\ 000 \times \frac{0.003\ (190 - 66)}{190}$$

$$= 391 \text{ MPa} < f_y$$

Using Eq. 4.33,

$$\rho_{max} = \bar{\rho} + \rho' \frac{(\phi_s f'_s - 0.85\ \phi_c f'_c)}{\phi_s f_y}$$

$$= 0.0230 + \frac{1400}{250 \times 316} \times \frac{(0.85 \times 391 - 0.85 \times 0.6 \times 30)}{0.85 \times 400}$$

$$= 0.0395$$

$$\text{Actual } \rho = \frac{2900}{250 \times 316} = 0.0367 < \rho_{max}$$

If Eq. 4.33a is used, $\rho_{max} = \bar{\rho} + \rho' \dfrac{f'_s}{f_y}$

$$= 0.0230 + \frac{1400}{250 \times 316} \times \frac{391}{400} = 0.0403$$

and still $\rho < \rho_{max}$

$$\rho_{min} = \frac{1.4}{f_y} = 0.0035$$

Hence, $\rho_{min} < \rho < \rho_{max}$ OK

It is obvious here that at factored strength the compression reinforcement does not yield. (In the balanced section $f'_s < f_y$. The actual section being under-reinforced, its $c < c_{max}$ and hence $f'_s < f'_{s\ max}, < f_y$).

Assume that the actual depth to neutral axis is c. Equating $C_c + C_s = T$ yields (Eq. 4.45).

$$0.85\ \phi_c f'_c\ b\ \beta_1 c + A'_s \left[\phi_s E_s \times 0.003 \frac{c - d'}{c} - 0.85\ \phi_c f'_c \right] = A_s \phi_s f_y$$

Substituting for $f'_c = 30$ MPa, $b = 250$ mm, $\beta_1 = 0.85$, $A'_s = 1400$ mm², $E_s = 200$ GPa,

$d' = 66$ mm, $A_s = 2900$ mm^2 and $f_y = 400$ MPa, and solving the quadratic equation for c,

$$c = 174 \text{ mm}$$

$$f_s' = E_s \times 0.003 \frac{c - d'}{c} = 372 \text{ MPa} < f_y$$

The design strength is given by (Eq. 4.46).

$$M_r = \left[C_c \left(d - \frac{\beta_1 c}{2} \right) + C_s(d - d') \right]$$

$$= \left[0.85 \times 0.6 \times 30 \times 250 \times 0.85 \times 174 \times \left(316 - \frac{0.85 \times 174}{2} \right) \right.$$

$$\left. + 1400 (0.85 \times 372 - 0.85 \times 0.6 \times 30)(316 - 66) \right] \times 10^{-6}$$

$$= 242 \text{ kN·m}$$

4.11.4 ANALYSIS OF T-BEAMS

When the depth of stress block, a, is less than or equal to the flange thickness, h_f, the analysis is identical to that for a rectangular beam. For $a > h_f$, the total compression area can be split up into a rectangular area, $b_w\, a$, corresponding to the web and an outstanding flange area, $(b - b_w)\, h_f$, as shown in Fig. 4.29. The average stress in compression is $0.85\phi_c f_c'$ and equating $C = T$, the actual value of a can be evaluated. The lever arm for the compression in the flange is $d - h_f/2$ and for the compression in the web portion is $d - a/2$. The relevant equations are

$$C_f + C_w = T$$
$$0.85\ \phi_c f_c'[(b - b_w)h_f + b_w a] = A_s \phi_s f_y$$

and hence:

$$a = \frac{A_s \phi_s f_y - 0.85\ \phi_c f_c'\ (b - b_w)\ h_f}{0.85\ \phi_c f_c' b_w} \tag{4.47}$$

Fig. 4.29 *Analysis of T-beam*

$$C_w = 0.85\,\phi_c f_c' b_w a \qquad\qquad C_f = 0.85\,\phi_c f_c'(b - b_w)\,h_f$$

The design strength is:

$$M_r = \left[C_f \left(d - \frac{h_f}{2} \right) + C_w \left(d - \frac{a}{2} \right) \right] \qquad (4.48)$$

──────────── **EXAMPLE 4.10** ────────────

A simply supported T-beam has a span of 8 m and the cross section dimensions shown in Fig. 4.30. Compute its factored resistance in flexure. Use $f'_c = 20$ MPa and $f_y = 400$ MPa.

Fig. 4.30 *Example 4.10*

SOLUTION

First check the effective flange width (CSA A23.3-M84(8.10). b shall not exceed $0.4 \times$ span $= 3200$ mm, and the overhang shall not exceed $12 \times h_f = 960$ mm nor one-half the clear distance to the next beam (not applicable here). Since the actual width is less than all these limitations, $b =$ actual width $= 1000$ mm.
Now check the reinforcement ratio:

$$\text{Actual } \rho = \frac{3500}{250 \times 500} = 0.028 > \rho_{\min} = \frac{1.4}{f_y} = 0.0035 \qquad \text{OK}$$

$\rho_{\max}$ may be computed using Eq. 4.35 or using the basic theory as follows:

$$\text{Maximum neutral axis depth } c_{\max} = \frac{600}{600 + f_y} d = 300 \text{ mm}$$

$$a_{\max} = \beta_1 c_{\max} = 0.85 \times 300 = 255 \text{ mm} > h_f$$

At maximum tension reinforcement conditions, the total concrete compression is:

$$C_{max} = 0.85 \ \phi_c f'_c \times (\text{concrete area over stress block depth})$$
$$= 0.85 \times 0.6 \times 20 \ [250 \times 255 + (1000 - 250) \times 100]$$
$$= 1.42 \times 10^6 \ \text{N}$$

Since
$$A_s \phi_s f_y = T = C$$
$$A_{s,max} = \frac{C}{\phi_s f_y} = \frac{1.42 \times 10^6}{0.85 \times 400} = 4180 \ \text{mm}^2$$

$$\text{Actual } A_s = 3500 \ \text{mm}^2 < A_{S,max} \qquad\qquad \text{OK}$$

The section is under-reinforced and the reinforcement will yield at the factored strength. The actual depth of stress block, a, is obtained by equating the tensile and compressive forces. Assuming $a > h_f$,

$$T = C = C_w + C_f$$

$$A_s \phi_s f_y = 0.85 \ \phi_c f'_c \ b_w \ a + 0.85 \ \phi_c f'_c \ (b - b_w) \ h_f$$

$$a = \frac{A_s \phi_s f_y - 0.85 \ \phi_c f'_c \ (b - b_w) \ h_f}{0.85 \ \phi_c f'_c \ b_w} \qquad\qquad (4.47)$$

$$= \frac{3500 \times 0.85 \times 400 - 0.85 \times 0.6 \times 20(1000 - 250)100}{0.85 \times 0.6 \times 20 \times 250}$$

$$= 167 \ \text{mm} > h_f, \text{ as assumed}$$

The design strength is given by Eq. 4.48,

$$M_r = M_w + M_f = \left[C_w \left(d - \frac{a}{2} \right) + C_f \left(d - \frac{h_f}{2} \right) \right]$$

$$= \left[0.85 \ \phi_c f'_c \ b_w a \left(d - \frac{a}{2} \right) + 0.85 \ \phi_c f'_c \ (b - b_w) h_f \left(d - \frac{h_f}{2} \right) \right]$$

$$= \left[0.85 \times 0.6 \times 20 \times 250 \times 167 \left(500 - \frac{167}{2} \right) \right.$$

$$\left. + \ 0.85 \times 0.6 \times 20(1000 - 250) \times 100 \left(500 - \frac{100}{2} \right) \right] \times 10^{-6}$$

$$= 522 \ \text{kN} \cdot \text{m}$$

4.11.5 TENSION REINFORCEMENT NOT YIELDED (OVER-REINFORCED BEAM)

All designs conforming to CSA A23.3-M84 have to be under-reinforced (see Section 4.9), so that in general the tension reinforcement would have yielded before the ultimate strength is reached. However, there may be special occasions when the nominal strength of an over-reinforced beam has to be evaluated.

In all beams at nominal ultimate strength the compressive strain at the extreme fibre in concrete will be $\epsilon_{cu} = 0.003.$ If the depth to the neutral axis, c, is also known, the steel strains and stresses can be evaluated. Knowing steel stresses and assuming the standard rectangular stress block for concrete in compression, the internal forces and the flexural strength can be evaluated from statics. Thus the problem reduces to finding the correct value of c. The value of c must be such that the equilibrium condition that the total compressive force is equal to the total tensile force is satisfied. Hence this condition can be used to evaluate c, either by a trial and error approach, or by setting up an algebraic equation in terms of c and solving it. After the correct value of c is determined, the strength can be evaluated using statics. Example 4.11 illustrates this procedure.

EXAMPLE 4.11

For the doubly reinforced section shown in Fig. 4.31 compute the design strength. Use $f'_c = 20$ MPa, $f_y = f'_y = 400$ MPa.

Fig. 4.31   *Example 4.11*

SOLUTION

First check whether the section is under-reinforced. The maximum reinforcement ratio can be found using Eq. 4.32 and 4.33. Equation 4.32 gives, for limiting reinforcement ratio:

$$f'_s = 600 \times \left(1 - \frac{400 + 600}{600} \times \frac{55}{500} \right) = 490 \text{ MPa} > f_y$$

So $f'_{s,\max} = 400$ MPa. From Table 4.1, for a rectangular section with tension reinforcement only, $\bar{p} = 0.0153$. The maximum reinforcement ratio (Eq. 4.33) is:

$$\rho_{\max} = \bar{\rho} + \rho' \frac{\phi_s f'_s - 0.85 \, \phi_c f'_c}{\phi_s f_y}$$

$$= 0.0153 + \frac{1400}{300 \times 500} \left(\frac{0.85 \times 400 - 0.85 \times 0.6 \times 20}{0.85 \times 400} \right) = 0.0244$$

actual $\rho = \dfrac{6000}{300 \times 500} = 0.04 > \rho_{\max}$

Hence the section is over-reinforced.

To compute M_r by the trial and error method, assume the tension reinforcement stress f_s as, say,

$$f_s = 400 \times \frac{0.0244}{0.04} = 244 \text{ MPa}$$

Note that assuming f_s is similar to assuming a value for c.

$$\epsilon_s = \frac{f_s}{E_s} = \frac{244}{200\ 000} = 0.00122, c = \frac{d \times 0.003}{\epsilon_s + 0.003} = 355 \text{ mm}$$

$$\epsilon_s' = \frac{0.003(c - d')}{c} = 0.00254 > \epsilon_y$$

so,

$$f_s' = f_y$$

$$a = \beta_1 c = 302 \text{ mm}$$

$$C = 0.85\ \phi_c f_c' \times 300 \times 282 + 1400\ (\phi_s\ 400 - 0.85\ \phi_c f_c') = 1.39 \times 10^6 \text{ N}$$

$$T = 6000 \times 0.85 \times 244 = 1.24 \times 10^6 \text{ N}$$

With the first trial $C > T$. Hence revise with a higher value for f_s. Try $f_s = 267$ MPa which gives a $T = 6000 \times 0.85 \times 267 = 1.36 \times 10^6$ N, which is slightly higher than the average of C and T obtained in the first trial. With $f_s = 267$ MPa, $\epsilon_s = 0.00134$, $c = 346$ mm, $\epsilon_s' > \epsilon_y'$, $f_s' = f_y = 400$ MPa, $a = 294$ mm, $C = 1.36 \times 10^6$ N. Since $C \simeq T$, this trial value may be assumed to be sufficiently accurate. The factored strength is:

$$M_r = \left[0.85\ \phi_c f_c'\ b\ a \left(d - \frac{a}{2} \right) + A_s'\ (\phi_s f_y - 0.85\ \phi_c f_c')(d - d') \right]$$

$$= \left[0.85 \times 0.6 \times 20 \times 300 \times 294 \left(500 - \frac{294}{2} \right) \right.$$

$$\left. + 1400(0.85 \times 400 - 0.85 \times 0.6 \times 20)(500 - 55) \right] \times 10^{-6}$$

$$= 523 \text{ kN} \cdot \text{m}$$

To solve the above example more exactly, assume the depth of neutral axis as c. Since the compression reinforcement is close to the top edge where the strain is 0.003, it may be assumed to have yielded so that $f_s' = f_y$ (and later confirmed). From the strain distribution,

$$f_s = E\ \epsilon_s = 200\ 000 \times \frac{0.003\ (d - c)}{c}$$

$$= \frac{600(500 - c)}{c}$$

Equating $C = T$,

$$0.85 \times 0.6 \times 20 \times 300 \,(0.85 \times c) + 1400 \,(0.85 \times 400 - 0.85 \times 0.6 \times 20)$$
$$= 6000 \times 0.85 \times \frac{600 \,(500 - c)}{c}$$

Solving this quadratic equation, $c = 346$ mm, $a = 294$ mm, $f_s = 267$ MPa, $\epsilon_s' = 0.00252 > \epsilon_y$, $f_s' = f_y$ (confirmed), and the factored strength is:

$$M_r = \left[0.85 \times 0.6 \times 20 \times 300 \,(294) \left(500 - \frac{294}{2} \right) \right.$$

$$\left. + \, 1400 \,(0.85 \times 400 - 0.85 \times 0.6 \times 20)(500 - 55) \right] \times 10^{-6}$$

$$= 523 \text{ kN} \cdot \text{m}$$

PROBLEMS

1. The rectangular beam shown in Fig. 4.32 has $f_c' = 30$ MPa and $f_y = 350$ MPa.
    a) Compute the stresses developed in concrete and steel under a service load moment of 175 kN·m. Check the computations using the flexure formula $f = Mc/I$.
    (b) If the service load stresses are to be limited to $0.4\, f_c'$ in concrete and 150 MPa in steel, compute the allowable service load moment for the section. Indicate whether the steel stress or the concrete stress controls.

Fig. 4.32   Prob. 1

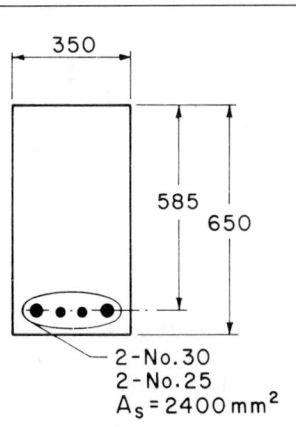

350

585

650

2-No.30
2-No.25
$A_s = 2400 \text{ mm}^2$

2. (a) For the rectangular beam shown in Fig. 4.33, $f_c' = 40$ MPa and $f_y = 400$ MPa. The service load stresses are not to exceed $0.4\, f_c'$ in concrete and 165 MPa in steel. Compute the allowable service load moment for the section.

Fig. 4.33 Prob. 2

6-No.30
$A_s = 4200\,mm^2$

(b) Repeat the problem at (a) with the flexural reinforcment taken as 3 — No. 30 bars ($A_s = 2100$ mm²).

3. A beam simply supported on a span of 6 m is subjected to the service loads (excluding self weight of the beam) shown in Fig. 4.34a. The cross-section dimensions of the beam are given in Fig. 4.34b. Compute the service load stresses in concrete and steel. Take $f'_c = 30$ MPa and $f_y = 300$ MPa.

Fig. 4.34 Prob. 3

3-No.35
$A_s = 3000\,mm^2$

4. For each of the beams in problems 1, 2, and 3 above, using basic principles, check whether the beams are under-reinforced or over-reinforced based on their behaviour at ultimate limit state. Check whether the beams satisfy CSA Code requirements for flexure. Also compute the factored resistance of each section.

5. The cross-section dimensions of a doubly reinforced beam are shown in Fig. 4.35. The beam is subjected to a service load moment of 70 kN·m.
   (a) Compute the stresses in concrete and steel under the specified loads. Take $f'_c = 25$ MPa and $f_y = 300$ MPa.
   (b) Compute the allowable service load moment if the concrete and steel stresses are not to exceed 0.4 f'_c and 140 MPa, respectively.

Fig. 4.35 Prob. 5

6. Compute the flexural strength of the beam in Problem 5 based on CSA A23.3-M84 requirements.

7. For the beam shown in Fig. 4.36, compute the factored resistance, M_r. Use $f'_c = 25$ MPa, $f_y = 400$ MPa and $f'_y = 300$ MPa (for compression reinforcement).

Fig. 4.36 Prob. 7

8. The cross-sectional dimensions of a T-beam are given in Fig. 4.37. Use $f'_c = 40$ MPa and $f_y = 400$ MPa.
   (a) Compute the steel and concrete stresses under a service load moment of 300 kN·m.
   (b) Compute the factored flexural resistance of the beam.

Fig. 4.37 Prob. 8

Fig. 4.38 Prob. 9

9. A simply supported T-beam with a span of 6 m has the cross-section dimensions shown in Fig. 4.38. Compute the flexural strength of the beam. Use $f'_c = 20$ MPa, $f_y = 350$ MPa.

10. A one-way slab having an overall thickness of 100 mm is reinforced with No. 10 bars placed at an effective depth of 75 mm and at a spacing of 125 mm c/c (Fig. 4.39). Material properties are $f'_c = 25$ MPa and $f_y = 400$ MPa. Compute the factored flexural resistance of the slab per metre width.

Fig. 4.39 Prob. 10

REFERENCES

4.1 Timoshenko, S.P., and Goodier, J.N., *Theory of Elasticity*, 2nd ed., McGraw-Hill, 1951, 506 pp.

4.2 Hognestad, E., Hanson, N.W., and McHenry, D., *Concrete Stress Distribution in Ultimate Strength Design*, J. of ACI, Vol. 52, Dec. 1955, pp. 455–479.

4.3 Popov, E.P., *Mechanics of Solids*, Prentice-Hall, Inc., Englewood Cliffs, New Jersey, 1976, 590 pp.

4.4 CSA Standard CAN3-S6-M78 — *Design of Highway Bridges*, Canadian Standards Association, Rexdale, Ontario, Canada, 1978, 272 pp.

4.5 Kaar, P.H., Hanson, N.W., and Capell, H.T., *Stress–Strain Characteristics of High-Strength Concrete*, Proc., Douglas McHenry Int. Symp., ACI Special Publication SP-55, 1978, pp. 161–185.

4.6 CSA Standard CAN3 A23.3 M84 — *Design of Concrete Structures for Buildings*, Canadian Standards Association, Rexdale, Ontario, Canada. 1984, 281 pp.

4.7 Mattock, A.H., Kriz, L.B., and Hognestad, E., *Rectangular Concrete Stress Distribution in Ultimate Strength Design*. J. of ACI, Vol. 57, Feb. 1961, pp. 875–928.

4.8 *Building Code Requirements for Reinforced Concrete*, ACI 318-83, American Concrete Institute, Detroit, Mich., 1983, 111 pp.

CHAPTER 5

DESIGN OF BEAMS AND ONE-WAY SLABS FOR FLEXURE

5.1 INTRODUCTION

In Chapter 4, behaviour of reinforced concrete beams in flexure was explained and procedures given for the analysis of sections. Analysis may involve calculations of stresses under known service loads, of allowable service load moments, and/or of factored flexural strength. From the latter two values, the allowable service loads and the factored loads in flexure may also be computed. In these computations, since the complete cross-sectional dimensions and the material properties are fixed (given) the results are unique, being dictated solely by the conditions of equilibrium of forces and compatability of strains (although simplifying assumptions may be used regarding material response).

The design problem is somewhat the reverse of the above. The external loads, material properties, and the skeletal dimensions (centreline dimensions such as span, height, etc.) are given and it is required to arrive at the cross-sectional dimensions including area of reinforcement necessary to carry all the load effects (including dead loads) satisfactorily. The dead load depends on the cross-sectional dimensions as yet unknown, and will have to be initially assumed and later verified and revised if need be. The flexural strength is dependent on the width and depth of sections, and on the area of reinforcement, and hence, there are several combinations of these that could satisfactorily carry a given moment. Thus, unless all but one of the variables affecting the strength are fixed, there is no unique solution to a design problem. Often one or more of these variables (or their ratios) may be limited to narrow ranges by consideration of aesthetics, economy, deflection control, crack control, ductility requirements, and other structural constraints (for example, b/d ratios, headroom limitations, reinforcement ratio limits, etc.).

In addition to determining the required overall cross-sectional dimensions and the area and disposition of flexural reinforcement at all sections of peak moments (which essentially follow in reverse the analysis procedures discussed in Chapter 4), a complete design of a beam involves the following considerations:

1. the detailing of longitudinal reinforcement in the section, with adequate concrete protection, without congestion, with proper bonding, and in a manner avoiding wide cracks;

2. detailing bar cut-off or bend points along the span, as the applied moment

decreases and some of the bars become no longer necessary to carry the reduced moment;

3. designing for effects of transverse shear force and torsion;
4. checking that serviceability requirements (deflection, crack width, etc.) are met.

Details of the design for shear and considerations of bond (development length) will be presented in Chapters 6, 8 and 9. Serviceability requirements and computations are detailed in Chapter 11. Design for torsion is presented in Chapters 7 and 8.

Methods for the computation of load effects on members such as factored bending moments, shear forces, axial forces and torsion are given in standard books on structural analysis (Refs. 5.1, 5.2). However, some of the approximate methods and simplified procedures permitted by the Code are presented in Chapter 10. For the purpose of design examples in this Chapter, it will be assumed that the factored moments have already been computed and hence known or they will be calculated using standard formulas.

General requirements for placing of flexural reinforcement are reviewed in the next Section as these somewhat influence the choice of beam sizes. This is followed by design for flexure and bar cut-off procedures.

5.2 CONCRETE COVER AND SPACING REQUIREMENTS FOR FLEXURAL REINFORCEMENT

It is convenient in the construction of formwork for the overall sizes of beams and thickness of slabs to be selected in round figures, such as in multiples of 5 mm for slabs, 10 mm for beams, and 20 mm for deeper girders. The effective depth of the member used in flexural computations depends on the reinforcing bar details and the amount of clear concrete cover. In order to protect the reinforcing bars from corrosion and fire, the Codes specify minimum concrete cover for all reinforcement, the amount of cover depending on the type of member and the exposure conditions. The minimum clear covers specified by CSA A23.1-M77 are given in Table 5.1.

Table 5.1 Minimum Specified Clear Concrete Cover for Reinforcement in Cast-in-Place Concrete, CSA A23.1-M77 (12.11)

Cast against and permanently exposed to earth	75 mm
Exposed to earth or weather:	
No. 20 through No. 55 bars	50 mm
No. 15 bars, 16 mm wire, and smaller	40 mm
Not exposed to weather or not in contact with the ground:	
Slabs, walls and joists:	
No. 45 and No. 55 bars	40 mm
No. 35 and smaller	20 mm
Beams, girders, and columns:	
Principal reinforcement, ties, stirrups, and spirals	40 mm
Shells and folded plate members:	
No. 20 bars and larger	20 mm
No. 15 bars, 16 mm wire, and smaller	15 mm

The Codes also specify minimum and maximum limits for the spacing between parallel reinforcing bars in a layer. The minimum limits are necessary to ensure that the concrete flows readily in between and around bars to facilitate proper placement. These limits can be met without difficulty in slabs in general because of the large width available and the relatively low area of reinforcement required for flexure. However, in selecting the width of beams, the minimum spacing between parallel bars is a major consideration because of the limited width within which bars can be placed, as stirrups have to be accommodated, and cover requirements have to be met at the two sides (Fig. 5.1). If the beam width is inadequate to accommodate

Fig. 5.1 *Limitations for flexural reinforcement placement (CSA A23.1, cl. 12)*

S_c = CLEAR SPACING $\geq (d_b$, 25 mm, 1.33 MAX. AGGR. SIZE)

d_c = 50 mm + d_b/2 FOR No. 10 STIRRUPS

   = 40 mm + DIA. OF STIRRUP + INSIDE RAD. OF STIRRUP BEND,
       FOR LARGER THAN No. 10

S = MIN. SPACING = S_c + d_b

MAX. NUMBER OF BARS = $\dfrac{b - 2d_c}{S}$ + 1

all the reinforcement in one layer with the necessary clearance between bars, the alternatives are: (1) to increase the beam width; (2) to place the reinforcement in two or more layers properly separated; and (3) to bundle groups of parallel bars in contact. In beams and girders, bundling of bars up to four in a bundle may be made, provided the bars are no larger than No. 35 and the bundle is enclosed by stirrups or ties. The complete requirements for fabrication and placement of reinforcement for building structures are given in CSA A23.1-M77 (12), and A23.3-M84(7). Those requirements that influence the flexural design are summarized below. The other requirements (such as for ties, splices, anchorage, etc.) will be considered later on when discussing associated design details. It may be noted that reinforcement placement requirements for bridge structures are somewhat different and are given in CSA S6-M78 (8.2).

CSA A23.1-M77 specifies a clear distance between parallel bars in a layer of not less than the diameter of bars, $1\frac{1}{3}$ times the maximum aggregate size, and 25 mm. When the reinforcement is placed in two or more layers, the bars in the upper layers must be placed directly above those in the bottom layer, with a clear distance between layers of at least 25 mm. These clear distance limitations also apply to the clear distance between the bars in a contact lap splice and adjacent parallel bars. Three and four bar bundles are arranged in triangular, L-shaped or square patterns. The bars in a bundle must be tied, wired or otherwise fastened together. For spacing and cover limitations, a unit of bundled bars is treated as a single bar of an equivalent diameter which would give the same total area. Cut off of individual bars in a bundle, within the span of flexural members should be staggered by at least 40 bar diameters. In beams and girders, compression reinforcement, if any, must be enclosed by ties or closed stirrups at least 6 mm in diameter, or welded wire fabric of equivalent area.

In slabs, the spacing of principal reinforcement should not exceed 3 times the slab thickness, (2 times the slab thickness at critical sections of two-way slabs), nor 500 mm. The spacing of shrinkage and temperature reinforcement in slabs should not exceed 5 times the slab thickness, nor 500 mm. The clearance requirements for beams and slabs are shown in Fig. 5.1.

In addition to these, the CSA A23.3-M84(10.6) provides for crack control by controlling the distribution of reinforcement in the zone of concrete tension. For a given area of reinforcement, providing several bars well-distributed over the zone of maximum concrete tension at a moderate spacing is more effective in controlling cracks and in improving bond than providing fewer bars of larger size. The two aspects related to detailing that affect crack width are the thickness of concrete cover, and the area of concrete in the zone of maximum tension surrounding each individual bar. The Code requirement in clause 10.6.4 translates into a maximum spacing limitation for one-way slabs and into a minimum required number of bars for a given width of beam stem. Here again, a bar bundle is treated as an equivalent single round bar giving the same area. While the basis for crack control is discussed in Chapter 11, the detailing requirements resulting from clause 10.6.4 are illustrated in Fig. 5.2.

It is important to note that the present cover and spacing requirements in the Code are primarily based on considerations of protection of reinforcement and

Fig. 5.2 *Minimum number N of tension reinforcement for crack control (CSA A23.3-M84, cl. 10.6.4)*

$$f_s \sqrt[3]{d_c A} \times 10^{-3} = Z \leq \begin{cases} 30 \, kN/mm \, (int. \, exp.) \\ 25 \, kN/mm \, (ext. \, exp.) \end{cases}$$

$$A = 2d_c b/N$$

$$N \geq 2bd_c^2 \left(\frac{f_s \times 10^{-3}}{Z} \right)^3$$

$$f_s \approx 0.6 \, f_y \, (MPa); \, b, d_c \text{ in mm}$$

N = number of bars

EFFECTIVE CONCRETE AREA
SURROUNDING TENSION BARS
$= 2d_c b$

efficient placement of concrete. However, the clear cover and bar spacing are important factors that influence the bond resistance of the bar (Chapter 9). Recent research indicates that if the clear cover on the main reinforcement is much less than about 50 mm, or the clear spacing between bars is less than about 100 mm, the present Code requirements for development length may be inadequate (unconservative), particularly for higher grade steels (see Section 9.10, and Ref. 5.3).

In relatively deep flexural members, a substantial portion of the web will be in tension. Properly distributed tension reinforcement will control the crack width at the level of the flexural tension reinforcement; however, higher up in the web wider cracks may develop. In order to control such cracks, CSA A23.3-M84(10.6.7) requires auxiliary longitudinal skin reinforcement to be placed near the web faces in the tension zone of members with web depth in excess of 750 mm.

5.3 GUIDELINES FOR SELECTION OF MEMBER SIZES

Further to the requirements for placement of reinforcement discussed in the previous Section, an additional major consideration that influences the overall cross-sectional dimensions is the need to control deflection. The effective flexural stiffness of a reinforced concrete section is dependent to a larger extent on the gross section dimensions that on the area of tension reinforcement (see Chapter 11). Table 5.2 gives minimum thickness to span ratios specified by CSA A23.3-M84 (9.5.2.1) for beams and one-way slabs. When these minimum thicknesses are provided, under normal situations in building structures, deflections will be satisfactory and the design will satisfy Code requirements. However, under unusual conditions, there is no guarantee that providing these minimum thicknesses will ensure satisfactory deflection behaviour. More elaborate recommendations for minimum thickness of beams and one-way slabs are given in Ref. 5.4

The deflection may also be controlled indirectly by using a relatively small tension

Table 5.2 Thickness Below Which Deflections Must Be Computed for Non-prestressed Beams or One-Way Slabs Not Supporting or Attached to Partitions or Other Construction Likely to Be Damaged by Large Deflections, CSA A23.3-M84 (9.5.2).

	Minimum Thickness, h*			
	Simply Supported	One End Continuous	Both Ends Continuous	Cantilever
Solid one-way slabs	$\ell/20$	$\ell/24$	$\ell/28$	$\ell/10$
Beams or ribbed one-way slabs	$\ell/16$	$\ell/18.5$	$\ell/21$	$\ell/8$

* The span length ℓ and the thickness h are both in mm.

+ The values given in this Table shall be used directly for non-prestressed reinforced concrete members made with *normal density concrete* ($w_c \geqslant 2400$ kg/m³) with reinforcement having a yield strength of 400 MPa. For other conditions, the values shall be modified as follows:

(a) For structural low density concrete and structural semi-low density concrete the values in the Table shall be multiplied by $1.65 - 0.0003\gamma_c$ but not less than 1.00 where γ_c is the mass density in kg/m³.

(b) For reinforcement having yield strengths other than 400 MPa the values in the Table shall be multiplied by $0.4 + f_y/670$.

Note: With the permission of CSA, this material is reproduced from CSA Standard CAN3-A23.3-M84, *Design of Concrete Structures*, which is copyrighted by the Canadian Standards Association and copies of which may be purchased from the Association, 178 Rexdale Boulevard, Rexdale, Ontario, M9W 1R3.

reinforcement ratio ($0.25\rho_b$ to $0.4\rho_b$) in the flexural design where ρ_b is the steel ratio obtained using Eq. 4.31 without the ϕ-factors. The low area of reinforcement leads to deeper members and hence higher flexural stiffness. Tension reinforcement ratios recommended for deflection control (Ref. 5.4) are presented in Table 5.3.

For a given reinforcement ratio, the resistance varies in direct proportion to the width b and to the square of the effective depth d (Eq. 4.38). Hence, an increase in depth is effective to one order higher than an increase in width. However, deeper beams result in loss of headroom or overall increase in the building height. For economy, in rectangular sections the proportion of overall depth to width is generally in the range of 1.5 to 2 for beams and may be higher (up to 3 or even more) for girders carrying heavy loads. These proportions also apply for the web portion of T-beams. The width and depth are also governed by the shear force on the section (shear stress limitation, see Chapter 6). The beam depth must also be sufficient to provide effective anchorage for the shear reinforcement (see Chapter 6). Often in T-beams the flange may be part of a slab in which case the flange dimensions are known.

The design sequence generally follows the progression of load transfer from the structure to the foundation. For example, in a slab-beam-girder floor system supported on columns, the design order will be slab, beam, girder, column and foundation as the member sizes of and loading on each of these elements are needed for the design of the subsequent element.

Table 5.3 Recommended Tension Reinforcement for Nonprestressed One-Way Construction So That Deflections Will Normally Be Within Acceptable Limits* (Ref. 5.4)

Members	Cross Section	Normal Density Concrete	Low Density Concrete
Not supporting or not attached to nonstructural elements, likely to be damaged by large deflections	Rectangular	$\rho \leqslant 35\% \; \rho_b$	$\rho \leqslant 30\% \; \rho_b$
	"T" or Box	$\rho_w \leqslant 40\% \; \rho_b$	$\rho_w \leqslant 35\% \; \rho_b$
Supporting or attached to nonstructural elements, likely to be damaged by large deflections	Rectangular	$\rho_w \leqslant 25\% \; \rho_b$	$\rho_w \leqslant 20\% \; \rho_b$
	"T" or Box	$\rho_w \leqslant 30\% \; \rho_b$	$\rho_w \leqslant 25\% \; \rho_b$

* For continuous members, the positive region steel ratios only may be used.
ρ_b refers to the balanced steel ratio corresponding to balanced strain conditions, computed without using ϕ-factors (i.e., $\phi_c = \phi_s = 1.0$) in Eq. 4.31 and is approximately 1.4 times ρ_{max}.
$\rho_w = A_s/b_w d$

5.4 DESIGN OF RECTANGULAR SECTIONS FOR BENDING-TENSION REINFORCEMENT ONLY

The problem generally is to determine the required section properties, b, d (hence h) and A_s, to carry a known design moment, M_f. The material properties, f_c' and f_y, are generally specified for the job, and if not, can be selected on the basis of availability and economy. For normal applications, the strengths used are 400 MPa for reinforcement and 20 to 40 MPa for concrete. For under-reinforced sections, the influence of f_c' on the factored flexural resistance is relatively small.

The Code requires that all sections be designed as under-reinforced (clause 10.3.3). For under-reinforced sections, the factored flexural resistance, M_r, is given by Eq. 4.38, (see Section 4.11). The design equation for flexure thus is (Eq. 3.13):

$$M_f = M_r$$
$$= \rho \; \phi_s f_y \left(1 - \frac{\rho \phi_s f_y}{1.7 \; \phi_c f_c'} \right) b d^2 \tag{5.1}$$

Obviously there are several combinations of ρ, b and d which will carry a given M_f and hence many different and satisfactory designs are possible. A general procedure for design when all section properties (b, d, h, A_s) are to be determined will be to select an appropriate reinforcement ratio ρ and then compute the other values (b, d, h) accordingly. This procedure is explained in Section 5.4.1.

There are also occasions where b and h (and hence d) are known and it is only necessary to compute the required area of reinforcement to carry the given moment. For example, in continuous beams, the beam section is first designed for the absolute maximum bending moment in the beam, which normally occurs at one of the support sections where the beam is continuous. The gross concrete section dimensions, $b \times h$, so determined may be provided for the entire length of the beam, since a prismatic member is often desired and may be more economical. Hence, in the design of the beam at other sections of peak moments, such as at midspan and at other support sections, the only unknown is the required area of reinforcement. The design in this case is simpler and may be made as explained in Section 5.4.2.

5.4.1 COMPLETE DESIGN TO DETERMINE b,d,h, A$_s$

Equation 5.1 can be rearranged as:

$$\frac{M_r}{bd^2} = K_r = \rho\phi_s f_y \left(1 - \frac{\rho\phi_s f_y}{1.7\ \phi_c f'_c} \right) \tag{5.2}$$

Step 1

Select a suitable ρ, within the limits of ρ_{max} and ρ_{min}. These limits are given in Sections 4.9 and 4.10 and also in Table 4.1. For economy and control of deflections, ρ selected should preferably be in the range of 0.35 ρ_{max} to 0.50 ρ_{max}. For the selected ρ, the right hand side of Eq. 5.2 gives the strength factor K_r.

$$K_{r,trial} = \rho_{trial}\ \phi_s f_y \left(1 - \frac{\rho_{trial}\ \phi_s f_y}{1.7\ \phi_c f'_c} \right)$$

Step 2

Compute required bd^2 from

$$bd^2 = M_r/K_{r,trial}$$

Now a suitable combination of b and d can be selected to yield the required bd^2. The value of b is taken as a convenient round figure, and d is selected such that the resulting h/b ratio is in the economic range (1.5 to 2 for beams and up to 3 for girders). For slabs, since design is made for a typical 1 m wide strip, $b = 1000$ mm and d can be directly calculated. The choice of d may also be guided by the minimum thickness limitations in Table 5.2 (deflection control).

Step 3

Select actual b, h, d. The overall depth h is selected as a round figure, and such that with allowances made for clear cover, stirrup size and bar size, an effective depth slightly in excess of the required value computed in Step 2 is obtained.

Step 4

Compute actual reinforcement area required. Since the actual d provided in Step 3 may differ slightly from the value computed in Step 2, a revised value for ρ is now computed. Using the dimensions b and d selected,

$$\text{actual } K_r = \frac{M_r}{bd^2}$$

ρ required is given by:

$$K_r = \rho\phi_s f_y\left(1 - \frac{\rho\phi_s f_y}{1.7\ \phi_c f'_c}\right)$$

or

$$\rho = \frac{0.85\ \phi_c f'_c}{\phi_s f_y}\left(1 - \sqrt{1 - \frac{2\ K_r}{0.85\ \phi_c f'_c}}\right) \tag{5.3}$$

Required $A_s = \rho bd$

Step 5

Check the design. Select size and number of bars required and locate the bars in one or more layers to meet the clearance requirements. Check whether the actual effective depth (distance from the compression face to the centroid of reinforcement area) is satisfactory. Note that even a slight reduction in d may be satisfactory if the bar selection gives an area correspondingly greater than the A_s computed in Step 4 (provided $\rho < \rho_{max}$). Probably the best way to check the design is to use the final dimensions selected to compute the factored resistance, M_r, of the section using Eq. 5.1 and ensure that this is greater than the design moment. If the design is not satisfactory, it is revised once again.

Frequently, during the analysis of the structure to compute the design moments, the self-weight of the member itself is initially based on an assumed member size. In such cases at Step 2 it can be checked whether the assumptions regarding self-weight are satisfactory, and if not, the moment calculation can be revised taking the value of b and h obtained in Step 2 as a guide.

EXAMPLE 5.1

A rectangular beam is simply supported on two walls 240 mm thick and 6 m apart (centre to centre). The beam has to carry, in addition to its own weight, a distributed live load of 15 kN/m and a dead load of 8 kN/m. Design the beam section for maximum moment at midspan. Use $f'_c = 30$ MPa, and $f_y = 400$ MPa.

SOLUTION

For $f'_c = 30$ MPa, $\beta_1 = 0.85$. The effective span length (CSA A23.3-M84(8.7)) will in this case be taken as the distance between supports $= 6$ m. Assuming a trial cross-section of 250 mm by 400 mm, the self-weight of the beam per metre length is.

$$0.25 \times 0.4 \times 2400 \times \frac{9.81}{1000} = 2.35 \text{ kN}$$

With the load factors specified in Section 3.6, the factored load is:

$$w_f = \alpha_D D + \gamma\psi\alpha_L L = 1.25(8 + 2.35) + 1.5 \times 15 = 35.4 \text{ kN/m}$$

$$M_f = \frac{w_f \ell^2}{8} = \frac{35.4 \times 6^2}{8} = 159 \text{ kN} \cdot \text{m}$$

The section will first be designed with the maximum allowable reinforcement ratio. The maximum allowable reinforcement ratio is (Eq. 4.31):

$$\rho_{max} = \frac{600}{600 + 400} \times \frac{0.85 \times 0.85 \times 0.6 \times 30}{0.85 \times 400} = 0.0230$$

For $\rho_{max} = 0.0230$,

$$\frac{M_r}{bd^2} = K_r = \rho\phi_s f_y \left(1 - \frac{\rho\phi_s f_y}{1.7\,\phi_c f'_c}\right)$$

$$= 0.0230 \times 0.85 \times 400 \left(1 - \frac{0.0230 \times 0.85 \times 400}{1.7 \times 0.6 \times 30}\right)$$

$$= 5.82 \text{ MPa}$$

Required $bd^2 = M_r/K_r$

$$= \frac{159 \times 10^6}{5.82} = 27.3 \times 10^6 \text{ mm}^3$$

Possible combinations of b and d are:

$b = 200$ mm	$d = 369$ mm
220 mm	352 mm
250 mm	330 mm

Select $b = 250$ mm and $d = 330$ mm. The minimum overall depth required allowing for 40 mm clear cover, No. 10 stirrups (diameter $= 11.3$ mm) and No. 30 bars (diameter $= 29.9$ mm) is $h_{min} = 396 + 40 + 11.3 + 29.9/2 = 396$ mm. Since the ρ selected is the maximum, the depth cannot be decreased. Hence select an overall member size of 250×400 mm. Actual effective depth then will be $d = 400 - (40 + 11.3 + 29.9/2) = 334$ mm.

$$\text{Actual } K_r = \frac{M_r}{bd^2} = \frac{159 \times 10^6}{250 \times 334^2} = 5.70 \text{ MPa}$$

$$= \rho \phi_s f_y \left(1 - \frac{\rho \phi_s f_y}{1.7 \phi_c f_c'} \right)$$

Solving for ρ, $\qquad\qquad \rho = 0.0223 < \rho_{max}$

$$\text{Area } A_s = \rho bd = 0.0223 \times 250 \times 334 = 1860 \text{ mm}^2$$

Fig. 5.3 Example 5.1

Two No. 30 bars and one No. 25 bar (Fig. 5.3) give an area of 1900 mm². The clear spacing between the bars is 1/2 (250 − 2 × 40 − 2 × 11.3 − 2 × 29.9 − 25.2) = 31.2 mm. The minimum clearance is d_b = 29.9 mm, 25 mm or 4/3 × 20 = 26.7 mm (assuming 20 mm size aggregate). The clearances are adequate. The cross-sectional dimensions designed are shown in Fig. 5.3. For this section, to check the actual M_r,

$$a = A_s \phi_s f_y / (0.85 \ \phi_c f_c' b) = 1900 \times 0.85 \times 400 / (0.85 \times 0.6 \times 30 \times 250)$$
$$= 169 \text{ mm}$$

$$M_r = A_s \phi_s f_y \left(d - \frac{a}{2} \right)$$
$$= 1900 \times 0.85 \times 400 \ (334 - 169/2) \times 10^{-6}$$
$$= 161 \text{ kN·m}$$
$$> 159 \text{ kN·m} \qquad\qquad\qquad\qquad\qquad\qquad\qquad\qquad \text{OK}$$

EXAMPLE 5.2

The beam in Example 5.1 will now be redesigned with a moderate reinforcement ratio.

SOLUTION

In order to improve the ductility and economy, and to reduce deflection, select a reinforcement ratio $\rho = 0.5$, $\rho_{max} = 0.5 \times 0.0230 = 0.0115$. This is greater than $\rho_{min} = 1.4/f_y = 0.0035$. From Eq. 5.2, the value of K_r, for $\rho = 0.0115$ is:

$$K_r = 0.0115 \times 0.85 \times 400 \left(1 - \frac{0.0115 \times 0.85 \times 400}{1.7 \times 0.6 \times 30} \right) = 3.41 \text{ MPa}$$

$$= M_r/bd^2$$

Required $bd^2 = M_r/K_r$

$$= 159 \times 10^6/3.41$$

$$= 46.6 \times 10^6 \text{ mm}^3$$

Trial sizes are:

b = 240 mm	d = 440 mm
250 mm	432 mm
260 mm	423 mm
280 mm	408 mm

The trial sizes make it clear that the self-weight has been underestimated. It appears that an overall size of 250×500 mm is adequate. This would give an effective depth close to the 432 mm indicated above. The larger moment due to higher self-weight will result in a reinforcement ratio slightly higher than the value of 0.5 ρ_{max}, which is acceptable.

Corrected self-weight = $0.25 \times 0.5 \times 24 = 3$ kN/m (unit weight of concrete ≈ 2400 $\times 9.81/1000 \approx 24$ kN/m³)

$w_f = 1.25(8 + 3) + 1.5 \times 15 = 36.3$ kN/m
$M_f = 36.3 \times 6^2/8 = 163$ kN·m
with $h = 500$ mm, $d = 500 - (40 + 11.3 + 29.9/2) = 434$ mm

Actual $M_r/bd^2 = K_r = 163 \times 10^6/(250 \times 434^2)$
$$= 3.46 \text{ MPa}$$

For $K_r = 3.46$ MPa, using Eq. 5.3, reinforcement ratio required is:

$$\rho = \frac{0.85 \times 0.6 \times 30}{0.85 \times 400} \left(1 - \sqrt{\frac{2 \times 3.46}{0.85 \times 0.6 \times 30}} \right) = 0.0117$$

Required $A_s = 0.0117 \times 250 \times 434 = 1270$ mm²
Two No. 30 bars give $A_s = 1400$ mm², with adequate clearances

To check M_r of the section,

$$a = \frac{A_s\phi_sf_y}{0.85\,\phi_cf_c'b} = \frac{1400 \times 0.85 \times 400}{0.85 \times 0.6 \times 30 \times 250} = 124 \text{ mm}$$

$M_r = A_s\phi_sf_y(d - a/2) = 1400 \times 0.85 \times 400 (434 - 124/2) \times 10^{-6}$
$$= 177 \text{ kN·m} > 163 \text{ kN·m} \qquad\qquad\qquad \text{OK}$$

5.4.2 DESIGN WHEN THE OVERALL CONCRETE SECTION DIMENSIONS ARE KNOWN

In this case b and h are known and from the latter d can be estimated, and it is only necessary to determine the area of steel A_s. The design then involves only the Steps 4 and 5 of the complete design process detailed in Section 5.4.1. The reinforcement obtained in Step 5 has to be checked against the maximum and minimum limits.

─────────────── **EXAMPLE 5.3** ───────────────

Design the beam in Example 5.1 if the beam size is fixed as 250 mm × 450 mm from architectural considerations.

SOLUTION

$$\text{Self-weight} = 0.25 \times 0.45 \times 2400 \times \frac{9.81}{1000} = 2.65 \text{ kN/m}$$

$$w_f = 1.25(8 + 2.65) + 1.5 \times 15 = 35.8 \text{ kN/m}$$

$$M_f = 35.8 \times 6^2/8 = 161 \text{ kN·m}$$

Assuming No. 10 stirrups and No. 30 bars,

$$d = 450 - (40 + 11.3 + 29.9/2) = 384 \text{ mm}$$

$$\frac{M_r}{bd^2} = K_r = \frac{161 \times 10^6}{250 \times 384^2} = 4.37 \text{ MPa}$$

$$K_r = \rho \phi_s f_y \left(1 - \frac{\rho \phi_s f_y}{1.7 \, \phi_c f_c'} \right)$$

Solving, $\qquad \rho = 0.0155, \; A_s = \rho bd = 1490 \text{ mm}^2$

Three No. 25 bars give $A_s = 1500 \text{ mm}^2$. Clear distance between bars $= 1/2(250 - 2 \times 40 - 2 \times 11.3 - 3 \times 25.2) = 35.9 \text{ mm}$, which is adequate.

With No. 25 bars, $d = 450 - (40 + 11.3 + 25.2/2) = 386 \text{ mm}$

$$\rho = \frac{1500}{250 \times 386} = 0.0155$$

$$\rho_{max} = 0.0230, \rho_{min} = 1.4/f_y = 0.0035$$

$$\rho_{min} < \rho < \rho_{max}$$

To check ultimate strength:

$$a = \frac{A_s \phi_s f_y}{0.85 \, \phi_c f_c' b} = \frac{1500 \times 0.85 \times 400}{0.85 \times 0.6 \times 30 \times 250} = 133 \text{ mm}$$

$$M_r = A_s \phi_s f_y \left(d - \frac{a}{2} \right) = 1500 \times 0.85 \times 400 \left(386 - \frac{133}{2} \right) \times 10^{-6}$$

$$= 163 \text{ kN} \cdot \text{m} > 161 \text{ kN} \qquad\qquad \text{OK}$$

5.5 DESIGN AIDS FOR RECTANGULAR BEAMS

The design procedure explained in Section 5.4 and the previous examples show that the design essentially involved the solution of Eq. 5.2, either to get a required $bd^2 = M_r/K_r$ for a selected reinforcement ratio, or to get a required ρ for a selected bd^2 from $K_r = M_r/(bd^2)$. The solution can thus be simplified by expressing the relation between K_r and ρ, either graphically or in a tabular form, for commonly used values of f'_c and f_y. Table 5.4 presents such a relation between K_r in MPa and $100\,\rho$. Table 5.4 is similar to Table 4.1 except that it is rearranged in terms of selected K_r values at closer intervals. This table also indicates the maximum and minimum limits for ρ. For design using the table, a suitable ρ is first selected and the resulting K_r read off from the table using the column for the specified values of f'_c and f_y. Knowing K_r, the required $bd^2 = M_r \times 10^6/K_r$, where b and d are in mm and M_r in kN·m. Conversely, when b and d are known or appropriately selected, the value $K_r = M_r \times 10^6/(bd^2)$ is computed first. For this K_r, the required value of ρ is read off from Table 5.4. For values in between those given in Table 5.4 linear interpolation may be used.

--------- **EXAMPLE 5.4** ---------

Design the beam in Example 5.1 using Table 5.4

SOLUTION

As in Example 5.1, required $M_r = M_f = 159$ kN·m. ...
Selecting $\rho = \rho_{\max} = 0.0230$, from Table 5.4, $K_r = 5.82$ MPa.

Required $bd^2 = 159 \times 10^6/5.82 = 27.3 \times 10^6$ mm^3
Select $b = 250$ mm,

$$\text{Required } d = \sqrt{\frac{27.3 \times 10^6}{250}} = 330 \text{ mm}$$

Select $h = 400$ mm.
Allowing for cover, actual $d = 400 - (40 + 11.3 + 29.9/2) = 334$ mm
With $b = 250$ mm, $d = 334$ mm

$$\text{Actual } K_r = \frac{M_r}{bd^2} = \frac{159 \times 10^6}{250 \times 334^2} = 5.70 \text{ MPa}$$

Table 5.4 Reinforcement Percentage, 100 ρ, for Resistance Factor, K_r

$M_r = K_r bd^2 \times 10^6$; $K_r = \rho \phi_s f_y [1 - \rho\phi_s f_y/(1.7\phi_c f_c')]$; $\rho = A_s/bd = 0.85\ \phi_c f_c'[1 - \sqrt{1 - 2K_r/(0.85\phi_c f_c')}]/(\phi_s f_y)$;
$\phi_c = 0.6$; $\phi_s = 0.85$. Units of M_r in kN·m; b and d in mm, K_r in MPa

f_y (MPa)	300					350					400				
f_c' (MPa)	20	25	30	35	40	20	25	30	35	40	20	25	30	35	40
$\rho_{min, slab}$ (%)	0.200					0.200					0.200				
$K_{r, min, slab}$ MPa	0.497	0.500	0.502	0.503	0.504	0.578	0.581	0.583	0.585	0.586	0.657	0.662	0.665	0.667	0.669
$\rho_{min, beam}$ (%)	0.467					0.400					0.350				
$K_{r, min, beam}$ MPa	1.12	1.13	1.14	1.15	1.16	1.12	1.13	1.14	1.15	1.16	1.12	1.13	1.14	1.15	1.16
↓ K_r 100 ρ →															
0.4	0.160	0.159	0.159	0.159	0.158	0.137	0.137	0.136	0.136	0.136	0.120	0.120	0.119	0.119	0.119
0.6	0.243	0.241	0.240	0.239	0.239	0.208	0.207	0.206	0.205	0.205	0.182	0.181	0.180	0.180	0.179
0.8	0.327	0.324	0.322	0.321	0.320	0.280	0.278	0.276	0.275	0.274	0.245	0.243	0.242	0.241	0.240
1.0	0.414	0.409	0.406	0.404	0.402	0.354	0.350	0.348	0.346	0.345	0.310	0.307	0.304	0.303	0.302
1.2	0.502	0.495	0.491	0.488	0.485	0.430	0.424	0.421	0.418	0.416	0.377	0.371	0.368	0.366	0.364
1.4	0.593	0.583	0.577	0.572	0.569	0.508	0.500	0.494	0.491	0.488	0.445	0.437	0.433	0.429	0.427
1.6	0.686	0.673	0.664	0.658	0.654	0.588	0.577	0.569	0.564	0.561	0.515	0.505	0.498	0.494	0.491
1.8	0.782	0.764	0.753	0.746	0.740	0.671	0.655	0.646	0.639	0.634	0.587	0.573	0.565	0.559	0.555
2.0	0.881	0.858	0.844	0.834	0.827	0.756	0.735	0.723	0.715	0.709	0.661	0.643	0.633	0.625	0.620
2.2	0.984	0.954	0.936	0.924	0.915	0.843	0.817	0.802	0.792	0.784	0.738	0.715	0.702	0.693	0.686

2.4	1.09	1.05	1.03	1.01	1.00	0.934	0.902	0.882	0.870	0.861	0.817	0.789	0.772	0.761	0.753
2.6	1.20	1.15	1.13	1.11	1.09	1.03	0.988	0.964	0.949	0.938	0.900	0.864	0.844	0.830	0.821
2.8	1.31	1.26	1.22	1.20	1.19	1.13	1.08	1.05	1.03	1.02	0.985	0.942	0.917	0.901	0.889
3.0	1.43	1.36	1.32	1.30	1.28	1.23	1.17	1.13	1.11	1.10	1.07	1.02	0.992	0.972	0.959
3.2	1.56	1.47	1.42	1.39	1.37	1.34	1.26	1.22	1.19	1.18	1.17	1.10	1.07	1.05	1.03
3.4	1.69	1.58	1.53	1.49	1.47	1.45	1.36	1.31	1.28	1.26	1.27	1.19	1.15	1.12	1.10
3.6	1.83	1.70	1.63	1.59	1.56	1.57	1.46	1.40	1.37	1.34	1.37	1.28	1.23	1.19	1.17
3.8	1.98	1.82	1.74	1.70	1.66	1.70	1.56	1.49	1.45	1.43	1.49	1.37	1.31	1.27	1.25
4.0	2.14	1.95	1.86	1.80	1.76	1.84	1.67	1.59	1.54	1.51		1.46	1.39	1.35	1.32
4.2		2.08	1.97	1.91	1.86		1.78	1.69	1.63	1.60		1.56	1.48	1.43	1.40
4.4		2.22	2.09	2.02	1.97		1.90	1.79	1.73	1.69		1.66	1.57	1.51	1.48
4.6		2.36	2.21	2.13	2.07		2.02	1.90	1.82	1.78		1.77	1.66	1.60	1.55
4.8		2.51	2.34	2.24	2.18		2.16	2.00	1.92	1.87		1.89	1.75	1.68	1.63
5.0		2.68	2.47	2.36	2.29		2.30	2.12	2.02	1.96			1.85	1.77	1.72
5.2			2.60	2.48	2.40			2.23	2.12	2.06			1.95	1.86	1.80
5.4			2.75	2.60	2.51			2.35	2.23	2.15			2.06	1.95	1.88
5.6			2.89	2.73	2.63			2.48	2.34	2.25			2.17	2.05	1.97
5.8			3.05	2.86	2.75			2.61	2.45	2.35			2.29	2.14	2.06
6.0			3.21	2.99	2.87			2.75	2.57	2.46				2.24	2.15
6.2			3.39	3.13	2.99				2.68	2.56				2.35	2.24
6.4				3.28	3.12				2.81	2.67				2.46	2.34
6.6				3.43	3.25				2.94	2.78					2.44
6.8				3.58	3.38					2.90					2.54
7.0				3.75	3.52					3.02					2.64
7.2					3.66					3.14					2.75
7.4					3.81					3.26					
7.6					3.96										
ρ_{max} (%)	2.27	2.83	3.40	3.78	4.11	1.84	2.30	2.76	3.07	3.33	1.53	1.91	2.30	2.55	2.77
$K_{r,max}$	4.14	5.18	6.21	7.04	7.78	4.01	5.01	6.01	6.80	7.51	3.88	4.84	5.81	6.57	7.25

From Table 5.4, for $K_r = 5.6$ MPa, $100 \rho = 2.17$
and $K_r = 5.8$ MPa, $100 \rho = 2.29$

Interpolating, for $K_r = 5.7$ MPa, $100 \rho = 2.23$
$A_s = \rho b d = 0.0223 \times 250 \times 334 = 1860$ mm² as before

The rest of the design follows that in Example 5.1.

5.6 DESIGN OF ONE-WAY SLAB

Slabs are designed considering a design strip of one metre width, so that $b = 1000$ mm. Otherwise, the design of a slab is identical in procedure to that of a beam of rectangular section. Slabs in general are not reinforced for shear, as shear stresses are kept in check by providing adequate depth. Hence, no allowance for stirrups need be made in the design of slabs. Since the one metre wide strip is only a typical design strip, there is no sense in specifying a certain number of reinforcing bars for the design section of one metre width. Instead what is required is the selection of a suitable size of bar and its spacing necessary to provide the calculated steel area. If A_s is the calculated reinforcement area required per one metre wide strip, and A_b is the area of one bar of the selected size, the required spacing s in mm is found from:

$$A_s = \frac{1000}{s} A_b, \qquad s = \frac{A_b}{A_s} \times 1000$$

To facilitate the selection of bar size and spacing for slabs, Table 5.5 gives the area, A_s, per metre width obtained by providing different size bars at different spacings. The principal reinforcement in slabs must be no less than that required for shrinkage and temperature stresses (CSA-A23.3-M84(7.8)). When the principal reinforcement extends only in one direction, as in one-way slabs, reinforcement required for shrinkage and temperature stresses is provided normal to the principal reinforcement.

EXAMPLE 5.5

The plan of a beam and slab floor is shown in Fig. 5.4a. The specified floor loading consists of a live load of 9 kN/m² and a dead load (due to floor finish, partitions, etc.) of 1 kN/m², in addition to the self-weight. Concrete has a strength $f'_c = 20$ MPa and $f_y = 400$ MPa. Design the slab thickness and the reinforcement area required at all critical sections.

SOLUTION

The slab is one-way and continuous and integral with the beam supports spaced at 3 m clear. A one metre wide design strip can be treated as a 4 span continuous beam.

Table 5.5 Area of Reinforcing Bars in Slabs in mm² Per Metre Width

Bar Size →	10	15	20	25	30	35
Nominal diameter mm →	11.3	16.0	19.5	25.2	29.9	35.7
Area of 1 Bar mm² →	100	200	300	500	700	1000
Bar Spacing ↓ mm						
60	16 667	3333	5000	8333	11 667	16 667
80	1 250	2500	3750	6250	8 750	12 500
100	1 000	2000	3000	5000	7 000	10 000
120	833	1667	2500	4167	5 833	8 333
140	714	1429	2143	3571	5 000	7 143
150	667	1333	2000	3333	4 667	6 667
160	625	1250	1875	3125	4 375	6 250
180	556	1111	1667	2778	3 889	5 556
200	500	1000	1500	2500	3 500	5 000
220	455	909	1364	2273	3 182	4 545
240	417	833	1250	2083	2 917	4 167
250	400	800	1200	2000	2 800	4 000
260	385	769	1154	1923	2 692	3 846
280	357	714	1071	1786	2 500	3 571
300	333	667	1000	1667	2 333	3 333
320	313	625	938	1563	2 188	3 125
340	294	588	882	1471	2 059	2 941
350	286	571	857	1429	2 000	2 857
360	278	556	833	1389	1 944	2 778
380	263	526	789	1316	1 842	2 632
400	250	500	750	1250	1 750	2 500
420	238	476	714	1190	1 667	2 381
440	227	455	682	1136	1 591	2 273
450	222	444	667	1111	1 556	2 222
460	217	435	652	1087	1 522	2 174
480	208	417	625	1042	1 458	2 083
500	200	400	600	1000	1 400	2 000

The design moments at all critical sections (at midspan for positive moments and at supports for negative moments) can be obtained by the analysis of this continuous beam, bearing in mind that the dead loads act on all spans, whereas the live loads have to be placed so as to cause the worst effect at the section considered. Such a detailed analysis will be discussed and illustrated in Chapter 10. In lieu of such an analysis the CSA A23.3-M84(8.3) permits the use of moment coefficients when the structure and loading meet the limitations in clause 8.3.3. This approximate procedure will be used in this example for computing design moments.

To compute the self-weight of the slab an estimate for the slab thickness may be taken as the minimum thickness specified in Table 5.2. (CSA A23.3-M84(9.5.2.1)).

Fig. 5.4 *Example 5.5*

(a) Plan of beam and slab floor

(b) Design strip for slab moments
by CSA A23·3–M84(8·3)

Considering the end span, which is more critical as one end is discontinuous (note that since the slab is monolithic with the spandel beam at the end, there is in fact some fixity at this end as well, but this will be discounted in estimating h_{min}),

$$h_{min} = \frac{\ell}{24}$$

The span ℓ for this purpose is defined in CSA A23.3-M84(8.7) and is the lesser of the clear span plus the depth of slab (say 3000 mm + 100 mm) and the distance between centres of supports = 3300 mm.

$$h_{min} = \frac{3100}{24} = 130 \text{ mm}$$

(Note that CSA A23.3-M84(8.7.4) permits the use of $\ell = 3$ m in this case.)

$$\text{Self-weight of slab} = 1 \times 1 \times 0.130 \times 2400 \times \frac{9.81}{1000}$$
$$= 3.06 \text{ kN/m}^2$$

Total dead load $= 1 + 3.06 = 4.06$ kN/m²,
  live load $= 9$ kN/m²

Since there are four equal spans and the factored live load is less than three times the factored dead load, the requirements of CSA A23.3-M84(8.3.3) are met and the moment coefficients may be used.

Total factored load per metre length for the one metre wide strip is:

$$w_f = 1.25 \times 4.06 + 1.5 \times 9 = 18.6 \text{ kN/m}$$

Clear span $\ell_n = 3$ m

The design moments at critical sections are (see Fig. 5.4b):

(i)   positive moment in end span (discontinuous end integral with supporting beam):

$$\frac{1}{14} w_f \ell_n^2 = \frac{18.6 \times 3^2}{14} = 12.0 \text{ kN} \cdot \text{m}$$

(ii) positive moment in interior spans:

$$\frac{1}{16} w_f \ell_n^2 = 10.5 \text{ kN} \cdot \text{m}$$

(iii) negative moment at faces of interior supports (for slab with span ≤ 3 m):

$$\frac{1}{12} w_f \ell_n^2 = 14.0 \text{ kN} \cdot \text{m}$$

(iv) negative moment at interior face of exterior support (integral beam support):

$$\frac{1}{24} w_f \ell_n^2 = 6.98 \text{ kN} \cdot \text{m}$$

The slab depth is selected for the largest moment which is the negative moment at supports $= 14.0$ kN·m.

Selecting a steel ratio of 0.4 ρ_{max}, from Table 4.1,
$$\rho = 0.4 \rho_{max} = 0.00612, \text{ for which } K_r = 1.87 \text{ MPa}$$
$$\text{Required } bd^2 = M_r/K_r = 14 \times 10^6/1.87$$
$$= 7.49 \times 10^6 \text{ mm}^2$$
$$d = \sqrt{7.49 \times 10^6/1000} = 86.5 \text{ mm}$$

Assuming No. 10 bars and 20 mm clear cover, the required overall thickness $= 86.5 +$

Select an overall thickness of 120 mm for the slab throughout. The initial self-weight computation is slightly on the safe side and hence satisfactory. The selected thickness is slightly below the computed minimum for deflection control, of 130 mm. However, considering the fixity at the discontinuous end, $h = 120$ mm should be adequate. The actual effective depth is:

$$d = 120 - 20 - 11.3/2 = 94.4 \text{ mm}$$

The area of reinforcement and bar spacing are computed for the interior support sections below. Similar computations will give the bar details required at all critical sections. At interior supports negative $M_r = 14.0$ kN·m

$$\text{Actual } K_r = \frac{M_r}{bd^2} = \frac{14 \times 10^6}{1000 \times 94.4^2} = 1.57 \text{ MPa}$$

For $K_r = 1.57$, $\rho = 0.00505$ (Eq. 5.3. or Table 5.4)
$A_s = 0.00505 \times 1000 \times 94.4 = 477 \text{ mm}^2$

$$\text{Spacing} = \frac{1000 \, A_b}{A_s} = \frac{1000 \times 100}{477} = 210 \text{ mm}$$

No. 10 bars at 200 mm is adequate. The bar spacing may also be selected from Table 5.5 which shows that No. 10 bars at 200 mm gives $A_s = 500$ mm² compared to the required 477 mm².

The calculations for other critical sections are set forth in Table 5.6. It is necessary also to check that the ρ provided is not less than the specified minimum for slabs and the spacing of bars does not exceed the maximum specified (CSA A23.3-M84 (7.4.1.2)).

In a one-way slab such as this with principal reinforcement designed for flexure only in one direction (bars placed along the short span), secondary reinforcement must be provided at right angles to the main reinforcement to take care of temperature and shrinkage stresses (A23.3-M84(7.8)). The minimum temperature and shrinkage reinforcement in this example = $0.002 \times$ gross area = $0.002 \times 1000 \times 120 = 240$ mm² per metre width. No. 10 bars at 400 mm give an area of 250 mm² per metre which is adequate. The maximum spacing for this is the lesser of 500 mm and $5 h = 600$. The spacing, 400 mm, is satisfactory. The actual arrangement of slab reinforcement is discussed in Section 5.7.

5.7 ARRANGEMENT OF FLEXURAL REINFORCEMENT IN CONTINUOUS ONE-WAY SLABS

The positive and negative moment reinforcement required in the span and over the support regions may be provided in one of two ways, shown in Fig. 5.5a and b. In the first method, separate reinforcement is detailed for the positive moments and the negative moments (Fig. 5.5a). Alternatively, the negative reinforcement over a support region may be provided in part by bending over to the top some of the positive moment reinforcement from the spans on either side as the bottom reinforcement becomes no longer needed in the support region (Fig. 5.5b). Detailing of bar cut-off, bending and extensions are discussed in Section 5.9. These details are dependent on the moment envelope, giving the upper and lower limits of the moment for the entire span. Determination of moment envelopes for continuous spans is explained in Chapter 10. However, for uniformly loaded continuous slabs of nearly equal spans, such detailed calculation is often unnecessary, and cut-off and bending points may be approximately related to span dimensions. Typical bar bending details for one-way solid slabs recommended in Ref. 5.5 are shown in Fig. 5.6. Based on this, two alternative arrangements of bars for the slab in Example 5.5 are given at Fig. 5.4c.

Fig. 5.4(c) *Arrangement of bars for the slab, Example 5.5*

TRANSVERSE SECTION OF SLAB
AT A A

ALTERNATIVE ARRANGEMENT
WITH BENT UP BARS

Fig. 5.5 *Arrangement of main reinforcement for continuous slabs*

Fig. 5.6 *Recommended typical bending details*

SINGLE SPAN, SIMPLY SUPPORTED

STANDARD HOOK
(TILT FROM VERTICAL IF NECESSARY
TO MAINTAIN 20 CLEARANCE)
25 COVER
0.25 L
20 CLEAR
SYMMETRICAL ABOUT ℄
TEMPERATURE BARS
100 MIN.
20 CLEAR
EXTEND ALL BOTTOM
BARS INTO SUPPORT
25 COVER MIN.
150 MAX.
L = CLEARSPAN**
BAR SPACING
SLAB THICKNESS

INTERIOR SPAN, CONTINUOUS

0.3L OR 0.3L₁
GREATER
20 CLEAR
0.3L OR 0.3L₁
GREATER
SYMMETRICAL ABOUT ℄
TEMPERATURE BARS
C
0.25L
20 CLEAR
L₁
150 MIN.
L = CLEARSPAN
SPACING
SLAB THICKNESS

END SPAN, SIMPLY SUPPORTED END

STANDARD HOOK
(TILT FROM VERTICAL IF NECESSARY
TO MAINTAIN 20 CLEARANCE)
0.25L
20 CLEAR
ALTERNATE BARS
BENT UP
0.3L OR 0.3L₁
GREATER
20 CLEAR
0.3L OR 0.3L₁
GREATER
SLAB THICKNESS
TEMPERATURE BARS
25 COVER
100 MIN.
20 CLEAR
0.25L
EXTEND ALL BOTTOM
BARS INTO SUPPORT
150 MAX.
150 MIN
L₁
25 COVER MIN.
L = CLEARSPAN**
SPACING

INTERIOR SPAN, CONTINUOUS
(WHEN STRAIGHT BARS ARE USED)

0.3L OR 0.3L₁
GREATER
20 CLEAR
0.3L OR 0.3L₁
GREATER
SLAB THICKNESS
0.125L
150 MIN.
L₁
L = CLEARSPAN
SPACING
TEMPERATURE BARS

Table 5.6 Computation for Flexural Reinforcement at All Critical Sections, Example 5.5

Location of Critical Section	End Span			Interior Span			Code A23.3 Ref. Clause
	End Support	Span	First Int. Support	Int. Support	Int. Span	Int. Support	
Moment Sign and Location of Reinforcement (top/bottom)	Negative	Positive	Negative				
Moment coefficient	$-1/24$	$+1/14$	$-1/12$	$-1/12$	$+1/16$	$-1/12$	8.3
Design moment, $M_r = $ Coeff $\times 18.6 \times 3^2$; kN·m	6.98	12.0	14.0	14.0	10.5	14.0	
Actual $K_r = \dfrac{M_r}{bd^2} = \dfrac{M_r \times 10^6}{1000 \times 94.4^2}$; MPa	0.783	1.35	1.57	1.57	1.18	1.57	
Required ρ (Table 5.4 or Eq. 5.3)	0.00240	0.00428	0.00505	0.00505	0.00370	0.00505	
ρ_{min}, on gross area	0.002						7.8.1
Required area $A_s = \rho \times 1000 \times 94.4$; mm²	227	404	477	477	349	477	
$A_{s,min} = \rho_{min} \times$ gross area	240	$0.002 \times 1000 \times 120 = 240$ mm², $A_s > A_{s,min}$					
Required spacing of No. 10 bars, $\dfrac{1000 \times 100}{A_s}$; mm	417	248	210	210	287	210	
Maximum allowable spacing, lesser of $3h$ or 500, mm	360						7.4.1.2
Select No. 10 bars at (mm)	350	240	200	200	250	200	
Area provided; mm²	286	417	500	500	400	500	

5.8 DESIGN OF RECTANGULAR BEAMS WITH COMPRESSION REINFORCEMENT

When the cross-sectional dimensions of a beam are so limited that even with the maximum allowable reinforcement ratio of ρ_{max} the factored resistance of the section is inadequate to carry the moment, M_f, due to the factored loads, the section is designed with compression reinforcement and additional tension reinforcement to carry the extra bending moment. All compression reinforcement must be enclosed by ties or closed stirrups conforming to CSA A23.3-M84(7.6.6), to prevent their buckling. The design of doubly reinforced rectangular beams is illustrated with Example 5.6 below.

--- **EXAMPLE 5.6** ---

If the beam in Example 5.1 is limited in size to $b = 250$ mm and $h = 400$ mm, and it has to carry, in addition to the loads already mentioned, a concentrated live load of 30 kN, design the reinforcement.

SOLUTION

Self-weight $= 0.25 \times 0.4 \times 2400 \times 9.81 \times 10^{-3} = 2.35$ kN/m
Factored distributed load $= 1.25(8 + 2.35) + 1.5 \times 15$
$\qquad\qquad\qquad\qquad\qquad = 35.4$ kN/m

Moment due to distributed load

$$= \frac{w_f \ell^2}{8} = \frac{35.4 \times 6^2}{8} = 159 \text{ kN·m}$$

The concentrated load is to be placed at midspan for maximum moment. This moment is:

$$\frac{P_f \ell}{4} = (1.5 \times 30) \times \frac{6}{4} = 67.5 \text{ kN·m}$$

Total design moment, $M_f = 227$ kN·m

The maximum strength of the section with tension reinforcement only, M_1, is checked first. As computed in Example 5.1, or from Table 4.1,

$$\rho_{max} = \bar{\rho} = 0.023$$
$$\text{for which } K_r = 5.81 \text{ MPa}$$

Assuming that tension reinforcement has to be provided in two layers with No. 30 bars,

$$d \approx 400 - (40 + 11.3 + 29.9 + 25/2) = 306 \text{ mm}$$
$$A_{s1} = 0.0230 \times 250 \times 306 = 1760 \text{ mm}^2$$
$$M_1 = K_r b d^2 = 5.81 \times 250 \times 306^2 \times 10^{-6} = 136 \text{ kN·m}$$

Fig. 5.7 *Example 5.6*

(a) Section (b) Couple M_1 (c) Strain (d) Couple M_2 (e) Bar details

Since $M_1 < M_f$, the section has to be additionally reinforced in compression and tension to carry the excess moment $M_2 = M_f - M_1 = 227 - 136 = 91$ kN·m. The compression reinforcement will be placed at a depth $d' = 40 + 11.3 + 29.9/2 \simeq 65$ mm. The internal resisting couples are shown in Fig. 5.7. The additional tension reinforcement area A_{s2} required is given by:

$$M_2 = A_{s2}\phi_s f_y (d - d')$$

$$A_{s2} = \frac{91 \times 10^6}{0.85 \times 400(306 - 65)} = 1110 \text{ mm}^2$$

Total tension reinforcement required $= 1760 + 1110$
$$= 2870 \text{ mm}^2$$

if the compression reinforcement yields $A'_s = A_{s2}$.

To check this, the strain at the compression reinforcement level is computed using the strain distribution shown in Fig. 5.7.

For the moment M_1, $\rho = \rho_{max}$, $c/d = c_{max}/d = 0.60$ as given in Table 4.1, and $c = 0.6 \times 306 = 184$ mm. Alternatively, equating $C_c = T_1$,

$$0.85 \, \phi_c f'_c \, ba = A_{s1}\phi_s f_y$$

$$a = \frac{A_{s1}\phi_s f_y}{0.85 \, \phi_c f'_c b} = \frac{1760 \times 0.85 \times 400}{0.85 \times 0.6 \times 30 \times 250} = 156.4 \text{ mm}$$

$$c = a/\beta_1 = 156.4/0.85 = 184 \text{ mm}$$

$$\epsilon'_s = \frac{0.003 \, (c - d')}{c} = \frac{0.003 \, (184 - 65)}{184} = 0.00194$$

$$\epsilon'_s < \epsilon_y = \frac{400}{200\,000} = 0.002$$

The compression steel does not yield and the stress in it is $f'_s = E_s \, \epsilon'_s = 200\,000 \times 0.00194 = 388$ MPa. To find A'_s, equating C_s to T_2,

$$A_s' \phi_s f_s' = A_{s2} \phi_s f_y$$

$$A_s' = A_{s2} \frac{f_y}{f_s'} = \frac{1110 \times 400}{388} = 1144 \text{ mm}^2$$

$$\rho' = A_s'/(bd) = 0.0150$$

Note that if allowance is made for the concrete area in compression which is displaced by A_s', the net effective steel stress will only be $(\phi_s f_s' - 0.85 \phi_c f_c')$ and A_s' will be obtained as:

$$A_s' = \frac{A_{s2} \phi_s f_y}{(\phi_s f_s' - 0.85 \phi_c f_c')} = 1200 \text{ mm}^2$$

The tension reinforcement ratio in this example can be expressed as:

$$\rho = \frac{A_{s1}}{bd} + \frac{A_{s2}}{bd}$$

$$= \bar{\rho} + \rho' \frac{f_s'}{f_y} = 0.0375$$

The required reinforcement areas can be provided by three No. 25 plus two No. 30 bars placed in two layers on the tension side ($A_s = 2900 \text{ mm}^2$) and two No. 30 bars on the compression side ($A_s = 1400 \text{ mm}^2$) as shown in Fig. 5.7. This arrangement meets the minimum clearance requirements.

An exact evaluation of the actual ultimate strength of this section can be made by the procedure outlined in Section 4.11.3. Such an analysis (see Example 4.9) shows that the actual strength is 242 kN·m > M_f = 227 kN·m. (The actual strength is slightly higher because with the reinforcement arrangement used, d = 316 mm rather than the 306 mm assumed in design.)

5.9 DESIGN OF T-BEAMS

The complete design of an isolated T-beam (such as may be used for a foot bridge) involves the determination of the flange and web dimensions and the area of reinforcement. In such a case, the overall flange width is normally specified and the flange thickness and flange reinforcement (transverse) required may be computed considering the transverse bending of the overhanging flange as a cantilever slab. However, the most common use of T-beams is in the monolithic slab-beam floor system where the slab dimensions are already known from the results of the slab design which precedes the T-beam design. Thus, the effective flange width and the flange thickness are generally known and the T-beam design involves the determination of web dimensions and the area of reinforcement. (Code Specifications for the effective flange width of T- and L-shapes are given in CSA A23.3-M84(8.10)).

In a continuous T-beam, the section over the support is under negative moment with the flange of the T being on the tension side of the section, and here the beam acts in effect as a rectangular section with compression in the bottom fibres of the web. In such cases, the rectangular section dimensions determined for the maximum negative moment (which generally exceeds the maximum positive moment) also control the web size in the positive moment region, and only the area of reinforcement in the T-beam remains to be determined.

In simply supported T-beams, however, the web dimensions must also be designed. The width of web b_w of a T-beam must be sufficient to provide the tension reinforcement in one or more convenient layers, and the depth, d, must be adequate to provide the area $b_w \times d$ to resist shearing forces (Chapter 6). The depth must also be adequate to control the deflection, and the minimum thicknesses specified in Table 5.2 may be used as guidance (although with caution as the loads on T-beams vary considerably depending on the type of structure — buildings, bridges, etc.)

In T-beam design procedure, the web dimensions are first assumed (if not already known) and the area of reinforcement required is computed. If the area so determined is beyond allowable limits or cannot be placed conveniently, the design is revised with modified web dimensions. Because of the very large compressive concrete area contributed by the flange, in T-beams of usual proportions the depth of stress block a will be less than the flange thickness h_f so that the T-beam effectively acts as a rectangular beam of width b. Actual T-beam action ($a > h_f$) occurs only when the flange dimensions are small while the beam is very deep and very heavily reinforced, as is likely with closely spaced bridge girders in a T-beam bridge.

Flanges of T-beams must have adequate transverse reinforcement provided near the top of the flange. Such reinforcement is usually present in the form of the negative moment reinforcement for the slab which forms the flange of the T-beam and generally spans across the T-beams. When this is not the case, the required transverse reinforcement must be designed by considering the over-hanging portion of the flange as a cantilever slab carrying the design loads applied on that portion of the slab. The spacing of this reinforcement is limited to the lesser of 5 times the flange thickness or 500 mm (A23.3-M84(8.10.5)).

At negative moment regions in a T-beam the flange is in tension and the beam is designed as a rectangular section having the width of the web. However, placing all the tension reinforcement in the web will leave the integral flange unprotected against wide tensile cracking. Hence A23.3-M84(10.6.6) specifies that a part of such tension reinforcement be distributed over a width equal to the lesser of the effective flange width or 1/10 span. If the effective width exceeds 1/10 span, the outer portions of the flange must be protected by some additional longitudinal reinforcement.

EXAMPLE 5.7

Design the interior beam in the floor system in Example 5.5.

SOLUTION

The slab is one-way, spanning the beams. Each interior beam carries a load from a width of 3.3 m of the slab (1.65 m on either side).

Dead load from slab including slab weight

$$= 3.3 \text{ m} \times \left(1 + 0.12 \times 2400 \times \frac{9.81}{1000}\right) = 13 \text{ kN/m}$$

The overall depth will be assumed as $\ell/16$ for deflection control,

$$= \frac{8000}{16} = 500 \text{ mm}$$

Self-weight of rib $= 0.3 \times (0.500 - 0.120) \times 2400 \times \dfrac{9.81}{1000}$

$$= 2.68 \text{ kN/m}$$

Live load $= 3.3 \text{ m} \times 9 \text{ kN/m}^2 = 29.7 \text{ kN/m}$

Total factored load $= 1.25 (13 + 2.68) + 1.5 (29.7) = 64.2 \text{ kN/m}$

$$\text{Design moment} = 64.2 \times \frac{8^2}{8} = 514 \text{ kN·m}$$

The effective width of flange is the least of
  (1) $0.4 \, \ell = 0.4 \times 8000 = 3200 \text{ mm}$
  (2) $24 \, h_f + b_w = 24 \times 120 + 300 = 3180 \text{ mm}$ (controls)
  (3) spacing $= 3300 \text{ mm}$

Assuming that reinforcement will be provided in two layers, take the effective depth $d \simeq 400 \text{ mm}$.

The steel area limitations will be computed first. The maximum allowable reinforcement ratio is given either by Eq. 4.33 or may be computed from first principles as follows:

$$c_{max} = \frac{600}{600 + f_y} \, d = 0.6 \times 400 = 240 \text{ mm}$$

$$a_{max} = 0.85 \times 240 = 204 \text{ mm} > h_f$$

$$C_{max} = [120(3180 - 300) + 300 \times 204] \times 0.85 \times 0.6 \times 20 = 4.15 \times 10^6 \text{ N}$$

$$A_{max} = \frac{4.15 \times 10^6}{400} = 12\ 200 \text{ mm}^2$$

$$A_{s,max} = \frac{1.4}{f_y} \, b_w d = \frac{1.4}{0.85 \times 400} \times 300 \times 400 = 420 \text{ mm}^2$$

The procedure for calculation of reinforcement area will depend on whether the depth of stress block for the given moment is less than h_f or not (whether rectangular beam action or T-beam action is to be considered). This may be checked by computing the factored moment resistance for $a = h_f$ and comparing this with the applied moment M_f.

If $a = h_f$,

$$M_r = bh_f \times 0.85 \, \phi_c f_c' \left(d - \frac{h_f}{2}\right)$$

$$= 3180 \times 120 \times 0.85 \times 0.6 \times 20 \left(400 - \frac{120}{2}\right) \times 10^{-6}$$

$$= 1326 \text{ kN·m} >> M_f = 514 \text{ kN·m}$$

Hence the depth $a < h_f$ and the beam acts like a rectangular beam of width $b = 3180$ mm and $d = 400$ mm. Following the design procedure for rectangular beams:

$$K_r = \frac{M_r}{bd^2} = \frac{514 \times 10^6}{3180 \times 400^2} = 1.01 \text{ MPa}$$

$$= \rho\phi_s f_y \left(1 - \frac{\rho\phi_s f_y}{1.7\ \phi_c f'_c}\right)$$

$$\rho = 0.00313$$

$$A_s = 0.00313 \times 3180 \times 400 = 3980 \text{ mm}^2$$

This is within the allowable limits. Provide six No. 30 bars in two layers, giving an area of 4200 mm^2 as shown in Fig. 5.8. The actual strength of this section may be computed as $M_r = 548$ kN·m > 514 kN·m. (See Example 5.10 for computations.)

Fig. 5.8 *Example 5.7*

6-No.30 BARS

$f'_c = 20$ MPa

$f_y = 400$ MPa

Alternative Procedure

The method of solution illustrated in Example 5.7 above gives the exact value of A_s required directly. However, the solution can be very much simplified by a trial-and-revision procedure as explained and illustrated below.

For T-beams of normal proportions, the depth of stress block, a, is normally within the flange. The lever arm, z, between the resultant compression in concrete, C, and tension in reinforcement, T, can be approximately estimated initially as the larger of $d - h_f/2$ and $0.9d$. With this value of z, the area of reinforcement necessary to carry a design moment M_r may be estimated as:

$$A_s = \frac{M_r}{\phi_s f_y z}$$

Using this approximate value of A_s, the depth of stress block, a, may be computed by equating $C = T$. Usually $a < h_f$, and even if $a > h_f$, in normal situations it is admissible to ignore the compressive concrete area below the flange and assume $a = h_f$.

Thus equating $C = T$ yields,

$$0.85 \, \phi_c f'_c \, b \times (a \leqslant h_f) = A_s \phi_s f_y$$

$$a = \frac{A_s \phi_s f_y}{0.85 \, \phi_c f'_c b} \leqslant h_f$$

and revised $z = d - \dfrac{a}{2}$

$$\text{Revised } A_s = \frac{M_r}{\phi_s f_y z}$$

Usually this revised value of A_s is very close to the exact value obtained by the detailed procedure.

To illustrate the procedure, the problem in Example 5.7 will be solved by this method.

$$\text{Approximate } z = d - \frac{h_f}{2} \text{ or } 0.9d$$

$$= 400 - \frac{120}{2} = 340 \text{ mm or } 0.9 \times 400 = 360 \text{ mm}$$

$$= 360 \text{ mm}$$

$$\text{Approximate } A_s = \frac{514 \times 10^6}{0.85 \times 400 \times 360} = 4199 \text{ mm}^2$$

$$a = \frac{4199 \times 0.85 \times 400}{0.85 \times 0.6 \times 20 \times 3180} = 44.0 \text{ mm}$$

$$\text{Revised } z = d - \frac{a}{2} = 378 \text{ mm}$$

$$\text{Revised } A_s = \frac{514 \times 10^6}{0.85 \times 400 \times 378} = 4000 \text{ mm}^2$$

compared to the value of 3980 mm² obtained by the detailed procedure.

In continuous T-beams, the negative moments at supports are usually greater than the positive moments in the span. Furthermore, under negative moments, the flange is on the tension side and the beam effectively acts as a rectangular section. In such cases the overall rib dimensions of the T-beam will be established first by the design of the beam section at the support. When the latter section is designed as an under-reinforced section (with tension reinforcement only), the T-section under the lesser positive moment in the span will naturally be under-reinforced and there is no need to check this.

EXAMPLE 5.8

A continuous T-beam has the cross-sectional dimensions shown in Fig. 5.9. The web dimensions have been determined from considerations of negative moment at supports and the shear strength requirements. The spans are 10 m, and the design moment at midspan under factored loads is 500 kN·m. Determine the flexural reinforcement requirement at midspan. Take $f'_c = 20$ MPa and $f_y = 400$ MPa.

Fig. 5.9 *Example 5.8*

$$M_1 = C_w \left(d - \frac{a}{2}\right) \qquad M_2 = C_f \left(d - \frac{h_f}{2}\right)$$

$$A_s \phi_s f_y = T = T_1 + T_2 = C_w + C_f$$

SOLUTION

The effective flange width is the least of:

$24\,h_f + b_w = 2700$ mm

actual width $= 1000$ mm ... (least)

$0.25 \times$ span $= 2500$ mm

Assuming that two layers of reinforcement are required, the effective depth is estimated as $d \simeq 600 - (40 + 11.3 + 29.9 + 25/2) = 506$ mm. First it will be checked whether the depth of stress block $a < h_f$ or not.

If $a = h_f = 100$ mm, the moment capacity will be given by:

$$M = 0.85\ \phi_c f_c'\, b\, h_f \left(d - \frac{h_f}{2}\right)$$

$$= 0.85 \times 0.6 \times 20 \times 1000 \times 100\ (506 - 100/2) \times 10^{-6}$$

$$= 465\ \text{kN·m}$$

Since design moment $M_r = 500$ kN·m is larger than this, a must be greater than h_f.

To find the correct value of a, equate M_r to the moment of the compressive forces in the web, C_w, and in the overhanging flange, C_f (see Fig. 5.9).

$$M_r = C_w\left(d - \frac{a}{2}\right) + C_f\left(d - \frac{h_f}{2}\right)$$

$$= 0.85\ \phi_c f_c'\, b_w\, a\left(d - \frac{a}{2}\right) + 0.85\ \phi_c f_c'\, (b - b_w)\, h_f\left(d - \frac{h_f}{2}\right)$$

$$500 \times 10^6 = 0.85 \times 0.6 \times 20 \left[300 \times a\left(506 - \frac{a}{2}\right) + (1000 - 300)100\left(506 - \frac{100}{2}\right)\right]$$

Solving this quadratic equation, $a = 129$ mm.

The steel area can now be obtained from

$$T = A_s\phi_s f_y \quad = C_w + C_f$$

$$A_s = \frac{C_w + C_f}{\phi_s f_y} = \frac{0.85 \times 0.6 \times 20\,[300 \times 129 + (1000-300) \times 100]}{0.85 \times 400}$$

$$= 3261 \text{ mm}^2$$

Now check whether this area is within allowable limits.

$$\rho_{min} = \frac{1.4}{f_y}, \text{ where } \rho = \frac{A_s}{b_w d}$$

$$A_{s,min} = \frac{1.4}{400} \times 300 \times 506 = 531 \text{ mm}^2$$

Maximum reinforcement such that $\dfrac{c}{d} \leq \dfrac{600}{600 + f_y} = 0.6$

Actual $c/d = a/(\beta d) = 129/(0.85 \times 506) = 0.30 << 0.6$ \qquad\qquad OK

The area designed, $A_s = 3261$ mm^2, is within the allowable limits and hence satisfactory.

5.10 CUT-OFF OR BENDING OF FLEXURAL REINFORCEMENT

5.10.1 INTRODUCTION

In simply supported beams, the maximum bending moment is at or near the midspan and the flexural design of the cross-section is first worked out for this section. Similarly, in continuous spans the maximum negative moment is at a support section and the maximum positive moment near midspan. Hence the required concrete section dimensions and the respective flexural reinforcement areas are determined by designing these sections. (For negative moment, CSA A23.3-M84(8.7.3) permits the use of the moment at the face of the support for design of beams.) At sections away from these critical design sections, the moment progressively decreases; however, the same overall section dimensions are usually provided throughout the length of the beam. Such a prismatic member is usually more economical than a member of varying cross-section because of the greater fabrication costs for the latter. Thus, as one moves away from the midspan section, the positive moment decreases and as a result the flexural reinforcement required also decreases. Hence some of the bars provided at the midspan section can be terminated as they become no longer necessary to resist bending moment or, alternatively, in continuous beams, bent over to the top side of the beam and continued to the support to take care of the negative moment requirement over the supports. In the same manner, as one moves away from the face of the support towards midspan in a continuous beam, some of the bars provided for the peak negative moment at the support section may be cut off or bent down.

5.10.2 THEORETICAL CUT-OFF POINTS CONSIDERING MOMENT ONLY

In a prismatic beam, as the bending moment decreases along the length of the member, the required flexural tension reinforcement decreases approximately in direct proportion to the moment. This can be illustrated as follows.

If M is the moment and A_s is the tension reinforcement area required at the critical section (Fig. 5.10b, c),

$$A_s = \frac{M}{\phi_s f_y z}$$

Fig. 5.10 *Theoretical bar cut-off (or bend) points*

At a section where the moment decreases to M_1, the required reinforcement area is:

$$A_{s1} = \frac{M_1}{\phi_s f_y z_1}$$

and

$$\frac{A_{s1}}{A_s} = \frac{M_1}{M} \frac{z}{z_1} \simeq \frac{M_1}{M}, \text{ since } \frac{z}{z_1} \simeq 1$$

In fact as the moment decreases, the lever arm z increases slightly as is evident from Fig. 5.10c; however, it is sufficiently accurate (and slightly conservative) to assume $z/z_1 = 1$. Hence the steel area requirement varies in proportion to the variation of the moment. Thus at a section where the moment is, say, 60 percent of M, the reinforcement area required is only 60 percent of the designed area, A_s, and the remaining 40 percent may be cut off as far as flexural requirement is concerned.

In practice, it may be that the actual reinforcement area, A_s, provided at the critical design section is in excess of the exact design requirement for M, as bars are available only in discrete sizes and can be provided only in full numbers. Further, bars to be cut off are selected by numbers of bars rather than percentage of areas. Hence it would be more appropriate to use the factored resistance, M_r, of the critical section, and the strength per bar M_r/N (where N is the total number of bars at the design section, assuming all of them to be of the same size) to detail bar cut-off. For example, the first two bars are no longer required for flexure where the moment reduces to $M = (N-2) \times M_r/N$; a third bar also becomes no longer needed where $M = (N-3) \times M_r/N$, and so on, as is illustrated in Fig. 5.10d. In determining the theoretical bar cut-off locations in this manner, the bending moment diagram must

Fig. 5.11 *Moment envelope for continuous spans*

represent the possible maximum at each section (moment envelope). Further, in continuous spans, the loading patterns for the maximum negative moment at supports and for the maximum positive moment in the span are different (see Chapter 10). Hence the points of inflection for the negative and the positive moment diagrams are usually different (see Fig. 5.11). (Also note that the bending moment diagram will be plotted on the side on which it causes tension).

5.10.3 ACTUAL POINT OF CUT-OFF

Although the theoretical location where a reinforcing bar is no longer required for flexure can be determined as above, the actual point of cut-off is selected more conservatively in order to allow for other factors such as possible diagonal tension cracks, unexpected shift in maximum moments, development length requirements, possible deterioration of shear strength, etc.

When a bar is terminated in the tension zone at the theoretical point of cut-off, the discontinuity effects tend to open up flexural cracks which could develop into diagonal tension cracks (see Chapter 6). Such a diagonal tension crack in a flexural member without shear reinforcement is shown in Fig. 5.12. Neglecting any dowel action in the reinforcing bars and any aggregate interlock forces, the equilibrium of forces indicated on the freebody in Fig. 5.12c shows that the reinforcement force at section b–b depends on the moment at section a–a. Thus the reinforcement area necessary to provide for the moment M_1 at section a–a is in fact needed up to section b–b.

Fig. 5.12 *Influence of diagonal tension cracks on reinforcement stress*

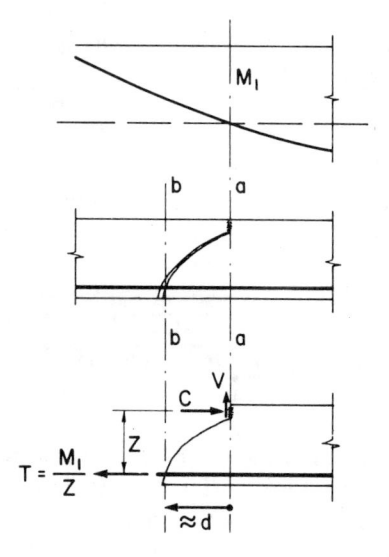

A similar situation exists near points of inflection in a continuous beam, as illustrated in Fig. 5.13. The development of flexural cracks and inclined flexure-shear

Fig. 5.13 *Reinforcement stresses at inflection points in cracked member*

(a) Bending moment diagram

(b) Tied-arch action in cracked beam

(c) Tensile stress in longitudial bars

cracks transforms the load transfer mechanism in a beam loaded on its top face and supported on the bottom into one approaching tied-arch action; with an inclined compression strut formed by the concrete between the load and reaction points, as shown in Fig. 5.13b. As a result, the zone of tension in the longitudinal reinforcement may extend beyond the point of inflection as shown in Fig. 5.13c, although the bending moment diagram would indicate zero tensile stress in the longitudinal bars at the inflection point.

Further, the theoretical bending moment diagrams are only idealizations or best estimates and the actual bending moments may differ because of the approximations involved in the calculation of loads and load effects, possible changes in loading, yielding of supports, lateral load effects, etc. In order to allow for these factors, CSA A23.3-M84(12.10.3) requires that the flexural reinforcement bars be extended beyond the theoretical point of cut-off for a distance equal to the larger of the effective depth d or 12 times the bar diameter d_b. Individual bars in a bundle, if cut off, must be terminated at different points at least 40 bar diameters apart. Bars for cut-off are preferably selected such that the continuing bars form an arrangement symmetric with respect to the vertical centreline of the beam cross-section. When bars are placed in two or more layers, the curtailment starts with bars in the innermost layer.

Instead of cutting off tension reinforcement in the zone of tension, it is frequently anchored into the compression zone by bending it across the web and making it continuous with the reinforcement on the opposite face of the member, or anchoring it there (CSA A23.3-M84(12.10.1)). Such bent bars restrain the spread of diagonal tension cracks and, furthermore, add to the shear resistance of the section (see Chapter 6). The discontinuity effects discussed in Sections 5.10.3 (Fig. 5.12) and 5.10.5 below are less severe for bars which are bent compared to bars cut off. Hence a requirement that bend points be extended beyond the theoretical points of cut-off by the larger distance of d or $12d_b$ may be too conservative. It has been suggested (Ref. 5.6) that for bar extension purposes, a bent bar may be considered effective up to the section where the bar crosses the mid-depth of the beam. This would reduce the extension required by $d/2$. However, for bar extension, the Code does not distinguish between cut off bars and bent bars.

5.10.4 CSA A23.3-M84 REQUIREMENTS ON BAR EXTENSION

The requirement of bar extension beyond the theoretical cut-off point by a distance equal to the larger of d or $12d_b$ (CSA A23.3-M84(12.10.3)) has already been explained above. Figure 5.12 shows that if a bar which is no longer required for moment at section a–a is cut off at section b–b, the remaining bars are stressed to their yield stress at section b–b. In order to develop this yield stress, the continuing bars must extend beyond section a–a for an embedment length of at least one full development length ℓ_d, plus the effective depth of the member or $12d_b$, whichever is greater (CSA A23.3-M84(12.10.4)). Development length is discussed in detail in Chapter 9, but stated simply it is the minimum length of embedment required beyond the point of peak bar stress (critical section) so that the bar force at the critical section ($A_s f_y$) can be fully transferred to the surrounding concrete through bond stress along the interface. This minimum extension by the length ℓ_d (or equivalent) beyond a critical section where the reinforcement stress is f_y is required for *all* bars (see Section 9.12). Requirements of clauses 12.10.3, 12.10.4 and 12.10.5, (see Section 5.10.5 below) do not apply if both the stirrups and the longitudinal reinforcement are designed using the General Method for shear given in CSA A23.3-M84(11.4), as the latter accounts for these effects directly.

To further provide for the possible shifting of the moment diagram, the Code (clauses 12.11 and 12.12) specifies that at least one-third of the positive moment reinforcement in simply supported members (and one-fourth in continuous members) be continued straight and into the supports for a length of at least 150 mm. However, if the member is part of a lateral load resisting system, in order to provide some strength and ductility under load reversals, the anchorage of this minimum positive moment reinforcement continued into the support must be adequate to develop the full yield stress f_y. (This does not apply for reinforcement continued into the support in excess of the minimum requirements.) Similarly at least one-third of the negative moment reinforcement required over a support must be continued beyond the point of inflection of the negative moment diagram for a distance equal to the largest of the effective depth, d, 12 times the bar diameter, or 1/16 of the clear span. These Code requirements concerning curtailment of flexural reinforcement are summarized in Fig. 5.14. For completeness, Fig. 5.14 also shows the development length requirement beyond all critical sections (minimum embedment, ℓ_d, beyond critical

Fig. 5.14 *Code requirements for flexural reinforcement cut-off*

Note: Code cl. 12·10·3, 12·10·4 & 12·10·5 do not apply if shear reinforcement is designed by cl. 11·4

The need for an embedment equal to a development length of ℓ_d (ℓ_d is a function of bar diameter d_b as described in Chapter 9) beyond a critical section at which the reinforcement stress is f_y has already been mentioned. For positive moment reinforcement continued beyond the section of zero moment (centre of simple supports and point of inflection for continuous spans), the portion ℓ_d (see Eq. 5.4) of the actual embedment beyond the point of zero moment that is considered to effectively contribute to ℓ_d is limited to:

1. at simple supports, the embedment length beyond the centre of support, and
2. at a point of inflection, the larger of the effective depth or $12d_b$, but not exceeding the actual bar extension beyond the point of inflection.

sections). However, the detailed consideration of bond and development and the methods for computing ℓ_d are given in Chapter 9.

CSA A23.3-M84(12.11.3) specifies that (except for reinforcement terminating in a standard hook extending ℓ_{db} beyond the centreline of simple supports or a mechanical anchorage at least equivalent to a standard hook) the total effective embedment length thus available, taken as $\ell_a + M_t/V_u$, must satisfy (see Section 9.12 for explanation):

$$\ell_d \le \ell_a + M_r/V_f \qquad (5.4)$$

Here M_r is the factored moment resistance of the section for the reinforcement that is continued, V_f is the factored shear force at the point of zero moment (see Fig. 5.15) and ℓ_{db} is the development length of standard hook in tension. If Eq. 5.4 is not satisfied, either smaller size bars have to be used so that the required ℓ_d is reduced or the area of reinforcement at the section has to be increased to increase M_r.

Fig. 5.15 *Positive reinforcement bar-size limitation at zero moment location (CSA A23.3-M84(12.11.3))*

Note: This does not apply if the General Method of shear design (cl. 11·4) is used (cl. 11·4·6·5)

5.10.5 EFFECT OF BAR CUT-OFF ON SHEAR STRENGTH

Cutting off bars in the tension zones lowers substantially the shear strength and ductility of beams (Ref. 5.7). This results from the premature flexural cracks initiated by the discontinuity effects near the cut bar, which may develop into diagonal tension cracks if the shear stress at this section also happens to be relatively high. To safeguard against this, CSA A23.3-M84(12.10.5) permits termination of flexural reinforcement in a tension zone only if one of the following three conditions is satisfied:

1. the shear at the section does not exceed 2/3 the permitted value;
2. excess stirrups, having an area

$$A_v \geq \frac{b_w s}{3 f_y}$$

are provided over a distance $0.75d$ from the cut-off point, at a spacing, s, not exceeding

$$\frac{d}{8\beta_b}$$

where β_b is the ratio of area of bars cut off to the total area of bars at the section;

3. with No. 35 and smaller size, the continuing bars provide at least twice the area required for flexure and the shear is no more than 3/4 the value permitted.

Bending the bars over to the opposite face instead of cutting them off generally avoids this detrimental effect on shear strength. Clause 12.10.5 does not apply if shear reinforcement is designed by the General Method for shear design (Cl. 11.4).

Examples of bar cut-off calculations given below consider only the moment requirement. The consideration of shear strength at bar cut-off points is illustrated in Examples 6.1 and 6.2 in Chapter 6.

──────────── **EXAMPLE 5.9** ────────────

In the beam designed in Example 5.3, locate the point where one of the three longitudinal reinforcement bars can be terminated.

SOLUTION

The bending moment diagram is parabolic with a central ordinate of 161 kN·m as shown in Fig. 5.16b.

The flexural capacity with all three bars is = 163 kN·m. If one bar is terminated, the flexural capacity with the remaining two bars is $\simeq 2/3 \times 163 = 109$ kN·m. In the beam, the location x where the bending moment is this value is obtained as:

$$M_x = w_f \frac{\ell}{2} x - w_f \frac{x^2}{2} = 109 \text{ kN·m}$$

$$= 35.8 \frac{6}{2} x - 35.8 \frac{x^2}{2} = 109 \text{ kN·m}$$

Solving $\qquad\qquad x = 1.29 \text{ m (or 4.71 m)}$

Fig. 5.16 *Example 5.9*

The middle bar is no longer required for moment at a section 1.29 m from the support. However, the bar should be continued beyond this point a distance equal to the larger of

$$d = 386 \text{ mm} \qquad \text{(controls)}$$
$$12d_b = 12 \times 25.2 = 302 \text{ mm}$$

Actual point of termination $= 1290 - 386 = 904$ mm

The middle bar may be terminated at 900 mm from the centre of the support. Since the bar is terminated in a tension zone, the requirements in CSA A23.3-M84(12.10.5) concerning the shear strength at this section must be satisfied. This is investigated in Example 6.1, Chapter 6.

Check bar size limitation at support (CSA A23.3-M84(12.11.3)).

$$\left(1.3 \frac{M_r}{V_f} + \ell_a\right) \text{ must exceed } \ell_d$$

For No. 25 bars, with $f_y = 400$ MPa and $f'_c = 30$ MPa, (From Table 9.1, Chapter 9)

$$\ell_d = 694 \text{ mm}$$

$$V_f = 35.8 \times \frac{6}{2} = 107 \text{ kN}$$

For two bars continued, $M_r = 109$ kN·m and $\ell_a = 120 - 40$ (cover) = 80 mm (Fig. 5.16)

$$1.3 \frac{M_r}{V_f} + \ell_a = \frac{1.3 \times 109 \times 10^6}{107 \times 10^3} + 80 = 1404 \text{ mm}$$

$$> \ell_d = 694 \text{ mm} \qquad \text{OK}$$

Check development length:
1. The cut off bar must have a length at least = ℓ_d beyond midspan (critical section).
   Length provided = 3000 − 900 = 2100 mm > ℓ_d = 594 mm   OK
2. The continuing bar shall have an embedment length of at least $\ell_d + d$ (on $\ell_d + 12d_b$) beyond the theoretical point of cut-off.

$$\text{Length provided} = 1290 + 80 = 1370 \text{ mm}$$
$$\ell d + d = 694 + 386 = 1080 \text{ mm} > 1370 \text{ mm} \qquad \text{OK}$$

EXAMPLE 5.10

Work out the bar cut-off details for the beam designed in Example 5.7, considering bending moments only. Fifty percent of the bars are to be terminated, in two stages.

SOLUTION

From the design in Example 5.7,

$$A_s = 4200 \text{ mm}^2, d = 406 \text{ mm}, f_c' = 20 \text{ MPa}, f_y = 400 \text{ MPa}$$
$$\text{Total factored load} = w_f = 64.2 \text{ kN/m}$$
$$M_{max} = 64.2 \times 8^2/8 = 514 \text{ kN·m}$$

The flexural strength of the section is first computed.

$$a = \frac{A_s \phi_s f_y}{b_f \times 0.85 \, \phi_c f_c'} = \frac{4200 \times 0.85 \times 400}{3180 \times 0.85 \times 0.6 \times 20} = 44 \text{ mm}, b_f = 120 \text{ mm}$$

$$M_r = A_s \phi_s f_y \left(d - \frac{a}{2} \right) = 4200 \times 0.85 \times 400 \left(406 - \frac{44}{2} \right) \times 10^{-6}$$

$$= 548 \text{ kN·m} > 514 \text{ kN·m} \qquad \text{OK}$$

The bending moment diagram is shown in Fig. 5.17b. The bars will be terminated in two stages, first the middle bar in the top layer at section 1–1 and secondly the two remaining bars in the top layer at location 2–2 (Fig. 5.17c). The flexural strength with the continuing bars at sections 1–1 and 2–2 can be taken approximately (and conservatively) as:

Fig. 5.17 *Example 5.10*

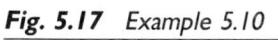

$$M_1 = \frac{5}{6} \times 548 = 457 \text{ kN·m}$$

and

$$M_2 = \frac{3}{6} \times 548 = 274 \text{ kN·m}$$

The location of sections 1–1 and 2–2 where the bending moments have values of $M_1 = 457$ kN·m and $M_2 = 274$ kN·m may be determined graphically, or by computation as follows:

At section 1–1, distant x_1 from A,

$$M_1 = \frac{w_f \ell}{2} x_1 - \frac{w_f x_1^2}{2}$$

$$457 = \frac{64.2 \times 8 \times (x_1)}{2} - \frac{64.2 \times (x_1^2)}{2}$$

Solving, $x_1 = 2.67$ m

Similarly for section 2–2, $x_2 = 1.27$ m

The bars must be continued beyond the above theoretical points for a distance equal to the larger of:

$$\text{effective depth} = 406 \text{ mm} \qquad \text{(controls)}$$
$$12 \times d_b = 12 \times 29.9 = 359 \text{ mm}$$

Actual points of termination are (see Fig. 5.17d):

for bar 1, $2670 - 406 = 2264 \simeq 2260$ mm from centre of support

and for bars 2 and 3, $1270 - 406 = 864 \simeq 860$ mm from centre of support.

Checks

1. The bars continuing must have an embedment length of at least $\ell_d + d$ beyond theoretical point of cut-off (clause 12.10.4).

   For No. 30 bars, $f_y = 400$ MPa, $f'_c = 20$ MPa, $\ell_d = 1190$ mm (see Chapter 9, Table 9.1).

   From Fig. 5.17c showing bar details,

Minimum provided at section 1–1 = $2670 - 860 = 1810$ mm
$$\ell_d + d = 1190 + 406 = 1596 \text{ mm} < 1810 \text{ mm} \qquad \text{OK}$$
Minimum provided at section 2–2 = 1350 mm
$$\ell_d + d = 1596 \text{ mm} \qquad \text{not OK}$$

   The straight length of embedment for bars continuing beyond secton 2–2 is inadequate to meet the requirement in clause 12.10.4. Hence these bars must be terminated in a standard hook. The development length, ℓ_{dh}, measured from the critical section to outside end of hook, required for standard hook in tension in this case (No. 30 bar, $f'_c = 20$ MPa, $f_y = 400$ MPa) is $\ell_{dh} = 669$ mm (see Section 9.11) and $1350 > 669 + 406 = 1075$ mm $\qquad$ OK

2. The bar that is cut off shall have an embedment length of at least ℓ_d beyond midspan section.

$$\text{Minimum length given (bar 1)} = 4000 - 2260 = 1740 \text{ mm} > \ell_d \qquad \text{OK}$$

3. Check maximum size of bar at support (clause 12.11.3).

   At support, flexural strength of bars continued to support (three bars),

$$= M_r = 274 \text{ kN·m}$$

Shear at support, $$V_f = \frac{64.2 \times 8}{2} = 257 \text{ kN}$$

The bar diameter must be small enough so that the development length

$$\ell_d \leq \left(1.3 \frac{M_r}{V_f} + \ell_a \right) \text{ (see Fig. 5.17d)}$$

$$1.3 \frac{M_r}{V_f} + \ell_a = 1.3 \times \frac{274 \times 10^6}{257 \times 10^3} + 80$$

$$= 1466 \text{ mm} > \ell_d = 1190 \text{ mm} \qquad\qquad \text{OK}$$

4. The shear strength at the bar cut-off points should be checked to conform with clause 12.10.5. This is done in Chapter 6, Example 6.2.

PROBLEMS

---

1. A rectangular beam of span 8 m (centre to centre) is to carry specified load moments of 120 kN·m due to live load and 100 kN·m due to dead load (in addition to its self-weight). Using material properties of $f_c' = 25$ MPa and $f_y = 350$ MPa, design the beam with (a) maximum permissible reinforcement and (b) with a moderate reinforcement ratio of $\rho = 0.4\, \rho_{max}$. Design using first principles and check using Table 5.3.

2. Design the beam in Example 5.1 if the overall beam size is restricted to $b = 350$ mm and $h = 500$ mm.

3. The floor plan of a building is shown in Fig. 5.18. The specified floor loading consists of a live load of 10 kN/m² (including floor finish, partitions, etc.), in addition to the self-weight. The material properties are $f_c' = 40$ MPa and $f_y = 400$ MPa. Design the slab thickness and the reinforcement area required at all critical sections.

4. Design the interior beam of the floor in Problem 5.3, considering the beam as simply supported.

5. Design the edge beam (L-beam) of the floor system in Example 5.3 with the restriction on stem width to be $b_w = 250$ mm and considering the beam as simply supported. Neglect the torsional effect.

6. A simply supported T-beam of 9 m span is subjected to a dead load of 20 kN/m (including self-weight) and a live load of 30 kN/m. The overall size of the beam is shown in Fig. 5.19. Design the beam for tension reinforcement and detail the bar cut-off. Assume that 50 percent of bars are to be cut off. Use $f_c' = 20$ MPa and $f_y = 400$ MPa and width of support as 300 mm.

7. The beam of rectangular section with overhang at both ends, shown in Fig. 5.20a is subjected to a dead load, including estimated self-weight, of 30 kN/m and a live load of 25 kN/m. The factored moment envelope is determined to be as shown in Fig. 5.20b. Design the beam and detail the bar cut-off. Assume that 50 percent of bars are to be cut off for both positive and negative moment reinforcement. Take $f_c' = 30$ MPa, $f_y = 350$ MPa and width of support as 300 mm.

Fig. 5.18 *Problems 5.3–5.5*

FLOOR PLAN

Fig. 5.19 *Prob. 5.6*

Fig. 5.20 *Prob. 5.7*

(a) Beam

(b) Moment envelope

REFERENCES

5.1 Wang, C.K., *Statically Indeterminate Structures*, McGraw-Hill, New York, 1953.

5.2 Weaver, W., and Gere, J.M., *Matrix Analysis of Framed Structures*, 2nd ed., Van Nostrand Company, New York, 1980, 492 pp.

5.3 Ferguson, Phil M., *Small Bar Spacing or Cover — A Bond Problem for the Designer*, J. of ACI, Vol. 74, Sept. 1977, pp. 435–439.

5.4 ACI Committee 435, *Proposed Revisions by Committee 435 to ACI Building Code and Commentary Provisions on Deflections*, J. of ACI, Vol. 75, No. 6, June 1978, pp. 229–238.

5.5 *Reinforcing Steel — Manual of Standard Practice*, Reinforcing Steel Institute of Ontario, Willowdale, Ontario, 1975, (3rd edition, 1983).

5.6 Winter, G., and Nilson, A.H., *Design of Concrete Structures*, 9th ed., McGraw-Hill Book Co., New York, 1979.

5.7 ASCE-ACI Task Committee 426, *The Shear Strength of Reinforced Concrete Members*, ASCE, Journal of Struct. Div., Vol. 99, No. ST6, June 1973, pp. 1091–1187.

CHAPTER 6

DESIGN FOR SHEAR

6.1 INTRODUCTION

The ultimate limit state considered in Chapters 4 and 5 dealt with failure in bending. Bending moments are generally accompanied by shear forces and sometimes by axial loads and torsion as well. Premature failure in shear can reduce the strength of members below their capacity in flexure and reduce the ductility of the elements considerably.

Failure in shear of reinforced concrete takes place under combined stresses resulting from an applied shear force, bending moments and, where applicable, axial loads and torsion as well. Because of the non-homogeneity of material, non-uniformity and nonlinearity in material response, presence of cracks, presence of reinforcement, combined load effects, etc., the behaviour of reinforced concrete in shear is very complex, and the current understanding of and design procedures for shear effects are, to a large measure, based on analysis of results of extensive tests and simplifying assumptions rather than on an exact universally acceptable theory.

Prior to its 1984 revision, CSA Standard A23.3 recommended a method for Shear and Torsion design based on the traditional method adopted by the ACI Code (Ref. 6.1). The background for this procedure, and the basic concepts, review of extensive test results and bibliographic references on shear strength of reinforced concrete beams are presented in Ref. 6.2 and 6.3. The 1984 revision of the Canadian Standard, CSA A23.3-M84, recommends two alternative methods for Shear and Torsion design. The first of these, termed the "Simplified Method" (CSA A23.3-M84(11.3)) is a shortened version of the traditional method followed by ACI and previous Canadian Codes. In the simplified method, the transverse reinforcement is designed for the combined effect of shear and axial load if any, while the longitudinal reinforcement is designed for the combined effect of flexure and axial load. The effects of shear on the longitudinal reinforcement (see Section 5.10.3) are taken care of by bar detailing requirements. Transverse and longitudinal reinforcement required for coexisting torsion, if any, are designed separately and added on to that already determined for flexure, shear and axial loads. The design by the simplified method is dealt with in this Chapter for shear and in Chapter 7 for torsion. Accordingly much of the information on shear strength that follows in this Chapter is based on Ref. 6.1–6.3 and gives the basis for the Simplified Method of CSA A23.3-84(11.3).

The second method, termed "General Method" for Shear and Torsion design

Fig. 6.1 *Stress distribution in rectangular beams*

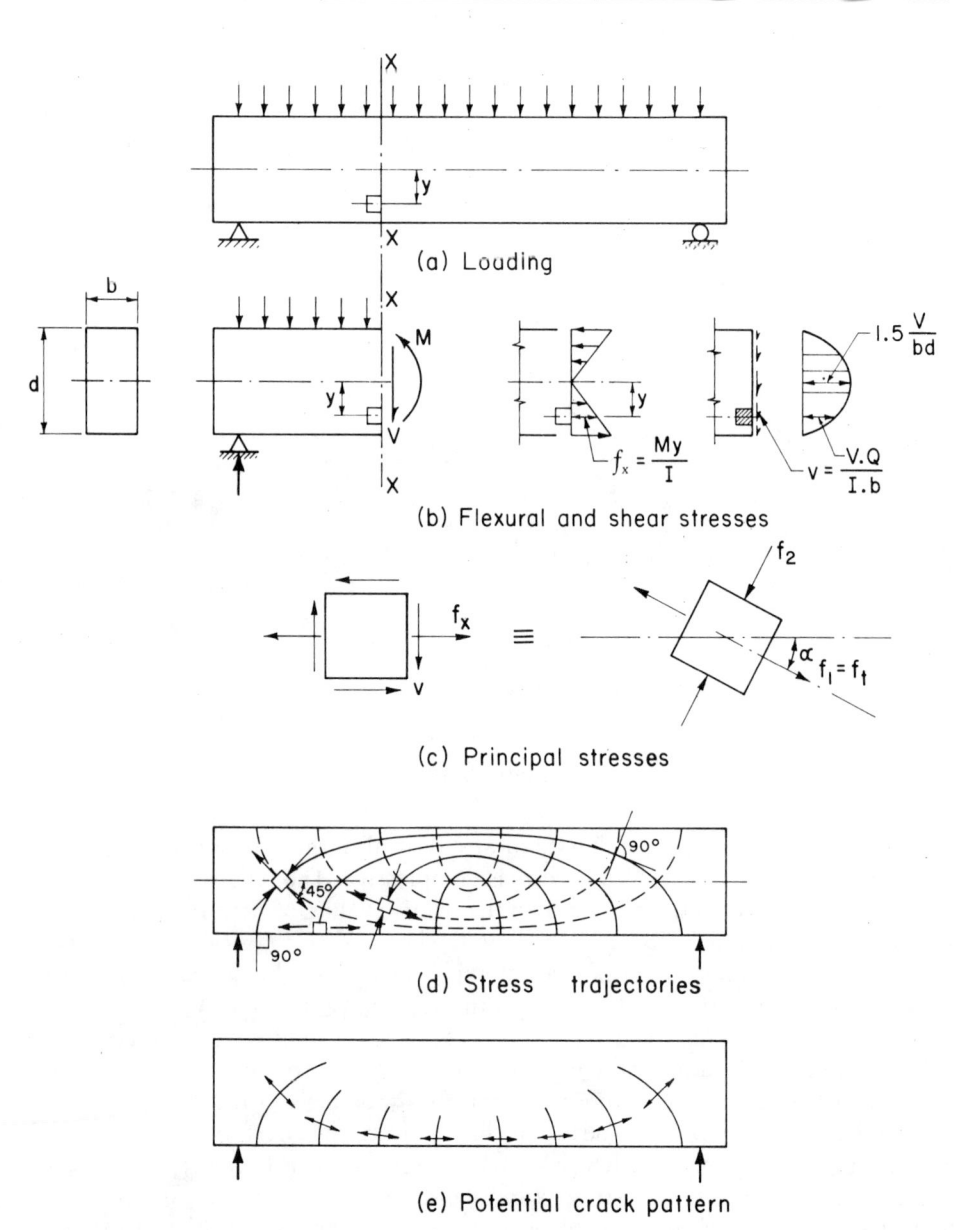

(a) Loading

(b) Flexural and shear stresses

(c) Principal stresses

(d) Stress trajectories

(e) Potential crack pattern

(CSA A23.3-M84(11.4)) is presented in Chapter 8. In this method, the member is designed for the combined effects of flexure, shear, axial load and torsion. The Code requires that deep beams, parts of members with deep shear span, brackets and corbels, and regions with abrupt changes in cross-section (such as regions of web openings in beams) be designed by the General Method only.

In order to gain an insight into the causes of shear failure in reinforced concrete, the stress distribution in a homogeneous elastic beam of rectangular section will be reviewed briefly. In such a beam, loaded as shown in Fig. 6.1a, a transverse section, X–X, is subjected, in general, to a bending moment M and a shear force V.

As described in texts on mechanics of materials (see Ref. 6.4), the flexural stress, f_x, and the shear stress, v, at a point in the section distant y from the neutral axis are given by:

$$f_x = \frac{My}{I}$$

$$v = \frac{VQ}{Ib} \qquad (6.1)$$

where Q = the first moment of the area of section above the depth y, about the neutral axis, and I = the moment of inertia of the section. The distribution of these stresses is as shown in Fig. 6.1b. Considering an element at depth y (Fig. 6.1c), the flexural and shear stresses can be resolved into equivalent principal stresses, f_1 and f_2, acting on orthogonal planes inclined at an angle α, where

$$f_{1,2} = \tfrac{1}{2} f_x \pm \sqrt{\left(\tfrac{1}{2} f_x\right)^2 + v^2} \qquad (6.2)$$

$$\text{and,} \quad \tan 2\,\alpha = \frac{2v}{f_x}$$

In general (neglecting the influence of any vertical normal stresses due to loads), the stress f_1 is tensile (say $= f_t$) and f_2 is compressive. The relative magnitudes of f_1 and f_2 and their directions depend on the relative values of f_x and v. In particular $f_t = f_x, f_2 = 0$, and $\alpha = 0$ at points along the bottom face; $f_t = v_{max}, f_2 = -v_{max}$ and $\alpha = 45°$ at points along the neutral axis. Considering the variation of stresses f_x and v over the depth at a section and the variations of M and V (and hence of f_x and v) along the span, the principal stress trajectories indicating the directions of principal stresses will follow the pattern shown in Fig. 6.1d. (Stress trajectories are a set of orthogonal curves, whose tangent at any point is in the direction of the principal stress at that point.) The broken lines indicate the directions of the principal tensile stresses. In a material which is weak in tension, tensile cracks would occur at right angles to the tensile stresses and hence the compressive stress trajectories indicate *potential* crack patterns (depending on the magnitude of the tensile stress) as shown in Fig. 6.1e. (Note that if in fact a crack is developed, the stress distributions assumed here are no longer valid in that region and redistribution of the internal stresses takes place.) The location of the absolute maximum principal tensile stress will depend on the variation of f_x and v, which in turn depends on the shape of the cross section and on the span and loading. For instance, in a relatively deep beam with a thin web (such as an I-section) and short span subjected to high shear force and low bending moment, the peak shear stress, v_{max}, may far exceed the peak flexural stress $f_{x,max}$, and hence the maximum principal tensile stress, $f_{t,max}$, may be at the neutral axis level at an inclination of $\alpha = 45°$. In contrast, for a shallow rectangular beam with high moment and low shear the maximum principal tensile stress may be the flexural stress $f_{x,max}$ in the outer fibre in the peak moment regions.

It is seen that the general influence of shear is to induce tensile stresses on an inclined or diagonal plane. Concrete is weak in tension. Further, as shown in Fig. 1.8, a biaxial state of combined tension and compression stresses, as exists in the beam shown in Fig. 6.1, significantly reduces both tensile and compressive strengths of concrete. Hence the tensile strength of the concrete in a reinforced concrete beam subjected to combined bending and shear is likely to be even less than the uniaxial tensile strength of concrete. Thus one can naturally expect some inclined tension cracks to develop in reinforced concrete beams subjected to any appreciable shear force. Failure of concrete beams in shear is triggered by the development of these inclined cracks (referred to as "diagonal tension" cracks) under combined stresses. To avoid a failure of the concrete in compression, it is also necessary to ensure that the principal compressive stress, f_2, is less than the compressive strength of concrete under the biaxial state of stress.

Although several theories of failure have been proposed for concrete under combined stresses, for the traditional method of shear and torsion design, the principal tensile stress theory has been followed.

6.2 NOMINAL SHEAR STRESS

The maximum shear stress at the neutral axis level in the homogeneous elastic beam of rectangular section is given in Fig. 6.1b as:

$$v_{max} = 1.5\,\frac{V}{bd} = k\,v_{av} \tag{6.3}$$

where the average shear stress,

$$v_{av} = \frac{V}{\text{area}}$$

The value of the constant, k, in Eq. 6.3 depends on the cross-sectional shape. Equations 6.1 and 6.3 are not applicable for a composite, cracked and non-linear material such as reinforced concrete, and the determination of the distribution and the maximum value of the shear stress in a reinforced concrete member is complex. Hence, as a means of comparing and correlating with test results, and as a parameter to aid design and control shear stress, the average shear stress $v = V/(bd)$, is generally used as a *measure* of the shear stresses resulting from the shear force, V.

6.3 EFFECT OF SHEAR ON REINFORCED CONCRETE BEAM BEHAVIOUR

If a reinforced concrete beam is subjected to bending only as in Fig. 6.2a, the principal tensile stresses will be parallel to the axis of the member and hence tension cracks will develop from the outer fibres where the tensile stress is maximum and spread inwards in a direction at right angles to the beam axis. The addition of shear stresses alters the magnitude and direction of the principal tensile stresses so that inclined tension cracks, called *diagonal tension cracks* may develop.

6.3.1 MODES OF INCLINED CRACKING

Inclined cracks have been classified as *web-shear cracks* and *flexure-shear cracks* (Fig. 6.2). When the shear stress, v, is large in relation to the flexural stress, f_x, the maximum principal tensile stress is near the neutral axis level at an inclination of nearly 45° to the beam axis. When this stress reaches a limiting value, cracks may initiate in the web, independent of any flexural cracks. This type of crack is termed *web-shear* crack and may occur in relatively deep beams with thin webs and low M/V ratios (Fig. 6.2b) or near inflection points of continuous beams. In general, web-shear cracks are extremely rare in reinforced concrete beams of usual proportions and are likely only in prestressed beams with thin webs. The web-shear cracking load may be computed with reasonable accuracy by equating the principal tensile stress in the web at or near the neutral axis to the tensile strength of the concrete.

In reinforced concrete beams of normal proportions inclined cracks usually form as an extension of the flexural cracks. The flexural cracks generally form first, but these are controlled and the tensile stresses carried by the tension reinforcement. However, these cracks lead to increased shear stresses and hence diagonal tensile stresses at the head of the crack, resulting in the extension of a flexure crack into a *diagonal tension* crack. This type of inclined crack is called a flexure-shear crack (Fig. 6.2c). Often, combined with flexure-shear cracks, secondary cracks (sometimes termed splitting cracks) may develop, extending from the inclined crack and along the tension reinforcement towards the region of lower moment (Fig. 6.2d). These cracks may be due to the splitting forces generated by the wedging action of the tension bar deformations and due to the vertical downward forces exerted on the cover concrete by the tension bars in transferring shear across the crack by dowel action (Fig. 6.2e).

6.3.2 SHEAR TRANSFER MECHANISMS IN REINFORCED CONCRETE

There are several mechanisms by which shear is transmitted between two planes in a concrete member. The prominent amongst these are identified in Fig. 6.3 which shows the freebody of one of the segments of a reinforced concrete beam separated by an inclined crack. The applied bending moment is resisted by the couple formed by flexural forces, C and T. The major components contributing to the shear resistance are: (1) the shear strength, V_{cz}, of the uncracked concrete; (2) the vertical component, V_{ay}, of the *interface shear*, V_a; (3) the dowel force, V_d, in the longitudinal

Fig. 6.2 *Shear cracks*

(a) Flexural Cracks

(b) Web-Shear Crack

(c) Flexure-Shear Crack

(d) Secondary Cracking

(e) Dowel Force

reinforcement; and (4) the shear, V_s, carried by the shear reinforcement, if any. The interface shear, V_a (also called aggregate interlock), is a tangential force transmitted along the plane of the crack, resulting from the resistance to relative movement (slip) between the two rough interlocking surfaces of the crack, much like frictional resistance. So long as the crack is not too wide, the force V_a may be very significant. The dowel force in the longitudinal tension reinforcement is the transverse force developed in these bars functioning as a dowel across the crack, resisting relative transverse displacements between the two segments of the beam. In deep beams, part of the applied load may be transmitted to the reaction points by tied-arch action, thereby reducing the effective shear force at a section. The equilibrium of vertical forces in Fig. 6.3 results in the relation

$$V_{ext} = V_{cz} + V_{ay} + V_d + V_s \tag{6.4}$$

Fig. 6.3 *Internal forces acting at inclined crack (adapted from Ref. 6.3)*

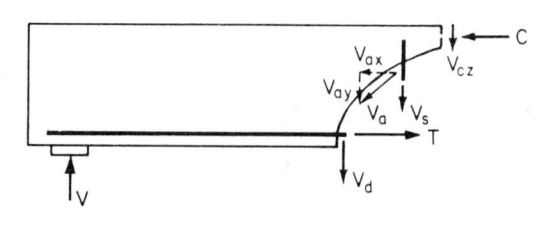

Different mechanisms dominate in members of different types and, further, their relative magnitudes also depend on the loading stage. Thus prior to flexural cracking, the applied shear is resisted almost entirely by the uncracked concrete. At flexural cracking, there is a redistribution of internal stresses and some measure of interface shear and dowel action develop. Prior to diagonal cracking, however, there is little tensile stress developed in shear reinforcement. At diagonal cracking, the shear reinforcement intercepted by the crack receives a sudden increase in tensile stress, and all the components in Eq. 6.4 now effectively contribute to the shear strength. The post-inclined cracking behaviour, the failure mode and the ultimate strength all depend on the variations in and the breakdown of these individual components and on how far these forces can be redistributed successfully.

In beams without shear reinforcement, the breakdown of any of the shear transfer mechanisms may soon cause failure. In such beams there are no stirrups enclosing the longitudinal bars and restraining them against splitting failure and the value of V_d is usually small. The component V_{ay} also decreases progressively due to the unrestrained opening up of the crack. The spreading of the crack into the compression zone decreases the area of uncracked concrete section contributing to V_{cz}. However, in relatively deep beams, tied-arch action may develop following inclined cracking, which in turn could transfer part of the load to the supports thereby reducing the effective shear force at the section. Because of the uncertainties in all these effects, it is difficult to predict precisely the behaviour and strength beyond diagonal cracking of beams without shear reinforcement.

In members with moderate amounts of shear reinforcement, the shear resistance continues to increase even after inclined cracking until the shear reinforcement yields and V_s can increase no more. Any further increase in applied shear force leads to increases in V_{cz}, V_d, and V_{ay}. With progressively widening crackwidth (which is now accelerated by the yielding of the shear reinforcement), V_{ay} begins to decrease forcing V_{cz} and V_d to increase at a faster rate until either a splitting (dowel) failure occurs, or the concrete in the compression zone fails under the combined shear and compression forces. Thus, in general, the failure of shear reinforced members is more gradual (ductile).

6.3.3 SHEAR FAILURE MODES

The magnitude and direction of the maximum principal tensile stress (see Eq. 6.2) and hence the development and growth of inclined cracks are affected by the relative magnitudes of the flexural stress, f_x, and the shear stress, v. As an approximation, stresses f_x and v can be considered proportional respectively to $M/(bd^2)$ and $V/(bd)$ where M is the applied bending moment and V the shear force at the section, b is the width of section, and d is the effective depth. The stress ratio f_x/v can be expressed as:

$$\frac{f_x}{v} = \frac{F_1 M/(bd^2)}{F_2 \, V/(bd)} = F_3 \frac{M}{Vd} \qquad (6.5)$$

where F_1, F_2, F_3 are constants of proportionality. Equation 6.5 shows that the dimensionless parameter $M/(Vd)$ strongly influences the development and propagation of inclined cracks, as well as the shear strength at inclined cracking. For beams subjected to concentrated loads, as in Fig. 6.4a, the ratio M/V is the so-called *shear span*, a, and $M/(Vd) = a/d$ is the shear span to depth ratio.

Typical shear failure modes of reinforced concrete beams, and the influence of the a/d ratio, are illustrated in Fig. 6.4 with reference to a simply supported rectangular beam subjected to symmetrical two-point loading (Refs. 6.3 and 6.5).

In very deep beams ($a/d < 1$) without web reinforcement, inclined cracking transforms the beam into a tied-arch (Fig. 6.4b). The tied-arch can fail by either a breakdown of its tension element, namely, the longitudinal reinforcement (by yielding, fracture or failure of anchorage), or by a breakdown of the concrete compression chord by crushing (Fig. 6.4b).

In relatively short beams, with a/d in the range of 1 to 2.5, the failure is initiated by an inclined crack, usually a flexural-shear crack. The actual failure may take place by (1) crushing of the reduced concrete section above the head of the crack under combined shear and compression (termed *shear-compression* failure, Fig. 6.4c)

Fig. 6.4 *Shear failure modes (adapted from Ref. 6.3)*

Tied-arch action in deep beams Modes of failure for deep beams

(b) a/d < 1

Fig. 6.4 *continued*

Shear-tension failure   **(c)** 1< a/d < 2.5   Shear-compression failure

Diagonal tension failure   **(d)** 2.5 < a/d <≈ 6

(e) Web-crushing failure

or (2) secondary cracking along the tension reinforcement resulting in loss of bond and anchorage of the tension reinforcement (*shear-tension* failure, Fig. 6.4c). This type of failure usually occurs before the flexural strength of the section is attained.

In beams with relatively short shear spans (*a/d* less than about 2.5) and with loads and reactions applied on the top (compressive) and bottom surfaces of the beam respectively, the effect of the vertical compressive stresses between the load and reaction may be significant and result in substantial increases in shear strength above the inclined cracking strength. Such beams are specially treated as *deep beams* (See Section 6.18.2).

Normal beams have *a/d* ratios in excess of about 2.5. Such beams may fail in shear or in flexure. The limiting *a/d* ratio above which flexural failure occurs depends on the tension reinforcement ratio, yield strength of reinforcement and concrete strength, and is in the neighbourhood of 6. For beams with *a/d* ratios in the range of 2.5 to 6, flexural tension cracks develop early on; however, before the ultimate flexural strength is reached the beam may fail in shear by the development of inclined flexure-shear cracks, which, in the absence of web reinforcement, rapidly extend right through the beam as shown in Fig. 6.4d. This type of failure is usually sudden and without warning and is termed *diagonal-tension* failure. Addition of web reinforcement in such beams leads to a shear-compression failure or a flexural failure.

In addition to these different modes, thin webbed members such as I-beams with web reinforcement may fail by crushing of the concrete in the web portion between inclined cracks under the diagonal compression forces (Fig. 6.4e).

6.4 SHEAR STRENGTH OF BEAMS WITHOUT SHEAR REINFORCEMENT

All shear failures result from or are affected by inclined cracks which precede them. In beams without shear reinforcement, the shear failure load may equal or exceed the load at which inclined cracks develop, depending on several variables such as ratio $M/(Vd)$, thickness of web, influence of vertical normal stresses, concrete cover and resistance to splitting (dowel) failure, etc. Further, the margin of strength beyond diagonal cracking is subject to considerable fluctuation. Hence, for beams of normal proportions ($M/(Vd) >$ about 2.5), as a design criterion, the shear force, V_{cr}, causing formation of the first inclined crack is generally considered as the usable ultimate strength for beams without shear reinforcement (Ref. 6.2).

6.5 SHEAR STRENGTH AT INCLINED CRACKING

(a) Web-Shear Cracking Load

Inclined cracks in concrete beams are of two types and they develop in different ways. *Web-shear* cracks are initiated in the web portion before any flexural cracks develop in its vicinity. Hence, assuming the principal tensile stress theory of failure, the web-shear cracking load may be estimated by relating the principal tensile stress in the uncracked web to the tensile strength of concrete. In the zone of potential web-shear cracking, which is close to the neutral axis level, the flexural stress, f_x, is very small and hence the maximum principal tensile stress is primarily dependent on the shear stress v. This concept together with correlation to extensive tests results (Ref. 6.2) has resulted in an equation for the average shear stress at web-shear cracking, v_{cw}, as:

$$v_{cw} = \frac{V_{cw}}{bd} = 0.3 \sqrt{f_c'} \text{ MPa} \tag{6.6}$$

where V_{cw} is the shear force at web-shear cracking and f_c' is in MPa.

(b) Flexure-Shear Cracking Load

The prediction of the *flexure-shear* cracking load is complicated due to the presence of prior flexural cracking. A semi-rational expression developed in Ref. 6.2 is the basis for the current ACI Code (Ref. 6.1) and the simplified method in CSA A23.3-M84(11.3), and the derivation of this equation is briefly described below.

If it is assumed that, at the point of inclined cracking, the principal tensile stress is equal to the tensile strength of concrete, f_t, using Eq. 6.2:

$$f'_t = \frac{1}{2} f_x + \sqrt{\left(\frac{1}{2} f_x\right)^2 + v^2} \qquad (6.7)$$

If the flexural tensile stress f_x is assumed proportional to the steel stress f_s, and the shear stress v is assumed proportional to the average shear stress,

$$f_x \propto \frac{f_s}{n} = F_1 \frac{M}{n \rho b d^2}$$

$$v = F_2 \frac{V}{bd}$$

where F_1 and F_2 are constants of proportionality, $n = E_s/E_c$, and ρ is the tension reinforcement ratio. Substituting for f_x and v, and recognizing that both the tensile strength f_t and modulus E_c are generally considered proportional to $\sqrt{f'_c}$, Eq. 6.7 can be expressed as a relation between two dimensionless parameters:

$$X = \frac{\rho\,Vd}{\sqrt{f'_c}\,M}, \text{ and } Y = \frac{V}{bd\,\sqrt{f'_c}}$$

The importance of the three major variables, $M/(Vd)$, ρ, and $\sqrt{f'_c}$, has also been-observed in tests. Having established, in this manner, the relevance of the two parameters X and Y, they are correlated to extensive test data on an $X - Y$ plot, yielding a valid relationship between X and Y. A simple expression so developed in Ref. 6.2, giving a lower bound estimate of the shear strength, V_{cr}, at flexure-shear cracking for beams without shear reinforcement, expressed in SI units is:

$$\frac{V_{cr}}{bd\,\sqrt{f'_c}} = 0.16 + (17) \frac{\rho}{\sqrt{f'_c}} \frac{Vd}{M} \qquad (6.8)$$

where f'_c and $\sqrt{f'_c}$ are both in MPa, and the constant 17 has the unit of MPa.

Combining Eq. 6.6 and 6.8, the nominal shear strength of reinforced concrete beams without shear reinforcement considered equal to the strength corresponding to inclined cracking is given by

$$V_{cn} = \left(0.16\,\sqrt{f'_c} + 17\,\rho\,\frac{Vd}{M}\right) bd \leq 0.3\,\sqrt{f'_c}\,bd \qquad (6.9)$$

where the term within the brackets has the unit of MPa.

The equivalent of Eq. 6.9, in imperial units, with rounding off of the numerical factors is:

$$V_{cn} = \left(1.9\,\sqrt{f'_c} + 2500\,\rho\,\frac{Vd}{M}\right) \leq 3.5\,\sqrt{f'_c}\,bd \qquad (6.10)$$

where V_{cn} is in pounds, f'_c in psi, b and d in inches, and ratios ρ and Vd/M are dimensionless.

Equation 6.10 is the expression recommended by the ACI Code (Ref. 6.1) for the *nominal shear strength* carried by concrete. Since the value of the moment M is zero at points of inflection, in order to limit V_{cn} near such points, the value of Vd/m in Eq. 6.10 is limited to a maximum of 1.0. As a convenient simplification useful for most designs, the ACI Code also permits the use of an approximate value for the second term within brackets of Eq. 6.10 of $0.1 \sqrt{f'_c}$, so that the shear strength provided by concrete may be computed as:

$$V_{cn} = 2 \sqrt{f'_c} \, bd \tag{6.10a}$$

where the various terms have the same units as in Eq. 6.10.

The 1977 revision of CSA A23.3 recommended Eq. 6.9 and, as a simplification, the SI units equivalent of Eq. 6.10a, namely $V_{cn} = 0.17 \sqrt{f'_c} \, b_w d$ for the nominal shear strength carried by the concrete. The V_{cn} so calculated was multiplied with a ϕ-factor of 0.85 to get the factored resistance. With the factor $\phi_c = 0.6$ applied to the material strength in CSA A23.3-M84, the coefficient 0.17 and related numbers in clauses 11.2 and 11.3 have been increased to 0.2 to partially offset the decrease in computed resistance on account of the reduced ϕ_c value. Introducing the material resistance factor ϕ_c and the factor λ to account for low density concrete (see Section 1.6), CSA A23.3-M84 (11.2.5.2 and 11.3.4.1) recommends Eq. 6.11 for computing the *factored shear resistance*, V_c, provided by tensile stresses in the concrete for members subjected to flexure and shear only.

$$V_c = 0.2 \, \lambda \, \phi_c \sqrt{f'_c} \, b_w d \tag{6.11}$$

where V_c is in Newtons, f'_c in MPa, and the web width b_w and effective depth d are in mm.

6.6 BEHAVIOUR OF BEAMS WITH SHEAR REINFORCEMENT

Shear reinforcement (also known as web reinforcement) in reinforced concrete members may consist of: (a) stirrups perpendicular to the member axis; (b) inclined stirrups making an angle of 45° or more with the longitudinal tension bars in order to intercept potential diagonal cracks; (c) welded wire fabric with transverse wires located at right angles to the member axis, provided that the transverse wires can undergo a minimum elongation of 4 percent measured over a gauge length of at least 100 mm, which includes at least one cross-wire; (d) longitudinal bars (No. 35 or smaller) bent at an angle of 30° or more with the longitudinal tensile bars to cross potential diagonal cracks, with the centre three-fourths of the inclined portion being considered effective; (e) spirals; or (f) a combination of stirrups and bent longitudinal bars (see Fig. 6.5). By far the most common type of shear reinforcement for beams is the U-shaped stirrup placed perpendicular to the axis of the member.

Fig. 6.5 *Types of shear reinforcement*

Forms of stirrups

Traditionally, the action of web reinforcement in reinforced concrete beams has been explained with the aid of the *truss analogy*, the simplest form of which is shown in Fig. 6.6. In this model, a reinforced concrete beam with inclined cracks is replaced with a pin-jointed truss, of which the compression chord represents the concrete compression zone at the top, the bottom tension chord represents the tension reinforcement, the web members in tension represent the stirrups and the diagonal web members in compression represent the concrete in compression between the inclined cracks. More complex forms of the truss, as well as other analogies (such as arch analogy, frame analogy) have also been proposed to better represent the complex stress field in a real reinforced concrete beam (Ref. 6.5). The truss model is a helpful tool to visualize the nature of stresses in the stirrups and in the concrete, and to base simplified design concepts and methods on. It may also be used to derive equations for the design of shear reinforcement. However, it does not recognize fully the actual action of web reinforcement and its effect on the various types of shear transfer mechanisms identified in Fig. 6.3.

Fig. 6.6 *Classical truss analogy for action of web reinforcement*

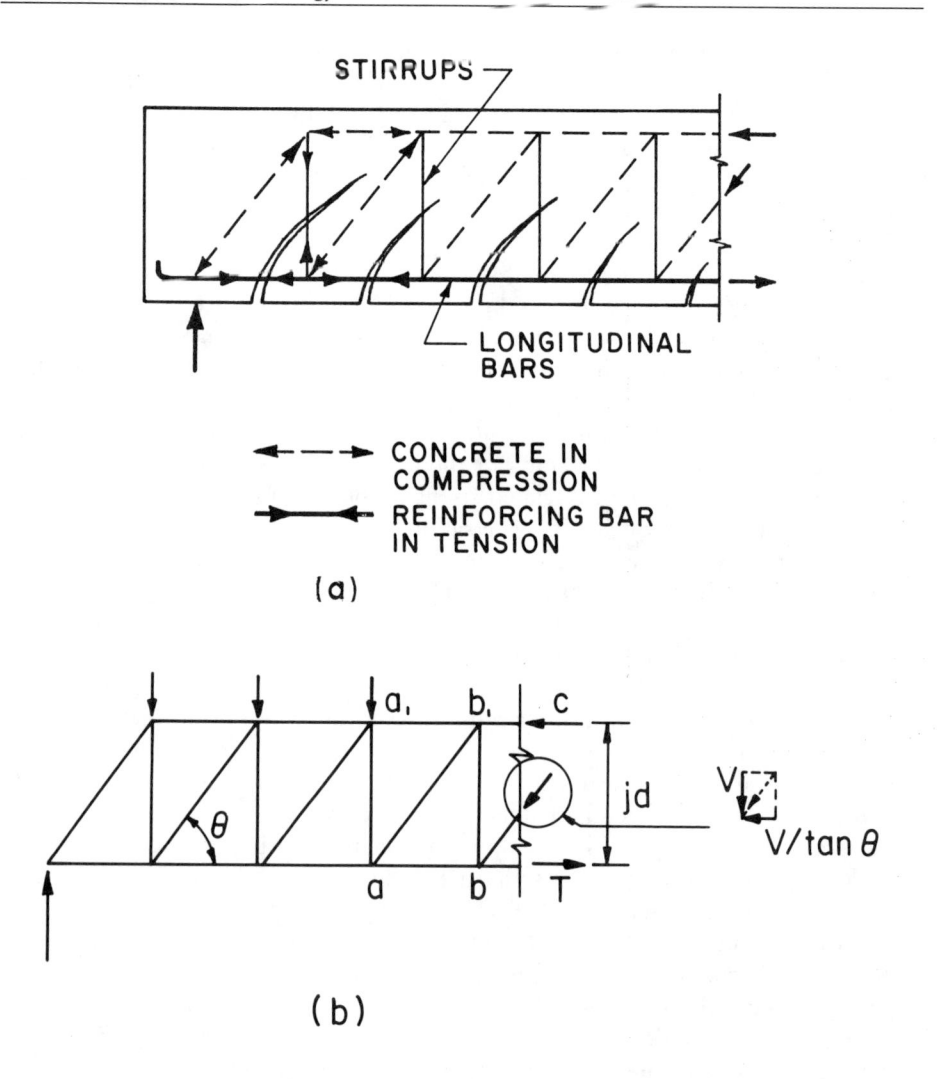

(a)

(b)

The necessity for extension of flexural tension reinforcement beyond the theoretical point of cut-off (see Section 5.10.3 and Fig. 5.12) can also be explained with the aid of the truss analogy. For the truss shown isolated in Fig. 6.6b, the tensile force in any bottom chord such as ab is dependent on the bending moment at section b_1b whereas the compression in the top chord a_1b_1 is dependent on the moment at section a_1a. Therefore, the tension reinforcement designed for section b_1b is in fact needed up to section a_1a. Thus flexural tension reinforcement must extend for a length $\approx jd/\tan\theta$ beyond the theoretical point of cut-off. From the equilibrium of forces at a section, it can also be seen that due to the shear force V, the tension T in a panel of the truss (or at any section of the reinforced concrete beam) exceeds the compressive force C by an amount equal to $V/\tan\theta$.

6.7 ACTION OF WEB REINFORCEMENT

Measurements of strain in web reinforcement have shown that prior to the formation of inclined tension cracks, there is little or no tensile stress developed in them. Hence, web reinforcement does not appreciably influence the location of inclined cracking or the inclined cracking load V_{cr}, and the latter can be computed with reasonable accuracy by the methods described in Section 6.5 for members without web reinforcement.

Following the development of inclined cracking, a portion of the shear is resisted by the web reinforcement intercepted by the crack. In this phase all the major shear transfer mechanisms are effective and the overall shear resistance can be expressed by Eq. 6.4. The web reinforcement contributes significantly to the overall shear strength by increasing or maintaining all the individual components of Eq. 6.4. Thus, first, there is the direct contribution V_s to the strength resulting from the tensile stress induced in the web reinforcement. Secondly, web reinforcement crossing the inclined crack restricts the widening of the crack and thereby helps maintain the aggregate interlock and the associated strength contribution V_{ay}. Further, by containing the crack and reducing its penetration deep into the compression zone, the shear resistance of the uncracked portion of the concrete, V_{cz}, is enhanced. Thirdly, the web reinforcement in the form of stirrups (which wrap around the longitudinal tension bars) restrains the longitudinal tension bars against vertical (dowel) displacement and thereby improves their contribution by dowel action, V_d. Furthermore, when sufficient web reinforcement is present, the shear failure is gradual with adequate warning in the form of large increases in the width of the inclined cracks as the web reinforcement reaches its yield point. Other favourable influences attributed to web reinforcement include improvement in concrete compressive strength and ductility due to the lateral confinement offered by the stirrups, transfer of some force across the crack by developing dowel action in the web reinforcement crossing the crack, and providing vertical support for the concrete between inclined cracks to facilitate arch action (see Fig. 6.7).

As described in Section 6.3b, the ultimate shear strength of a web reinforced section is reached soon after the yielding of this reinforcement. As stirrups yield, V_s remains constant and on further loading, the accelerated widening and spreading of the crack which follows soon leads to the breakdown of one or more of the remaining components of shear transfer (V_{ay}, V_d and V_{cz}) and hence to failure. Thus the shear strength and mode of shear failure of beams with web reinforcement are influenced by the amount, distribution and yield strength of web reinforcement, in addition to the variables f_c', ρ and $M/(Vd)$ which were shown to influence the cracking load and failure mode of beams without web steel. If the web reinforcement is so small that the sudden increase in stress in it at the formation of the inclined crack is sufficient to cause its yielding, the strength and behaviour of the beam may be little different from those of a similar beam without web reinforcement. For instance, a sudden diagonal tension failure may occur in beams of low $\rho(Vd)/M$ ratios. This suggests the need for a minimum amount of web steel in members where the sudden formation of inclined cracking may lead to distress. On the other hand, if a member is heavily reinforced with web steel, particularly in short deep members, the compres-

Fig. 6.7 *Secondary effects of web reinforcement*

Stirrup action as
tied-arch support

Confinement of concrete
in compression zone by
closed stirrups

Dowel action in web
reinforcement

sion zone of the concrete may fail in shear-compression before the web steel reaches yield point. (Shear-compression failure before yielding of web steel may also occur under dynamic loading.) Such failure may occur without warning and hence an upper limit on the shear strength with web reinforcement is also desirable. A shear-compression failure before the yielding of web steel could also occur if the web steel has a very high yield strength. Within the two extremes considered above, for moderate amounts of web reinforcement there is an increase in shear strength beyond diagonal cracking due to the web reinforcement effects. Further, the shear failure of such beams follows the yielding of web reinforcement and is more gradual with sufficient warning of impending failure in the form of widening cracks.

6.8 SHEAR STRENGTH OF BEAMS WITH WEB REINFORCEMENT

The overall shear strength of beams with web reinforcement was expressed by Eq. 6.4. The magnitudes of the strength components in this equation due to individual shear transfer mechanisms vary widely in members of different types, and further they are also interdependent. For simplicity, the ultimate nominal shear strength, V_n, is generally expressed as

$$V_n = V_{cn} + V_{sn} \tag{6.12}$$

where V_{cn} is referred to as the *shear carried by the concrete* at ultimate and includes all the components V_{cz}, V_{ay}, and V_d of Eq. 6.4; and V_{sn} is the shear carried by the web reinforcement. Although the relative magnitudes of the components of V_{cn} vary depending on the stage of loading and the state of cracking, Ref. 6.3 recommends that their aggregate value represented by V_{cn} can be assumed to be constant and equal to the shear strength at diagonal cracking, V_{cr}. Such an approximation was also found to result in good correlation with test results. V_{cr}, the shear strength at diagonal cracking, is independent of the web reinforcement and, hence, may be computed using Eq. 6.9. An expression for the shear, V_{sn}, carried by web reinforcement is derived below on the basis of some simplifying assumptions.

Fig. 6.8 *Shear strength due to web reinforcement*

(a) Inclined stirrups   (b) Vertical stirrups

Figure 6.8a shows one segment of a beam separated by a diagonal tension crack. This is an idealizaton of Fig. 6.3 by assuming that the diagonal crack is straight and at an angle θ to the beam axis, and that it extends over the full depth of the beam. In Fig. 6.8a only the forces in the web reinforcement at the section which constitute the strength V_{sn} are shown. The web reinforcement is assumed to be placed at an angle α with the beam axis and spaced s apart along the beam axis. If A_v is the total cross-section area of one stirrup (considering all legs) and f_v is the stress in it, the vertical component of the tensile force in one stirrup is $A_v f_v \sin \alpha$. The total number, N, of stirrups intercepted by the diagonal crack is seen from Fig. 6.8a to be:

$$N = \frac{d(\cot \theta + \cot \alpha)}{s}$$

The total vertical component of the stirrup forces across the crack, which make up the shear strength contribution. V_{sn} of web steel is:

$$\begin{aligned} V_{sn} &= N A_v f_v \sin \alpha \\ &= \frac{A_v f_v d}{s} \sin \alpha (\cot \theta + \cot \alpha) \end{aligned} \qquad (6.13)$$

Two further simplifying assumptions are usually made. First, the diagonal crack is assumed to occur at 45° to the beam axis; and secondly, the web reinforcement is assumed to yield as the *ultimate strength* is attained. The latter can be ensured by limiting the amount of web reinforcement (or alternatively limiting the ultimate strength including the web reinforcement) and by specifying an upper limit for the yield strength of web steel. With these two assumptions, at ultimate strength Eq. 6.13 reduces to:

$$V_{sn} = A_v f_y \frac{d}{s} (\sin \alpha + \cos \alpha) \qquad (6.14)$$

Equation 6.14 gives an expression for the nominal shear strength for web reinforcement placed at an angle α.

When vertical stirrups are used as web reinforcement, as in Fig. 6.8b, $\alpha = 90°$, and the shear strength is given by:

$$V_{sn} = \frac{A_v f_y d}{s} \qquad (6.15)$$

An equation similar to Eq. 6.15 could also be derived from the truss analogy. Because of this, the design procedure using Eq. 6.12 and 6.14 (which is the basis for both ACI 318-83 and CSA A23.3-M84(11.3)) is often referred to as the *modified truss analogy* procedure.

Web reinforcement is also provided in the form of bent-up longitudinal bars. When there is a series of parallel bent up bars at different distances from the support, the shear strength can be computed, as for inclined stirrups, using Eq. 6.14. However, when a single bar or a group of parallel bars are all bent-up at the same location at an angle α, the vertical component of the tension in these bars gives:

$$V_{sn} = A_v f_y \sin \alpha \qquad (6.16)$$

For bent-up bars, only the centre three-fourths of the inclined portion is considered effective as shear reinforcement.

Note that Eqs. 6.14 to 6.16 give the *nominal ultimate shear strength* provided by shear reinforcement. The corresponding *factored shear resistance* provided by shear reinforcement, V_s, is obtained by multiplying f_y in these equations with ϕ_s, the resistance factor for reinforcing bars. The three equations then respectively yield:
(i)   for inclined stirrups,

$$V_s = A_v \phi_s f_y \frac{d}{s} (\sin \alpha + \cos \alpha) \qquad (6.17)$$

(ii)   for shear reinforcement perpendicular to the axis of the member,

$$V_s = A_v \phi_s f_y \, d/s \qquad (6.18)$$

(iii)   for a single bar or a group of parallel bars, all bent up at the same location,

$$V_s = A_v \phi_s f_y \sin \alpha \qquad (6.19)$$

6.9 CSA A23.3–M84 REQUIREMENT FOR SHEAR DESIGN OF REINFORCED CONCRETE BEAMS BY SIMPLIFIED METHOD (CLAUSE 11.3)

(a) Design Basis

The strength criterion in Eq. 3.13 applied to shear design leads to the requirement:

Factored shear force at section ≤ factored shear resistance of the section

$$V_f \leq V_r \tag{6.20}$$

where V_f is the applied shear force under factored load combinations and V_r is the factored shear resistance.

The simplified design procedure for shear in CSA A23.3-M84(11.3) for members subjected to flexure and shear only is based on the assumptions that (1) the factored shear resistance of a reinforced concrete member can be taken as the sum of the factored shear resistance, V_c, provided by the concrete and the factored shear resistance, V_s, provided by the shear reinforcement; (2) the strength, V_c, provided by the concrete can be taken as equal to the shear resistance at inclined cracking (see Section 6.5, Eq. 6.11); and (3) V_s can be computed by the equations developed in Section 6.8 (Eqs. 6.17 to 6.19).

On this basis, the Code (Clause 11.3) recommends Eqs. 6.20 to 6.25 for the shear design of members subjected to shear and flexure only. (Flexural members subjected to axial load are dealt with in Section 6.14 and torsion design is considered in Chapter 7.) The factored shear resistance, V_r, is given by:

$$V_r = V_c + V_s \tag{6.21}$$

The factored shear resistance provided by tensile stresses in the concrete, V_c, is given by:

$$V_c = 0.2 \, \lambda \, \phi_c \, \sqrt{f'_c} \, b_w d \tag{6.22}$$

where V_c is in Newtons, f'_c in MPa, and the web width, b_w, and effective depth, d, are in mm. The factored shear resistance provided by shear reinforcement, V_s, is given by:

(i)   for shear reinforcement perpendicular to the axis of the member (vertical stirrups),

$$V_s = \frac{\phi_s A_v f_y d}{s} \tag{6.23}$$

(ii)   for inclined stirrups,

$$V_s = \frac{\phi_s A_{vi} f_y (\sin \alpha + \cos \alpha) d}{s} \tag{6.24}$$

(iii) for a single bar or a single group of parallel bars, all bent-up at the same section,

$$V_s = \phi_s A_{vi} f_y \sin \alpha \leq 0.3 \lambda \phi_c \sqrt{f'_c} b_w d \tag{6.25}$$

In Eqs. 6.23 to 6.25,

A_v = cross-sectional area of shear reinforcement perpendicular to the axis of the member within a distance of s

A_{vi} = cross-sectional area of inclined shear reinforcement within a distance s, or cross-sectional area of bars bent up at the same location

b_w = web width or diameter of circular section

d = distance from the extreme compression fibre to the centroid of the longitudinal tension reinforcement

s = spacing of shear reinforcement measured parallel to the member axis

α = angle between inclined stirrups or bent-up bars and longitudinal axis of the member

Where shear reinforcement consists of a series of parallel bent-up bars or groups of parallel bent-up bars, bent at different distances from the support, the shear strength V_s is to be taken as three-fourths of the value given by Eq. 6.24, but not greater than $0.5 \lambda \phi_c \sqrt{f'_c} b_w d$.

It should be noted that A_v represents the total area of shear reinforcement within the spacing s, and thus for a U-shaped stirrup with 2 legs, A_v is equal to twice the cross-sectional area for the bar size used. When more than one type of web reinforcement is used in the same portion of the web, the total shear strength V_s is taken as the sum of the individual shear strengths computed for each type.

In an "analysis" problem, Eq. 6.21 can be applied directly to evaluate the factored shear strength of a given section (see Example 6.1). In a "design" problem (to design the required web reinforcement for a given V_f) the factored resistance provided by the concrete, V_c, is first computed using Eq. 6.22 and the minimum required shear reinforcement contribution V_s can be obtained from Eq. 6.21. Knowing V_s, Eq. 6.23 or 6.24 can be used to determine the amount and distribution of shear reinforcement (see Example 6.2).

(b) Limitations Set in the Code

1. Yield Strength f_y The Code limits the yield strength used in design calculations of web reinforcement for shear and torsion to 400 MPa. The shear strength computations are based on the assumption that web reinforcement yields. The width of the inclined crack is a function of the stirrup strain. The yield strength limitation ensures that the diagonal crack widths are not excessive. Further, higher strength reinforcement may be brittle at sharp bends which often occur in stirrups.

For welded wire fabric used as shear reinforcement, the Code (cl. 11.2.2.1b) also specifies a minimum elongation requirement of 4 percent over a gauge length of at least 100 mm, which is not covered by the material standards CSA G30.5 and CSA G30.15. In order to mobilize the shear resistance, V_s, of the reinforcement, the transverse wires have to undergo substantial strains. The elongation requirement is to guard against premature failure of the transverse wires at or between the cross-wire intersection.

2. Minimum Shear Reinforcement When the shear force, V_f, does not exceed $V_c/2$ (that is, the nominal shear stress $V_f/(b_w d) \leq 0.1 \lambda \phi_c \sqrt{f_c'}$), beams need not be reinforced for shear. As explained in earlier sections, in some members, formation of an inclined crack may lead to sudden failure. In order to restrain the spread of inclined cracking and thereby provide increased ductility, and to obtain a gradual failure with adequate warning of distress, the Code specifies a minimum area of shear reinforcement in reinforced concrete flexural members whenever the factored shear force, V_f, exceeds $0.5 V_c$ (where V_c, in lieu of more detailed calculations, may be computed by Eq. 6.22), except in slabs and footings, concrete joist floors and certain classes of relatively wide and shallow beams. The minimum shear reinforcement is:

$$A_{v,min} = 0.35 \frac{b_w s}{f_y} \qquad (6.26)$$

Smax = d/2 > 75 mm

use d/3 on 8 ... mm

In satisfying the requirement for $A_{v,min}$ in Eq. 6.26, inclined reinforcement and transverse reinforcement used to resist torsion may be included.

The factored shear resistance of shear reinforcement corresponding to the minimum A_v required in Eq. 6.26 is obtained by substituting $A_v = A_{v,min}$ in Eq. 6.23 as:

$$V_s = 0.35 \phi_s b_w d$$

Hence, for the case of stirrups placed perpendicular to the beam axis, the minimum reinforcement controls as long as $V_s \leq 0.35 \phi_s b_w d$.

3. Maximum shear stress or minimum size of cross-section Members heavily reinforced with web steel may fail in shear-compression before the web steel yields and such failures are usually sudden and brittle. Hence, a limitation is placed on the shear strength provided by shear reinforcement. Accordingly, A23.3-M84(11.3.6.6) limits V_s to:

$$V_s \leq 0.8 \lambda \phi_c \sqrt{f_c'} \, b_w d \qquad (6.27)$$

CSA A23.3-M84(11.3.7.9) also limits $V_f/(b_w d)$ to a maximum of $0.25 \lambda \phi_c f_c'$. However, in most cases, Eq. 6.27 is more restrictive and hence controls. For several reasons (see Ref. 6.3) bent-up bars used as shear reinforcement give lower shear strength and wider cracks than similar beams with stirrups. As a result, the Code (cl. 11.3.6.3) limits the shear that can be carried by bent-up bars by restricting V_s in Eq. 6.25 to:

$$V_s \leq 0.3 \, \lambda \phi_c \, \sqrt{f'_c} \, b_w d \qquad (6.28)$$

The limitation on V_s given in Eq. 6.27 also serves as a serviceability limit for the control of width of inclined cracks. The latter is a function, among other things, of the stirrup strain, which in turn is a function of V_s.

Furthermore, in thin webbed members, failure may occur by crushing of the web prior to yielding of stirrups. To avoid such failures, the diagonal compression stress in the web and hence the maximum shear stress $v_f = V_f/(b_w d)$ must be limited. Such limits for v_f have been suggested in the range of $0.2 f'_c$ (Ref. 6.2) without the inclusion of the ϕ-factors. Equations 6.22 and 6.27 together give the upper limit for $V_f/(b_w d)$ as $1.0 \lambda \phi_c \sqrt{f'_c}$, which is a more conservative limit than $v_f < 0.2 f'_c$. Hence Eq. 6.27 also covers the requirement to prevent crushing of the web. The minimum size of cross-section that is required from considerations of shear is in effect determined by Eq. 6.27.

4. Spacing limits for shear reinforcement Web reinforcement becomes effective when it is crossed by inclined cracks. Further, dowel action will depend on the longitudinal bars being restrained by closely spaced stirrups. The Code limitation on spacing is such that every 45° line extending toward the reaction from mid-depth of the member to the longitudinal tension bars (representing a potential inclined crack) is crossed by at least one line of effective shear reinforcement (Fig. 6.9). For stirrups placed

Fig. 6.9 *Maximum spacing of web reinforcement*

(a) Vertical stirrups

(b) Inclined stirrups

(c) Bent up bars

perpendicular to the member axis, this leads to a maximum spacing of $0.5d$, which is further restricted to 600 mm. For inclined stirrups (Fig. 6.9b) the limitation leads to a spacing of $s \leq \frac{1}{2}d(1 + \cot \alpha)$. For bent-up bars (Fig. 6.9c) where the centre three-fourths of the inclined portion only is effective, $s \leq \frac{3}{8}d(1 + \cot \alpha)$. When the factored shear resistance contributed by shear reinforcement, V_s, exceeds $0.4\lambda\phi_c\sqrt{f_c'}b_w d$, the above maximum spacings are reduced by half. The minimum spacing between stirrups should be adequate for effective placement of concrete and is usually in excess of about 75 mm.

In summary, the shear reinforcement area is controlled by the following requirements:

$$V_f \leq V_c/2 \quad \text{— No reinforcement required.}$$

$$V_f > V_c/2, \text{ and } V_s \leq 0.35 \; \phi_s b_w d \quad \text{— Provide minimum reinforcement area as in Eq. 6.26.}$$

$$0.4 \; \lambda\phi_c \; \sqrt{f_c'} \; b_w d > V_s > 0.35 \; \phi_s b_w d \quad \text{— Provide reinforcement for } V_s = V_f - V_c, \text{ Eqs. 6.23–6.25, spacing limitations, cl. 11.3.8.1, 11.3.8.2.}$$

$$0.8\lambda\phi_c \; \sqrt{f_c'} \; b_w d > V_s > 0.4\lambda \; \phi_c \; \sqrt{f_c'} \; b_w d \quad \text{— Provide reinforcement for } V_s, \text{ Eqs. 6.23–6.25, maximum spacing limitations to be one-half of previous case (cl. 11.3.8.3).}$$

$$V_s > 0.8\lambda \; \phi_c \; \sqrt{f_c'} \; b_w d \quad \text{— Provide larger beam cross section (see Eq. 6.27).}$$

6.10 CRITICAL SECTION FOR SHEAR DESIGN NEAR SUPPORTS

Usually the maximum shear force in a member occurs at the face of supports and near concentrated loads. When a support reaction introduces transverse compressive stresses in the end region of the member, the shear strength of this region is enhanced, and inclined cracks do not generally form for a distance from the face of the support equal to the effective depth, d. Accordingly, for such cases, the Code permits sections

Fig. 6.10 *Critical sections for shear at support*

located less than a distance d from the face of the support to be designed for the same factored shear force, V_f, as that computed for the section at a distance d. Critical sections to be used in computing V_f for shear design for a few typical support conditions are shown in Fig. 6.10.

When a load is applied to the side of a member, through brackets, ledges or cross beams (Fig. 6.11), additional full depth transverse reinforcement is necessary in both members (CSA A23.3-M84 (11.3.9)) in the vicinity of the interface, for "hanging up" a portion $(1 - h_b/h_1)$ of the interface shear. The transverse reinforcement is designed to transfer a tensile force of $V_f(1 - h_b/h_1)$. (See Fig. 6.11 for notation.) The effective region in each member within which this additional reinforcement should be placed in order to be effective is also shown in Fig. 6.11. For the supporting member, this is a zone within a distance of h_b from the shear interface, and for the supported member it is a zone within a distance of $d/4$ on each side of the shear interface. The value of h_b need not be taken as less than 75 mm.

Fig. 6.11 *Hanging up bars for indirect support*

Section of girder

Section of cross beam

(a)

(b) Effective region for supporting member

(c) Effective region for supported member

6.11 ANCHORAGE OF SHEAR REINFORCEMENT

Web reinforcement is designed on the assumption that the yield stress ($f_y \leq 400$ MPa) is developed at ultimate strength. Hence, in order to be fully effective, such reinforcement must be adequately anchored at both ends so as to develop the yield strength. Web reinforcement is carried as close to the compression and tension surfaces as practical, consistent with minimum cover requirements.

Minimum requirements for effective anchorage are given in CSA A23.3-M84(12.13). For the most common type of web reinforcement, the U-stirrup, the Code requirements are shown in Fig. 6.12. The maximum stress in the stirrup, equal to design

Fig. 6.12 C.S.A. A23.3-M84 requirements for anchorage of U-stirrups

f_y, is assumed to be developed at a depth $d/2$ from the compression face. In Fig. 6.12, ℓ_d represents the development length (see Chapter 9). When the standard hook (Fig. 6.12c) is used, in order to satisfy the requirement in Fig. 6.12a, a reduction in stirrup bar size and spacing, or alternatively an increase in the effective depth, d, may become necessary. The minimum required depth, d, for effective anchoring of web reinforcement with standard hooks is tabulated in Table 6.1 for No. 10 and No. 15 bar stirrups. Between the anchored ends, each bend in the continuous portion of the stirrup must enclose a longitudinal bar.

Table 6.1 Minimum Effective Depth of Beam for Effective Anchorage of Vertical Stirrups with Standard Hooks, mm

Hook Type	Bar Size	$f_y = 300$ MPa		350 MPa		400 MPa	
		ℓ_d	d_{min}	ℓ_d	d_{min}	ℓ_d	d_{min}
90°	10	197	345	229	377	262	410
	15	278	454	325	501	371	547
135°	10	–	No limit	–	299	–	321
	15	–	No limit	–	391	–	421

Based on: $f'_c = 20$ MPa, minimum cover $d_c = 40$ mm
$d_{min} = \ell_d + 6d_b + 2d_c$ for 90° bend
$\quad\quad = 0.66\,\ell_d + 6d_b + 2d_c$ for 135° bend
(Refer Fig. 6.12 a, b, c)

6.12 ADDITIONAL COMMENTS ON SHEAR REINFORCEMENT DESIGN

Bent-up bars generally give lower shear strength and wider cracks than stirrups. Hence, unless there is a series of such bars bent up at relatively close spacings, there is not much economy resulting from accounting for their shear strength contribution. In the usual situations where there are only a few isolated bent-up bars scattered widely along the span, their shear strength contribution is ignored and stirrups are designed to carry the full excess shear $V_s = V_f - V_c$.

Inclined stirrups are most effective in reducing crack width, and are desirable when full depth transverse cracks are likely, as in the case of beams with high axial tensile forces. However, inclined web reinforcement may become ineffective if the direction of shear force is reversed when the potential inclined cracks parallel the inclined web reinforcement. In general, smaller diameter, closely spaced stirrups (for example, welded wire fabric stirrups) give better crack control than stirrups of larger size bars, placed farther apart.

Although the Code does not require shear reinforcement in portions of beams where $V_f < V_c/2$, it is good practice to provide the minimum web reinforcement in this region also, in order to provide ductility and restrain cracks in the event of accidental overloading. In beams subjected to reversed bending to the full design capacities, stirrups may be provided for the full factored shear force, neglecting the shear strength provided by concrete; and such stirrups should be in the form of closed ties.

Shear reinforcement is designed usually by selecting a suitable stirrup bar size (which is kept the same for the entire span) and then computing the required spacings. In general, the required spacing will vary continuously along the length of the beam due to the variation in shear force, V_f. However, stirrups are usually arranged with the spacing kept uniform over portions of the span, the spacing varying between these portions as necessitated by strength and limiting spacing considerations. The first stirrup is placed at not more than one-half space from the face of support. Between the anchored ends, each bend in the continuous portion of a stirrup must enclose a longitudinal bar.

Termination of flexural reinforcement in the tension zone may lower the shear strength of beams (see Section 5.10.5). Hence, such sections may also be critical and have to be checked for shear and, if necessary, provided with additional shear reinforcement to satisfy the requirement in CSA A23.3-M84 (12.10.5) (see Example 6.1).

6.13 SHEAR DESIGN EXAMPLES

--------------------------------- **EXAMPLE 6.1** ---------------------------------

The simply supported beam in Examples 5.3 and 5.9 is provided with web reinforcement consisting of 6 mm plain bar U-stirrups at a uniform spacing of 190 mm, as shown in Fig. 6.13a. Check the adequacy of the shear design. If necessary revise the design.

SOLUTION

1. The factored loads are:
   Dead load $=$ 1.25 (10.65) $=$ 13.3 kN/m
   Live load $=$ 1.5 $\times$ 15 $=$ 22.5 kN/m
2. The shear force envelope giving the maximum factored shear force at each location along the span is first drawn for one half of the span. The placement of live load to give the maximum shear force at a section, X, on the left half of the span is as shown in Fig. 6.13b. Dead load generally acts on the entire span.

   The actual shape of the shear envelope is curvilinear. However, it is sufficiently accurate and conservative to consider the shear force due to live load to vary as a straight line between the maximum values computed for the support and for the midspan. In this problem,

$$V_{f,\text{support}} = \frac{(13.3 + 22.5)6}{2} = 108 \text{ kN}$$

$$V_f \text{ at } \mathbb{C} = 0 + \frac{22.5 \times 3 \times 1.5}{6} = 16.9 \text{ kN} = \frac{w \frac{\ell}{2} \frac{\ell}{4}}{\ell}$$

The critical section for shear (cl. 11.3.3.2) is at a distance $d = 386$ mm from face of support, or $120 + 386 = 506$ mm from centre of support.

Fig. 6.13 Example 6.1

Factored shear force at critical section:

$$V_f = 16.9 + \frac{(108 - 16.9)(3000 - 506)}{3000} = 92.6 \text{ kN}$$

The factored shear force diagram is shown in Fig. 6.13c.

3. The factored shear resistance provided by concrete, V_c, using Eq. 6.22 is:

$$V_c = 0.2\lambda\phi_c \sqrt{f'_c} \, b_w d$$
$$= 0.2 \times 1.0 \times 0.6 \times \sqrt{30} \times 250 \times 386 \times 10^{-3} = 63.4 \text{ kN}$$

4. Check adequacy of section
   Maximum permitted $V_s = 0.8\lambda\phi_c\sqrt{f'_c}b_w d = 4 \times 63.4 = 254$ kN
   Actual required $V_s = 92.6 - 63.4 = 29.2$ kN < maximum permitted.
   Section size is adequate.

5. Check adequacy of shear reinforcement

Since $V_f > V_c/2$, shear reinforcement is required.

Minimum shear reinforcement required is:

$$A_{v,min} = 0.35\frac{b_w s}{f_y} = \frac{0.35 \times 250 \times 190}{400} = 41.6 \text{ mm}^2$$

$$A_v \text{ provided} = 2 \times 28.3 = 56.6 > \text{minimum} \qquad \text{OK}$$

Required spacing for the V_s is obtained from Eq. 6.23:

$$\text{Required } s = \frac{\phi_s A_v f_y d}{V_s} = \frac{0.85 \times 56.6 \times 400 \times 386}{29.2 \times 10^3} = 254 \text{ mm}$$

$$\text{Provided } s = 190 \text{ mm} < \text{required} \qquad \text{OK}$$

Alternatively, the factored resistance of the section, V_r, is:

$$V_r = V_c + V_s = 63.4 + \frac{\phi_s A_v f_y d}{s}$$

$$= 63.4 + \frac{0.85 \times 56.6 \times 400 \times 386}{190} \times 10^{-3} = 63.4 + 39.1$$

$$= 103 \text{ kN} > V_f = 92.6 \text{ kN} \qquad \text{OK}$$

6. Check spacings

To check condition in Code clause 11.3.8.3

$$0.4\lambda\phi_c\sqrt{f_c'}\, b_w d = 2 \times 63.4 = 126.8 \text{ kN}$$

Required $V_s = 29.2 \text{ kN} < 0.4\lambda\phi_c\sqrt{f_c'}\, b_w d$

Hence the maximum spacings in clause 11.3.8.1 need not be halved.

Maximum spacing limit is the least of:

$$0.5d = 0.5 \times 386 = 193 \text{ mm} \qquad \text{(controls)}$$
$$\text{or } 600 \text{ mm}$$
$$s_{max} = 193 \text{ mm}$$
$$s_{provided} = 190 \text{ mm} < s_{max} \qquad \text{OK}$$

7. Check anchorage

The stirrups are bent around the bars, at an angle of 90°. For 6 mm bars and $f_c' = 30$ MPa and $f_y = 400$ MPa, the development length, $\ell_d = 139$ mm (clause 12.2.2). In order to have adequate anchorage, minimum required depth of beam (Fig. 6.12 a, c) is:

$$d_{min} = \ell_d + 6d_b + 2d_c$$
$$= 139 + 6 \times 6 + 2 \times 40 = 255 \text{ mm}$$

Since $d = 386 \text{ mm} > d_{min}$, the anchorage is adequate.

8. Check shear strength at bar cut-off point

One longitudinal tension reinforcement is terminated in the tension zone, at distance of 900 mm from support. Hence this section should be checked for shear to conform with requirements in clause 12.10.5. Loading condition for maximum shear at this section is shown in Fig. 6.13d.

Factored shear at cut-off point, $V_f = 76.7$ kN

Shear resistance of this section:

$$V_r = V_c + V_s = 63.4 + 39.1 = 103 \text{ kN}$$

$$\frac{2}{3} \text{ shear resistance} = \frac{2}{3} \times 103 = 68.3 \text{ kN}$$

Since $V_f > \frac{2}{3}$ that permitted $= 68.3$ kN, clause 12.10.5a is not satisfied and additional stirrups must be provided. Here, adding two additional stirrups along the last portion of the cut-off bar, as shown by broken lines in Fig. 6.13a, will reduce the spacing to $190/2 = 95$ mm. Then the shear strength at the cut-off point is:

$$V_c + V_s = 63.4 + 2 \times 39.1 = 142 \text{ kN}$$

$$V_f < \frac{2}{3} \text{ permitted} = \frac{2}{3} \times 142 = 94.4 \text{ kN} \qquad \text{OK}$$

The requirement in clause 12.10.5a is now satisfied. Alternatively, the additional stirrups can be shown to satisfy the requirements in Clause 12.10.5b.

The reinforcement provided corresponds to the minimum spacing. Since the design is safe at the critical section for shear, it is safe at other points along the beam.

According to the Code, shear reinforcement is not required where $V_f < V_c/2$ or

$$V_f < 63.4/2 = 31.7 \text{ kN}$$

Referring to Fig. 6.13c,

$$V_f < V_c/2$$

for a distance of 487 mm from centre of span, and the stirrups in this region (five in all for the entire span) are not essential.

EXAMPLE 6.2

Design the shear reinforcement for the T-beam in Examples 5.7 and 5.10.

SOLUTION

The section properties are given below and shown in Fig. 6.14a.

$b = 3180$ mm, $b_w = 300$ mm, $h = 500$ mm

d at support $= 434$ mm, d at midspan $= 406$ mm

A_s at support $= (3 \text{ No. } 30 \text{ bars}) = 2100 \text{ mm}^2$

A_s at midspan = (6 No. 30 bars) = 4200 mm^2
Span centre to centre of support = 8 m
Width of bearing = 240 mm
f'_c = 20 MPa, f_y = 400 MPa

1. Factored loads are:

   Dead load = 1.25 × 15.7 = 19.6 kN/m
   Live load = 1.5 × 29.7 = 44.6 kN/m
   Total = 64.2 kN/m

2. Compute shear force envelope

$$\text{Maximum shear at support} = \frac{64.2 \times 8}{2} = 257 \text{ kN}$$

   Maximum shear at midspan (live load only on one half of span)

$$= 0 + \frac{1}{4} \frac{44.6 \times 8}{2} = 44.6 \text{ kN}$$

   Shear force diagram may be assumed conservatively as a straight line shown in Fig. 6.14b.
   Critical section at support is at a distance d away from the face of support, or 120 + 434 = 554 mm from the centre of support.

$$\text{Shear at critical section} = 44.6 + \frac{(257 - 44.6)(4 - 0.554)}{4}$$

$$= 228 \text{ kN}$$

3. Factored shear resistance of concrete, V_c:

$$V_c = 0.2\lambda\phi_c\sqrt{f'_c}\, b_w d$$

   At support section,

$$V_c = 0.2 \times 1.0 \times 0.6 \times \sqrt{20} \times 300 \times 434 \times 10^{-3} = 69.9 \text{ kN}$$

   At midspan section, d = 406 mm and

$$V_c = 0.2 \times 1.0 \times 0.6 \times \sqrt{20} \times 300 \times 406 \times 10^{-3} = 65.4 \text{ kN}$$

Variations of V_f, V_c and V_s are shown in Fig. 6.14c
   Note that the effective depth, d, varies along the span as tension reinforcement bars are terminated. Here, the lower value of V_c = 65.4 kN computed for midspan section will be considered as the factored shear resistance contributed by concrete for the entire length of the beam. Consideration of variation of d along span is not generally warranted by normal levels of accuracy desired in practice.
   The excess shear to be carried by shear reinforcement, V_s, is represented by the

shaded area in Fig. 6.14c. The location of the section where $V_f = V_c$ is calculated from similar triangles, in Fig. 6.14c as:

$$\frac{\ell_c}{3446} = \frac{65.4 - 44.6}{228 - 44.6}; \qquad \ell_c = 391 \text{ mm}$$

4. Check adequacy of overall depth
Maximum permitted $V_s = 0.8\lambda\phi_c\sqrt{f'_c}b_w d = 4 \times 65.4 = 262$ kN
Actual maximum $V_s = 228 - 65.4 = 163$ kN < 262 kN   OK

5. The zones with different shear reinforcement requirements are calculated below and identified in Fig. 6.14c

Zone I:

$$V_f < V_c/2 = 32.7 \text{ kN}$$

No shear reinforcement required

In this example $V_{f,\min} = 44.6$ kN $> V_c/2$. Hence, no such zone in this beam.

Zone II:

$$V_f > \frac{V_c}{2} \text{ and } V_s \leq 0.35 \ \phi_s b_w d = 0.35 \times 0.85 \times 300 \times 406 \times 10^{-3} = 36.2 \text{ kN}$$

or

$$V_f \leq 65.4 + 36.2 = 101.6 \text{ kN}$$

Minimum reinforcement requirement controls. Location for this is:

$$\ell_2 = \left(\frac{101.6 - 44.6}{228 - 44.6}\right) \times 3446 = 1070 \text{ mm}$$

Zone II is from midspan to 1070 mm.

Zone III:

$$0.35 \ \phi_s b_w d < V_s < 0.4\lambda\phi_c\sqrt{f'_c}b_w d = 2 \times 65.4 = 131 \text{ kN}$$

Reinforcement designed for excess shear V_s, and maximum spacing limitations as in CSA A23.3-M84(11.3.8.1). The upper limit to this is at $V_f = V_c + 131 = 196$ kN, and its location is given by:

$$\ell_3 = \left(\frac{196 - 44.6}{228 - 44.6}\right) \times 3446 = 2845 \text{ mm from midspan}$$

Zone IV:

$$0.4\lambda\phi_c\sqrt{f'_c}b_w d < V_s < 0.8\lambda\phi_c\sqrt{f'_c}b_w d = 4 \times 65.4 = 262 \text{ kN}$$

Fig. 6.14 *Example 6.2*

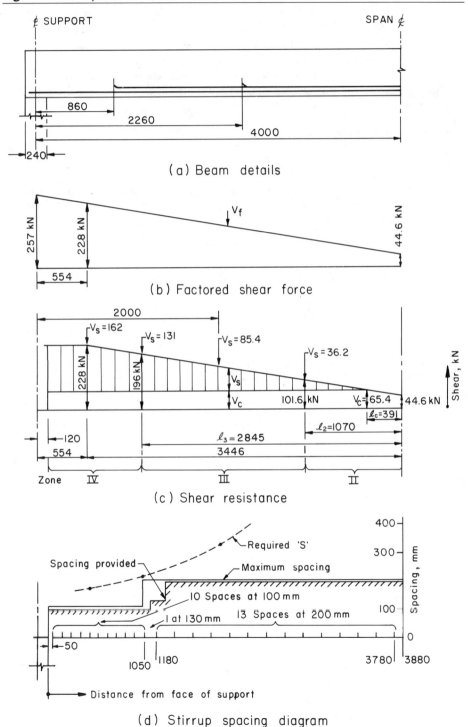

(a) Beam details

(b) Factored shear force

(c) Shear resistance

(d) Stirrup spacing diagram

Reinforcement designed for excess shear V_s, but maximum spacing limitations reduced to one-half of that for Zone III. The region from $\ell_3 = 2845$ mm to face of support falls in this category.

Note that the zones have been identified here for illustration purposes only. For designing shear reinforcement, such a detailed zoning is not necessary and it is sufficient to locate the region where shear reinforcement is necessary ($V_f > V_c/2$) and the section ($V_s = 0.4\lambda\phi_c\sqrt{f_c'}b_w d$) beyond which the closer spacing requirement controls.

6. Design of shear reinforcement
   Try No. 10 bar U-stirrups placed vertically.
   $A_v = 2 \times 100$ (for 2 legs) $= 200$ mm^2.
   Check adequacy of beam depth for anchorage of stirrups.
   For No. 10 bar stirrups, minimum depth required for effective anchorage with standard 90° hook is (Table 6.1) $= 410$ mm; $d = 434$ mm at support where shear is maximum.   OK

The spacing requirement is given by (Eq. 6.23),

$$s = \frac{\phi_s A_v f_y d}{V_s} = \frac{0.85 \times 200 \times 400 \times 406}{V_s} = \frac{27\ 600 \times 10^3}{V_s}\ \text{mm}$$

The required stirrup arrangement can be determined by plotting a spacing curve giving the variation of required s as computed by the above equation wherein the spacing limitations are also superimposed. Having done this, convenient bar spacings are selected which are on the safe side of the requirement. The calculations for the spacing curve for this example are set up in tabular form below.

Distance of section from support, mm	V_s kN	Required s = 27 600/V_s, mm
Critical section, 554	162	170
1155	131	211
2000	85.4	323
For min. reinf. 2930	36.2	763

The maximum allowable spacing in Zones I to III is the least of:

$$0.5d = 0.5 \times 406 = 203\ \text{mm} \quad \text{(controls)}$$
or $$600\ \text{mm}$$

Maximum allowable spacing in Zone IV is:

$$\tfrac{1}{2}(0.5d) = \tfrac{1}{2}(0.5 \times 434) = 109\ \text{mm}$$

(Note that in Zone IV where top layer tension reinforcement is cut off, d = 434 mm.)

The curve for required spacing is shown in Fig. 6.14d, wherein the maximum spacing limitations are also indicated. The solid line in this figure represents the spacing requirements. A stirrup layout meeting these requirements is detailed along the base of Fig. 6.14d and the corresponding spacing diagram is shown by the shaded area. In this layout, the first stirrup is placed at 50 mm from face of support and this is followed in turn by 10 spaces (stirrups) at 100 mm, one at 130 mm and 13 at 200 mm making up one-half of the span.

This spacing can be selected as final if no flexural reinforcement bar is terminated in the tension zone. In this Example, since the main flexural reinforcement is terminated in the tension zone, the shear at these locations must be checked to satisfy CSA A23.3-M84(12.10.5)

7. Check shear strength at bar cut-off points
    The cut-off point closest to support, which is more critical, will be checked first. The section is 0.86 m from centre of support. Since the shear here may be critical and in order to avoid too much conservatism, the actual loading condition for maximum shear (shown in Fig. 6.15) and the effective depth at support will be used to check the strength of this section.

    From Fig. 6.15, at the cut-off point

$$V_f = 204 \text{ kN}$$

Fig. 6.15 *Stirrup layout, Example 6.2*

(a) Loading for maximum shear at cut-off point

(b) Stirrup layout

Shear carried by concrete for $d = 434$ mm is:

$$V_c = 69.9 \text{ kN}$$

Shear carried by shear reinforcement (at spacing of 100 mm provided at this location) is:

$$V_s = \frac{\phi_s A_v f_y d}{s} = \frac{0.85 \times 200 \times 400 \times 434}{100} \times 10^{-3} = 295 \text{ kN}$$

Permitted shear without additional stirrups (CSA A23.3-M84(12.10.5)) is:

$$\tfrac{2}{3} (V_c + V_s) = \tfrac{2}{3} \times (69.9 + 295) = 243 \text{ kN}$$

$$V_f = 204 < \text{permitted} \qquad\qquad \text{OK}$$

Shear strength at cut-off point is adequate for the arrangement of stirrups worked out at step 6 above.

A similar evaluation at the cut-off point at 2260 mm from the support will show that the shear strength at this location is adequate with the stirrup spacing just designed. Hence a final satisfactory stirrup layout is as shown in Fig. 6.15b.

6.14 SHEAR STRENGTH OF FLEXURAL MEMBERS SUBJECTED TO AXIAL LOADS

6.14.1 INCLINED CRACKING LOAD

Frequently, flexural members in frames are subjected to axial loads in addition to bending moment and shear. Axial tension may also result from restraint of deformations due to temperature changes, creep and shrinkage. In determining shear resistance, effects of axial forces, in particular of tensile forces, must be considered wherever applicable.

Inclined cracking is related to the attainment of a limiting principal tensile stress in the concrete. The addition of an axial force alters the stress distribution and hence the shear strength at inclined cracking. In particular, axial compression decreases the normal tensile stress, f_x. Furthermore, it decreases the shearing stress, v, by decreasing the height of flexural cracks and thereby increasing the effective concrete area resisting shear. Both these effects reduce the principal tensile stress and, hence, tend to increase the shear strength at diagonal cracking. In contrast, axial tension will have just the opposite effects.

For beams with axial loads but without web reinforcement, the shearing force, V_{cr}, which causes the first diagonal cracking is taken as the shear strength, V_c, of the beam. The expressions for computing $V_c = V_{cr}$ when there is axial load, N, are given below.

6.14.2 AXIAL COMPRESSION

Following the general derivation of Eq. 6.9, Ref. 6.2 has shown that the effect of an axial compression, N, can be accounted for conservatively by replacing the moment, M, in Eq. 6.9 by a modified moment M_m given by:

$$M_m = M - N\left(\frac{4h - d}{8}\right) \tag{6.29}$$

where h is the overall thickness of the member, d is the effective depth, and the axial load N is considered positive if compressive. Correspondingly, the upper limit $V_{cn} = 0.3 \sqrt{f_c'}\, bd$ is modified as

$$V_{cn} \leqslant 0.3 \sqrt{f_c'}\left(\sqrt{1 + 0.3 \frac{N}{bd}}\right)bd \tag{6.30}$$

where $N/(bd)$ is in MPa. With these modifications to Eq. 6.9, the nominal shear strength of concrete beams without shear reinforcement when subjected to axial compression, N, is,

$$V_{cn} = \left(0.16 \sqrt{f_c'} + 17\rho \frac{Vd}{M_m}\right)bd \leqslant 0.3 \sqrt{f_c'}\left(\sqrt{1 + 0.3 \frac{N}{bd}}\right)bd \tag{6.31}$$

where

$$M_m = M - N\left(\frac{4h - d}{8}\right)$$

For computing the nominal shear strength provided by concrete of members subjected to axial compression, the ACI Code (Ref. 6.1) specified Eq. 6.32, which is the equivalent in imperial units of Eq. 6.31.

$$V_{cn} = \left(1.9 \sqrt{f_c'} + 2500\, \rho \frac{Vd}{M_m}\right)bd \tag{6.32}$$

$$\leqslant 3.5 \sqrt{f_c'}\left(\sqrt{1 + \frac{N}{500A_g}}\right)bd$$

where

$$m_m = M - N\left(\frac{4h - d}{8}\right)$$

Here A_g is the gross area of the section and all the terms are in Imperial units. In Eq. 6.32, Vd/M_m is not limited to 1.0 as it was in Eq. 6.10.

A simplified expression for computing V_{cn}, given in Ref. 6.1, is:

$$V_{cn} = 2\sqrt{f'_c}\left(1 + \frac{N}{2000A_g}\right)bd \tag{6.33}$$

where the terms are in Imperial units.

The S.I. unit versions of Eqs. 6.32 and 6.33 were specified in CSA A23.3-M77. However, CSA A23.3-M84 (11.3) gives Eq. 6.34 for computing the factored shear resistance of concrete of members subjected to axial compression.

$$V_c = 0.2\,\lambda\,\phi_c\,\sqrt{f'_c}\left(1 - \frac{3N_f}{A_g f'_c}\right)b_w d \tag{6.34}$$

where N_f is the factored axial load, in Newtons, occurring simultaneously with V_f and includes effects of tension due to creep and shrinkage, and is considered positive if tensile; A_g is the gross area of concrete cross-section in mm^2; and the other terms have previously been defined. Equation 6.34 is based on the recommendations of ACI–ASCE Committee 426 (Refs. 6.6, 6.7) and gives approximately a lower bound to the test data for axially compressed beams. The factor within brackets in Eq. 6.34, giving the increase in shear resistance due to axial compression (compare with Eq. 6.22), is expressed in terms of the nondimensional ratio $N_f/(A_g f'_c)$, because the depth of the compression zone at ultimate strength will be a function of f'_c.

6.14.3 AXIAL TENSION

On the basis of the development, Eq. 6.32 should also be applicable when the axial force is tensile. However, comparison with test results (Ref. 6.8) has shown the equation to be unsafe for beams with axial tension.

CSA A23.3-M84 (11.3) specifies a simple straight line interaction equation, Eq. 6.35, for computing the factored shear resistance of concrete beams when subjected to axial tension. This equation gives the same value for V_c as Eq. 6.22 when the axial force is zero, and $V_c = 0$ when the axial tension, N_f, is such that it, together with the simultaneously occurring bending moment, causes yielding of the longitudinal reinforcement. In between these two limits, Eq. 6.35 gives a linear variation for V_c.

$$V_c = 0.2\,\lambda\,\phi_c\,\sqrt{f'_c}\left(1 - \frac{N_f}{N_r}\right)bd \tag{6.35}$$

where N_f = the factored axial load occurring simultaneously with V_f, including effects of tension due to creep and shrinkage, considered positive when tensile; and N_r = factored axial tensile resistance of the member, taking into account the co-existing bending moment.

In general, N_r has to be computed by a "plane section analysis" of the member. If the coexisting moment at the section is equal to M_r of the section, then $N_r = 0$. At the other extreme, if the moment at the section is zero, $N_r = A_s\phi_s f_y$ if the longitudinal reinforcement, A_s, is symmetrically placed. As a conservative approximation, V_c can be taken equal to zero, as the axial tension approaches the tensile cracking load of the section. On this basis, as an alternative to detailed plane section analysis, the Code permits the use of Eq. 6.36 for computing N_r.

$$N_r = 0.6 \lambda \, \phi_c \, \sqrt{f_c'} \, A_g \tag{6.36}$$

where A_g is the gross area of concrete section.

6.15 WEB REINFORCEMENT IN BEAMS WITH AXIAL FORCE

It has been reported (Ref. 6.3), based on test results, that the presence of axial forces does not influence the effectiveness of the web reinforcement. It is further suggested that the design procedure taking the factored shear resistance, $V_r = V_c + V_s$, leads to conservative designs. Here, V_c is the shear *carried by the concrete*, assumed equal to the inclined cracking shear in the presence of the axial load, discussed in Section 6.14, and V_s is the shear carried by the web reinforcement (Eqs. 6.23–6.25).

When the axial force is tensile and relatively high, transverse tension cracks may traverse the entire beam depth. In such situations, inclined stirrups which intercept these transverse cracks may be preferable to vertical stirrups, in order to prevent shearing displacement along such cracks.

6.16 STRUCTURAL LOW DENSITY CONCRETE

A study of diagonal tension test results of structural low density concrete beams has shown (Ref. 6.9) that the general principles of shear strength discussed in Sections 6.3–6.5 for normal density concrete apply equally well to low density concrete members. Once again, the major variables which affect the inclined cracking strength of such beams are the ratio $M/(Vd)$, ρ, and tensile strength of concrete; and further, these variables can be grouped into the same parameters as used in the development of Eq. 6.10. However, the tensile strength of low density concrete is highly dependent on the type of aggregate, and is somewhat less than that of normal density concrete of the same specified compressive strength, f_c'. As a simplification, it can be assumed that for a given compressive strength of concrete, the tensile strength of lightweight concrete is a fixed proportion of the tensile strength of normal density concrete (Ref. 6.9). Thus, CSA A23.3-M84(11.2.3) has introduced a modification factor, λ, for all values of $\sqrt{f_c'}$ affecting shear resistance, in order to account for the density of concrete. The factor λ has values of unity for normal density concrete, 0.85 for structural semi-low density concrete and 0.75 for structural low density concrete.

6.17 MEMBERS OF VARYING DEPTH

In flexural members of varying depth, as shown in Fig. 6.16, the vertical component of the flexural tensile force may affect the shear force for which the section is to be

Fig. 6.16 *Design shear force in members of variable depth*

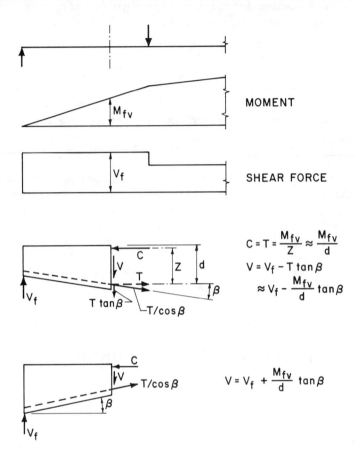

designed. Assuming the horizontal component of the tension in the flexural reinforcement as $T = M_{fv}/z \approx M_{fv}/d$, it can be seen from Fig. 6.16 that the net shear force for which the section should be designed is:

$$V_{net} = V_f \pm \frac{M_{fv}}{d} \tan \beta \qquad (6.37)$$

where V_f and M_{fv} are the applied shear force and moment at the section, β is the inclination of the tension reinforcement to the beam axis; and the negative sign applies where M_{fv} increases in the same direction as depth, d, and the positive sign applies where M_{fv} decreases in this direction.

A similar adjustment to the shear, V_f, is necessary when the flexural compression is inclined to the beam axis. CSA A23.3-M84 (11.2.1.4) specifies that when the effect of the vertical component is unfavourable (that is, additive to V_f), it should be taken into account in designing for shear. The effect may also be considered when favourable.

6.18 SHEAR STRENGTH OF SPECIAL MEMBERS

6.18.1 GENERAL

CSA A23.3-M84 presents separate provisions for the shear design of special types of members such as deep beams, brackets and corbels, and walls, and for the shear design of regions adjacent to supports, concentrated loads or abrupt changes in cross section. These members are relatively short and deep (low $M/(vd)$ ratio) and their shearing failure may differ considerably from that in normal beams. The inclined cracks in these types of members are much steeper than in normal beams and as a result the relative importance of the various shear transfer mechanisms, as well as the relative effectiveness of vertical and horizontal (transverse and longitudinal) shear reinforcement is altered (Fig. 6.17). Further, because of their relatively small size, these members often require special detailing considerations for bond, anchorage, bearing and crack (spalling) control.

Special provisions are also made for design by the shear-friction method, where appropriate, and for the design of slabs and footings for two-way shear.

The Code requires that deep beams (and parts of members having deep shear span) and brackets and corbels be designed by the General Method of clause 11.4, and these are discussed in Chapter 8. For walls, design for shear forces in the plane of the wall is made as for beams, and for shear forces due to loads acting perpendicular to the face of the wall the design is as for slabs and footings. Design for shear strength at construction joints in walls is based on the shear-friction theory (see Section 6.18.2). Considerations of one- and two-way shear in slabs are discussed in Chapter 14.

Fig. 6.17 Typical crack patterns in short, deep members

(a) Deep beam   (b) Corbel

6.18.2 SHEAR-FRICTION

There are situations where shear has to be transformed across a plane, nearly parallel to the shear force and along which slip could occur. Examples are planes of existing or potential cracks, interface between dissimilar materials, and interface between

Fig. 6.18 *Typical cases where shear friction is applicable (adapted from Ref. 6.10)*

(a) Corbel

(b) Precast beam seat

(c) Column face plate

concrete placed at different times (Fig. 6.18). In such cases failure is in direct shear rather than in diagonal tension and hence it would be appropriate to consider shear resistance developed along such planes in the form of resistance to the tendency to slip. The *shear-friction* concept is a method to do this.

When two bodies are in contact with a normal reaction, R, across the surface of contact, the frictional resistance, F, acting tangential to this surface and resisting relative slip is known to be $F = \mu R$ where μ is the coefficient of friction (Fig. 6.19a). Figure 6.19b shows a cracked concrete specimen loaded in shear. In such a specimen, a clamping force between the two faces of the crack can be induced by providing reinforcement perpendicular to the crack surface (shear-friction reinforcement, A_{vf}). Any slip between the two faces of the rough irregular crack causes the faces to ride upon each other which opens up the crack. This in turn induces tensile forces in the reinforcement which ultimately yields (Fig. 6.19c, d). If the area of reinforcement is A_{vf} and yield stress f_y, at ultimate, the clamping force between the two faces is $R = A_{vf}f_y$, and the frictional resistance is $A_{vf}f_y\mu$.

In reality, the actual resistance to shear, V_r, is composed of this frictional force, the resistance to shearing off of the protrusions on the irregular surface of the crack, and the dowel force developed in the transverse reinforcement. In the simple shear-friction model of behaviour, it is assumed that all the shear resistance is due to the friction between the crack faces, and then the factored shear resistance, V_r, is given by Eq. 6.38; however, the coefficient of friction, μ, is adjusted upwards to yield computed strengths in agreement with test results. Other less simple methods of

Fig. 6.19 *Shear-friction analogy*

(a) Frictional force

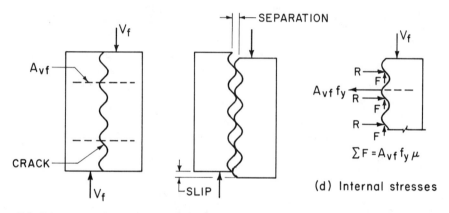

(b) Direct shear crack   (c) Slip and separation

calculation have been proposed (Refs. 6.11, 6.12) which result in predictions of shear transfer resistance in substantial agreement with comprehensive test results, and such methods are also permitted by the Code (clause 11.7.3).

For shear-friction reinforcement placed perpendicular to the shear plane,

$$V_r = \phi_s A_{vf} f_y \mu \qquad (6.38)$$

CSA A23.3-M84(11.7.4.3) specifies values for μ of
$\qquad \mu = 1.25\lambda$ for concrete cast monolithically,
$\qquad = 0.9\lambda$ for concrete placed against hardened concrete, with the surface intentionally roughened as specified in clause 11.7.9,
$\qquad = 0.5\lambda$ for concrete placed against hardened concrete *not* intentionally roughened, and
$\qquad = 0.6\lambda$ for concrete placed against as-rolled structural steel
where λ is the factor to account for low density concrete.

Shear-friction reinforcement may also be placed at an angle α_f to the shear plane such that the shear force produces tension in the shear-friction reinforcement as shown in Fig. 6.20a, b (i.e., $\alpha_f \leq 90°$). As the shear-friction reinforcement yields,

Fig. 6.20 *Inclined shear-friction reinforcement*

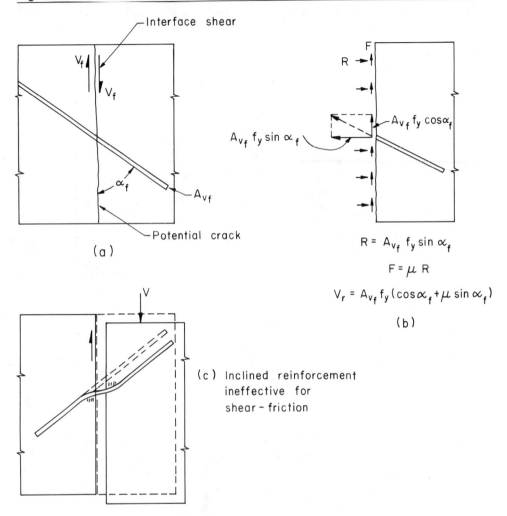

(a)

$R = A_{v_f} f_y \sin \alpha_f$

$F = \mu R$

$V_r = A_{v_f} f_y (\cos \alpha_f + \mu \sin \alpha_f)$

(b)

(c) Inclined reinforcement ineffective for shear-friction

the tensile force in the reinforcement is $A_{vf}f_y$, which has a component parallel to the shear plane of $A_{vf}f_y \cos \alpha_f$, and a component normal to the plane equal to $A_{vf}f_y \sin \alpha_f$ and the latter produces the clamping force. The total force resisting shear is then obtained as $A_{vf}f_y \cos \alpha_f + \mu.A_{vf}f_y \sin \alpha_f$, and the factored shear resistance is given by:

$$V_r = \phi_s A_{vf}f_y (\cos \alpha_f + \mu \sin \alpha_f) \tag{6.39}$$

Note that reinforcement inclined at an angle $\alpha_f > 90°$ is mostly ineffective in resisting interface shear because as relative slip between the two parts occurs and the reinforcement deforms, the effect is to separate the two parts farther apart rather than introducing any clamping forces, as shown in Fig. 6.20c.

For calculating the factored shear resistance, V_r, the yield strength of the shear-friction reinforcement is limited to a maximum of 400 MPa. An upper limit on the shear resistance is also specified equal to $0.25\ \phi_c\ f'_c\ A_{cv} \leqslant 6.5\ \phi_c\ A_{cv}$, where A_{cv} is the area of concrete section resisting shear transfer. This limit is to prevent congestion of reinforcement and also because of the unconservative nature of Eqs. 6.38 and 6.39 at high values of V_f (Ref. 6.10).

Any direct tension, N_f, across the shear plane must be provided for by additional reinforcement having an area equal to $N_f/(\phi_s f_y)$. Such tensile forces may be caused by restraint of deformations due to temperature change, creep and shrinkage, etc. Although the beneficial effect of a permanently occurring net compressive force across the shear plane may be taken advantage of in reducing the amount of shear-friction reinforcement required, it is prudent to ignore this effect. When there is a bending moment acting on the shear plane, the flexural tensile and compressive forces balance each other, and the ultimate compressive force across the plane (which induces the frictional resistance) is equal to $A_s f_y$. Hence, the flexural reinforcement area, A_s, can be included in the area A_{vf} for computing V_r. When there is no bending moment acting on the shear plane, the shear-friction reinforcement is best distributed uniformly along the shear plane in order to minimize crack widths. When a bending moment also exists, most of the shear-friction reinforcement is placed closer to the tension face to provide the required effective depth. Since it is assumed that the shear-friction reinforcement yields at the ultimate strength, it must be anchored on both sides of the shear plane so as to develop the specified yield strength in tension. Although the ACI Code (Ref. 6.1) (and the former CSA 23.3) permits the shear design of brackets and corbels by the shear-friction method, CSA A23.3-M84 requires that these members be designed by the General Method in clause 11.4.

PROBLEMS

1. A simply supported beam of 6 m span, shown in Fig. 6.21, is to carry a uniform dead load of 25 kN/m (including beam weight) and a uniform live load of 40 kN/m. The width of the supporting wall is 250 mm.

Fig. 6.21 *Problem 6.1*

a) Determine the adequacy of the No. 10 U-stirrups as shear reinforcement.
b) If the shear reinforcement is to be provided in the form of No. 10 bar stirrups inclined at 60° to the beam axis, determine the required spacing.
c) If two of the tension reinforcement bars are terminated at 300 mm from the centre of the support, check the shear strength at the bar cut-off point. Use $f'_c = 30$ MPa and $f_y = 350$ MPa.

2. A simply supported T-beam of 9 m span is subjected to a dead load (including self weight) of 20 kN/m and a live load of 30 kN/m. Details of the section and bar cut-offs are shown in Fig. 6.22a and b. Design and detail the shear reinforcement using vertical stirrups. Use $f'_c = 20$ MPa and $f_y = 400$ MPa.

Fig. 6.22 *Problem 6.2*

(a)

(b) Section at midspan

3. A simply supported beam with overhangs at both ends shown in Fig. 6.23 is subjected to a dead load (including self weight) of 30 kN/m and a live load of 25 kN/m. Design and detail the shear reinforcement using vertical stirrups. Use $f'_c = 25$ MPa and $f_y = 350$ MPa. The beam is symmetric about the centreline.

4. Figure 6.24 shows a uniformly loaded cantilever beam. The dead load including self-weight of the beam is 20 kN/m and the live load is 50 kN/m. Design the shear reinforcement using vertical stirrups. The bar cut-off details are as shown. Assume $f'_c = 25$ MPa and $f_y = 400$ MPa.

Fig. 6.23 Problem 6.3

(a)

(b) Section at support

(c) Section at midspan

Fig. 6.24 Problem 6.4

(a)

(b) Section at support

REFERENCES

6.1 *Building Code Requirements for Reinforced Concrete (ACI 318–83)*, American Concrete Institute, Detroit, Michigan, 111 pp.

6.2 ACI-ASCE Committee 326 Report, *Shear and Diagonal Tension*, J. of ACI, Vol. 59, Jan., Feb., and Mar., 1962, pp. 1–30, 277–334 and 352–396.

6.3 ASCE-ACI Committee 426, *The Shear Strength of Reinforced Concrete Members*, ASCE Journal, Struc. Div., Vol. 99, No. ST 6, June 1973, pp. 1091–1187.

6.4 Popov, E.P., *Mechanics of Materials*, 2nd ed., Prentice Hall, Inc., Englewood Cliffs, New Jersey, 1978.

6.5 Bresler, B., and MacGregor, J.G., *Review of Concrete Beams Failing in Shear*, ASCE Journal, Struc. Div., Vol. 93, No. ST 1, Feb. 1967, pp. 343–372.

6.6 ACI-ASCE Committee 426, *Suggested Revisions to Shear Provisions for Building Codes*, J. ACI, Vol. 74, Sept. 1977, pp. 458–469.

6.7 MacGregor, J.G., and Gergely, P., *Suggested Revisions to ACI Building Code Clauses Dealing with Shear in Beams*, J. ACI, Vol. 74, Oct. 1977, pp. 493–500.

6.8 MacGregor, J.G., and Hanson, J.M., *Proposed Changes in Shear Provisions for Reinforced and Prestressed Concrete Beams*, J. of ACI, Vol. 66, April 1969, pp. 276–288.

6.9 Hanson, J.M., *Tensile Strength and Diagonal Tension Resistance of Structural Lightweight Concrete*, J. of ACI, Vol. 58, No. 1, July 1961, pp. 1–37.

6.10 *Commentary on Building Code Requirements for Reinforced Concrete (ACI 318–77)*, American Concrete Institute, Detroit, Michigan, 132 pp.

6.11 Mattock, A.H., *Shear Transfer in Concrete Having Reinforcement at an Angle to the Shear Plane*, Shear in Reinforced Concrete, SP-42, American Concrete Institute, Detroit, 1974, pp. 17–42.

6.12 Mattock, A.H., Li W.K., Wang, T.C., *Shear Transfer in Lightweight Reinforced Concrete*, J. PCI, Vol. 21, No. 1, Jan.-Feb. 1976, pp. 20–39.

CHAPTER 7

DESIGN FOR TORSION

7.1 GENERAL

In this chapter, the basic concepts of the behaviour of concrete members under torsion are briefly discussed. The basis and the methods for design for torsion by the Simplified Method of CSA A23.3-M84(11.3) are also dealt with here.

Torsional moments are developed in elements of reinforced concrete structures in two distinct forms as follows:

1. Equilibrium torsion In this case the torsional moment (torque) has to be developed to maintain equilibrium, no matter what the magnitude of the stiffness of the members. The torque can be determined from statics alone. In this case the member has to be designed for the full torsion, as no redistribution to other members is possible. Two examples of this type of torsion are shown in Fig. 7.1. In usual structures, this type of torsion is relatively rare.

Fig. 7.1 Equilibrium torsion

subjected to
equilibrium torsion

beam subjected
to equilibrium
torsion

2. Compatibility torsion Most of the torsional moments which arise in reinforced concrete structural members are of this type. Here the torsion is induced in the member because it is restrained and has to undergo a twist in order to maintain compatibility of deformations. The amount of torsional moment in the member depends on its torsional stiffness relative to the stiffness of the connecting members. An example of compatibility torsion is the spandrel beam ABC in Fig. 7.2. The flexure of the beam BD introduces a rotation, θ, at the end B. Since both ABC and BD are monolithically connected at B, this introduces a twist equal to θ in the beam ABC, and hence torsion. Here, changes in relative stiffnesses due to cracking and yielding can lead to redistribution and reduction of the torsional moment. At torsional cracking, reinforced concrete members undergo a fairly large twist. Furthermore, with only moderate amounts of torsional reinforcement, the post-cracking torsional stiffness is only a small fraction of the pre-cracking value. Both these factors lead to a redistribution of forces and moments, resulting usually in a reduction of the twisting moment in the torsional member. Although heavy torsional reinforcement can increase the torsional strength, such a high strength can be attained only with very large twist and accompanying deformations of the structure. Thus, in most practical situations, the maximum torsional moment in a reinforced concrete member under compatibility torsion is likely to be the value corresponding to torsional cracking of the member.

Fig. 7.2 *Compatibility torsion*

The preceding discussion indicates that torsion is not a serious problem in usual concrete situations. Equilibrium torsion occurs rarely. Compatibility torsion, when it occurs, is usually limited to the cracking torque. The latter can be predicted with reasonable accuracy, and has a relatively low value.

7.2 TORSION FORMULAS

The theory of torsion (St. Venant torsion) of prismatic members having circular, noncircular and thin-walled tubular cross-sections is given in detail in books on mechanics of materials (Refs. 7.1, 7.2). For rectangular sections under elastic behaviour, the distribution of shear stress over the cross-section is as shown in Fig. 7.3. The maximum shear stress and angle of twist are given by (Ref. 7.1):

$$v_{t,\max} = \frac{T}{\alpha x^2 y} \tag{7.1}$$

$$\theta_t = \frac{T}{\beta G \, x^3 y} \tag{7.2}$$

where
G = modulus of rigidity of the material
T = applied torsional moment
$v_{t,\max}$ = maximum shear stress, which occurs at the middle of the wider face
x = length of shorter side of the cross section
y = length of longer side of the section
α, β = constants, whose values depend on the ratio, y/x (α varies from 0.208 to $\frac{1}{3}$ and β from 0.141 to $\frac{1}{3}$ (Refs. 7.1, 7.2))
θ_t = angle of twist per unit length of the member

The torsional stiffness of the member, represented by GC, is the torque per unit twist, T/θ_t, and is obtained from Eq. 7.2 as:

$$GC = G \, \beta x^3 \, y \tag{7.3}$$

where C is a property of the section having the same relationship to torsional stiffness as does the polar moment of inertia for a circular section. An approximate expression for C has been obtained as (Ref. 7.2):

$$C = (1 - 0.63 \, x/y) \, x^3 \, y/3 \tag{7.4}$$

For sections composed of rectangular elements, such as T-, L- and ⊏ shapes, C may be computed by summing up the individual values for each of the component rectangles as given by Eq. 7.5. The value of C so obtained will always be less than the correct value (as shown by the membrane analogy, Ref. 7.1), so that the subdivision of the section into component rectangles may be made such as to yield the highest possible value for C. It should be noted that Eq. 7.5 is used in the computation of C values for torsional members in Chapter 14.

Fig. 7.3 *Torsional shear stresses in a beam of rectangular section*

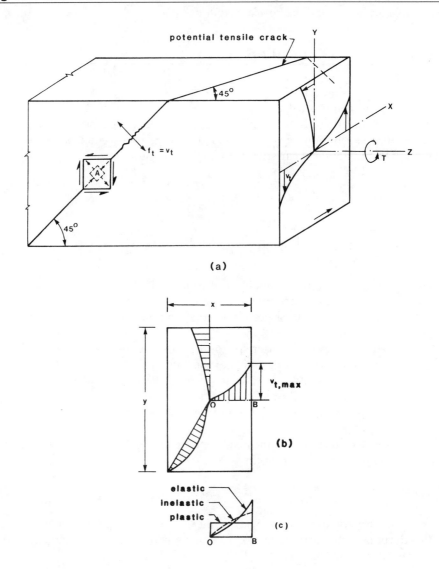

For a section composed of several rectangular elements, x by y,

$$C = \Sigma(1 - 0.63 \, x/y) \, x^3 \, y/3 \qquad (7.5)$$

For thin-walled tubular sections of any shape, as shown in Fig. 7.4, the torsion formulas are (Ref. 7.1):

$$q = v_t t = \frac{T}{2A_o} \qquad (7.6)$$

Fig. 7.4 *Torsion of thin-walled tubular section*

$$v_t = \frac{T}{2A_o t} \tag{7.7}$$

$$\theta_t = \frac{T}{G4A_o^2} \oint \frac{ds}{t} \tag{7.8}$$

$$GC = \frac{T}{\theta_t} = G\frac{4A_o^2}{\oint \dfrac{ds}{t}}$$

where A_o = area enclosed by the centreline of the wall
         q = the constant shear flow across the wall thickness
         t = thickness of the wall, at element ds along the perimeter
         v_t = torsional shear stress at the wall section having thickness t
and $\oint$ = indicates integration around the perimeter of the tube

For a tube of uniform wall thickness,

$$\oint \frac{ds}{t} = p_o/t$$

where p_o is the perimeter of the tube measured along the wall centreline (Fig. 7.4). With this substitution, Eq. 7.8 for torsional stiffness reduces to:

$$GC = G4 A_o^2 t/p_o \tag{7.9}$$

7.3 STRENGTH AND BEHAVIOUR OF CONCRETE MEMBERS IN TORSION

7.3.1 GENERAL BEHAVIOUR

For simplicity, the behaviour of concrete members subjected to pure torsion is discussed with reference to a member of rectangular cross section. A plain concrete member under pure torsion develops shear stresses in the cross section with a distribution in the early stages of loading close to that given by the elastic theory, and shown in Fig. 7.3. The maximum shear stress occurs at the middle of the wider face. The state of pure shear develops direct tensile and compressive stresses along the diagonal directions as shown in the element at *A* in Fig. 7.3a. The principal tensile and compressive stress trajectories spiral around the beam in opposite directions at 45° to the beam axis. One such line, on which at all points the principal tensile stress is at right angles (and compressive stress tangential), is shown in Fig. 7.3a. This is a potential line of crack. In a concrete beam, such a crack would develop when the diagonal tensile stress reaches the tensile strength of concrete. The bulk of the torsional strength of the section is contributed by the material in the outer portions of the cross section. (The stresses in the outer fibres are the largest and have the greatest lever arm.) Hence, in a plain concrete member, the diagonal torsional cracking in the outer fibres would immediately lead to a sudden failure of the entire section.

The distribution of shear stress in the cross section may not be elastic, particularly as the failure strength is reached. An inelastic stress distribution will result in somewhat larger stresses in the interior fibres, and a fully plastic behaviour will give uniform shear stress distribution over the entire cross-section (see Fig. 7.3c).

A typical torque-twist relation for a plain concrete section is shown by curve *a* in Fig. 7.5 (Ref. 7.3). Torque and twist increase nearly linearly up to failure, which is sudden and brittle, and occurs immediately after the formation of the first torsional crack. The failure strength of a plain concrete member, T_{cr}, is generally computed by equating a theoretical nominal maximum shear stress, $v_{t,max}$ (which is a measure of the diagonal tension), to the tensile strength of concrete. Expressions for T_{cr} have been derived based on (a) elastic theory, (b) plastic theory, (c) skew bending theory, and (d) the equivalent tube analogy (Refs. 7.4, 7.5). Each of these theories gives slighly different expressions for T_{cr}, and hence has to be correlated, experimentally, with an appropriate measure of the tensile strength of concrete to be used with it. CSA A23.3-M84 requirements are based on the equivalent thin-walled tube approach (Ref. 7.5). The derivation of an expression for T_{cr} on this basis is presented later.

The failure of a plain concrete member in torsion is caused by the torsional cracking due to the diagonal tensile stresses. Hence the ideal form of reinforcing for torsion is in the form of a spiral along the direction of principal tensile stresses. However, this is often impractical, and the usual form of torsional reinforcement consists of a combination of (a) longitudinal reinforcement distributed around the cross section, and (b) closed stirrups placed perpendicular to the member axis. A combination of both types is needed to resist the inclined tensile stresses. Further-

Fig. 7.5 *Typical torque-twist curves for concrete members in pure torsion*

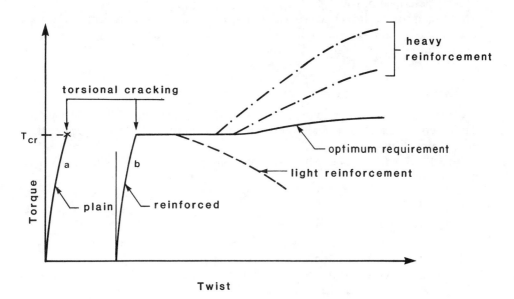

more, these tensile stresses develop on all faces of the beam and hence the stirrups must be closed. Alternatively, a closed cage of welded wire fabric with the transverse wires located perpendicular to the member axis can be used.

Members with torsional reinforcement behave similarly to the plain section until the formation of the first torsional crack, at a torque, identified as the cracking torque, T_{cr}, as shown in Fig. 7.5 (Ref. 7.6). The value of T_{cr} is insensitive to the amount of torsional reinforcement, and is very nearly the same as the failure strength of an identical plain section. When cracking occurs, there is a large increase in twist under nearly constant torque (Fig. 7.5). Beyond this, however, the strength and behaviour depend on the amount of torsional reinforcement.

For very small amounts of torsional reinforcement, the failure occurs soon after the first crack in a brittle manner. Increasing the torsional reinforcement will increase the ultimate torsional strength, but this is accompanied by larger twist. The torsional stiffness after cracking is primarily dependent on the amount of torsional reinforcement, and is usually only a small fraction (0 to 10 percent) of its value before cracking.

7.3.2 STRENGTH IN TORSION

(a) Cracking Torque Using Equivalent Tube Concept
As already indicated, the failure strength of a plain concrete member and the strength at torsional cracking, T_{cr}, of a reinforced member are essentially the same. The latter is also essentially independent of the amount of reinforcement. Although several methods have been used to compute T_{cr}, only the equivalent thin-walled tube ap-

proach (Ref. 7.5, 7.10) which forms the basis of the CSA A23.3-M84 requirements is briefly described below.

In this approach, the actual cross section is replaced by an equivalent thin-walled tube having the same external dimensions as the actual section (Fig. 7.6). The uniform wall thickness, t_c, of the equivalent tube is taken as:

$$t_c = 0.75 \, A_c/p_c \tag{7.10}$$

where p_c = the external perimeter of the actual section, and A_c = the area enclosed within perimeter p_c. For actual hollow sections, t_c must not be taken greater than the actual wall thickness.

Using Eq. 7.6 for thin-walled tubular sections, the tourque, T, in the equivalent tube is:

$$T = 2 \, A_{oc} \, t_c \, v_t \tag{7.11}$$

where T = the torque, and A_{oc} = area enclosed within the centreline of the wall thickness, and v_t is the torsional shear stress.

Assuming that diagonal cracking occurs when the torsional shear stress, v_t, reaches $0.4 \sqrt{f_c'}$, at cracking, Eq. 7.11 reduces to:

$$T_{cr} = 2 \, A_{oc} \, t_c \times 0.4 \sqrt{f_c'} \tag{7.12a}$$

For design purposes, for rectangular sections of usual proportions the approximation $A_{oc} \simeq 2/3 \, A_c$ may be used. With this, and substituting Eq. 7.10 for t_c,

Fig. 7.6 *Equivalent tube, before cracking*

(a) Actual section   **(b) Equivalent tube**

$$T_{cr} = A_c^2/p_c \times 0.4 \sqrt{f_c'} \qquad (7.12b)$$

Now introducing the material resistance factor ϕ_c and the λ-factor to allow for low density concrete, the factored torsional cracking resistance in pure torsion is:

$$T_{cr} = (A_c^2/p_c)\, 0.4\lambda\, \phi_c \sqrt{f_c'} \qquad (7.12c)$$

prior to cracking

CSA A23.3-M84 (11.2.4) recommends Eq. 7.12c computing T_{cr}.

The Code clause 11.2.4 allows torsion to be neglected if the design torque does not exceed $0.25\ T_{cr} = (A_c^2/p_c)\, 0.1\lambda\, \phi_c\, \sqrt{f_c'}$. This corresponds to a nominal shear stress of $0.1\lambda\, \phi_c\, \sqrt{f_c'}$. Torsion of such magnitude does not cause any significant reduction in the strength in either flexure or shear (Ref. 7.7).

The torsional stiffness before cracking, GC_{gross}, may be computed by the elastic theory, Eq. 7.4. For this, the modulus of rigidity, G, for concrete may be taken as (assuming the Poisson ratio, $\mu = 0$):

$$G = 0.5E_c$$

where E_c is the elastic modulus.

Alternatively, the uncracked torsional stiffness may be computed using the tubular section formula (Eq. 7.9) applied to the equivalent tube in Fig. 7.6. On this basis,

$$GC_{\text{gross}} = G\, \frac{4\, A_{oc}^2\, t_c}{p_{oc}} \qquad (7.13)$$

where p_{oc} is the perimeter of the tube measured along the wall centreline.

(b) Torsional Strength of Cracked, Reinforced Concrete Members

Several theories have been proposed for the computation of the torsional strength of reinforced concrete members, notable amongst them being (1) the ultimate equilibrium method, (2) the skew bending theory, and (3) the space truss analogy (Refs. 7.8, 7.9, 7.10).

CSA A23.3-M84(11.3) requirements for design of reinforced concrete members for torsion are based on the space truss analogy and its further developments (Refs. 7.5, 7.10, 7.11). In this approach, after torsional cracking, the longitudinal and transverse torsional reinforcement and surrounding layer of concrete in the shape of a thin-walled tube (Fig. 7.7) may be considered as a space truss. The truss is made up of the corner longitudinal bars as stringers, the closed stirrup legs as transverse ties, and the concrete between diagonal cracks as compression diagonals. In the development of the general equations in this section, the tube wall is assumed to have a thickness t_o as shown in Fig. 7.7. However, it should be noted that at higher loads the concrete cover will spall off, and hence the CSA A23.3-M84 equations are based on the assumption that the effective outer dimensions of the tube coincide with the centreline of the exterior closed transverse torsion reinforcement (see Fig. 7.7d).

Fig. 7.7 *The space truss model*

(a)

(b) DETAIL 'X' (ENLARGED)

Fig. 7.7 *concluded*

(c)

(d) effective outer dimensions used
in CSA A23.3 – M84

Considering a transverse section of the tube, the inclined diagonal compressive force C_d, in the concrete has a tangential component providing the shear stress, v_t, and the shear flow, $q = v_t t_o$, necessary to resist the applied torque, T. The normal component of the diagonal compression is balanced by the tensile forces in the longitudinal bars. Considering a transverse section, such as the face AB of the element shown in Fig. 7.7b, the normal component of the diagonal compression is balanced by the stirrup tension and the tangential component provides the complementary shear v_t. In the thin-walled tube, the area enclosed by the shear flow path is A_o and the perimeter of area A_o is p_o as shown in Fig. 7.7c. (Compare with Fig. 7.4.)

Assuming that at failure both the longitudinal and stirrup reinforcement yield, the equilibrium of the axial and tangential forces on the cross section in Fig. 7.7a. and of the forces normal to section AB in Fig. 7.7b gives:

$$A_\ell f_y = \frac{q}{\tan \alpha} p_o \tag{7.14}$$

$$T = 2 A_o q \tag{7.15}$$

$$A_t f_y = s q \tan \alpha \tag{7.16}$$

and the shear stress, v_t, is

$$v_t = q/t_o = T/(2 A_o t_o)$$

where A_ℓ = total area of symmetrically placed longitudinal reinforcement

A_t = area of one leg of the closed stirrup
A_o = area enclosed by the shear flow path
p_o = the perimeter of area A_o
q = shear flow across thickness of the tube
s = spacing of stirrups
T = torsional resistance
v_t = torsional shear stress

Eliminating q (and α) in Eqs. 7.14 and 7.16, the relation between A_t and A_ℓ and the torsional resistance can be obtained as:

$$\frac{A_t}{s \tan \alpha} = \frac{A_\ell \tan \alpha}{p_o} \tag{7.17}$$

$$T = \frac{2 A_o A_t f_y}{s \tan \alpha} \tag{7.18a}$$

$$= 2 A_o f_y \sqrt{\frac{A_t A_\ell}{s p_o}} \tag{7.18b}$$

Note that the same expression for T as Eq. 7.18b can be obtained by using the thin-walled tube concept (Eq. 7.6a), with an equivalent steel tube having wall thickness, t_s, given by:

$$t_s = \sqrt{\frac{A_t A_\ell}{s p_o}} \tag{7.19}$$

It is generally assumed, and this indeed is the case in pure torsion, that $\alpha = 45°$. In that case, Eq. 7.17 simplifies to Eq. 7.20 giving the condition that the volume of longitudinal torsional reinforcement be equal to the volume of closed transverse stirrups, and the torsional resistance is given by Eq. 7.21:

$$\frac{A_t}{s} = \frac{A_\ell}{p_o} \qquad (7.20)$$

$$T = \frac{2 A_o A_t f_y}{s} \qquad (7.21)$$

Introducing the material resistance factor, ϕ_s, the factored torsional resistance, T_r, is obtained as:

$$T_r = \frac{2 A_o A_t \phi_s f_y}{s} \qquad (7.22)$$

after crack.

and the corresponding total area, A_ℓ, of symmetrically distributed longitudinal reinforcement required (from Eq. 7.20) is

$$A_\ell \geqq A_t\, p_o/s \qquad (7.23a)$$

In adopting Eqs. 7.21 and 7.23a, CSA A23.3-M77 had taken A_o as the area enclosed by a line connecting the centres of longitudinal bars at the corners of the stirrups (i.e., $A_o = b_o \times h_o$ in Fig. 7.7a). The CEB Code (Ref. 7.18) also uses a similar definition for the area A_o.

CSA A23.3-M84(11.3) specifies Eqs. 7.22 and 7.23b below for the design for torsion.

$$A_\ell = A_t\, p_h/s \qquad (7.23b)$$

where p_h = perimeter of the centreline of closed transverse torsion reinforcement, and the area A_o in Eq. 7.22 may be taken as $0.85\,A_{oh}$, A_{oh} being the area enclosed by perimeter p_h.

If a member is not to fail suddenly in a brittle manner after the development of torsional cracks (see Fig. 7.5), the torsional strength of the cracked section must at least equal the cracking torque T_{cr} given by Eq. 7.12c. For this condition to hold,

$$T_r = \frac{2 A_o A_t \phi_s f_y}{s} \geqslant \frac{A_c^2\, 0.4\lambda\, \phi_c \sqrt{f_c'}}{p_c}$$

$$\frac{A_t}{s} = \frac{A_\ell}{p_o} \geqslant \frac{A_c^2}{A_o p_c} \frac{0.2\lambda\, \phi_c \sqrt{f_c'}}{\phi_s f_y}$$

min reinf so T_r will not be reached before T_{cr}

The torsional stiffness of a cracked reinforced concrete member, GC_{cr}, is only a small fraction (usually less than 10 percent) of its value before cracking. Expressions for the postcracking stiffness have been derived on the basis of the space truss model (Ref. 7.12). However, a much simpler relation is obtained by using the equivalent thin-walled tube concept (Ref. 7.5). On this basis, assuming a steel tube of wall thickness, t_s, given by Eq. 7.19, and substituting in Eq. 7.9,

$$GC_{cr} = G_s\, 4\, A_o^2\, t_s\, p_o$$

$$\simeq \frac{E_s}{2} \frac{4\, A_o^2\, t_s}{p_o} \tag{7.24}$$

7.4 COMBINED LOADINGS

In practice, torsion exists in reinforced concrete members mostly in combination with bending moments and transverse shear. Several forms of moment-torque interaction curves and moment-shear-torque interaction surfaces have been developed, and a summary of the early theories is presented in Ref. 7.13. For the most part, these theories are quite complex and awkward for design applications.

For design, CSA A23.3-M84(11.3) adopts the simple and conservative procedure of merely adding the reinforcement required for torsion to that required for flexural, axial and shear loads that act simultaneously. This procedure is simple and has been found to give satisfactory designs (Refs. 7.10, 7.14). However, the general method in clause 11.4 takes an interactive approach for the design of longitudinal reinforcement required for shear and torsion coexisting with flexure and axial force.

7.5 CSA CODE APPROACH TO ANALYSIS AND DESIGN FOR TORSION

7.5.1 GENERAL

The Simplified Method in CSA A23.3-M84(11.3) is based on the 45° space truss model. When the factored torsion, T_f, exceeds 25 percent of the pure torsional cracking resistance, T_{cr}, given by Eq. 7.12c, the torsional reinforcement, consisting of closed stirrups perpendicular to the member axis and longitudinal bars distributed symmetrically around the section, is designed to resist the full torsion, T_f, using Eq. 7.22 and 7.23b. This reinforcement is provided in addition to the reinforcement required for shear, flexure, and axial loads that act in combination with the torsion. However, in the flexural compression zone, the area of longitudinal torsion steel required may be reduced by an amount equal to $M_f/(0.9\, d\, f_y)$, where M_f is the factored moment at the section acting simultaneously with the torsion. This is because the flexural compression counteracts the required tension due to torsion. The Code also requires that the cross-sectional dimensions of the member be such that,

$$\frac{V_f}{b_w d} + \frac{T_f\, p_h}{A_{oh}^2} \leq 0.25\, \lambda\, \phi_c\, f_c' \tag{7.25}$$

This limitation is to ensure that the concrete in the web does not fail by crushing due to the diagonal compression prior to the development of full torsional resistance, T_{cr}, with yielding of reinforcement. The maximum diagonal compression is dependent on the maximum nominal web shear stress due to shear and torsion combined, represented by the left hand side of Eq. 7.25. In order to control diagonal crack width at specified loads, the yield strength of torsion reinforcement used in design calculations is limited to a maximum of 400 MPa.

The Code requires a minimum area of torsion reinforcement whenever $T_f >$ 0.25 T_{cr}, except in slabs and footings, concrete joist floors, and certain classes of relatively wide and shallow beams (cl. 11.2.5.1). The minimum transverse reinforcement, as in the case of shear (Eq. 6.26), is given by:

$$A_v = 0.35\, b_w s/f_y \qquad \qquad (7.26)$$

— code minimum

For hollow sections, the distance measured from the centreline of the transverse torsion reinforcement to the inside face of the wall should be greater than or equal to 0.5 A_{oh}/p_h. Sections located less than a distance d from the face of supports may be designed for the same torsion, T_f, as that at a distance d, provided that the support reaction introduces compression into the end region of the member and that no concentrated load occurs within the distance d from the face of support.

7.5.2 COMPATIBILITY TORSION

This type of torsion arises out of a requirement to twist, and hence its magnitude depends on the torsional stiffness of the member. The design considerations are therefore directly related to the torque-twist relationship of the reinforced concrete member such as shown in Fig. 7.5. (Although members may be subjected to combined loadings, the torsional stiffness characteristics are not seriously altered.) The major factors related to this curve which affect the analysis and design procedures are:

1. The uncracked stiffness, GC_{gross}, is effective prior to torsional cracking and the maximum torque that can be reached with this stiffness is T_{cr}.
2. At cracking, a large twist occurs at nearly constant torque, resulting in redistribution of force in the structure (Refs. 7.14, 7.15, 7.16). Furthermore, torsional cracking drastically reduces the torsional stiffness in relation to the flexural stiffness. For practical ranges of twist, a previously cracked member, on subsequent loading, will develop torque far less than that at the initial loading.
3. Although increasing the torsional reinforcement will increase the torsional capcity of the section, often it may not be possible to utilize the increased strength as it can be attained only with very large twist and associated cracking.
4. For the torsional member to behave in a ductile manner after cracking, torsional reinforcement must be provided sufficient to develop a strength equal to T_{cr}.

Recognizing these factors in situations where redistribution of internal forces can occur, the Code clause 11.2.4.2 permits the reduction of the maximum factored torsion, T_f, to 0.67 T_{cr}, provided corresponding adjustments are made to the bending moments, shear forces and torques in the member and in the members framing into it.

In many practical situations, the compatibility torsion on members in concrete frames may be extremely low, and there is no possibility of torsional cracking. In such cases, the minimum torsional reinforcement requirement may be omitted. One example of such a case is the negligible torsion introduced in an interior beam such as beam *EF* in Fig. 7.2 due to the bending of the column at its end. The torque below which no torsional reinforcement need be provided is $0.25\ T_{cr}$. However, in this range the effective torsional stiffness is GC_{gross}, and must be used in the analysis for torsion.

7.5.3 DETAILING OF REINFORCEMENT

In members having significant torque, the torsional reinforcement should not be less than that given in Eq. 7.26. For satisfying this requirement, the closed stirrups provided to resist transverse shear may be included.

The Code requirements for detailing torsion reinforcement are given in CSA A23.3-M84(11.3.8) and shown in Fig. 7.8. These requirements are meant to avoid premature brittle failures and to control crack widths. The functions of the reinforcement are evident from the equilibrium conditions of the internal forces in the space truss shown in Fig. 7.7; their relation to detailing requirements are explained in Ref. 7.17.

Fig. 7.8 *Details of torsion reinforcement*

Because stirrup legs are needed as ties on all sides of the member for torsion, they must be provided in the closed form. To ensure proper anchorage even after the spalling of the concrete cover, the free ends of the closed stirrups must be bent into the concrete within the stirrups with 135° hooks. In regions where the concrete surrounding the anchorage is restrained against spalling, standard hooks as in Fig. 6.12c may be used. The stirrup spacing must not exceed $p_h/8$ and 300 mm. The longitudinal bars are distributed around the perimeter of the stirrups, at a spacing not exceeding 300 mm. The spacing limitations are intended to control the crack widths. Inside each corner of the closed stirrup at least one longitudinal bar, having a diameter not less than $s/16$, must be provided and anchored to provide full development in tension. This is necessary to distribute the resultant stirrup force at the corners where the stirrup changes its direction. These stirrup spacing and corner bar size requirements have been found to be adequate to prevent the outward buckling of the corner bar between the stirrups, due to the outward thrust exerted by the compression diagonal of the truss.

EXAMPLE 7.1

Reinforced concrete member, subjected to pure torsion, has the cross-sectional dimensions shown in Fig. 7.9. $f'_c = 30$ MPa and $f_y = 400$ MPa. Compute (1) the torsional cracking resistance, (2) the torsional stiffness prior to cracking, (3) the factored torsional resistance of the section, and (4) the torsional stiffness after cracking.

Fig. 7.9 *Example 7.1*

SOLUTION

1. Torsional cracking resistance
    Area, $A_c = 300 \times 500 = 150 \times 10^3$ mm^2
    Perimeter $= p_c = 2 (300 + 500) = 1600$ mm
    Using Eq. 7.12c:

    $$T_{cr} = \frac{A_c^2}{p_c} \, 0.4 \, \lambda \, \phi_c \, \sqrt{f'_c}$$

    $$= \frac{(150 \times 10^3)^2}{1600} \times 0.4 \times 1.0 \times 0.6 \times \sqrt{30} \times 10^{-6} = 18.5 \text{ kN·m}$$

2. Precracking stiffness
    If the elastic theory is applied to the gross concrete section, the property C (Eq. 7.4) is:

$$C = \left(1 - 0.63 \times \frac{300}{500}\right) \frac{300^3 \times 500}{3} = 2.80 \times 10^9 \text{ mm}^4$$

$$G = E_c/2 = 5000 \sqrt{30}/2 = 13.7 \times 10^3 \text{ MPa}$$

$$GC_{\text{gross}} = 13.7 \times 10^3 \times 2.8 \times 10^9 = 38.3 \times 10^{12} \text{ N·mm}^2$$

Alternatively, using the equivalent tube formula, Eq. 7.13, thickness of equivalent tube, before cracking, is,

$$t_c = \frac{0.75 \, A_c}{p_c} = \frac{0.75 \times 150 \times 10^3}{1600} = 70.3 \text{ mm}$$

$$b_{oc} = 300 - 70.3 = 229.7 \text{ mm}, \, h_{oc} = 500 - 70.3 = 429.7 \text{ mm}$$

$$A_{oc} = 229.7 \times 429.7 = 98.7 \times 10^3 \text{ mm}^2$$

$$p_{oc} = 2(229.7 + 429.7) = 1319 \text{ mm}$$

$$GC_{\text{gross}} = G \frac{4 \, A_{oc}^2 t_c}{p_{oc}}$$

$$= 13.7 \times 10^3 \times \frac{4 \, (98.7 \times 10^3)^2 \times 70.3}{1319} = 28.5 \times 10^{12} \text{ N·mm}^2$$

3. Factored torsional resistance, T_r

Width between stirrup centrelines, $b_h = 180 + 16 + 11.3 = 207.3$ mm
Depth between stirrup centrelines, $h_h = 380 + 16 + 11.3 = 407.3$ mm
Area enclosed by stirrup centrelines, $A_{oh} = 207.3 \times 407.3 = 84.4 \times 10^3$ mm^2
Perimeter of $A_{oh} = p_h = 2(207.3 + 407.3) = 1229$ mm
Area enclosed by shear flow path, $A_o \simeq 0.85 \, A_{oh} = 71.7 \times 10^3$ mm^2
Area of transverse reinforcement, $A_t = 100$ mm^2

Using Eq. 7.22,

$$T_r = \frac{2 \, A_o A_t \phi_s f_y}{s} = \frac{2 \times 71.7 \times 10^3 \times 100 \times 0.85 \times 400}{150} \times 10^{-6} = 32.5 \text{ kN·m}$$

For this to be valid, check:
(i) longitudinal steel area required,

$$A_\ell = A_t p_h/s = 100 \times 1229/150 = 819 \text{ mm}^2$$

Area provided $= 4 \times 200 + 2 \times 100 = 1000$ mm^2 OK

(ii) minimum area of transverse reinforcement required (Eq. 7.26),

$$A_{v,\text{min}} = \frac{0.35 \times 300 \times 150}{400} = 39.4 \text{ mm}^2$$

A_v provided $= 2 \times 100 = 200$ mm^2 OK

(iii) adequacy of section dimensions (Eq. 7.25),

$$0.25\lambda \, \phi_c f_c' = 0.25 \times 1.0 \times 0.6 \times 30 = 4.5 \text{ MPa}$$

$$\frac{T_r p_h}{A_{oh}^2} = \frac{32.5 \times 10^6 \times 1229}{(84.4 \times 10^3)^2} = 5.61 \text{ MPa} > 4.5 \text{ MPa} \quad \textbf{not OK}$$

The condition in Eq. 7.25 is not satisfied, and the nominal shear stress is excessive. Hence the nominal shear stress (diagonal compression failure) in the concrete controls. In order to satisfy Eq. 7.25,

$$T_r \leqslant 0.25 \, \lambda \, \phi_c f'_c \left(\frac{A_{oh}^2}{p_h} \right)$$

$$\leqslant 4.5 \times \frac{(84.4 \times 10^3)^2}{1229} \times 10^{-6} = 26.1 \text{ kN} \cdot \text{m}$$

Factored torsional resistance, $T_r = 26.1$ kN·m.

4. Stiffness after cracking
    Wall thickness, t_s, of equivalent steel tube, using Eq. 7.19, is:

$$t_s = \sqrt{\frac{A_t}{s} \frac{A_\ell}{p_o}} = 0.772 \text{ mm}$$

$$G_s \simeq E_s/2 = 100 \text{ GPa}$$

Using Eq. 7.24:

$$GC_{cr} = G_s \frac{4 A_o^2 t_s}{p_o}$$

$$= 100 \times 10^3 \times \frac{4 (68.4 \times 10^3)^2 \times 0.772}{1120}$$

$$= 1.29 \times 10^{12} \text{ N} \cdot \text{mm}^2$$

Note: The member in this example has torsional reinforcement only slightly in excess of the minimum required to resist the cracking torque. It is seen that the post-cracking stiffness is negligible, compared to the precracking value. This member, if used to resist compatibility torsion, will develop very little torque after it has cracked. To compare, the unit twist, θ_t, when $T = T_{cr}$ during the first time loading (uncracked stiffness) is:

$$\theta_{t,\text{gross}} = \frac{T_{cr}}{GC_{\text{gross}}} = \frac{18.5 \times 10^6}{28.5 \times 10^{12}}$$

$$= 0.649 \times 10^{-6} \text{ rad/mm}$$

This kind of twist is in the probable range. If the member is cracked, $GC = GC_{cr}$, and in order to develop a torque equal to T_{cr}, the unit twist has to be,

$$\theta_t = \frac{T_{cr}}{GC_{cr}} = \frac{18.5 \times 10^6}{1.29 \times 10^{12}}$$

$$= 14.3 \times 10^{-6} \text{ rad/mm}$$

This is 22 times the precracking value and is unusually large. Hence the procedure of analyzing the structure assuming $GC = 0$ is a satisfactory one and is realistic, once the torsional member is cracked. However, the member may be subjected to $T = T_{cr}$ before cracking has occurred, and hence the minimum reinforcement to resist $T = T_{cr}$ is essential to avoid a sudden brittle failure at the formation of the first crack. Though the failure strength for this member is 26.1 kN·m, it can seldom be realized in compatibility torsion.

EXAMPLE 7.2

Figure 7.10 shows a continuous beam supporting a monolithically cast cantilever slab. The specified live load on the slab is 2.4 kN/m². In addition, the beam supports a dead load of 15 kN/m, applied directly over the beam width; $f'_c = 30$ MPa and $f_y = 400$ MPa. Design the reinforcement for the beam at a critical section near the first interior support.

Fig. 7.10 *Example 7.2*

SOLUTION

1. Factored load effects (Fig. 7.10b)
   The loads on the overhanging portion of the slab per metre length of beam are:

$$DL = 1.25 \times 0.125 \times 1.50 \times 24 = 5.63 \text{ kN/m}$$
$$LL = \qquad\quad 1.50 \times 1.5 \times 2.4 = 5.40 \text{ kN/m}$$
$$\text{Total} = 11.0 \text{ kN/m}$$

Assuming a beam size of 350×500 mm, the torque due to the loads on slab is:

$$T = 11.0 \left(\frac{1.5}{2} + 0.175 \right) = 10.2 \text{ kN·m/m length}$$

Dead load directly on the beam, including the beam weight, is:

$$1.25 \, (15 + (0.35 + 0.5) \times 24) = 24.0 \text{ kN/m}$$

Total load on beam $= 11.0 + 24.0 = 35.0$ kN/m
Maximum moment in the beam at the face of supporting column, using the CSA moment coefficients, is:

$$M_f = \frac{w_f \ell_n^2}{10} = \frac{35.0 \times 7^2}{10} = 172 \text{ kN·m}$$

$$\text{Maximum shear in the beam} = 1.15 \, \frac{w_f \ell_n}{2}$$

$$V_{\text{max}} = 1.15 \times \frac{35.0 \times 7}{2} = 141 \text{ kN}$$

Maximum torque in beam, at face of support:

$$T_{\text{max}} = 10.2 \times \frac{7}{2} = 35.7 \text{ kN·m}$$

The shear and torque vary linearly. The design shear and torque are taken at a section a distance d from face of the support. Assuming 40 mm cover, No. 10 stirrups and No. 25 bars, $d = 500 - 64 = 436$ mm. At the critical design section:

$$\text{Factored shear, } V_f = 141 - 35.0 \times 0.436 = 126 \text{ kN}$$
$$\text{Factored torque, } T_f = 37.5 - 10.2 \times 0.436 = 31.3 \text{ kN·m}$$

2. Check shear resistance and torsion resistance requirements and adequacy of overall section size
   Factored shear resistance provided by concrete, V_c, is (Eq. 6.22),

$$V_c = 0.2\lambda \, \phi_c \sqrt{f_c'} \, b_w d$$
$$= 0.2 \times 1.0 \times 0.6 \times \sqrt{30} \times 350 \times 436 \times 10^{-3} = 100 \text{ kN}$$

$$V_f = 126 \text{ kN} > 0.5 \, V_c = 50 \text{ kN, shear reinforcement is required}$$

$$V_s = V_f - V_c = 126 - 100 = 26 \text{ kN}$$

$$0.4\lambda \, \phi_c \sqrt{f_c'} \, b_w d = 0.4 \times 1.0 \times 0.6 \sqrt{30} \times 350 \times 436 \times 10^{-3} = 201 \text{ kN}$$

Since $V_s < 0.4\lambda\,\phi_c\,\sqrt{f_c'}\,b_w d$, normal spacing requirements control and also the section is adequate for shear. Pure torsional cracking resistance, T_{cr}, is obtained as (Eq. 7.12c),

$$T_{cr} = \frac{A_c^2}{p_c} 0.4\lambda\,\phi_c\,\sqrt{f_c'}$$

$$A_c = 350 \times 500 = 175 \times 10^3\ \text{mm}^2, p_c = 2(350 + 500) = 1700\ \text{mm}$$

$$T_{cr} = \frac{(175 \times 10^3)^2}{1700} \times 0.4 \times 0.6\,\sqrt{30} \times 10^{-6} = 23.7\ \text{kN·m}$$

$T_f > 0.25\,T_{cr}$, and hence torsion reinforcement is required.

The total nominal shear stress, considering V_f and T_f, must be such that Eq. 7.25 is satisfied, otherwise the section dimensions must be increased. Assuming 40 mm clear cover all around and No. 10 stirrups, the centreline dimensions of the stirrup cage (Fig. 7.10c) are 258.7×408.7 mm.

$$A_{ob} = 408.7 \times 258.7 = 106 \times 10^3\ \text{mm}^2, p_h = 2(408.7 + 258.7) = 1335\ \text{mm}$$

$$\frac{V_f}{b_w d} + \frac{T_f p_h}{A_{ob}^2} = \frac{126\,000}{350 \times 436} + \frac{31.3 \times 10^6 \times 1335}{(106 \times 10^3)^2} = 4.54\ \text{MPa}$$

$$0.25\lambda\,\phi_c\,f_c' = 0.25 \times 1.0 \times 0.6 \times 30 = 4.5\ \text{MPa}$$

Eq. 7.25 is satisfied for all practical purposes and the cross-sectional dimensions are adequate.

Design of flexural reinforcement

$$K_r = \frac{M_f}{bd^2} = \frac{172 \times 10^6}{350 \times 436^2} = 2.59\ \text{MPa}$$

For this K_r, from Table 5.4:

$$\rho = 0.0084 << \rho_{max} = 0.0230$$
$$A_{sf} = 0.0084 \times 350 \times 436 = 1282\ \text{mm}^2.$$

Reinforcement for flexural shear

$$V_s = V_f - V_c = 126 - 100 = 26\ \text{kN}$$

Using Eq. 6.23, area of vertical stirrup, A_v, is given by:

$$\frac{A_v}{s} = \frac{V_s}{\phi_s f_y d} = \frac{26\,000}{0.85 \times 400 \times 436} = 0.175\ \text{mm}^2/\text{mm}$$

Minimum transverse shear reinforcement is given by Eq. 6.26 as:

$$\left(\frac{A_v}{s}\right)_{min} = \frac{0.35\,b_w}{f_y} = \frac{0.35 \times 350}{400} = 0.306 \text{ mm}^2/\text{mm}$$

For satisfying the minimum A_v requirement, Code (cl. 11.2.5.6) permits inclusion of transverse torsion reinforcement. Hence, the checking of $A_{v,min}$ will be done after designing torsion reinforcement.

Choosing two-legged closed stirrups, area of stirrup required per leg, A_{v1}, is given by:

$$\frac{A_{v1}}{s} = \frac{A_v/s}{2} = \frac{0.175}{2} = 0.0875 \text{ mm}^2/\text{mm}$$

5. Torsional reinforcement
    Area of stirrup required for torsion, using Eq. 7.22 is:

$$\frac{A_t}{s} = \frac{T_f}{2\,A_o\phi_s f_y}$$

$$A_o \simeq 0.85\,A_{oh} = 0.85 \times 106 \times 10^3 = 90.1 \times 10^3 \text{ mm}^2$$

$$\frac{A_t}{s} = \frac{31.3 \times 10^6}{2 \times 90.1 \times 10^3 \times 0.85 \times 400} = 0.511 \text{ mm}^2/\text{mm}$$

and total longitudinal torsional steel is (Eq. 7.23b):

$$A_\ell = \frac{A_t}{s}p_h = 0.511 \times 1335 = 682 \text{ mm}^2$$

6. Selection of stirrup and spacing
    Total area required for each leg, for resisting shear and torsion, is given by:

$$\left(\frac{A_t}{s}\right)_{total} = \frac{A_{v1}}{s} + \frac{A_t}{s} = 0.0875 + 0.511$$

$$= 0.599 \text{ mm}^2/\text{mm} \qquad 0.306/2$$

This is more than the minimum requirement of 0.153 mm²/mm and hence controls. Also note that the A_ℓ/p_h required by design, $= 0.511$, is also greater than the $(A_v/s)_{min} = 0.153$ and hence adequate.

Using No. 10 stirrups, $A_v = 100$ mm², and the required spacing is:

$$s = \frac{A_v}{0.599} = \frac{100}{0.599} = 167 \text{ mm}$$

Spacing must not exceed:

For shear, $d/2 = 436/2 = 218$ mm
For torsion, $p_h/8 = 1335/8 = 167$ mm and 300 mm

Select a spacing of 160 mm. Since this is only slightly smaller than the maximum of 167 mm, the same spacing (160 mm) may be provided throughout the length of the beam even though the theoretical requirement is less.

7. Selection of longitudinal reinforcement

Spacing of longitudinal reinforcement is limited to a maximum of 300 mm. Since $b_b = 408.7$ mm, longitudinal reinforcement will be required at mid-depth. Thus A_ℓ will be distributed in three layers. Providing two No. 10 bars at mid-depth, one on each side (see Fig. 7.10c), the balance of A_ℓ to be provided between the top and bottom layers is $682 - 200 = 482$ mm², at $482/2 = 241$ mm² each.

Combined longitudinal reinforcement on the flexural tension side (top face) is:

$$A_s = A_{sf} + A_{\ell,\text{top}}$$
$$= 1282 + 241 = 1523 \text{ mm}^2$$

Provide five No. 20 bars, giving $A_s = 1500$ mm². The longitudinal reinforcement on the flexural compression side is (bottom face here):

$$A_s' = A_{\ell,\text{bottom}} - \frac{M_f}{0.9\, d\, f_y}$$
$$= 241 - \frac{172 \times 10^6}{0.9 \times 436 \times 400} = -797 \text{ mm}^2$$

The tension due to torsion is more than balanced by flexural compression. However, inside each corner of the stirrup, at least one longitudinal bar having a diameter not less than $s/16$ ($160/16 = 10$ mm) is a minimum requirement (clause 11.2.2.4). Hence, provide two No. 10 bars on the bottom side, as shown in Fig. 7.10c. In practice, the two longitudinal corner bars at the bottom may be provided by extending two of the positive moment bars from the midspan section into the support. The longitudinal bars required for torsion must be anchored into the support so as to develop the yield strength at the face of support.

The net longitudinal tension reinforcement, $A_s - A_s' = 1500 - 200 = 1300$ mm², is well within the maximum allowable limit of $\rho_{\max} = 0.023$.

PROBLEMS

1. The critical section of a balcony beam has the cross-sectional dimensions shown in Fig. 7.11, and is subjected to the following actions:

$$M_f = +76 \text{ kN·m}, \ T_f = 35 \text{ kN·m}, \ V_f = 52 \text{ kN}.$$

Check the adequacy of the section. Take $f_c' = 30$ MPa and $f_y = 400$ MPa.

2. A continuous bridge girder, curved in plan, has the following actions acting at the critical section, a distance d from the face of support: $M_f = -39.5$ kN·m, $V_f = 53.5$ kN, and $T_f = 34.2$ kN·m. Design the girder section and reinforcement. Take $f_c' = 30$ MPa, $f_y = 350$ MPa.

3. Figure 7.12 shows the overall dimensions of a sign post. The post is a reinforced concrete member of square cross section. The net wind load is estimated as 1.5 kN/m². Design the post, assuming:

Fig. 7.11 Problem 1

No. 10

No. 10 at 100

3 - No. 20

Fig. 7.12 Problem 3

post

a) the effects of vertical loads are negligible compared to wind load effects; and
b) the weight of the sign board is 5 kN and gravity load effects are to be included.

REFERENCES

7.1 Timoshenko, S., and Goodier, J.N., *Theory of Elasticity*, 3rd ed., McGraw-Hill, New York, 1970.

7.2 Popov, E.P., *Mechanics of Materials*, 2nd ed., Prentice-Hall, Englewood Cliffs, New Jersey, 1978.

7.3 Cowan, H.J., *Reinforced and Prestressed Concrete in Torsion*, Edward Arnold Ltd., London, 1965.

7.4 Hsu, T.T.C., *Torsion of Structural Concrete — Plain Concrete Rectangular Sections*, Torsion of Structural Concrete, ACI Publication SP-18, American Concrete Institute, Detroit, 1968, pp. 207–238.

7.5 Collins, M.P., *The Torque-Twist Characteristics of Reinforced Concrete Beams*, Inelasticity and Non-linearity in Structural Concrete, SM Study No. 8, University of Waterloo Press, Waterloo, 1972, pp. 211–232.

7.6 Hsu, T.T.C., *Torsion of Structural Concrete — Behaviour of Reinforced Concrete Rectangular Members*, ACI Publication SP-18, American Concrete Institute, Detroit, 1968, pp. 261–306.

7.7 ACI Committee 438, *Tentative Recommendations for the Design of Reinforced Concrete Members to Resist Torsion*, ACI Journal, Vol. 66, No. 1, Jan. 1969, pp. 1–8.

7.8 Lyalin, I.M., *The Ultimate Equilibrium Method*, Chapter 6 in Ref. 7.3, pp. 81–106.

7.9 Hsu, T.T.C., *Ultimate Torque of Reinforced Rectangular Beams*, ASCE Journal, Struct. Div., Vol. 94, No. ST2, Feb. 1968, pp. 485–510.

7.10 Lampert, P., and Collins, M.P., *Torsion, Bending, and Confusion — An Attempt to Establish the Facts*, ACI Journal, Vol. 69, Aug. 1972, pp. 500–504.

7.11 Mitchell, D., and Collins, M.P., *Diagonal Compression Field Theory — A Rational Method for Structural Concrete in Pure Torsion*, ACI Journal, Vol. 71, Aug. 1974, pp. 396–408.

7.12 Lampert, P., *Post Cracking Stiffness of Reinforced Concrete Beams in Torsion and Bending*, Analysis of Structural Systems for Torsion, ACI Publication SP-35, American Concrete Institute, Detroit, 1973, p. 385–433.

7.13 Zia, Paul, *What Do We Know About Torsion in Concrete Members?* ASCE Journal, Struct. Div., Vol. 96, No. ST6, June 1970, pp. 1185–1199.

7.14 Collins, M.P. and Lampert, P., *Redistribution of Moments at Cracking — The Key to Simpler Torsion Design?* Analysis of Structural Systems for Torsion, ACI Publication SP-35, American Concrete Institute, Detroit, 1973, pp. 343–383.

7.15 Hsu, T.T.C., and Burton, K.T., *Design of Reinforced Concrete Spandrel Beams*, ASCE Journal, Struct. Div., Vol. 100, No. ST1, Jan. 1974, pp. 209–229.

7.16 Hsu, T.T.C., and Hwang, C.S., *Torsional Limit Design of Spandrel Beams*, ACI Journal, Vol. 74, American Concrete Institute, Detroit, Feb. 1977, pp. 71–79.

7.17 Mitchell, D. and Collins, M.P., *Detailing for Torsion*, ACI Journal, Vol. 73, American Concrete Institute, Detroit, Sept. 1976, pp. 506–511.

7.18 CEB-FIP, *Model Code for Concrete Structures*, CEB-FIP International Recommendations, 3rd ed., Comité Euro-International du Béton, Paris, 1978.

CHAPTER 8

SHEAR AND TORSION DESIGN — GENERAL METHOD

8.1 INTRODUCTION

As in the case of the Simplified Method dealt with in Chapter 6, the General Method for shear design also uses the truss analogy. However, the influence of diagonal tension cracking on the diagonal compressive strength of concrete, and the influence of shear on the design of longitudinal reinforcement are accounted for in a more direct manner.

In order to identify the various parameters to be considered in analysing for shear strength, consider the compressive stress trajectories in a beam subjected to a bending moment, M, an axial force, N (considered positive if tensile), and a shear force, V, as shown in Fig. 8.1. Note that this is similar to the trajectories shown in Fig. 6.1d, except that in Fig. 8.1b the influence of axial force is also included. At any cross section such as section 1–1, the magnitude and direction of the principal compressive stresses and principal compressive strains will vary over the depth of the section. At the top face the inclination θ will be 0° and at the bottom face θ will be 90°. The shear stress distribution over the depth of section will also be non-uniform. Considering a small element such as at A at a depth y, the stresses, strains and the corresponding Mohr's circles are as shown in Fig. 8.1h, j. In drawing these it is assumed that concrete has no tensile strength, and that the directions of principal stresses coincide with the directions of principal strains.

For a correct analysis, three parameters are required to be known/computed at each point over the depth of the section. These may be considered, for instance, as the principal strains ϵ_1 and ϵ_2 and the angle θ. The stress–strain relation for concrete and reinforcing steel are also necessary. With these known, the principal stress, f_2, can be computed from the stress–strain relations for concrete and hence the normal stress, f_{cx}, and the tangential stress, v, at all points over the depth of the section. The longitudinal steel strain, ϵ_{sx}, and hence the stress in longitudinal steel, f_s, can also be computed assuming that reinforcing steel carries only axial forces. Thus, the distributions of axial and tangential stresses over the cross section can be obtained as shown in Fig. 8.1d and e. The resultant of the forces obtained as the product of these stresses and the cross-sectional area integrated over the entire depth of section gives the stress resultants M, N and V (Fig. 8.1f). The strain in the transverse direction, ϵ_t, determines the tensile stress, f_v, in the transverse shear reinforcement,

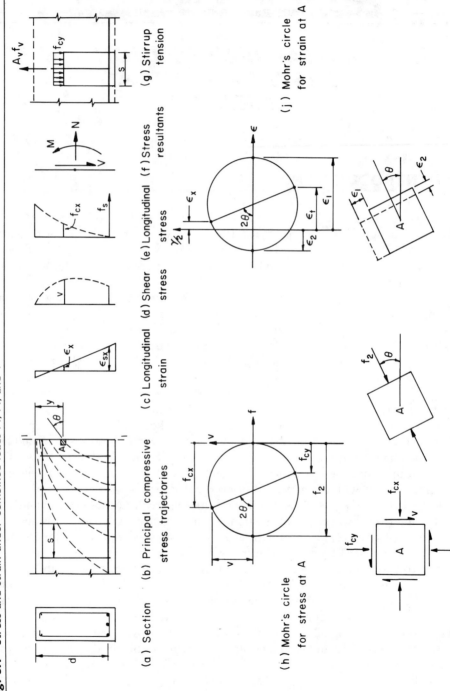

Fig. 8.1 Stress and strain under combined loads M, N, and V

(a) Section   (b) Principal compressive stress trajectories   (c) Longitudinal strain   (d) Shear stress   (e) Longitudinal stress   (f) Stress resultants   (g) Stirrup tension   (h) Mohr's circle for stress at A   (j) Mohr's circle for strain at A and the tension in this reinforcement balances the transverse tensile stress in the concrete, f_{cy}, over the area tributary to it (Fig. 8.1g).

The strain distribution must be compatible with the geometry of deformation. Thus, assuming that beam deformations are such that plane sections of the beam remain plane, the distribution of ϵ_1, ϵ_2 and θ must be such that the associated longitudinal strain, ϵ_x, varies linearly over the depth of the member, as shown in Fig. 8.1c.

An analysis problem may be the computation of the maximum shear that a given section can carry concurrently with given axial load and moment or the computation of the axial load and moment capacity of the section in the presence of a given shear force. Because of the large number of unknowns involved, a direct solution to these problems is not possible and a trial and error procedure has to be resorted to. Two parameters that may be assumed initially are the shear stress distribution and the longitudinal strain distribution. This gives v and ϵ_x at all points. Taking trial values of a third parameter also, such as ϵ_1, and by successive iterations to satisfy equilibrium, compatibility and stress–strain relations, the appropriate values of f_{cx} and v at element A could be found. Such a procedure to compute the shear strength of a given section acted on by given moment M and axial force N is presented in Ref. 8.1. However, such procedures, although permissible according to CSA A23.3-M84 (11.4.1.2), are lengthy and tedious and seldom necessary in practice. Accordingly, an approximate procedure is specified in CSA A23.3-M84 (11.4.1.5) and described in Section 8.4

8.2 STRESS–STRAIN RELATION FOR DIAGONALLY CRACKED CONCRETE

The necessity to know the stress–strain relationship for steel and concrete has already been mentioned. The state of stress in element A shown in Fig. 8.1h indicates that the maximum (principal) compressive stress in concrete, f_2, is inclined at an angle θ to the axis of the member. Associated with this stress is a compressive strain, ϵ_2, along f_2 and a tensile strain, ϵ_1, at right angles to the direction of f_2. Because of the low tensile strength of concrete (which is neglected here), tensile cracks will develop early along the direction of f_2, and the concrete in between these cracks acts as the parallel compression diagonals in the truss analogy. Hence the concrete carrying the diagonal compressive stress can be represented as shown in Fig. 8.2a.

Biaxially strained concrete, as in Fig. 8.2a, with compression in one direction and a concurrent transverse tensile strain is weaker than concrete in uniaxial compression as in a cylinder test (Fig. 8.2b) where the lateral strain is only due to the Poisson effect. Based on results of tests on reinforced concrete panels subjected to in-plane shear and normal stresses (Ref. 8.2) the maximum compressive strength, $f_{2,\text{max}}$, of concrete in the presence of transverse tensile strain, ϵ_1, has been expressed by Eq. 8.1:

$$f_{2,\text{max}} = \lambda \phi_c f_c' / (0.8 + 170\epsilon_1)$$
$$\leq \lambda \phi_c f_c' \tag{8.1}$$

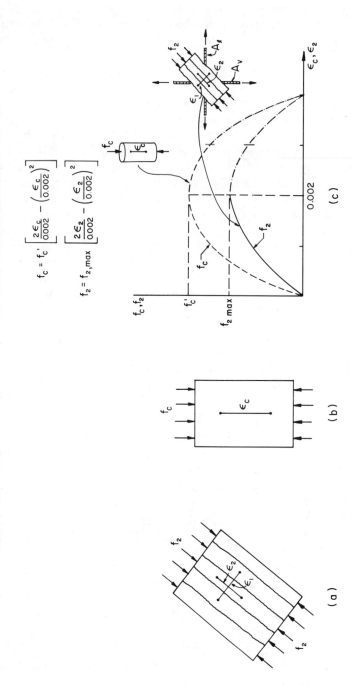

Fig. 8.2 *Stress–strain relationship for diagonally cracked concrete*

where $f_{2,max}$ = compressive strength of concrete in presence of transverse tensile strain ϵ_1

f_c' = specified compressive strength of concrete

ϵ_1 = transverse tensile strain

The material resistance factor ϕ_c and the concrete density factor λ are included in Eq. 8.1. It is also important to note that the test panels forming the basis for the derivation of Eq. 8.1 were reinforced with uniform and closely spaced reinforcement along the longitudinal and transverse directions.

While Eq. 8.1 gives $f_{2,max}$, the *maximum strength* of concrete under transverse tensile strain ϵ_1, in order to compute the stress f_2 corresponding to a principal compressive strain ϵ_2 (concurrent with transverse tensile strain ϵ_1), a corresponding stress–strain relation for biaxially strained concrete (Fig. 8.2a) is also necessary. For this, it may be assumed that the general shape of this stress–strain relation remains the same as for uniaxial compression. One such parabolic relationship proposed in Ref. 8.2 is given in Eq. 8.2. Equation 8.2 is also shown in Fig. 8.2c where it is compared with the parabolic stress–strain diagram for uniaxial compression:

$$f_2 = f_{2,max} \left[\frac{2\epsilon_2}{0.002} - \left(\frac{\epsilon_2}{0.002} \right)^2 \right] \tag{8.2}$$

where f_2 is the compressive stress corresponding to compressive strain ϵ_2, in presence of transverse tensile strain ϵ_1, and

$f_{2,max}$ is given by Eq. 8.1.

8.3 APPROXIMATE PROCEDURE FOR ANALYSIS (CLAUSE 11.4.1.5)

8.3.1 ASSUMPTIONS AND CONTROLLING EQUATIONS

In this procedure, the response to shear force is evaluated separately and superposed on the strength under bending and axial load computed assuming shear to be zero. The plane section theory for flexure is described in Chapters 4 and 5, and for flexure and axial load in Chapter 15. Figure 8.3a represents the results of such an analysis for flexure and axial load, assuming the shear force to be zero. The stress resultants, M_p and N_p, are taken with reference to the axis of $d_v/2$ for convenience in combining with those due to shear. In the analysis for shear only, the following simplifying assumptions are made:

(a) the shear stress v is uniformly distributed over the web, having a width b_v and depth d_v, so that $v = V/(b_v d_v)$;

(b) the longitudinal strain, ϵ_x, and the inclination, θ, of the principal compressive stress remain constant over the depth, d_v, and the former may be taken as the longitudinal strain at mid-depth of the member determined from a plane section analysis for flexure and axial load; and

(c) the tensile strength of the concrete is ignored.

Fig. 8.3 *Approximate method of analysis for M, N, and V*

(i) (ii) Strain (iii) Stress (iv) Stress resultants

(a) Plane section analysis for bending moment and axial force only

$$= \frac{V}{b_v d_v}$$

$$f_2 = \frac{v}{\sin\theta \cdot \cos\theta}$$

$$D = \frac{V}{\sin\theta}$$

$$= v/\tan\theta$$

$$N_v = V/\tan\theta$$

(i) (ii) Assumed longitudinal strain (iii) Assumed uniform shear stress distribution (iv) Diagonal compression stresses (v) Longitudinal stresses due to shear (vi) Stress f_{cy} (vii) Stress resultants

(viii) Mohr's circle of stress

(ix) Mohr's circle of strain

(x) Forces in stirrups

(xi) Directions of direct strains

(b) Analysis for shear only

With the above simplifying assumptions, the internal forces, stress and strain distributions and the stress resultants at a section subjected to shear only (such as at a point of contraflexure) are as shown in Fig. 8.3b. The Mohr's circles for stress and strain states at all points on the section are as shown in Fig. 8.3b(viii) and (ix). From the equilibrium of forces on an element, or from the Mohr's circle for stresses, the following relations can be worked out:

$$v = \frac{V}{b_v d_v} \tag{8.3}$$

$$f_2 = \frac{v}{\sin\theta\cos\theta} = \frac{V}{b_v d_v}\left(\tan\theta + \frac{1}{\tan\theta}\right) \tag{8.4}$$

$$f_{cx} = \frac{v}{\tan\theta} \tag{8.5}$$

$$f_{cy} = v\tan\theta \tag{8.6}$$

The equilibrium of forces across the section gives:

$$N_v = b_v d_v \frac{v}{\tan\theta} = \frac{V}{\tan\theta} \tag{8.7}$$

This force N_v must be properly accounted for both in respect of the longitudinal reinforcement requirement and the maximum compressive strains in the concrete in the plane section analysis for bending plus axial force (see Section 8.3.2). Combining the two stress resultants, in Fig. 8.3a(iv) and b(vii), the following net values are obtained for the resistance including shear effects:

$$M = M_p \tag{8.8}$$

$$\begin{aligned} N &= N_p - N_v \\ &= N_p \frac{V}{\tan\theta} \end{aligned} \tag{8.9}$$

If M_p and N_p are computed for the ultimate conditions, the resistance of the section in the presence of shear, V, can be expressed as $M_r = M_p$ and $N_r = N_p - V/\tan\theta$.

The force in the transverse reinforcement balances the vertical compressive stresses f_{cy} in the concrete. Considering equilibrium of stirrup forces and vertical forces in the concrete area tributary to a stirrup as shown in Fig. 8.3b(x):

$$\begin{aligned} A_v f_v &= b_v s f_{cy} \\ &= b_v s v \tan\theta \\ &= \frac{Vs}{d_v}\tan\theta \end{aligned} \tag{8.10}$$

The strain in the transverse reinforcement, ϵ_t, and the stress f_v are also related by the stress–strain relation for steel.

The strain sets ϵ_x, ϵ_t and γ and ϵ_1, ϵ_2 and θ are interrelated as indicated by the Mohr's circle of strain. From this circle, the following relations can be obtained:

$$\tan^2 \theta = \frac{\epsilon_x + \epsilon_2}{\epsilon_1 - \epsilon_x} = \frac{\epsilon_1 - \epsilon_t}{\epsilon_t + \epsilon_2} = \frac{\epsilon_x + \epsilon_2}{\epsilon_t + \epsilon_2} \tag{8.11a}$$

$$\epsilon_1 = \epsilon_x + (\epsilon_x + \epsilon_2)/\tan^2 \theta \tag{8.11b}$$

$$\epsilon_t = \epsilon_1(1 - \tan^2 \theta) + \epsilon_x \tan^2 \theta \tag{8.11c}$$

Equations 8.3 to 8.10 establish the applicable relations between the internal stresses and applied forces M, N and V. Equations 8.11a, b and c give the compatibility conditions for the strain. Equations 8.1 and 8.2 establish the stress–strain relationship for concrete in diagonal compression. When f_2 reaches $f_{2,\max}$ failure will occur by diagonal crushing of the concrete. For reinforcement the usual stress–strain relationship of $f_s = E_s\epsilon_s \leqslant f_y$ is applicable. Once again, for the analysis of a section to find, say, the shear resistance V_r, in the presence of bending moment M_f and axial force N_f will involve a process of trial and error. An algorithm for a trial and error solution is presented in Ref. 8.1. It still involves choosing trial values for three parameters, say, ϵ_1, θ and f_v, assumed uniform over the section and adjusting these until all conditions are satisfied. The process is "impossibly tedious" for hand computation and is workable only with a computer. It should also be pointed out that the axial strain ϵ_x assumed for shear analysis is the strain at depth $d_v/2$ obtained by plane section analysis under the combination of the applied moment M and an axial force equal to the applied axial force (net axial force) plus the axial force N_v caused by shear (see Eq. 8.9a).

8.3.2 INFLUENCE OF SHEAR ON LONGITUDINAL REINFORCEMENT

The analysis for shear only and associated stress resultants (Fig. 8.3b(vii)) show that there is an axial compression, N_v, developed due to shear. While the plane section analysis for flexure and axial load is done for moment M_p and axial force N_p, the final net resistances in the presence of shear V are only $M_r = M_p$ and $N_r = N_p - N_v$. Conversely, in a design for M_f, N_f and V_f, the longitudinal reinforcement has to be designed for a moment M_f, and an axial force $N = N_f + N_v = N_f + V_f/\tan \theta$ (CSA A23.3-M84(11.4.6.1)). This is represented schematically in Fig. 8.4a–c. Reference 8.3 recommends that instead of designing the section for the additional axial force N_v, the section may be designed for an additional moment of $0.5N_v d_v$ (see Fig. 8.4d). Even though this procedure gives essentially the same additional longitudinal tension reinforcement, the two procedures are not identical, particularly in respect of associated compressive forces in concrete. It is also to be noted that when the applied axial force N_f is compressive and relatively high, the axial compressive stress in concrete is enhanced by the stresses, f_{cx}, due to shear (Fig. 8.4e). In such cases the concrete section should also be sufficient to carry the moment M_f together

Fig. 8.4 *Schematic representation of influence of shear on longitudinal reinforcement*

(a) Design forces   (i) flexure + axial force   (ii) shear force

(b) Analysis

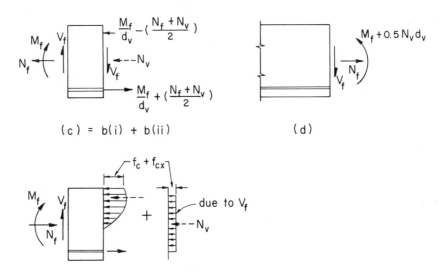

(c) = b(i) + b(ii)   (d)

(e) Case of high axial compression

with a total axial compressive force of $N_f + N_v$ (CSA A23.3-M84(11.4.6.4)).

When the general method is adopted for the design of longitudinal reinforcement, the flexural reinforcement extension requirements to account for shear (see Sections 5.10.3, 5.10.4, 5.10.5 and 6.12 and Code clauses 12.10.3, 12.10.4, 12.10.5 and 12.11.3) may be waived (CSA A23.4-M84 (11.4.6.5)).

8.3.3 COMPARISON OF THE APPROXIMATE PROCEDURE WITH TRUSS ANALOGY EQUATIONS

The expressions for stresses given in Section 8.3.1 can be derived directly using the truss analogy. Figure 8.5 shows the truss already described in Chapter 6 (Fig. 6.6). The compression diagonal is assumed to be along the direction of the principal compressive stress, inclined at θ to the axis of the beam. The lever arm is assumed

Fig. 8.5   *Truss analogy*

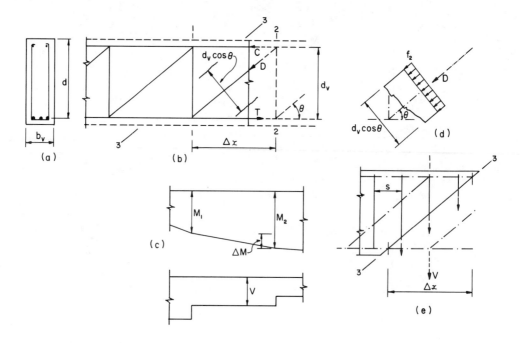

to be d_v. Equilibrium of forces at the section and considerations of geometry give the following results:

$$\Delta x = d_v / \tan \theta$$

$$C = \frac{M_1}{d_v}$$

$$T = \frac{M_2}{d_v}$$

$$T - C = \frac{M_2 - M_1}{d_v} = \frac{\Delta M}{\Delta x \tan \theta} = \frac{V}{\tan \theta}$$

$$D = \frac{V}{\sin \theta}$$

The concrete diagonal carrying force D has a width $d_v \cos \theta$ (Fig. 8.5b, d) and thickness b_v so that the diagonal compressive stress f_2 is given by:

$$f_2 = \frac{D}{b_v d_v \cos \theta} = \frac{V}{b_v d_v \sin \theta \cos \theta}$$

From the vertical equilibrium of forces in Fig. 8.5e:

$$V = \frac{\Delta x}{s} A_v f_v$$

$$= \frac{d_v}{s \tan \theta} A_v f_v$$

or,
$$A_v f_v = \frac{Vs}{d_v} \tan \theta$$

The equations for f_2 and $A_v f_v$ are identical with Eq. 8.4 and 8.10. However, the excess tension in the bottom chord over the compression in the top chord is $T - C = V/\tan \theta$. Thus the longitudinal reinforcement representing the bottom chord must be designed for the applied flexure and tension and an additional tensile force due to shear of $V/\tan \theta$.

The method explained in Section 8.3.1 and 8.3.2 attempts to relate the stress f_2 to the principal strains in concrete, ϵ_1 and ϵ_2. In order to achieve this, however, several simplifying assumptions are required with regard to the strain distributions. (See assumptions in Sections 8.3.1 and 8.4.)

8.4 SHEAR AND TORSION DESIGN USING THE APPROXIMATIONS TO THE GENERAL METHOD

8.4.1 GENERAL

A Simplified General Method for Shear and Torsion design is recommended in CSA A23.3-M84(11.4.1.5) and detailed in clauses 11.4.1.5 to 11.4.6.5. The procedure is based on further simplifying assumptions (see Section 8.4.2) and essentially concentrates on the conditions at mid-depth of the beam. As in the traditional method of design detailed in Chapters 6 and 7, torsion can be neglected if $T_f \le 0.25 \, T_{cr}$. The total necessary transverse reinforcement is taken as the sum of that required for shear and that required for coexisting torsion, while the additional longitudinal reinforcement required is designed for an additional axial tensile force, N_v, equivalent to the combined effects of shear and torsion.

The design shear force is the factored shear force V_f. However, for variable depth members, components of flexural compression and tension in the directions of the applied shear must be taken into account as given by Eq. 6.37.

8.4.2 FURTHER ASSUMPTIONS AND DESIGN EQUATIONS FOR SHEAR

CSA A23.3-M84(11.4) recommends a Simplified General Method for shear design based on several simplifying assumptions. In addition to those given in Section 8.3.1, the following additional assumptions are made:

(a) the angle θ may be chosen to have any value between $15°$ and $75°$;

(b) the axial strain ϵ_x, assumed uniform over the depth for shear calculations may be taken as the longitudinal strain at mid-depth of the member when the section is subjected to M_f, N_f and N_v, determined by performing a plane section analysis. In lieu of such an analysis, ϵ_x may be taken as 0.002;

(c) the maximum value of the principal compressive strain, ϵ_2, is assumed to be 0.002;

(d) the stress–strain relation for diagonal compression is assumed to be given by Eq. 8.1 and 8.2; and

(e) the effective shear depth, d_v, may be taken as the distance, measured perpendicular to the neutral axis, between the resultants of the tensile and compressive forces due to flexure, but need not be taken less than $0.9d$ where d is the effective depth of the beam.

For design, the following procedure can be adopted. With ϵ_x, ϵ_2 and θ assumed, ϵ_1 can be obtained from Eq. 8.11b as:

$$\epsilon_1 = \epsilon_x + \frac{\epsilon_x + 0.002}{\tan^2 \theta} \tag{8.12}$$

Using this ϵ_1, $f_{2,\text{max}}$ can be computed from Eq. 8.1. To avoid diagonal crushing, the cross-sectional dimensions must be such that the stress f_2 given by Eq. 8.4 be less than $f_{2,\text{max}}$, or:

$$f_2 = \left(\tan \theta + \frac{1}{\tan \theta}\right) \frac{V_f}{b_v d_v} < f_{2,\text{max}} \tag{8.13}$$

$$< \lambda \phi_c f'_c / (0.8 + 170 \epsilon_1)$$

The requirement in Eq. 8.13, after substituting for ϵ_1 from Eq. 8.12, can be expressed as:

$$\frac{V_f}{\lambda \phi_c f'_c b_v d_v} < \frac{1}{\left(\tan \theta + \dfrac{1}{\tan \theta}\right)\left[0.8 + 170\left(\epsilon_x + \dfrac{\epsilon_x + 0.002}{\tan^2 \theta}\right)\right]} \tag{8.14}$$

The right hand side of Eq. 8.14, for $\epsilon_x = 0.002$, is plotted against θ in Fig. 8.6 and the conditions in Eq. 8.14 are satisfied for points below the solid line in Fig. 8.6. For a given beam, the value of θ corresponding to the shear stress ratio, $V_f/(\lambda \phi_c f'_c b_v d_v)$, read from the curve in Fig. 8.6 gives the minimum value of θ at which $f_2 = f_{2,\text{max}}$. Note that Eqs. 8.13 and 8.14 consider diagonal compressive stress due to transverse shear V_f only. When significant torsion, T_f, is also present the term $V_f/(b_v d_v)$ in these equations must be replaced by $[V_f/(b_v d_v) + T_f p_h/A_{oh}^2]$ (see Section 8.4.4).

In order to ensure a ductile behaviour, the Code clauses 11.4.1.3 and 11.4.3.1 require that the transverse shear reinforcement must yield before diagonal crushing of concrete, that is:

$$\epsilon_t > f_y/E_s$$

Assuming that yielding of transverse reinforcement is ensured (see below), the required amount of transverse reinforcement or the factored shear resistance, V_r, can be computed using Eq. 8.15, (Eq. 8.10 with material resistance factor ϕ_s):

$$V_r = \frac{A_v \phi_s f_y d_v}{s \tan \theta} \geq V_f \tag{8.15}$$

The transverse strain ϵ_t can be expressed from Eq. 8.11a as:

$$\epsilon_t = \epsilon_1 - \epsilon_x - \epsilon_2$$

$$= \frac{\epsilon_x + \epsilon_2}{\tan^2 \theta} - \epsilon_2$$

Thus the yielding requirement for transverse reinforcement can be expressed as:

$$\frac{\epsilon_x + \epsilon_2}{\tan^2 \theta} - \epsilon_2 \geq f_y/E_s$$

$$\tan^2 \theta < \frac{\epsilon_x + \epsilon_2}{\dfrac{f_y}{E_s} + \epsilon_2} \tag{8.16}$$

If the simplifying assumptions, (b) and (c) above, recommended by the Code are used, $\epsilon_x = 0.002$ (tensile) and $\epsilon_2 = 0.002$ (compressive), the maximum values of θ for $f_y = 400$ MPa and 300 MPa (with $E_s = 200$ GPa) are obtained as 45° and 46.9°, respectively. These limiting values for θ are also shown in Fig. 8.6. Obviously an angle for θ selected in excess of this, as recommended in Code clause 11.4.2.6, will not satisfy the condition in Code clause 11.4.3.1. Conversely, in order to satisfy the yield conditions in Eq. 8.16 with the value of $\theta = 75°$ permitted by the Code, the value of ϵ_x must be:

$$\epsilon_x > \left(\frac{f_y}{E_s} + \epsilon_2 \right) \tan^2 75° - \epsilon_2 \tag{8.16a}$$

which gives for $f_y = 400$ MPa and $\epsilon_2 = 0.002$:

$$\epsilon_x > 0.0537$$

This means that for the plane section analysis, limiting the maximum compressive strain in the concrete at the top fibre to 0.003 as required by the Code, the tensile strain in the flexural steel should be (see Fig. 8.7a) over 50 times the yield strain of the steel, $\epsilon_y = f_y/E_s$. Even for a lower value of $\epsilon_2 = 0.001$, Eq. 8.16a gives a tensile steel strain of about 38 ϵ_y. However, even for the minimum flexural steel specified

Fig. 8.6 *Limiting values for shear stress ratio and angle θ*

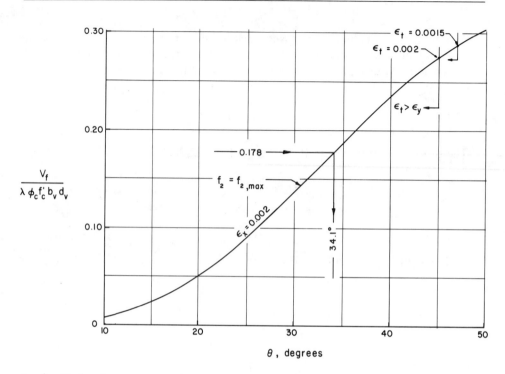

in the Code of $\rho_{min} = 1.4/f_y$, the tension steel strain will be only of the order of $9 \epsilon_y$ (see Fig. 8.7b).

8.4.3 USE OF INCLINED STIRRUPS

When inclined stirrups having an area of cross section A_{vi} and spaced s_i measured along the axis of the beam are provided, part of the shear is carried directly by these bars by diagonal tension. The net shear force at the section, to be resisted by diagonal compression in the concrete, can be computed as follows (see Fig. 8.8a):

Assuming stirrups to yield,   stirrup force $= A_{vi}\phi_s f_y$

$$\text{Number of bars crossing section} = \frac{d_v}{s_i \tan \alpha}$$

$$\text{Total diagonal tension force} = D_{si} = \phi_s A_{vi} f_y \frac{d_v}{s_i \tan \alpha}$$

Shear carried by inclined stirrups is

$$V_{si} = D_{si} \sin \alpha$$

$$= \phi_s A_{vi} f_y d_v \frac{\cos \alpha}{s_i}$$

Fig. 8.7 Strain diagram for yielding of transverse steel, with $\theta = 75°$ and $\epsilon_x = 0.002$ (CSA A23.3-M84(11.4))

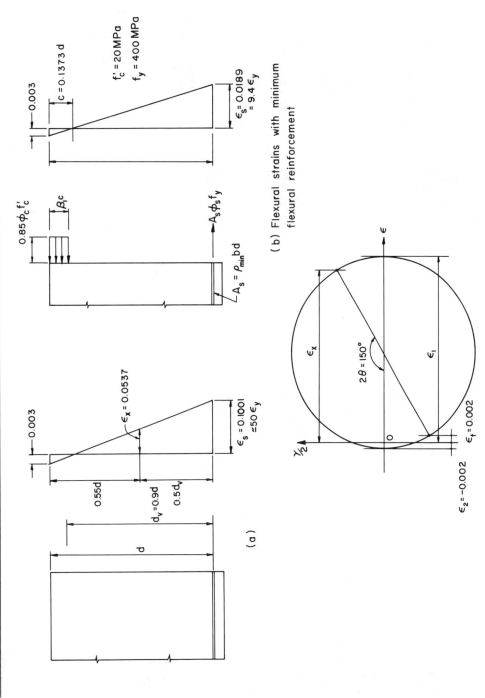

Fig. 8.8 *Shear carried by inclined stirrups*

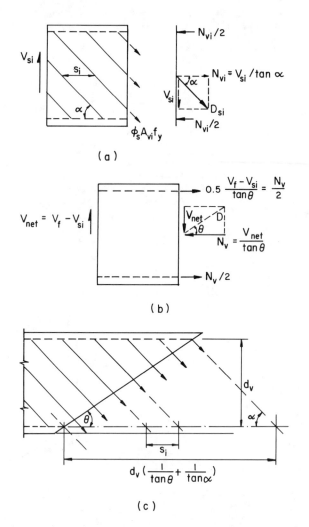

(a)

(b)

(c)

In thise case, while checking for the diagonal compressive stress, f_2, in Eq. 8.13, V_f can be replaced by the net shear to be resisted by diagonal compression (see Fig. 8.8b), namely:

$$V_{\text{net}} = V_f - V_{si}$$
$$= V_f - \phi_s A_{vi} f_y \, d_v \frac{\cos \alpha}{s_i} \tag{8.17}$$

CSA A23.3-M84(11.6.2.10) limits V_{net} to not less than 0.66 V_f. The shear resistance contributed by the inclined stirrups (in addition to the resistance, V_r, due to the transverse shear reinforcement provided perpendicular to the axis of the beam and given by Eq. 8.15) can be computed (see Fig. 8.8c) as:

$$V_{r,i} = \frac{A_{vi}\phi_s f_y d_v}{s_i}\left(\frac{\sin\alpha}{\tan\theta} + \cos\alpha\right) \tag{8.18}$$

Figure 8.8a also shows that there is an axial tensile force, N_{vi}, due to the force in the inclined stirrups which partly affects the compressive forces, N_v, due to diagonal compression shown in Fig. 8.8b. However, CSA A23.3-M84(11.4.6.4) requires longitudinal reinforcement to be designed for the N_v shown in Fig. 8.8b, $(V_f - V_{si})$ being limited to not less than $0.66\ V_f$.

In deriving the expressions given in Eq. 8.17 and 8.18, it was assumed that the inclined stirrups yielded. In order to ensure this, the strain ϵ_α, inclined at an angle α to the longitudinal axis of the member, must exceed the yield strain. Using Mohr's circle (see Fig. 8.3b ix, xi) and with $\epsilon_2 = 0.002$, this condition can be expressed as (see CSA A23.3-M84(11.4.3.3)):

$$\epsilon_\alpha = \frac{\epsilon_1 - 0.002}{2} - \frac{\epsilon_1 + 0.002}{2}\cos 2\,(\theta + \alpha) > \frac{f_y}{E_s} \tag{8.19}$$

where ϵ_1 may be computed from Eq. 8.12.

CSA A23.3-M84(11.4.4.5) recommends that a series of parallel bent up bars at different distances from the support may be treated as inclined stirrups, except that the term d_v in Eq. 8.18 be replaced by $0.75\ d_v$ and that the shear resistance $V_{r,i}$ given by this equation be limited to a maximum of $0.3\ \lambda\ \phi_c\ \sqrt{f_c'}\ b_v d_v$.

8.4.4 DESIGN FOR TORSION

The design for torsion by the General Method is also based on the assumption that after cracking, torsion is resisted by a hollow tubular space truss consisting of compression diagonals spiralling around the tube and tensile steel placed longitudinally and transversely. It is assumed that the concrete cover spalls off and the effective outer dimensions of the tube coincide with the centreline of the closed transverse (hoop) reinforcement (see Fig. 7.7.d), and this area enclosed by the hoop centreline is denoted by A_{oh} and its perimeter by p_h. Torsion may be ignored if $T_f < 0.25\ T_{cr}$ as already given in Chapter 7. When torsion is to be accounted for, the shear stress due to torsion is taken as:

$$v_t = T_f \frac{p_h}{A_{oh}^2}$$

Combining this shear with the shear stress due to coexisting transverse shear V_f (Eq. 8.3), in order to avoid diagonal crushing of the concrete, the total principal diagonal compressive stress, f_2, should be limited to $f_{2,max}$ as given by Eq. 8.20:

$$f_2 = \left(\tan\theta + \frac{1}{\tan\theta}\right)\left(\frac{V_f}{b_v d_v} + \frac{T_f p_h}{A_{oh}^2}\right) \leq f_{2,max} \tag{8.20}$$

where $f_{2,\max}$ is given by Eq. 8.1. The angle θ can vary as in the case of design for shear.

The Code recommends Eqs. 8.21 and 8.22 for the area enclosed by the torsional shear flow, A_o, and the torsional depth of compression, a_o (a_o is also the thickness of an equivalent tube with uniform shear stress across its thickness). These expressions are based on Ref. 8.4:

$$A_o = A_{oh} - a_o p_h / 2 \tag{8.21}$$

$$a_o = \frac{A_{oh}}{p_h}\left[1 - \sqrt{1 - \frac{T_f p_h}{0.7\,\lambda\phi_c f_c' A_{oh}^2}\left(\tan\theta + \frac{1}{\tan\theta}\right)}\,\right] \tag{8.22}$$

The area of transverse reinforcement, A_t, or alternatively the factored torsional resistance, T_r, can be computed from Eq. 8.23:

$$T_r = \frac{\phi_s A_t f_y}{s}\frac{2A_o}{\tan\theta} \tag{8.23}$$

where s is the spacing of transverse reinforcement and the design requirement is that $T_r \geq T_f$.

The additional axial force due to torsion alone can be calculated as (see Eq. 7.14 and 7.15):

$$N = \frac{T_f p_o}{2A_o \tan\theta}$$

When shear and torsion are present the Code recommends that longitudinal reinforcement be designed for an effective additional axial tensile load of:

$$N_v = \frac{1}{\tan\theta}\sqrt{v_f^2 + \left(\frac{T_f p_o}{2A_o}\right)^2} \tag{8.24}$$

where $p_o = p_h - 4a_o$

8.5 ADDITIONAL REQUIREMENTS IN CSA A23.3-M84

8.5.1 GENERAL

When the factored shear force, V_f, exceeds $\lambda\phi_c\sqrt{f_c'}\,b_w d$ or the factored torsion, T_f, exceeds $0.25\,T_{cr}$, the concrete cover down to the centreline of the outermost reinforcement should be assumed to have spalled off (unless the ends of the concrete compression diagonals are restrained against spalling) and the width of web, b_v, determined accordingly. Otherwise, spalling may be neglected. When spalling is to be allowed for, transverse reinforcement must not be anchored in the cover concrete. To provide ductility, in addition to the requirement that $\epsilon_t > f_y/E_s$ given in Section 8.4.2, when V_f exceeds $0.5 V_c$ or T_f exceeds $0.25\ T_{cr}$, a minimum area of shear reinforcement is required to be provided, as already given in Chapter 6, Eq. 6.26.

For beams supported on the bottom face and uniformly loaded on the top face, the Code permits the design of transverse reinforcement required within a length of $d_v/\tan\theta$ to be based on the lowest value of V_f and T_f occurring within this length. This concept is represented in Fig. 8.9. For vertical equilibrium, the total shear resistance offered by the transverse reinforcement bars within the length AB need only equal the shear $V_{f,B}$ at section B. However, the value of θ, stress f_2, and the strain limits discussed in Section 8.4.2 are to be based on the highest value of V_f and T_f occurring within this length. For hollow sections subjected to torsion, the distance from the centreline of the outermost transverse torsional reinforcement to the inside face of the wall should be greater than 1.5 a_o, where a_o is given by Eq. 8.22.

Fig. 8.9 *Design shear for beams uniformly loaded on top face*

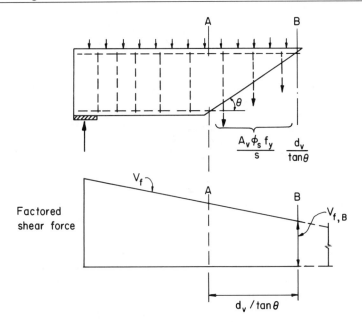

8.5.2 CONTROL OF DIAGONAL CRACKING

When the shear due to specified loads, V_{se}, is greater than the cracking shear, v_{cr}, diagonal cracking at specified loads must be investigated and controlled (CSA A23.3-M84(11.4.8)). Control of diagonal cracking may be considered adequate if:
   (a) the spacing of transverse reinforcement is ≤ 300 mm;
   (b) $V_f \leq 0.6\ \lambda\phi_c\ \sqrt{f_c'}\ b_w d$; and
   (c) $T_f \leq 0.25\ T_{cr}$.

Alternatively, the crack control may be considered adequate if the requirements of (a) and (b) and either (c) or (d), given below, are met.

(a) spacing of transverse reinforcement is ≤ 300 mm; and

(b) spacing between longitudinal reinforcement bars along the exposed sides is ≤ 300 mm; and

(c) calculated strain in transverse reinforcement under specified loads, ϵ_{se}, does not exceed 0.001 for interior exposure and 0.0008 for exterior exposure; or

(d) the value of f_y used in calculating transverse reinforcement is taken as not greater than 300 MPa.

To check the requirement at (c) above, ϵ_{se} may be calculated as:

$$\epsilon_{se} = \left[\frac{V_{se}s}{A_v E_s d_v} + \frac{T_{se}s}{1.6\,A_t E_s A_{oh}} \right] \sqrt{\tan\theta} \left[1 - \left(\frac{V_{cr}}{V_{se}} \right)^3 \right] \tag{8.25}$$

where A_v and A_t are the amounts of transverse reinforcement provided for shear and torsion respectively. The cracking shear, V_{cr}, may be calculated as follows:

(a) for members not subjected to torsion:

$$V_{cr} = 0.2\,\lambda\,\sqrt{f_c'}\,b_w d \left[\sqrt{1 - \frac{N_{se}}{A_g 0.4\,\lambda\,\sqrt{f_c'}}} \right] \tag{8.26}$$

(b) for members subjected to combined shear and torsion, the V_{cr} obtained by Eq. 8.26 may be divided by the factor:

$$\sqrt{1 + \left(\frac{p_c b_w d}{2 A_c^2} \frac{T_{se}}{V_{se}} \right)^2} \tag{8.27}$$

Apart from the usual notations, in Eq. 8.26 and 8.27 T_{se} and N_{se} are the torsion and axial tensile load, respectively, acting concurrently with the shear force, V_{se}, at the specified loads; p_c is the outside perimeter of the concrete cross-section, and A_c is the area enclosed by p_c.

When inclined stirrups are used as shear reinforcement, the strain in these at service loads may be computed by dividing the value obtained using Eq. 8.25 by $[\sin\alpha/\tan\theta + \cos\alpha]$, and with A_v replaced with A_{vi}.

8.5.3 SPACING OF TRANSVERSE REINFORCEMENT

Spacing of transverse shear reinforcement placed at right angles to the axis of the member must not be greater than $d_v/(3\tan\theta)$, d_v or 600 mm. Inclined stirrups and/ or bent up flexural reinforcement used as shear reinforcement must be so spaced that every potential diagonal tension crack line inclined at θ to the longitudinal axis of the member is crossed by at least three lines of effective shear reinforcement. Spacing of transverse torsion reinforcement must not exceed $p_h/(8\tan\theta)$.

8.6 SPECIAL PROVISIONS FOR REGIONS ADJACENT TO SUPPORTS, CONCENTRATED LOADS, OR ABRUPT CHANGES IN CROSS SECTION

The design procedures described in Section 8.4 assumed that the shear stresses are uniformly distributed over the depth d_v, and the angle θ is constant. In regions of discontinuities and of concentrated loads and reactions, as shown in Fig. 8.10a, this assumption does not apply and hence the design may be based on load transfer through a simple truss system consisting of concrete struts and reinforcing steel ties interconnected by nodal zones of concrete. It is in fact a very basic concept in structural design that, for transferring a system of loads to the supports, any stable skeletal framework such as a truss or grid, compatible with the actual deformation pattern, may be delineated and the members and their joints designed for the resulting forces thereon. Designers have always used this concept for design in special situations such as openings in webs of beams, corbels, end blocks in prestressed concrete beams, etc. The skeleton (or truss) may be implicit and embedded within a member, as in the case of the truss analogy for shear design of concrete beams and the truss analogy for plate girder design, or it may be explicit and externally visible as in a real truss. Thus for a deep beam, for instance, it is possible to conceive a truss, a tied arch, a strutted catenary and several other combinations (see Fig. 8.11) as the skeletal load transfer forms within the member and design accordingly. The optimum design is the one which is economic, has stability, adequate reserve strength and ductility and meets the serviceability conditions satisfactorily.

Truss models for load transfer schemes for the cases shown in Fig. 8.10a are shown in Fig. 8.10b. The forces in the truss members can be computed from statics. The truss members and their joints (nodal zones) must have adequate strength to carry these forces. The area of the tension tie is obtained as the tie force divided by $\phi_s f_y$. This tension tie reinforcement must be anchored to the nodal zones by appropriate embedment lengths, hooks, or mechanical devices (see Chapter 9).

The compressive stress in the concrete strut must not exceed $f_{2,max}$ given by Eq. 8.1. If a tension tie is crossing a compression strut, the fact that the strain along the direction of the bar is equal to the yield strain, $\epsilon_y = f_y/E_s$, must be recognized in evaluating the transverse strain ϵ_1 necessary for computing $f_{2,max}$. Thus referring to Fig. 8.12, if a tension tie reinforcement is crossing a compression strut at an angle α_s, the strain along the reinforcement is ϵ_y and with the maximum value of $\epsilon_2 = 0.002$ (compressive), the value of ϵ_1 can be computed using the Mohr's circle for the strain set up as:

$$\epsilon_1 = \epsilon_y + (\epsilon_y + 0.002)/\tan^2 \alpha_s$$

The nodal zones are the regions of concrete subjected to multidirectional compression, where the struts and ties of the truss meet. For the purpose of checking the stresses in the nodal zones, the tie reinforcement can be assumed as being anchored by bearing against the far side of the nodal zone, the effective bearing area being the area of concrete surrounding the tension tie reinforcement and having the same centroid as the reinforcement (Fig. 8.10c(ii)). The nodal zones are bounded by planes

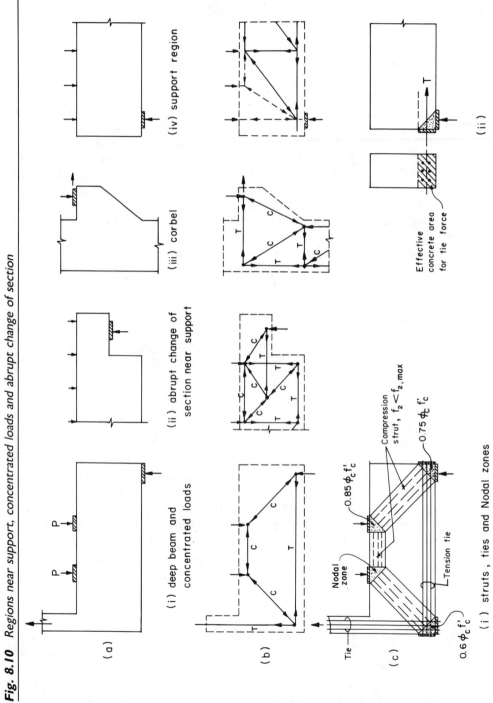

Fig. 8.10 Regions near support, concentrated loads and abrupt change of section

Fig. 8.11 *Alternative load transfer schemes*

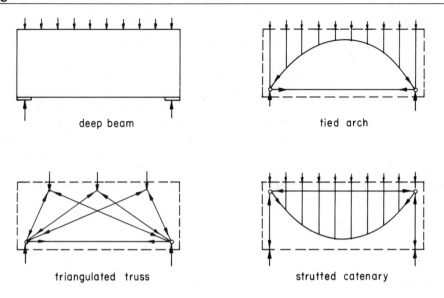

deep beam

tied arch

triangulated truss

strutted catenary

Fig. 8.12 *Strain conditions in concrete strut*

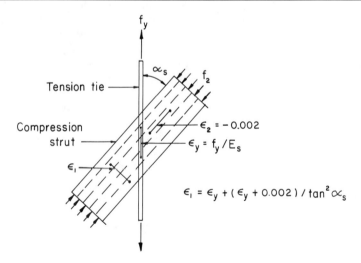

f_y

Tension tie

α_s

f_2

Compression strut

$\epsilon_2 = -0.002$

$\epsilon_y = f_y / E_s$

ϵ_1

$\epsilon_1 = \epsilon_y + (\epsilon_y + 0.002) / \tan^2 \alpha_s$

perpendicular to the lines of actions of the truss member forces at the joint and the load/reaction bearing area, if any, as shown in Fig. 8.10c.

Unless special confining reinforcement is provided, the compressive stresses in the concrete in the nodal zones are limited to the maximum values given below (CSA A23.3-M84(11.4.7.5)):

(a) in nodal zones bounded by compressive struts and bearing areas
    — $0.85\ \phi_c f_c'$;
(b) in nodal zones anchoring only one tension tie — $0.75\ \phi_c f_c'$; and
(c) in nodal zones anchoring tension ties in more than one direction
    — $0.60\ \phi_c f_c'$.

Examples of zones to which each of these limits apply are also identified in Fig. 8.10c(i). In most situations, since the compressive stress in the compression strut is limited to a maximum of $f_{2,max}$, the compressive stress on the face of the nodal zone bearing against a compression strut will be within safe limits. In regions near direct supports and concentrated loads in normal beams (Fig. 8.10c(ii)) it is only necessary to check the bearing stresses produced by the concentrated loads and the effective concrete stress due to the tie force (tie force divided by the effective area of concrete over which tie reinforcement is distributed), if any, against the above stress limits within the nodal zone.

EXAMPLE 8.1

Design the shear reinforcement for the T-beam in Example 5.7 by the General Method.

SOLUTION

1. Shear force diagram
   The factored shear force envelope (see Example 6.2) is shown in Fig. 8.13b. The shear force at the face of support is $44.6 + (257 - 44.6)\ (4 - 0.12)/4 = 251$ kN.

2. Check spalling of concrete cover
   To prevent spalling of the concrete cover, V_f should be less than $\lambda \phi_c \sqrt{f_c'}\ b_w d$. At the support section, where the shear is maximum, $d = 434$ mm:

   $$\lambda \phi_c \sqrt{f_c'}\ b_w d = 1.0 \times 0.6 \times \sqrt{20} \times 300 \times 434 \times 10^{-3} = 349 \text{ kN}$$

   Since maximum V_f is less than 349 kN, spalling does not occur. Hence $b_v = b_w = 300$ mm; $d_v = 0.9;\ d = 0.9 \times 434 = 391$ mm.

3. Select value of θ to avoid crushing of concrete
   The shear stress ratio, at face of support, where shear is maximum is:

   $$\frac{V_f}{\lambda \phi_c f_c' b_v d_v} = \frac{251 \times 10^3}{1.0 \times 0.6 \times 20 \times 300 \times 391} = 0.178$$

   Assuming $\epsilon_x = 0.002$, from Fig. 8.6, the lowest value of θ to avoid diagonal crushing of concrete is 34.1°. For ensuring the yielding of transverse steel, θ must also be less than 45° (see Fig. 8.6). Choose $\theta = 35°$. Note that since θ is arbitrarily chosen within the permissible range, and the spacing limit is related to $d_v/\tan \theta$, θ could be selected to give a convenient round figure for $d_v/\tan \theta$. For instance, a value of $\theta = 35.91°$ will give the spacing limit $d_v/(3 \tan \theta)$ of 180 mm, which controls in this problem. Also in uniformly loaded beams with the load applied on its top surface and supported on its bottom surface (see Fig. 8.9), the transverse

Fig. 8.13 *Example 8.1*

(a) beam

(b) factored shear, V_f ; kN

(c) spacing , s

(d) service load shear, V_{se} ; kN

(e) effective concrete area tor tensile tic force distribution

reinforcement required within a length of $d_v/\tan \theta$ may be determined for the lowest V_f within this length. In such cases it may also be convenient to select θ such that $d_v/\tan \theta$ is a suitable round figure. However, in this problem, θ will be taken as 35°.

4. Select transverse reinforcement

Transverse reinforcement, placed perpendicular to the beam axis should be such that:

$$V_r = \frac{\phi_s A_v f_y d_v}{s \tan \theta} \geq V_f$$

or the spacing and area of transverse reinforcement is given by:

$$s \leq \frac{\phi_s A_v f_y d_v}{V_f \tan \theta}$$

Choosing No. 10 U-stirrups with 135° hooks bent around the longitudinal bars and placed perpendicular to the beam axis:

$$s \leq \frac{0.85 \times 200 \times 400 \times 391}{V_f \tan 35°} = \frac{37\,970 \times 10^3}{V_f} \text{ mm}$$

The spacing diagram according to the above requirement is plotted in Fig. 8.13c.

The spacing is also controlled by the minimum transverse area requirement and the spacing limitations. Since in this case V_f exceeds 0.5 V_c (as shown in Example 6.2) the minimum shear reinforcement requirement is:

$$A_v > A_{v,\min} = 0.35 \frac{b_w s}{f_y}$$

or
$$s < \frac{A_v f_y}{0.35 \, b_w} = \frac{200 \times 400}{0.35 \times 300} = 762 \text{ mm}$$

The spacing limits for transverse reinforcement are:

$$s \leq d_v/(3 \tan \theta) = 391/(3 \tan 35°) = 186 \text{ mm} \qquad \text{(controls)}$$
$$d_v = 391 \text{ mm}$$
$$\text{and } 600 \text{ mm}$$

Thus the controlling limit is $s \leq 186$ mm and this is shown in Fig. 8.13c. It can be seen that except for a very short distance of 995 mm from the centreline of support (or 875 mm from the face of the support) the minimum spacing requirement controls. A stirrup placing arrangement which satisfies all the requirements is shown in Fig. 8.13a. The first stirrup is placed at 75 mm from face of support and this is followed by five stirrups at a spacing of 150 mm and 16 more stirrups equally spaced (average spacing = 185 mm) making up the distance to one-half spacing from centreline, giving a total of 44 stirrups for the entire span.

For satisfactory anchorage of No. 10 stirrups with 135° hooks, the minimum effective depth required is 321 mm (Table 6.1). As $d > 321$ mm, the anchorage requirement is satisfied.

5. Check control of diagonal cracking

Control of diagonal cracking must be investigated if the shear due to the specified loads, V_{se}, is greater than the shear, V_{cr}, causing diagonal cracking.

V_{se} at the face of the support for a total service load of $15.7 + 29.7 = 45.4$ kN/m is:

$$V_{se} = \frac{45.4 \times 7.88^2}{8 \times 2} = 176 \text{ kN}$$

$$V_{cr} = 0.2 \lambda \sqrt{f'_c} \, b_w d$$

$$= 0.2 \times 1.0 \times \sqrt{20} \times 300 \times 434 \times 10^{-3} = 116 \text{ kN}$$

$V_{se} > V_{cr}$ at support

The service load shear at midspan is (due to live load only):

$$V_{se,\mathcal{C}_L} = \frac{29.7 \times 4^2}{8 \times 2} = 29.7 \text{ kN}$$

Assuming a linear variation of V_{se} between 176 kN at the face of the support and 29.7 kN at midspan, (Fig. 8.13d), $V_{se} = V_{cr}$ at a distance of 2290 mm from the centre of the span. The region beyond this up to the face of the support must be checked for crack control.

Control of cracking is adequate (clause 11.4.8.2) if:
(a) spacing $s < 300$ mm;   OK
(b) torsion can be neglected; and   OK
(c) V_f does not exceed $0.6 \lambda \phi_c \sqrt{f'_c} \, b_w d$.

$$0.6 \lambda \phi_c \sqrt{f'_c} \, b_w d = 0.6 \times 1.0 \times 0.6 \times \sqrt{20} \times 300 \times 434 \times 10^{-3}$$
$$= 210 \text{ kN}$$

From Fig. 8.13b, $V_f = 210$ kN at a distance of 3110 mm from the centre of the span. Hence within this region crack control is adequate and further investigation is needed only from 3110 mm to face of support, a distance of 770 mm.

Add two additional longitudinal bars, one on each face, at mid-depth to a distance of 800 mm from the face of the support as shown in Fig. 8.13a. With this, the requirements in clause 11.4.8.3 (a) and (b) are met, namely:
(a) $s < 300$ mm; and
(b) spacing of longitudinal bars < 300 mm.

The one additional requirement to be satisfied is either the transverse reinforcement strain at service load, ϵ_{se}, be limited to 0.001 (or 0.0008 for exterior exposure), or the f_y used in calculating transverse reinforcement be not greater than 300 MPa (clause 11.4.8.3(c) or (d), respectively). As the actual spacing provided near the support is close to the required spacing calculated with $f_y = 400$ MPa, the latter requirement cannot be met and hence ϵ_{se} must be checked.

$$\epsilon_{se} = \frac{V_{se} s}{A_v E_s d_v} \sqrt{\tan \theta} \left[1 - \left(\frac{V_{cr}}{V_{se}} \right)^3 \right]$$

$$= \frac{176 \times 10^3 \times 150}{200 \times 200\,000 \times 391} \sqrt{\tan 35°} \left[1 - \left(\frac{116}{176} \right)^3 \right]$$

$$= 8.29 \times 10^{-4} < 0.001 \qquad\qquad\qquad \text{OK}$$

Hence the crack control requirements are satisfied.

6. Design of longitudinal reinforcement required for shear

   The additional tensile force to be carried is $N_v = V_f/\tan \theta$. Distributing this equally between the top and bottom chords, the additional tensile reinforcement needed is that required to carry a tensile force of $N_v/2$. [The tension of $N_v/2$ at the top chord only counteracts the flexural compression in this region and hence no tension reinforcement is required at the flexural compression level. This is the effect, as far as total tensile steel is concerned, of the proposal in Ref. 8.3 of designing for an additional moment of $0.5\ N_v d_v$.]

   Additional tensile reinforcement required at the support is

$$A_\ell = \frac{0.5\ N_v}{\phi_s f_y} = \frac{0.5 \times 251 \times 10^3}{0.85 \times 400 \times \tan 35°} = 780 \text{ mm}^2$$

and at midspan is

$$A_\ell = \frac{0.5 \times 44.6 \times 10^3}{0.85 \times 400 \times \tan 35°} = 138 \text{ mm}^2$$

   Example 5.7 shows that the flexural reinforcement area provided at midspan is 4200 mm² and a requirement of 3980 mm², giving an excess of 220 mm² which is ample to cover the additional requirement for shear of 138 mm². (Alternatively the increased design moment at midspan, accounting for shear is $M_f + 0.5\ N_v d_v$ = $514 + 0.5 \times (44.6/\tan 35°) \times 0.391 = 532$ kN·m,

and the actual moment of resistance $M_r = 548$ kN·m > 532 kN·m, hence OK.)

   At the support section the moment is zero and three No. 30 bars giving an area of 2100 mm² are extended into the support compared to a requirement for N_v of 780 mm². (Also the M_r of the section for three bars is $\simeq 274$ kN·m, whereas $0.5\ N_v d_v = 0.5 \times (251/\tan 35°) \times 0.391 = 104$ kN·m.)

   Similar checks at bar cut-off points at 860 mm and 2260 mm from the centre of support (see Fig. 5.17, Example 5.10) show that the longitudinal reinforcement provided meets the requirements for N_v. Alternatively a fresh bar cut-off can be detailed considering the tension reinforcement requirement for flexure and shear.

7. Check stresses in regions adjacent to supports

   The support reaction is 257 kN. Actual bearing stress:

$$= \frac{257 \times 10^3}{300 \times 240} = 3.57 \text{ MPa}$$

Allowable bearing stress = $0.75\ \phi_c f_c'$
$$= 0.75 \times 0.6 \times 20 = 9.00 \text{ MPa}$$
$$> 3.57 \qquad\qquad\qquad \text{OK}$$

   The effective area of concrete required to distribute tensile tie force:

$$= \frac{\text{tie force}}{0.75\ \phi_c f_c'}$$

Tie force at support region:

$$= 0.5 \, N_v = 0.5 \, \frac{V_f}{\tan \theta}$$

$$= \frac{0.5 \times 257}{\tan 35°} = 184 \text{ kN}$$

Area required $= \dfrac{184 \times 10^3}{0.75 \times 0.6 \times 20} = 20\,444 \text{ mm}^2$

Effective area provided (Fig. 8.13e):

$$= 300 \times 132 = 39\,600 \text{ mm}^2 \qquad\qquad \text{OK}$$

EXAMPLE 8.2

Design an interior span of the beam in Example 7.2 by the General Method for Shear and Torsion Design.

SOLUTION

1. Factored load effects

   Assuming a 350 mm × 500 mm cross section for the beam, the bending moment, shear force and torsion distributions along an interior span are as shown in Fig. 8.14a. (See Example 7.2 for loads. The bending moments shown are approximate and based on the CSA moment coefficients.)

2. Check whether torsional effects are to be considered

   Pure torsional cracking resistance is

$$T_{cr} = (A_c^2/p_c) \, 0.4 \, \lambda \phi_c \, \sqrt{f_c'}$$

$$= \frac{(350 \times 500)^2}{2(350 + 500)} \times 0.4 \times 1.0 \times 0.6 \times \sqrt{30} \times 10^{-6} = 23.7 \text{ kN·m}$$

$T_f > 0.25 \, T_{cr}$ and hence torsional effects are to be considered and concrete cover must be assumed to have spalled. To locate the region where shear reinforcement is not required:

$T_f = 0.25 \, T_{cr} = 5.93 \text{ kN·m}$ at 0.7 m from midspan;

and $\quad V_c - 0.2 \, \lambda \psi_c \, \sqrt{f_c'} \, b_w d$

$\qquad = 0.2 \times 1.0 \times 0.6 \times \sqrt{30} \times 350 \times 436 \times 10^{-3} = 100 \text{ kN}$

$V_f = V_c/2 = 50 \text{ kN}$ at 1.42 m from midspan

Hence shear reinforcement is not required for a distance of 0.7 m on either side of midspan.

Fig. 8.14 *Example 8.2*

(a) beam

(b) M_f, (kN·m)

(c) V_f, (kN)

(d) T_f, (kN·m)

(e) cross section

(f) longitudinal reinforcement

3. Cross-section properties

The cross-section properties are, assuming 40 mm cover all around, No. 10 U-stirrups and No. 25 bars for longitudinal flexural reinforcement,

$$b_v = 350 - 2 \times 40 - 11.3 \qquad = 259 \text{ mm}$$
$$d = 500 - 40 - 11.3 - 25.2/2 = 436 \text{ mm}$$
$$d_v \simeq 0.9d = 0.9 \times 436 \qquad = 392 \text{ mm}$$
$$d_b = 500 - 2 \times 40 - 11.3 \qquad = 409 \text{ mm}$$
$$A_{oh} = 259 \times 409 \qquad = 106\,000 \text{ mm}^2$$
$$p_h = 2 \times (259 + 409) \qquad = 1336 \text{ mm}$$

4. Select θ to avoid diagonal crushing
    θ is selected for the maximum V_f and T_f at face of support:

$$\text{Shear stress ratio} = \frac{v}{\lambda\phi_c f'_c} = \left(\frac{V_f}{b_v d_v} + \frac{T_f p_h}{A_{oh}^2}\right)\frac{1}{\lambda\phi_c f'_c}$$

$$= \left(\frac{116 \times 10^3}{259 \times 392} + \frac{29.6 \times 10^6 \times 1336}{106\,000^2}\right)\frac{1}{1.0 \times 0.6 \times 30}$$

$$= 0.259$$

Assuming $\epsilon_x = 0.002$, from Fig. 8.6 the minimum value of θ is 42.75°. Since the load is uniformly distributed and mostly applied on the top surface of the beam, the transverse reinforcement will be designed for the lowest values of V_f and T_f occurring within successive segments along the beam having a length of $d_v/\tan\theta = 420$ mm. This length of 420 mm chosen corresponds to $\theta = 43°$.

5. Design of transverse reinforcement
    Within the first segment of 420 mm from face of support, lowest $V_f = 103$ kN and $T_f = 26.1$ kN·m.
    The equivalent torsional depth of compression is (Eq. 8.22):

$$a_o = \frac{A_{oh}}{p_h}\left[1 - \sqrt{1 - \frac{T_f p_h}{0.7\,\lambda\phi_c f'_c A_{oh}^2}\left(\tan\theta + \frac{1}{\tan\theta}\right)}\right]$$

$$= \frac{106 \times 10^3}{1336}\left[1 - \sqrt{1 - \frac{26.1 \times 10^6 \times 1336}{0.7 \times 1.0 \times 0.6 \times 30 \times 106\,000^2}\left(\tan 43° + \frac{1}{\tan 43°}\right)}\right]$$

$$= 22.9 \text{ mm}$$

Area enclosed by shear flow path, A_o, is (Eq. 8.21):

$$A_o = a_{oh} - a_o p_h/2 = 106\,000 - 22.9 \times 1336/2 = 90\,700 \text{ mm}^2$$

Area and spacing of transverse reinforcement for torsion, A_t, is given by Eq. 8.23 as:

$$\frac{A_t}{s} = \frac{T_f \tan\theta}{2A_o\phi_s f_y} = \frac{26.1 \times 10^6 \times \tan 43°}{2 \times 90\,700 \times 0.85 \times 400} = 0.398 \text{ mm}^2/\text{mm}$$

Area and spacing of transverse reinforcement for shear is given by (Eq. 8.15):

$$\frac{A_v}{s} = \frac{V_f \tan\theta}{\phi_s f_y d_v} = \frac{103 \times 10^3 \times \tan 43°}{0.85 \times 400 \times 392} = 0.721 \text{ mm}^2/\text{mm}$$

Assuming closed two-legged stirrups, area of one leg and spacing is given by:

$$\frac{A_{total}}{s} = \frac{A_t}{s} + 0.5\frac{A_v}{s} = 0.398 + 0.721/2 = 0.758 \text{ mm}^2/\text{mm}$$

Using No. 10 stirrups, area per leg = 100 mm² and the required spacing is:

$$s = \frac{100}{0.758} = 132 \text{ mm}$$

Maximum allowable spacing is:

$$s \leqslant d_v/(3 \tan \theta) = 140 \text{ mm}$$
$$\leqslant p_h/(8 \tan \theta) = 179 \text{ mm}$$
$$\leqslant 600 \text{ mm}$$

No. 10 closed stirrups at 130 mm may be provided in the first segment. A similar procedure may be used in successive segments along the beam. However, in this example s_{min} = 140 mm and this will control for the remaining length of beam. The first stirrup may be placed at 60 mm from the face of the support, followed by three more at a spacing of 130 mm, and thereafter 21 stirrups may be provided at a uniform spacing of 140 mm.

6. Design of longitudinal reinforcement
    The longitudinal reinforcement may be designed for a moment of $M_f + 0.5\ N_v d_v$. At the support section, N_v is given by:

$$N_v = \frac{1}{\tan \theta} \sqrt{V_f^2 + \left(\frac{T_f p_o}{2A_o}\right)^2}$$

$$p_o = p_h - 4a_o = 1336 - 4 \times 22.9 = 1244 \text{ mm}$$

$$N_v = \frac{1}{\tan 43°} \sqrt{116^2 + \left(\frac{29.6 \times 10^3 \times 1244}{2 \times 90\ 700}\right)^2} = 251 \text{ kN}$$

Required moment of resistance at support,

$$M_r = 148 + 0.5 \times 251 \times 0.392 = 197 \text{ kN·m}$$

$$K_r = \frac{M_r}{bd^2} = \frac{197 \times 10^6}{350 \times 436^2} = 2.96 \text{ MPa}$$

From Table 5.4, for K_r = 2.96, ρ = 0.00977

$$A_s = 0.00977 \times 350 \times 436 = 1490 \text{ mm}^2$$

Three No. 25 bars give an area of 1500 mm². The increased moment allowing for 0.5 $N_v d_v$ and the moment of resistance are plotted in Fig. 8.14b. One bar may be cut off at a distance of 1.5 m from face of support. This will be fully effective at a distance of ℓ_d = 1.4 × 694 = 972 mm (see Chapter 9 and Table 9.1 for development length) from the point of cut-off, that is up to a point 1.5 − 0.972 = 0.528 m from the face of the support. Note that two bars are continued, though not required for flexure, to function as stirrup supports. Smaller size stirrup holder bars may be used near the midspan region.

Similarly the beam may be designed (as an ell section) for the increased positive moment, including the contribution $0.5\,N_v d_v$. In this case two No. 15 bars plus one No. 20 bar will be adequate. The two No. 15 bars may be continued into the support and the No. 20 bar cut off at a distance of 2.50 m from midspan. The positive moment resistance for this arrangement is also indicated in Fig. 8.14b. The Code places a maximum spacing limit of 300 mm for the longitudinal reinforcement for torsion when designed by the Simple Method in clause 11.3; however, no such limit is specified for design by the General Method in clause 11.4 (except the limitations applicable for flexure).

7. Check crack control
    At specified loads, loads from slab are

$$\text{D.L.} = 0.125 \times 24 = 3.00 \text{ kN/m}$$
$$\text{L.L.} = 1.5 \times 2.4 = \underline{3.60} \text{ kN/m}$$
$$6.60 \text{ kN/m}$$

Torque per metre length $= 6.60(0.175 + 1.5/2) = 6.11 \text{ kN·m/m}$
Load on beam is D.L. $= 15 + 0.35 \times 0.5 \times 24 = 19.20 \text{ kN/m}$
Torque at face of support, $T_{se} = 6.11 \times 3.5 = 21.4 \text{ kN·m}$
The cracking shear, V_{cr}, is (Eq. 8.26)

$$V_{cr} = \frac{0.2\,\lambda\,\sqrt{f_c'}\,b_w d}{\sqrt{1 + \left(\dfrac{p_c b_w d}{2 A_o^2}\dfrac{T_{se}}{V_{se}}\right)^2}}$$

$$= \frac{0.2 \times 1.0 \times \sqrt{30} \times 350 \times 436 \times 10^{-3}}{\sqrt{1 + \left(\dfrac{1700 \times 350 \times 436}{2 \times (350 \times 500)^2} \times \dfrac{21.4 \times 10^3}{90.3}\right)^2}} = 118 \text{ kN}$$

As $V_{se} < V_{cr}$, diagonal cracking is under control at specified loads.

8.7 SPECIAL PROVISIONS FOR DEEP SHEAR SPANS, BRACKETS AND CORBELS

8.7.1 DEEP SHEAR SPANS

Members or portions of members in which the distance from the point of zero shear to the face of the support is less than $2d$, or in which a load causing more than 50 percent of the shear at a support is located less than $2d$ from the face of the support, are considered as deep shear spans (see Fig. 8.15) and must be designed by the General Method in accordance with the considerations described in Section 8.6. In such spans, the area of transverse shear reinforcement, A_v, must be not less than $0.002\,b_w s$ and the spacing, s, should not exceed $d/5$ nor 300 mm. In addition, longitudinal bars must also be distributed over the depth, in pairs, with one bar

Fig. 8.15 *Examples for deep shear spans*

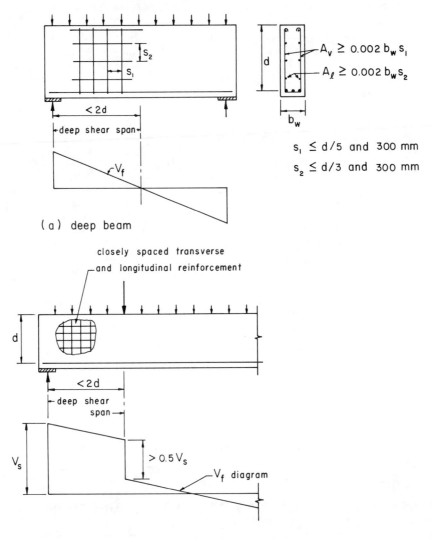

(a) deep beam

(b) deep shear span region in normal beam

near each face at a spacing, s, not to exceed $d/3$ nor 300 mm. Each such pair of bars must have a total area of $0.002b_w s$. These requirements amount to a minimum reinforcement, in each direction, of 0.2 percent of the gross concrete area, and this is provided in order to control diagonal cracking and to increase the ductility of the member.

8.7.2 BRACKETS AND CORBELS

Brackets and corbels are short cantilevered members, projecting beyond the face of another member such as a column (Figs. 6.17b and 8.16). They are used widely in

Fig. 8.16 *Requirements for brackets and corbels*

$$N_f \geq 0.2 V_f, \quad = 0 \text{ if } N_f \text{ eliminated by special provisions}$$

$$d \geq a \quad ; \quad d_o \geq 0.5 d$$

$$A_s / bd \geq 0.04 (f_c' / f_y) \quad ; \quad A_h \geq 0.5 A_s$$

precast construction as well as in cast-in-place construction for supporting members such as overhead crane runways. The loads applied on a corbel generally consist of a shear, V_f, which is the vertical reaction from the supported member acting at a distance, a, from the face of the support and a normal force, N_f. The latter may result from any longitudinal force on the supported member and from the restraint to horizontal deformations due to creep, shrinkage and temperature changes in the supported member. The member must be designed for the moment resulting from eccentricity of applied loads, shear and axial tension. When provisions are made to avoid the tensile force, N_f, the corbel is designed for effects of V_f only. Vertical stirrups are generally ineffective in resisting shear in corbels. Projecting members with an a/d ratio > 1.0 are not considered as corbels, and are designed for shear using the procedure described for deep beams or ordinary beams depending on the a/d ratio.

CSA A23.3-M84(11.8) requires brackets and corbels to be designed by the General Method using the truss (concrete struts, reinforcement ties and nodal zones) concept described in Section 8.6. Additional detailing requirements given in the Code are summarized below (also see Fig. 8.16). These detailing provisions are based on results of extensive tests, and background information on their development is contained in Ref. 8.5. The external tensile force, N_f, acting on the bearing area must be taken as not less than 0.20 V_f unless special provisions are made to avoid tensile forces. The effective depth, d, measured at the face of the support should not be less than the distance, a, between the load and the face of the support. In order to avoid premature failure due to a major diagonal tension crack propagating from below the bearing area to the outer sloping face of the corbel or bracket, a minimum depth of $0.5d$ is specified at the outer edge of the bearing area. The tension reinforcement ratio $\rho = A_s/bd$, calculated at the face of the support, should not be less than $0.04f_c'/f_y$, where A_s is the area of the primary tensile tie reinforcement. Closed stirrups or ties, parallel to the primary tensile tie reinforcement and having a total area A_h not less than $0.5\ A_s$, must be uniformly distributed within a depth of $2/3d$ adjacent to A_s.

Because of the limited dimensions of a corbel, bond, anchorage and bearing considerations are very important in their design (Ref. 8.6). The primary tensile tie reinforcement must be anchored close to the outer face of the corbel to develop the required force in the tie such as by welding to steel plate, angle or to a crossbar or by bending preferably into a horizontal loop. When the member is designed to carry an axial load N_f the bearing plate may be set into the concrete cover and welded to the tension reinforcement. The bearing area should not project beyond the straight portion of the tensile tie bars, nor beyond the interior face of transverse anchor bars, if any. The reinforcing details and the Code limitations are also presented in Fig. 8.15.

PROBLEMS

See Chapters 6 and 7 for problems on shear and torsion design.

REFERENCES

8.1 *Concrete Design Handbook*, Part II, Chapter 4, Canadian Portland Cement Association, Ottawa, 1985.

8.2 Vecchio, F.J. and Collins, M.P., *The Modified Compression-Field Theory for Reinforced Concrete Elements Subjected to Shear*, J. ACI, Vol. 83, March-April 1986, pp. 219–231.

8.3 Collins, M.P., Mitchell, D., and MacGregor, J.G., *Explanatory Notes on CSA Standard A23.3-M84*, Canadian Portland Cement Association, Ottawa, 1985, 140 pp.

8.4 Collins, M.P., and Mitchell, D., *Shear and Torsion Design of Prestressed and Non-Prestressed Concrete Beams*, J. Prestressed Concrete Institute, Vol. 25, No. 5, Sep/Oct 1980, pp. 32–100.

8.5 ASCE-ACI Committee 426, *The Shear Strength of Reinforced Concrete Members*, ASCE J. Struct. Div., Vol. 99, No. ST6, June 1973, pp. 1091–1187.

8.6 *Commentary on Building Code Requirements for Reinforced Concrete (ACI 318–83)*, American Concrete Institute, Detroit, Michigan, November 1983, 155 pp.

CHAPTER 9

BOND AND DEVELOPMENT

9.1 BOND STRESS

The transfer of axial force from a reinforcing bar to the surrounding concrete results in the development of tangential stress components along the contact surface (Fig. 9.1d). This stress acting parallel to the bar along the interface is called **bond stress** and will be denoted by u per unit area of bar surface. Bond stress results from a change in the bar force, T, along its length.

Bond resistance is made up of chemical adhesion, friction and mechanical interlock between the bar and surrounding concrete. In plain bars, only the first two of these components contribute to the bond strength. In contrast, with deformed bars, the surface protrusions or ribs (oriented transversely to the bar axis) interlocking with and bearing against the concrete key formed between the ribs contributes more positively to bond strength, and is the major reason for their superior bond effectiveness. It should be noted here that CSA A23.3-M84(3.1.2) permits only deformed bars as reinforcement except for spirals, and for stirrups and ties smaller than 10 mm size.

9.2 FLEXURAL BOND

Fig. 9.1b shows the flexural stresses at two adjacent sections of a beam, dx apart, between which the moment changes by dM. With the usual assumptions made in flexural design, the change in the bar force over the length dx is:

$$dT = \frac{dM}{z}$$

This unbalanced bar force is transferred to the surrounding concrete through the bond stress, u, developed along the interface (Fig. 9.1d). The bond stress, u per unit of bar surface, assuming it to be uniformly distributed over the surface, is obtained from equilibrium of forces as:

Fig. 9.1 *Bond stress in a beam*

(a)

(b)

(c)

(d) Flexural bond

(e) Development bond

$$u \, \Sigma o \, dx = dT$$

$$u = \frac{1}{\Sigma o} \frac{dT}{dx} \qquad (9.1\text{a})$$

$$= \frac{1}{\Sigma o} \frac{dM/dx}{z}$$

$$u = \frac{V}{\Sigma o \, z} \qquad (9.1\text{b})$$

where Σo is the total perimeter of the bars at the section, and $V = dM/dx$ is the transverse shear force. It may be noted from Eq. 9.1a that the bond stress is proportional to the change of the bar force.

The bond stress expressed by Eq. 9.1b is associated with a shear or variation of bending moment and is called **flexural bond**. Although Eq. 9.1b appears to give the exact value of the local bond stress, because of variations along the bar due to the influence of flexural cracking, local slip, splitting, and other secondary effects, in reality it represents only a measure of an average local bond stress.

9.3 ANCHORAGE OR DEVELOPMENT BOND

The anchorage zone CD of the tension reinforcement of a cantilever beam is shown in Fig. 9.2. The design bending moment and hence the reinforcement stress, f_s, are maximum at section D. At the end C of the bar, the stress is zero. The change in

Fig. 9.2 *Anchorage bond stress*

(a) Cantilever beam

(b) Anchorage bond

force between C and D, equal to $T = A_b f_s$, is once again transferred to the surrounding concrete through bond. If the bond stress is assumed to have an average uniform value of u, equilibrium of forces gives:

$$(\pi d_b L)\, u = A_b f_s$$

$$= \frac{\pi d_b^2}{4} f_s$$

$$u = \frac{A_b f_s}{\pi d_b L} = \frac{d_b f_s}{4L} \tag{9.2}$$

where A_b = area of bar
d_b = bar diameter
L = length of embedment CD
and f_s = maximum stress in bar (at D)

The *average* bond stress so calculated using the nominal bar diameter d_b gives the nominal bond stress. A similar situation exists in the segment CD of the reinforcement in Fig. 9.1, between the point of maximum tension and the end of the bar (see Fig. 9.1e). The bond stress in these situations can be viewed as providing anchorage for a critically stressed bar, or alternatively as helping develop over the length L the maximum stress f_s needed at the critical section, and is referred to as **anchorage bond** or **development bond**.

9.4 VARIATION OF BOND STRESS

Actual bond stress in a bar is not uniform along its length. In anchorage bond, the stress may be high near the loaded end and much lower and even zero near the outer end. In flexural members, in a constant moment region the shear force, V, is zero and Eq. 9.1b would indicate zero bond stress also. However, the steel force, T, does vary between flexural cracks and bond stresses are developed as shown in Fig. 9.3. At the flexural cracks the tension is carried by the reinforcement alone while in between the cracks concrete carries some tension and thereby partially relieves the steel bar force. Since the bond stress is proportional to the rate of change of bar force (Eq. 9.1a), local bond stresses do develop following a distribution somewhat like that shown in Fig. 9.3d, with the direction of bond stress reversing between the cracks (Ref. 9.1). The net bond force between the cracks will, of course, be zero. With large size bars, these local peak bond stresses may be large enough to cause local slip and/or splitting. In regions of flexural cracking where the moment varies, the distribution of bond stress will differ from that in Fig. 9.3d as now the net bond force between the cracks must equal the unbalanced bar force, dT. Beam tests show that longitudinal splitting cracks tend to initiate at flexural cracks.

Unlike tension bars, transverse cracking is not present generally when the bar is in compression. Further, the bar end bearing against the concrete will transfer part of the force to the concrete. Hence bond resistance of a compression bar is generally greater than that of a tension bar.

9.5 BOND FAILURE

The mechanisms that initiate bond failure may involve a breakup of adhesion between the bar and the concrete, longitudinal splitting of the concrete around the bar, crushing of the concrete in front of the bar ribs, shearing of the concrete keyed between the ribs along a cylindrical surface surrounding the ribs, or any combinations

Fig. 9.3 *Effect of flexural cracks on bond stress in constant moment region*

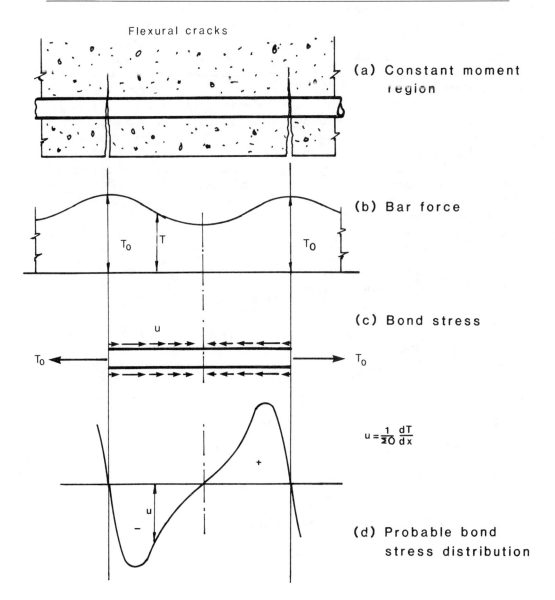

Flexural cracks

(a) Constant moment region

(b) Bar force

T_0 T T_0

(c) Bond stress

T_0 ← → T_0

u

$$u = \frac{1}{\Sigma 0}\frac{dT}{dx}$$

u

+

−

(d) Probable bond stress distribution

of these. Bond failures in beams occur generally when the reinforcement bar pulls loose following longitudinal splitting of the concrete along the bar embedment (Fig. 9.3). Occasionally failure can occur when the bar pulls out of the concrete leaving a circular hole and without extensive splitting of the concrete. The latter type of failure may occur with plain smooth bars placed with large cover, and with very small diameter deformed bars (wires) having large concrete covers. This is particularly true with top bars cast in low density concrete (Ref. 9.1). However, with deformed bars and with the normal amount of concrete cover used in ordinary beams, bond failure is usually a result of longitudinal splitting. Because of the inclination of the rib face, the bearing pressure between the rib and concrete is at an angle to the bar axis which introduces radial forces in the concrete (commonly termed wedging action), as shown in Fig. 9.5. The radial forces, p, cause circumferential tensile stresses in the concrete surrounding the bar (similar to the stresses in a pipe subjected to internal pressure) and tend to split the concrete along the weakest plane. Although the splitting tendency is present even in plain bars, it is apparent that splitting is generally more critical with deformed bars. If the analogy of the bursting of a pipe is applied to a failure cylinder as shown in Fig. 9.5c, it can be shown (Refs. 9.2, 9.3, 9.4) that for a bar subjected to an axial force only and not enclosed by stirrups, the average uniform bond stress at which splitting occurs is a function of the tensile strength of concrete (or $\sqrt{f_c'}$), the ratio C/d_b, and ratio d_b/L. Here, C is the thickness of the idealized failure cylinder, equal to the least of the bottom clear cover, side cover, and one-half the clear spacing between bars (Fig. 9.5c, d); d_b is the bar diameter; and L the embedment length. Splitting occurs along

Fig. 9.4 *Typical bond splitting crack patterns*

(a) Bottom splitting

SPLITTING CRACKS

$C_b \ll C_s$

CRACKS

C_s

C_b

$C_b < C_s$

CRACKS

$C_b > C_s$

(b) Bottom and side splitting cracks

the thinnest surrounding concrete section and the direction of the splitting crack (bottom splitting versus side splitting) will depend on the relative values of the bar spacing and concrete cover as shown in Fig. 9.4b.

Fig. 9.5 *Splitting forces with deformed bars*

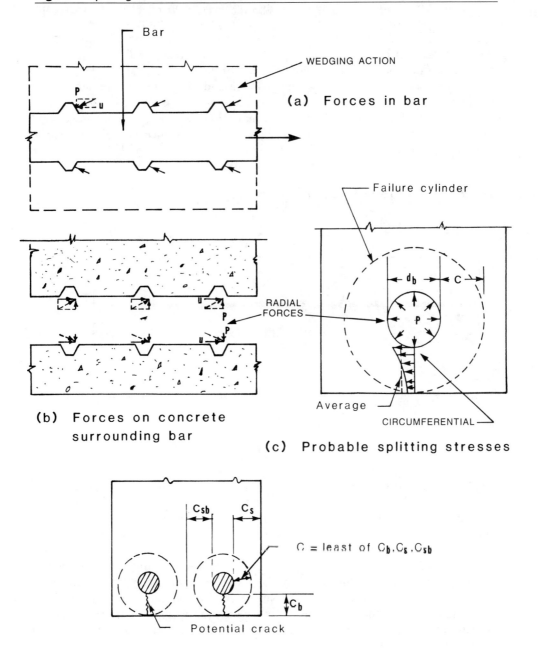

(a) Forces in bar

(b) Forces on concrete surrounding bar

(c) Probable splitting stresses

In flexural members, splitting cracks usually appear on the surface as starting from flexural and diagonal tension cracks, beginning in regions of high steel stress (regions where bond stress has local high values, Fig. 9.3), and with increased loads, progress along the length of embedment. Since the distribution of bond stress along the length of the bar is non-uniform, splitting is usually a gradual and progressive phenomenon, with local splitting at regions of high bond stress resulting in redistribution of bond stresses along the length of embedment. Local splitting does not affect the load carrying capacity of a beam, although it indicates bond distress. Splitting may develop over 60–75 percent of the bar length without loss of average bond strength. In beams without stirrups, however, the final failure is sudden, as the longitudinal split runs through to the end of the bar (Ref. 9.1).

A simple beam can act as a two-hinged arch and carry substantial loads even if the bond is destroyed over the length of the bar, as long as the tie bar is anchored at its ends (Fig. 9.6). Such end anchorage may be realized through anchorage bond over an adequate development length or by special devices (such as hooks and bends, nuts and bolts, welded plates, etc.) or by a combination of both. However, the deflection and crack widths of such a beam may be excessive.

Fig. 9.6 *Tied-arch action with bar anchorage only*

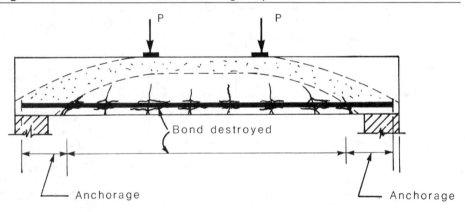

9.6 FACTORS INFLUENCING BOND STRENGTH

The bond strength and mode of bond failure are affected by several variables. The importance of the parameters C, d_b, L and the tensile strength of the concrete was indicated in the previous section. Any increase in confinement of the bar by the surrounding concrete, transverse reinforcement (stirrups, ties, and spirals) or bearing reaction increases bond strength and minimizes splitting. Confinement by the concrete is dependent on the cover and bar spacing and recent studies indicate that it can be appropriately accounted for by the parameter C/d_b (Refs. 9.2, 9.3). Stirrups and ties cross potential longitudinal splitting cracks at right angles (Fig. 9.7) and hence can resist, partially, the tensile force which causes splitting thereby delaying initiation of splitting. They can also restrain the width and propagation of splitting

Fig. 9.7 *Stirrup forces due to bond splitting*

Stirrups

cracks, once formed. Statistical analysis of test results indicates good correlation between the increase in bond stress due to stirrups and the parameter $A_t f_y / s d_b$, there being an upper limit for the increase that can be so obtained (Ref. 9.2, 9.3). Here, A_t, f_y, and s are the area, yield stress and spacing of the transverse reinforcement, respectively, and d_b is the diameter of the longitudinal reinforcement.

Formation of diagonal tension cracks leads to an increase in the stress in the tension reinforcement where it intercepts the cracks (Section 5.10.3 and Fig. 5.12) and hence higher bond stresses on both sides of the crack. In addition, the dowel force that may develop in the tension reinforcement at a diagonal crack due to shear adds to the splitting force due to bond. One conservative estimate, based on test results, puts the loss of bond capacity due to relatively high values of dowel forces at 20 percent (Ref. 9.3). However, closely spaced stirrups will reduce the adverse effect of dowel forces and improve the bond strength.

When flexural reinforcement is terminated in a tension zone, the discontinuity results in a greater rate of change of bar forces and hence larger bond stresses. In addition, premature flexural cracks develop at such points increasing the local bond stress concentrations.

Another factor that influences the bond strength of bars is the depth of fresh concrete below the bar during casting. Water and air which rises towards the top of the concrete mass tend to get entrapped beneath horizontally placed bars and weaken the bond as well as the concrete on the underside of the bar. Hence top reinforcement, which is classified in CSA A23.3 as horizontal bars placed with more than 300 mm of concrete cast in the member below the bar, has a lower bond resistance.

As already indicated, bond strength is proportional to the tensile strength of concrete and the latter, for normal density concretes, can be assumed to vary as the square root of the cylinder strength f_c'. However, for low density concrete the tensile strength is somewhat less than that of normal density concrete having the same compressive strength. Also, bond failure by pull out of the bar, following crushing or shear failure of the concrete between bar ribs, has been observed more often in low density concrete than normal density concrete.

The dimensions and shape of the ribs (deformations) of deformed bars may be expected to have some influence on bond resistance. Experiments show that several deformation patterns, all meeting the specifications for bar deformations, yield substantially the same bond resistance (Refs. 9.3, 9.5).

Because of the extreme non-uniformity in the bond stress distribution, and the large number of variables and uncertainties involved, there is as yet, no theoretical method for evaluating the bond strength of a reinforcing bar embedded in concrete. Hence, specifications and design practice have depended on bond test results to arrive at maximum allowable nominal bond stresses, computed as the bar force divided by the nominal surface area of the embedment length of the bar ($u = P/(\Sigma oL)$), or alternatively to arrive at the minimum development length required to develop a required stress in the reinforcement.

9.7 BOND TESTS

Bond strength is determined with either pull out tests or some sort of beam tests. In a typical pull out test, shown schematically in Fig. 9.8a, a bar embedded in a concrete cylinder or prism is pulled until failure by splitting, excessive slip, or pull out. The nominal bond stress is computed as $u = P/(\Sigma oL)$, where P is the pull at failure. Bond conditions in a pull out test do not ideally represent those in a flexural member. Factors such as flexural cracks, shear, diagonal tension cracks and dowel forces which lower the bond resistance of a flexural member are not present in a concentric pull out test. The concrete in the pull out specimen is in compression, and the friction at the bearing on the concrete offers some restraint against splitting.

Several forms of beam tests for bond have been developed, all designed to duplicate as best as practicable the actual bond conditions in a flexural member (Ref. 9.6). The set up for one such test is shown in Fig. 9.8b (Ref. 9.3). In addition to simulating the bond stress conditions in a beam, this set up is capable of allowing for dowel force effect at diagonal shear cracking.

Based on early bond test results, the maximum allowable ultimate bond stress was expressed as $u_u = K \sqrt{f_c'}/d_b$ (and $K' \sqrt{f_c'}$ for special large size deformed bars) in the 1970 edition of CSA A23.3 (and also in Ref. 9.7), where K and K' are appropriate constants specified in the Code. Although from the 1973 edition of the CSA A23.3 onwards the emphasis on design has shifted from an allowable bond stress to a required development length, ℓ_d, the basis for ℓ_d is still essentially the same allowable bond stress values specified in Ref. 9.7.

9.8 DEVELOPMENT LENGTH

The *development length* is the length necessary to introduce or develop a given stress, f_s, into the bar through bond or, viewed in reverse, the length necessary to take the maximum stress, f_s, out of the bar (length CD in Fig. 9.1 and 9.2). Sometimes

Fig. 9.8 *Bond test specimens*

(a) Pull out test set up

(b) Modified cantilever bond specimen

the term *anchorage length* is used for the development length, in situations where the embedment portion of the bar is not subjected to any flexural bond (Fig. 9.2). However, the distinction should be recognized that *anchorage* can be obtained by methods other than bond development such as by hooks and bends, mechanical fasteners, etc. The concept of development length is that a certain minimum length of bar is needed on either side of a point of maximum steel stress to prevent the bar from pulling out. However, when the bar cannot be provided with the required extension due to placement difficulties, or when it is more convenient to do so, the required development length may be provided by a combination of the equivalent embedment length of a hook or mechanical anchorage, plus additional embedment length of the bar.

Prior to the 1973 edition of CSA A23.3, design for bond required the consideration of both flexural bond stress and anchorage bond stress. Because of the unpredictable and extreme non-uniformity in the distribution of the actual bond stress, the flexural bond stress computed by Eq. 9.1b gives only an unreliable index. With such variations in bond stress, localized bond failures may occur at loads well below ultimate; however, these local failures do not impair the strength of the beam, provided the bars are still anchored at their ends (Fig. 9.6). Further, the allowable bond stress determined from bond tests represents an average bond resistance over the full embedment length. Thus the development of bar stresses through adequate anchorage is the primary requirement rather than the control of local bond stresses whose predicted values are unrealistic. Thus the 1973 and subsequent editions of CSA A23.3 consider essentially the development bond only in the form of a specified minimum required development length. This length is derived on the basis of a maximum permissible average bond stress over the length of embedment. The *basic development length* specified in CSA A23.3-M84 is the length required to develop a bar stress equal to f_y with a permissible average anchorage bond stress u_u essentially the same as used in the 1970 edition of the Code. This value of ℓ_d may be obtained from Eq. 9.2 as:

$$\ell_d = \frac{f_y \, d_b}{4u_u} \tag{9.3}$$

For example, with No. 35 and smaller bars placed at the bottom, CSA A23.3-70 specified an ultimate bond stress expressed in SI units, of:

$$u_u = \frac{19.7 \sqrt{f_c'}}{d_b} \qquad \text{or} \qquad 5.52 \text{ MPa*}$$

where u_u and f_c' are in MPa and d_b is in mm. Further, flexural bond stress could be ignored when anchorage bond stress was limited to $0.8 \, u_u$. Using these values in Eq. 9.3 gives:

$$\ell_d = \frac{f_y \, d_b}{4 \, (0.8 \times 19.7 \, \sqrt{f_c'}/d_b)} \qquad \text{or} \qquad \frac{f_y \, d_b}{4 \, (0.8 \times 5.52)}$$

---

*The Code had specified $9.5 \sqrt{f_c'}/d_b$ or 800 psi where f_c' is in psi and d_b in inch units.

$$= 0.02 \frac{A_b f_y}{\sqrt{f_c'}} \qquad \text{or} \qquad 0.057 \, d_b f_y$$

CSA A23.3-M84 specifies $0.019 \, A_b f_y \sqrt{f_c'}$ but not less than $0.058 \, d_b f_y$.

9.9 DEVELOPMENT LENGTH SPECIFIED IN CSA STANDARD

The development lengths specified in CSA A23.3-M84(12) are the same as those specified in the ACI Standard 318-83. The latter is based on the permissible bond stress specified in Ref. 9.7. In development length computations, no allowance need be made for the material resistance factors. The area of steel being developed is already computed during its design with an allowance for the ϕ factors. Hence the only requirement is to provide adequate development for the steel area so designed. The required development length must be provided on both sides of peak stress points. The Code (clause 12) requirements are summarized below.

Note that these requirements are used to compute the development length, ℓ_d, necessary to develop a stress of f_y in the bar at one end of this length. Conversely these relations may also be used to compute the reduced stress $f_s < f_y$ obtainable at one end of a given development length which is less than the minimum required for f_y.

9.9.1 TENSION BARS

For bars in tension, the development length, ℓ_d, in mm is computed as the product of a *basic development length*, ℓ_{db}, given in (a) below (see also Table 9.1) and a series of applicable modification factors to allow for top bars, low density concrete, wide spacing, etc., given in (b) and (c) below:

(a) Basic development length, ℓ_{db}, (Table 9.1):
   (i)   No 35 or smaller bars $\qquad\qquad\qquad 0.019 \, A_b f_y / \sqrt{f_c'} \geqslant 0.058 \, d_b f_y$
   (ii) No 45 bars $\qquad\qquad\qquad\qquad\qquad\qquad 26 \, f_y / \sqrt{f_c'}$
   (iii) No. 55 bars $\qquad\qquad\qquad\qquad\qquad\qquad 34 \, f_y / \sqrt{f_c'}$
   (iv) deformed wire $\qquad\qquad\qquad\qquad\qquad 0.36 \, d_b f_y / \sqrt{f_c'}$

(b) Modification factors greater than 1.0:
   (i)   for top bars (horizontal bars with more than 300 mm of concrete cast in the member below the bar): $\qquad\qquad\qquad\qquad\qquad\qquad\qquad\qquad 1.4$
   (ii) for $f_y > 400$ MPa: $\qquad\qquad\qquad\qquad\qquad\qquad\qquad 2 - \dfrac{400}{f_y}$
   (iii) Structural low density concrete: $\qquad\qquad\qquad\qquad\qquad 1.33$
         semi-low density concrete: $\qquad\qquad\qquad\qquad\qquad\quad 1.18$

Table 9.1 Basic Development Length ℓ_{db} in mm, for Bars in Tension

Bar no.	Area of one bar, A_b mm²	Dia of one bar, d_b mm	f'_c = 20 MPa			f'_c = 25 MPa			f'_c = 30 MPa			f'_c = 35 MPa			f'_c = 40 MPa		
			f_y = 300 MPa	350 MPa	400 MPa	300 MPa	350 MPa	400 MPa	300 MPa	350 MPa	400 MPa	300 MPa	350 MPa	400 MPa	300 MPa	350 MPa	400 MPa
10	100	11.3	197	229	262	197	229	262	197	229	262	197	229	262	197	229	262
15	200	16.0	278	325	371	278	325	371	278	325	371	278	325	371	278	325	371
20	300	19.5	383	446	510	342	399	456	339	396	452	339	396	452	339	396	452
25	500	25.2	638	744	850	570	665	760	520	607	694	480	562	642	451	526	601
30	700	29.9	893	1041	1190	798	931	1064	728	850	971	672	787	899	631	736	841
35	1000	35.7	1275	1488	1700	1140	1330	1520	1040	1214	1388	960	1124	1285	901	1052	1202
45	1500	43.7	1744	2035	2326	1560	1820	2080	1424	1661	1899	1318	1538	1758	1233	1439	1644
55	2500	56.4	2280	2661	3041	2040	2380	2720	1862	2173	2483	1724	2011	2299	1613	1882	2150

(i) For No. 35 or smaller bars, $\ell_{db} = 0.019\, A_b f_y / \sqrt{f'_c}$ but not less than $0.058\, d_b f_y$

(ii) For No. 45 bars, $\ell_{db} = 26\, f_y / \sqrt{f'_c}$

(iii) For No. 55 bars, $\ell_{db} = 34\, f_y / \sqrt{f'_c}$

Actual development length, ℓ_d, obtained by multiplying ℓ_{db} by all applicable modification factors shall be taken as not less than 300 mm.

(c) Modification factors less than 1.0:

Basic development length, ℓ_{db}, modified by factors at (b) may be reduced by multiplication by one or more of the following factors, with a cumulative value of not less than 0.6:

(i) reinforcement being developed in length under consideration and spaced at least 150 mm on centre and at least 75 mm from side face (Fig. 9.9):   0.8

(ii) flexural reinforcement in excess of that required: $\dfrac{A_{s,required}}{A_{s,provided}}$

(iii) bars enclosed in spirals which are at least 6 mm diameter and have a pitch of at most 100 mm: 0.75

Here f_y, f'_c and $\sqrt{f'_c}$ are in MPa, A_b is the area of bar in mm², and d_b the diameter of bar in mm. The spiral referred to in (c)(iii) may be an individual spiral around each bar or group of bars, or a column type spiral. In applying the factor in c(i) to the case in Fig. 9.9(b), both bars A and B have the same spacing, s, as bars A are developed in length XY while bars B are developed in length YZ (Ref.9.10). The development length, ℓ_d, shall not be taken less than 300 mm except for lapped splices in tension and for anchorage of web reinforcement.

Fig. 9.9 *Spacing considerations for modification factor in 9.9.1(c)(i) (Ref. 9.10)*

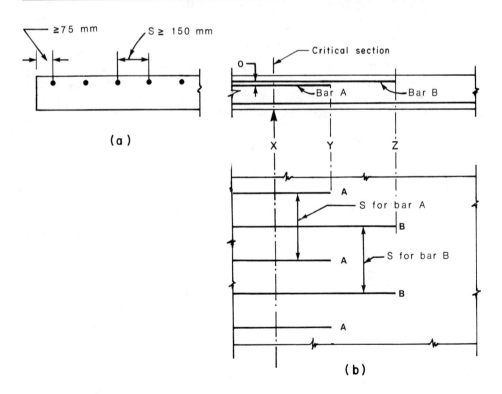

(a)

(b)

9.9.2 COMPRESSION BARS

For deformed bars in compression the development length, ℓ_d, is again computed as the product of the basic compression development length, ℓ_{db}, and the applicable modification factors. The basic compression development length, $\ell_{db} = 0.24\,d_b f_y/\sqrt{f_c'}$, but not less than $0.044\,d_b f_y$. The modification factors, with a cumulative value of not less than 0.6, are:

(a) When reinforcement in excess of that required by analysis is provided: $\dfrac{A_{s,required}}{A_{s,provided}}$

(b) Reinforcement enclosed within spiral reinforcement having a diameter of at least 6 mm and having a pitch no greater than 100 mm: 0.75

9.9.3 BUNDLED BARS

In three- and four-bar bundles, the perimeter of individual bars exposed to concrete is reduced and the bond resistance may not be fully mobilized in the core of the bundle. Hence the development length computed for the individual bars in a bundle is increased by 20 percent for a three-bar bundle and 33 percent for a four-bar bundle. Also, the cut-off points within the span of flexural members of individual bars in a bundle are staggered by at least forty bar diameters.

9.9.4 WELDED WIRE FABRIC IN TENSION

For smooth WWF, the yield strength is considered to be developed with the embedment of at least two cross wires, with the closer one at least 50 mm from the critical section (Fig. 9.10). However, the length, ℓ_d, measured from the critical section to the outermost cross wire must be not less than:

$$\ell_d = 3.3\,\frac{A_w}{s_w}\,\frac{f_y}{\sqrt{f_c'}}$$

Fig. 9.10 *Development of welded wire fabric*

(a) Smooth wire   (b) Deformed wire

where A_w is the area and s_w is the spacing of the individual wires to be developed. The length given by the above equation may be modified by the factors given in Sections 9.9.1 at (b)(iii) and (c)(ii) for low density concrete and excess area provided, respectively. However ℓ_d must be not less than 150 mm.

For deformed WWF, the development length, ℓ_d, measured from the critical section to the end of the wire (Fig. 9.10(b)) is taken as the product of the basic development length, ℓ_{db}, given below and applicable modification factors as given in Section 9.9.1 at (b) and (c). However, ℓ_d must not be taken less than 200 mm except in the development of web reinforcement.

The basic development length for deformed WWF is,

(a) when there is at least one cross wire
within the development length, $\ell_{db} = 0.36 \, d_b \, (f_y - 140)/\sqrt{f'_c}$

but not less than $\qquad 2.4 \dfrac{A_w}{s_w} \dfrac{f_y}{\sqrt{f'_c}}$

(b) when there are no cross wires within the development
length, ℓ_{db} is calculated as for deformed wires in Section 9.9.1,
that is: $\qquad \ell_{db} = 0.36 \, d_b f_y / \sqrt{f'_c}$

9.10 COMMENTS ON DEVELOPMENT LENGTH SPECIFIED IN THE CODE

Results of studies on development length (Refs. 9.2, 9.8, 9.9) indicate that the CSA A23.3-M84 specifications in this regard may not be conservative in some cases, particularly when the clear cover and bar spacings are minimal. The Code equations do not recognize the influence on bond resistance of clear cover, bar spacing, shear force, dowel force and shear reinforcement. The average permissible bond stress forming the basis for the Code equations was established experimentally where wide beams of large cross-section with widely spaced bars were often used. Thus the ill effects of small cover and close bar spacing may have been inadvertently eliminated from the Code formulas. Based on a preliminary evaluation of some of the recent test results, Ferguson has suggested that the development length calculated using the Code may be inadequate if the clear cover over the main bar is less than about 50 mm or the clear spacing between bars is less than about 100 mm (Ref. 9.9). In such situations one suggested remedy is to increase the development length computed, by a factor equal to the ratio of the above minimum values to the corresponding values actually provided. For example, if the clear cover provided is 40 mm, the correction factor is 50/40 = 1.25; and if the minimum clear spacing is 65 mm the correction factor will be 100/65 = 1.54. Obviously there are other options such as changing the bar size, arranging the bars in several layers, and modifying the beam size. Also the ill effect of small cover and spacing is partially offset by the improvement in strength due to stirrups which are normally present.

Recent studies have concentrated on predicting the average bond strength at splitting, accounting for the influence of bar cover and spacing, shear force, dowel force and transverse reinforcement (stirrups) (Refs. 9.2, 9.3, 9.4 and 9.8). Based on these, Refs. 9.2 and 9.3 have recommended Eqs. 9.4 and 9.5, respectively, for the average ultimate bond strength, u:

$$u = \left[0.1 + 0.25 \frac{C}{d_b} + 4.2 \frac{d_b}{\ell_s} + 0.024 \frac{A_{tr} f_{yt}}{s d_b} \right] \sqrt{f'_c} \qquad (9.4)$$

where the last term within the bracket is limited to 0.25

$$u = \left[0.55 + 0.24 \frac{C}{d_b} \right] \sqrt{f'_c} + 0.191 \frac{A_t f_{yt}}{s d_b} \qquad (9.5)$$

In Eq. 9.4 and 9.5, C is the lesser of clear cover and one-half bar spacing in mm, d_b is the diameter of the main bar in mm, A_{tr} is the effective area of transverse reinforcement (stirrups) per main bar in mm^2, A_t is the area of stirrup in mm^2, f_{yt} is the yield strength of stirrups in MPa, s is the spacing of stirrups in mm, and ℓ_s is the splice length or development length in mm. Reference 9.2 also concludes that the basic requirements for development length and lap splice length are identical.

Although proposals such as those in Eq. 9.4 and 9.5 have been made, no generally accepted recommendations have yet been evolved. In the meantime, the development length, ℓ_d, calculated by the Code equations may be modified as indicated earlier when the clear cover or bar spacings are small.

9.11 HOOKS AND MECHANICAL ANCHORAGES

There are situations where the straight distance available beyond a critical section may be inadequate to provide the required development length, ℓ_d (Fig. 9.11). Figure 9.11a shows a deep beam where inclined cracks cause the beam to behave like a tied-arch. The full yield strength may have to be developed at the face of the support and the available straight enbedment length, E, may be less than the length, ℓ_d, required. A similar situation exists at the top bars in a beam or bracket where it frames into a column (Fig. 9.11b). The column face is the critical section for the negative moment reinforcement and again the straight embedment length available within the column may be less than ℓ_d. In such situations CSA A23.3-M84 allows the development of the tension in the reinforcement through a combination of the straight lead-in length, ℓ_{ti}, available plus a standard hook or mechanical anchorage. The effectiveness of mechanical anchorage devices must be ascertained by tests. Hooks can be used in developing tensile stresses only and are ineffective in compression.

The proportions of a standard hook as defined in CSA 23.1-M77 are shown in Fig. 9.12 (proportions for stirrup and tie hooks were given in Chapter 6, Fig. 6.12). The development length, ℓ_{dh}, of a deformed bar terminating in a standard hook is the length measured from the outside of the hook to the critical section where the reinforcement yield stress, f_y, is developed, as shown in Fig. 9.12. The length, ℓ_{dh}, is computed as the product of a basic development length of a standard hook in tension, ℓ_{hb}, and applicable modification factors given below. However ℓ_{dh} is taken as not less than $8 d_b$ or 150 mm, whichever is greater.

Fig. 9.11 *Combination development length with hooks and embedment*

(a) Deep beam

(b) Beam–column connection

The basic development length for a hooked bar with $f_y = 400$ MPa is

$$\ell_{hb} = 100d_b/\sqrt{f'_c}$$

The values of ℓ_{hb} for various size bars and concrete strengths are given in Table 9.2. The modification factors for hook development length are:

(i) for f_y other than 400 MPa $f_y/400$

(ii) for No. 35 and smaller bars, with side cover (normal to plane of hook) not less than 60 mm, and for 90° hooks with cover on the bar extension beyond the hook of not less than 50 mm (see Fig. 9.12c)   0.7

Fig. 9.12 *Development length of standard hooks*

(a) 90° Hook

(b) 180° Hook

$\ell_{dh} = \ell_{hb} \times$ modification factor

(i) Mod. factor = 0.7

(ii) Mod. factor = 0.8

(iii) Ties required, 0.8 factor not applicable

(c) Modification factors for development length of hooks

(iii) for No. 35 and smaller bars, with hook enclosed vertically or horizontally within at least three ties spaced along a length at least equal to the inside diameter of the hook, at a spacing not greater than $3d_b$, d_b being the diameter of the hooked bar (Fig. 9.12c)   0.8

(iv) for reinforcement provided in excess of that required by analysis, for which development of f_y is not required $\dfrac{A_{s \text{ required}}}{A_{s \text{ provided}}}$

(v) for structural low density concrete   1.3

For bars being developed by a standard hook at the ends of members where the side cover and the top (or bottom) cover over the hook is less than 60 mm, the hook must be enclosed within a minimum of three ties spaced along a length at least equal to the inside diameter of the hook and at a spacing not greater than $3d_b$. In this case the factor of 0.8 at (iii) above will not apply (Fig. 9.12(c)(iii)).

Table 9.2 Basic Development Length, ℓ_{hb}, for Standard Hooks in Tension

Bar no.	Dia. of bar, d_b mm	ℓ_{hb}				
		$f'_c = 20$ MPa	$f'_c = 25$ MPa	$f'_c = 30$ MPa	$f'_c = 35$ MPa	$f'_c = 40$ MPa
10	11.3	253	226	206	191	179
15	16.0	358	320	292	270	253
20	19.5	436	390	356	330	308
25	25.2	563	504	460	426	398
30	29.9	669	598	546	505	473
35	35.7	798	714	652	603	564
45	43.7	977	874	798	739	691
55	56.4	1261	1128	1030	953	892

$\ell_{hb} = 100d_b/\sqrt{f'_c}$ for $f_y = 400$ MPa

Development length, $\ell_{dh} = \ell_{hb} \times$ modification factors, but not less than $8d_b$ and 150 mm

Studies indicate that tail extensions beyond those of a standard hook are not effective in providing development length. (Ref. 9.11). It may be emphasized that hooks are used for anchorage or development, and should not be used to replace the bar extensions discussed in Section 5.10.4.

9.12 GENERAL DEVELOPMENT REQUIREMENTS IN THE CODE

The computed stress at *every section* of a reinforcing bar must be developed on both sides of the section. Flexural tension reinforcement can be developed by bending it over to the compression face of the member and anchoring it or making it continuous with the compression reinforcement at this face. In flexural members the critical sections for development are: (1) the points of maximum stress in the bar (such as section C1 for positive moment reinforcement and section C2 for negative moment reinforcement, (see Fig. 5.14;) and (2) when part of the reinforcement is terminated (or bent) at points within the span, for the continuing reinforcement, the point where parts of the bars are terminated (or bent) (such as sections, X_1, Y_1 in Fig. 5.14). The development length and bar extension requirements at these critical sections in flexural members, specified in CSA A23.3-M84 (12.10), were detailed in Sections 5.10.4 and 5.10.5 and Figs. 5.14 and 5.15. When the flexural member is part of the primary lateral load resisting system, the minimum positive reinforcement required to be continued into the support (A23.3-M84 (12.11)) must be anchored to develop its yield strength in tension at the face of the support (at section C2 in Fig. 5.14). Likewise, if the bottom reinforcement at the face of the support is counted on as compression reinforcement at this section, it must be anchored to develop the stress, f_y.

In order to ensure that the computed bar stress is developed at every section, the positive moment region of beams with distributed loading requires special consideration. In such regions the moment diagram is nonlinear (and convex), but the bar stress development over the length ℓ_d is assumed to be linear. Hence, providing the minimum ℓ_d beyond a critical section may not always ensure the development of required stresses at other sections. Such a situation is illustrated in Fig. 9.13a. In this figure, assuming a linear variation of bar stress over the length, ℓ_d, the moment of resistance corresponding to the developed bar stress varies as shown by the chain line. At a section such as C, the moment that is developed (M_{dev}) is less than the applied moment. To avoid such a situation, the Code stipulates that (Fig. 9.13b):

$$\ell_a + \left(\frac{M_r}{V_f} \right) \geq \ell_d \qquad (9.6)$$

where M_r is the factored moment corresponding to the reinforcement extended beyond the section of zero moment, V_f is the factored shear force at the section of zero moment, and ℓ_a is limited to the larger of d or $12d_b$. V_f represents the slope of the moment diagram at the point of zero moment and the length $\ell_a + M_r/V_f$ is indicated in Fig. 9.13b. The moment developed with this requirement is shown by the dotted line in Fig. 9.13b. The requirement in Eq. 9.6 is to be satisfied at supports of simply supported members and at points of inflection of continuous beams. Although a similar situation could arise at bar cut-off points, if the bar size satisfies the requirement at the point of zero moment, usually the condition will be satisfied at cut-off points as well. When the General Method of shear design in CSA A23.3-M84 (11.4) is followed, the requirement given in Eq. 9.6 can be waived.

Special members such as sloped, stepped or tapered footings, deep beams, brackets, and members of variable depth in which the steel stress is not directly proportional to the moment require special consideration for the development of tension reinforcement. For the cases shown in Figs. 9.11a and 9.14, the steel stress is relatively high close to the end of the member and the development of the bar depends on providing adequate end anchorage.

For web reinforcement, the critical section for development is considered to be at mid-depth, $d/2$, of the beam. The anchorage requirements for web reinforcement were detailed in Section 6.11 and Fig. 6.12.

Fig. 9.13 *Bar size limitation at points of inflection*

9.13 FACTORED MOMENT RESISTANCE DIAGRAMS

The theoretical cut-off or bend points may often be determined by scaling from the factored moment envelope. Once the actual cut-off details are worked out, a convenient method to check their adequacy is to construct the corresponding factored moment resistance diagram (Fig. 9.15c). In determining the latter, a bar is assumed to develop its full yield strength, and hence its share of the moment resistance, linearly over a length ℓ_d from the *effective* point of cut-off (Fig. 9.15b). The Code requirement in clause 12.10.3 implies that at cut-off points the factored moment resistance diagram must have a horizontal offset from the applied moment diagram by the larger of d or $12d_b$, except at simple or exterior supports and at free ends of cantilevers.

Fig. 9.14 *Example of special end anchorage requirement*

9.14 DESIGN EXAMPLES

Calculations for the development requirements of flexural reinforcement in simply supported beams were illustrated in design Examples 5.9 and 5.10 in Chapter 5. Examples of a continuous beam and a beam-column connection are given below.

EXAMPLE 9.1

The moment envelope for an interior span of a continuous beam is shown in Fig. 9.16a. With a beam size of 300 × 450 mm and an effective depth of 390 mm, the flexural design gave three No. 25 bars as positive moment reinforcement at midspan and four No. 25 bars as negative moment reinforcement at the support. The beam is designed for shear using the Simplified Method in CSA A23.3-M84(11.3). That is, the effect of shear on longitudinal reinforcement has not been accounted for in flexural design.

One-third of the positive moment reinforcement and one-half of the negative moment reinforcement are to be terminated as they become no longer necessary for flexure. Design the bar cut-off details and check the development requirements by the Code. Take $f_y = 400$ MPa, and $f'_c = 30$ MPa. The factored load on the span is 48 kN/m.

Fig. 9.15 *Factored moment resistance diagram*

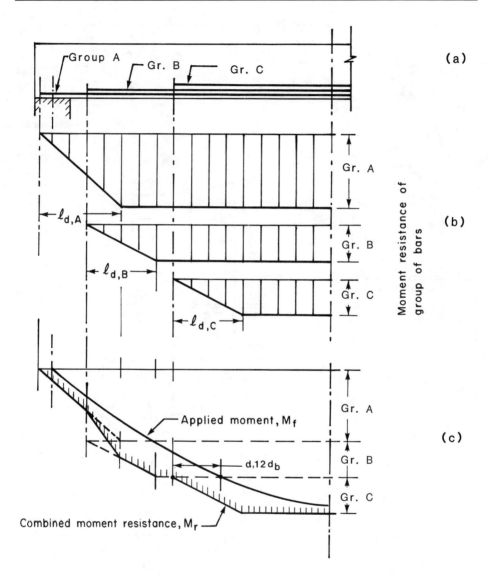

SOLUTION

(See Fig. 9.16b)

(a) Positive Moment Reinforcement

    1. Development length required

        The required development length for the bottom reinforcement is (none of the modification factors in 9.9.1b,c are applicable in this case):

Fig. 9.16 *Example 9.1*

$$\ell_d = 0.019\, A_b\, f_y / \sqrt{f_c'} \geq 0.058\, d_b f_y$$

$$= 0.019 \times 500 \times 400 / \sqrt{30} = 694 \text{ mm}$$

2. Location of bar cut-off

   The reinforcement ratio provided at midspan is $\rho = 3 \times 500/(300 \times 390) = 0.0128$. For this, from Table 5.4, $K_r = 3.73$ and the factored moment of resistance is $M_r = 3.73 \times 300 \times 390^2 \times 10^{-6} = 170$ kN·m. Then, one of the three bars is no longer needed for bending moment, at a section where

$$+ M = \frac{2}{3}(+ M_r) = \frac{2}{3} \times 170 = 113 \text{ kN·m}$$

   From the moment envelope in Fig. 9.16a, this section is located at A, a distance of 1190 mm from midspan. The bar must be continued beyond this point, a minimum distance equal to the larger of $d = 390$ mm or $12\, d_b = 302$ mm (CSA A23.3-M84(12.10.3)). Hence the one bar is continued a distance = $1190 + 390 \approx 1600$ mm from midspan.

3. Check development
    (i)   The cut-off bar has its critical section at midspan. The actual length of embedment beyond the critical section is:

    $$1600 \text{ mm} > \ell_d = 694 \text{ mm} \qquad \text{OK}$$

    (ii) The continuing bar has the critical section for development at section Λ, where the terminated bar is no longer required to resist flexure. With the two bars continued into the support a distance of 150 mm (CSA A23.3-M84(12.11.1)), the length of embedment provided is:

    $$3500 + 150 - 1190 = 2460 \text{ mm}$$
    $$> \ell_d + (d, \text{ or } 12d_b) = 694 + 390 = 1084 \text{ mm} \qquad \text{OK}$$

    (iii) To check bond stresses at the point of inflection (bar size limitation – CSA A23.3-M84(12.11.3)), the maximum additional embedment length beyond the point of inflection, $I1$, that can be considered effective, is

    $$\ell_a = \text{the larger of } d \text{ or } 12d_b$$
    $$= 390 \text{ mm}$$
    $$M_r = \text{flexural strength for the 2 bars}$$
    $$\simeq \frac{2}{3} \times 170 = 113 \text{ kN·m}$$

    Shear at face of support $= 48 \times 3.5 = 168 \text{ kN}$

    $$V_f = \text{shear at point of inflection}$$
    $$= 168 - 48 \times 1.025 = 119 \text{ kN}$$

    $$\ell_a + \frac{M_r}{V_f} = 390 + \frac{113 \times 10^6}{119 \times 10^3}$$
    $$= 1340 \text{ mm} > \ell_d \qquad \text{OK}$$

    (iv) In addition, since a flexural reinforcement bar has been terminated in a tension zone, the shear strength requirements at the cut-off location (CSA A23.3-M84(12.10.5)) must be checked. This was illustrated in Chapter 6 in Examples 6.1 and 6.2.

Note that the minimum positive moment reinforcement that must be continued into the support is only $0.25 A_s$, which in this example would be one bar. If the beam is part of a primary lateral load resisting system (by frame action, either alone or in combination with shearwalls, etc.), this minimum required reinforcement must be anchored into the support to develop stress f_y at the column face. This is to take care of possible stress reversals and to ensure ductility by yielding of bars at the connection. In view of the ductility requirement, it is *not* sufficient to develop the larger area continued into the support for a proportionately lower stress in lieu of providing full anchorage for the minimum required reinforcement.

(b) Negative Moment Reinforcement
1. Development length ℓ'_d for top reinforcement
   The negative moment reinforcement has more than 300 mm of concrete below it and hence will classify as "top reinforcement" and the modification factor of 1.4 is applicable.

$$\ell'_d = 1.4 \times 694 = 971 \text{ mm}$$

2. Location of cut-off
   With four No. 25 bars, $A_s = 2000$ mm², $\rho = 1.71\%$, and from Table 5.4, $K_r = 4.71$, $M_r = 4.71 \times 300 \times 390^2 \times 10^{-6} = 215$ kN·m. Two of the four bars are not needed for flexure at the section where

$$-M = \frac{1}{2}(-M_r) = 107 \text{ kN·m}$$

   Assuming for simplicity (and slightly conservatively) that the negative moment envelope is a straight line, $-M = 107$ kN·m at section B, a distance 835 mm from the face of the support. Continuing the bars for $d \geqslant 12d_b$ beyond B, two of the four bars can be terminated at $835 + 390 \simeq 1230$ mm from the face of the support.
   The remaining two bars (which is greater than the minimum of 1/3 required by the Code) are continued beyond the point of inflection $I2$ for a distance equal to the greater of d, $12d_b$, or 1/16 of clear span, $= 440$ mm. Hence these bars must be continued to $1670 + 440 = 2110$ mm, at least, from the face of the support.

3. Check development
   (i) For the two bars cut off, the actual embedment length beyond the critical section C (face of support) is:

$$1230 \text{ mm} > \ell'_d = 971 \text{ mm} \qquad\qquad \text{OK}$$

   (ii) For the two bars continued, the critical section for development is at B where the terminated bars are no longer required to resist flexure. Length provided beyond B, determined above, is:

$$2110 - 835 = 1275 \text{ mm}$$

   Length required in accordance with CSA A23.3-M84(12.10.4) is,

$$\ell'_d + (d \text{ or } 12d_b) = 971 + 390 = 1361 \text{ mm}$$

   This is more than the length determined above $= 1275$ mm. Hence the two bars continued must be taken to a length of $835 + 1361 \simeq 2200$ mm from the face of the support.

   (iii) All the negative moment reinforcement must also be fully developed to the left of section C by anchorage in or through the column with embedment length, hooks, or mechanical anchorage. At an interior column this is usually done by continuing the reinforcement into the adjacent span as part of the reinforcement in that span.

(iv) At the bar cut-off in the tension zone, the shear strength requirements must be checked as before.

EXAMPLE 9.2

Design the anchorage details for the negative moment reinforcement in the beam at the exterior beam-column connection shown in Fig. 9.17, using f_y = 400 MPa, and f'_c = 20 MPa.

Fig. 9.17 *Example 9.2*

SOLUTION

Required development length in tension, allowing the 1.4 factor for top reinforcement, is

$$\ell_d = 1.4 \times 0.019 \, A_b f_y / \sqrt{f'_c}$$
$$= 1.4 \times 0.019 \times 500 \times 400 / \sqrt{20} = 1190 \text{ mm}$$

Allowing the minimum clear cover of 40 mm, the maximum available straight length, ℓ, is $400 - 40 = 360$ mm $< \ell_d$. Since a straight embedment is inadequate, anchorage through the provision of a standard 90° hooked with a lead-in embedment length is considered. Basic development length for a standard hooked bar is

$$\ell_{hb} = 100 \, d_b / \sqrt{f'_c} = 100 \times 25.2 / \sqrt{20} = 563 \text{ mm}$$

The critical section is x at the face of support. With beam width = 300 mm and column width = 400 mm, the side cover is in excess of 60 mm. The hook will be placed with a cover of 51.3 mm on the vertical bar extension beyond the hook (that is, within column ties). Further the hook will be enclosed horizontally within at least three ties conforming to Code clause 12.5.3.4. With these, the modification factors applicable are 0.7 and 0.8, and the development length of the hook will be

$$\ell_{dh} = 0.7 \times 0.8 \times 563 = 315 \text{ mm}$$

The length available with the 51.3 mm cover is $400 - 51.3 = 348$ mm > 315, OK. Using No. 10 closed horizontal ties to confine the concrete in the anchorage zone, the maximum spacing is $3d_b = 3 \times 25.2 = 75.6$ mm. Use a spacing of 75 mm.

The minimum length over which the stirrups are to be placed is equal to the inside diameter of hook which for No. 25 bars with $f_y = 400$ MPa is 150 mm (CSA A23.1-M77, Table 6). Three No. 10 horizontal ties at a spacing of 75 mm may be provided as shown.

―――――――――――――― **EXAMPLE 9.3** ――――――――――――――

In the beam-column connection shown in Fig. 9.17, if the positive moment reinforcement continued into the support is considered as compression reinforcement in the design for negative bending moment at the column face, check the anchorage requirements at the support. Use $f_y = 400$ MPa and $f'_c = 20$ MPa.

SOLUTION

The basic development length for compression is:

$$\ell_{db} = 0.24 \, f_y d_b / \sqrt{f'_c} \geq 0.044 \, f_y d_b \text{ or } 200 \text{ mm}$$
$$= 0.24 \times 400 \times 25.2 / \sqrt{20} = 541 \text{ mm}$$

As no modification factore are applicable in this case,

$$\ell_d = \ell_{db} = 541 \text{ mm}$$

The maximum available straight length is 360 mm (Fig. 9.17), and hooks are ineffective for development of compressive stress. Hence the compressive stress in the bar at the column face has to be limited to the value that can be developed or alternatively smaller diameter bars must be used.

The stress that can be developed in the No. 25 bar with the available development length of 360 mm may be computed from:

$$\ell_{dc} = 0.24 \, f_s d_b / \sqrt{f_c'}$$

$$360 = 0.24 \times f_s \times 25.2 / \sqrt{20}, \qquad f_s = 266 \text{ MPa}$$

If the No. 25 bars are replaced by smaller size bars, the maximum size of bar which can develop a compressive stress of $f_y = 400$ MPa at the column face with the available development length may be computed from the above equation as:

$$360 = 0.24 \times 400 \times d_b / \sqrt{20}$$

$$d_b = 16.8 \text{ mm}$$

Thus No. 15 size bottom bars can develop a stress of f_y in compression at the column face.

---------------------------- **EXAMPLE 9.4** ----------------------------

The reinforcement details and the moment envelope for a continuous T-beam are given in Fig. 9.18. Draw the moment capacity diagram and check the development requirements at the inflection points. Use $f_c' = 20$ MPa, and $f_y = 300$ MPa. Assume that for bar extension requirements bent bars are effective from mid-depth of the beam. The beam is 300 mm wide and 600 mm deep overall, and the effective width of flange is 1850 mm. $h_f = 110$ mm. The maximum shear at the exterior face of the first interior support is 199 kN. Assume that design for shear is by the Simplified Method in CSA A23.3-M84 (11.3).

SOLUTION

The moment capacities are first computed. The reinforcement ratios at all sections are below the maximum permissible of $\rho_{max} = 0.0227$.
1. Section at exterior support:
   (i) With full reinforcement, $A_s = 1500$ mm^2

   $$d = 536 \text{ mm,}$$

   $$a = \frac{1500 \times 0.85 \times 300}{0.85 \times 0.6 \times 20 \times 300} = 125 \text{ mm}$$

$M_r = 0.85 \times 1500 \times 300 \times (536 - 125/2) \times 10^{-6} = 181 \text{ kN} \cdot \text{m}$

With two No. 25 bars, $M_r = \frac{2}{3} \times 181 = 121 \text{ kN} \cdot \text{m}$

With one No. 25 bar, $M_r \simeq \frac{1}{3} \times 181 = 60 \text{ kN} \cdot \text{m}$

ℓ_d for No. 25 top bars (Table 9.1) in this case is $= 1.4 \times 638 \simeq 900$ mm. The moment resistance diagram for the negative moment in this region is drawn in Fig. 9.18b.

Fig. 9.18 *Example 9.4*

(a) Beam B1 - B2 - B1

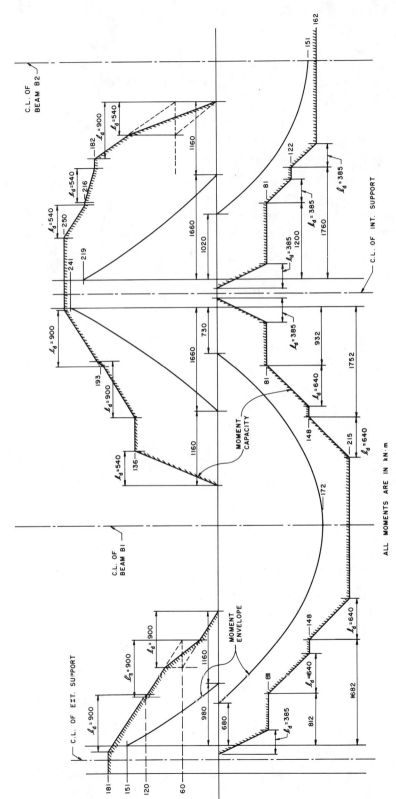

ALL MOMENTS ARE IN kN·m

(b) Moment resistance diagram

2. Section at interior support:
   (i)   For full reinforcement
         A_s = 2200 mm², d = 538 mm, a = 183 mm, M_r = 250 kN·m
   (ii) With four No. 20 and one No. 25,
         $A_s = 1700$ mm², $M_r = \dfrac{1700}{2200} \times 250 = 193$ kN·m
   (iii) With four No. 20 bars only
         A_s = 1200 mm², M_r = 136 kN·m
         ℓ_d for No. 20 top bars = 1.4 × 383 ≃ 540 mm
         ℓ_d for No. 25 top bars = 900 mm
   (iv) For interior span at this section
         For two No. 25 and three No. 20 bars, A_s = 1900 mm², M_r = 216 kN·m
         For two No. 25 bars and two No. 20 bars, A_s = 1600 mm², M_r = 182 kN·m
         For No. 20 bars only, A_s = 600 mm², M_r = 68 kN·m

3. Positive moment region, exterior span:
   (i)   Width of T-beam flange = 1850 mm
         d = 537 mm
   (ii) With full reinforcement, A_s = 1600 mm²

   $$a = \frac{1600 \times 0.85 \times 300}{0.85 \times 0.6 \times 20 \times 1850} = 21.6 \text{ mm} < h_f$$

   M_r = 0.85 × 1600 × 300 × (537 − 21.6/2) × 10^{-6} = 215 kN·m
   (iii) With one No. 25 bent, A_s = 1100 mm², M_r = 148 kN·m
   (iv) With two No. 25 bent, A_s = 600 mm², M_r = 81 kN·m
   (v) ℓ_d for No. 20 bottom bars = 385 mm
   (vi) ℓ_d for No. 25 bottom bars = 640 mm

4. Positive moment region, interior span:
   (i)   With full reinforcement
         A_s = 1200 mm², d = 539 mm, a = 16.2 mm, M_r = 162 kN·m
   (ii) With one No. 20 bent, M_r = ¾ × 162 = 122 kN·m
   (iii) With two No. 20 bent, M_r = ½ × 162 = 81 kN·m

The complete moment capacity diagram is shown in Fig. 9.18b.

The factored moment resistance developed at all sections exceeds the factored moment, M_f. The requirement of bar extension by the larger of d or $12d_b$ (= 539 mm here) specified in Code clauses 12.10.3 and 12.10.4 implies that the moment resistance diagram must have a horizontal clearance of at least d or $12d_b$ from the factored moment diagram at points of bar cut-off. This requirement is met by the moment resistance diagram in Fig. 9.18b. Also at least one-third of the negative moment reinforcement must extend beyond the point of inflection for a length equal to the greatest of d, $12d_b$ or $\ell_n/6$, equal to 1159 mm in this case. The extension provided is 1160 mm.

The No. 25 bars used as negative reinforcement at the exterior support must have adequate anchorage within the beam. Providing a standard 90° hook, the basic development length of hook is

$$\ell_{hb} = 100d_b/\sqrt{f'_c} = 563 \text{ mm}$$

Here the yield strength of reinforcement is f_y = 300 MPa. Further, the side cover to the hook is greater than 60 mm and the cover on the bar extension beyond the hook is 51.3 mm $>$ 50 mm. The applicable modificaton factors are $f_y/400$ and 0.7 (CSA 23.3-M84(12.5.3)). Hence the development length of the hook is

$$\ell_{dh} = 563 \times 0.7 \times 300/400 = 296 \text{ mm}$$

Length available within the girder is

$$450 - 51.3 = 399 \text{ mm} > 296 \text{ mm} \qquad\qquad \text{OK}$$

To check the bar diameter restriction of positive moment reinforcement at inflection points (Code clause 12.11.3), the critical section will be at the inflection point closest to the interior support of the outer span as the shear is greatest here. The shear at the inflection point may be taken conservatively as equal to the shear at the support = 199 kN. $\ell_a = d = 537$ mm.

$$\frac{M_r}{V_f} + \ell_a = \frac{81 \times 10^6}{944 \times 10^3} + 537$$

$$= 944 \text{ mm} > \ell_d = 385 \text{ mm} \qquad\qquad \text{OK}$$

9.15 SPLICING OF REINFORCEMENT

9.15.1 GENERAL

Splices are required when reinforcement has to be extended beyond the available length of individual bars or when bars are placed short of the required length, in several stages (such as in columns) for convenience in construction. Splicing of reinforcement is done by butt-welding, mechanical connection or overlapping the bars to develop their strength through bond.

CSA A23.3 requires that for a full welded splice, the bars be butted and welded to develop in tension at least 125 percent the specified yield strength of the bar. The 25 percent increase above f_y ensures yielding and a ductile failure and is adequate for a compression splice as well. All welding of reinforcing steel must conform to CSA Standard W186 – *Welding of Reinforcing Bars in Reinforced Concrete Construction*.

A mechanical connection must also be capable of developing in tension or compression, at least 125 percent of the specified yield strength of the bar. However, in regions where the area of reinforcement provided is at least twice that required, by analysis, welded or mechanical splices of strength less than 1.25 f_y are permitted as specified in CSA A23.3-M84 (12.15.4).

9.15.2 LAP SPLICE

In general, No. 35 and smaller size reinforcing bars are spliced by lapping the bars (Fig. 9.19a). The cracking and splitting behaviour observed in lap splice tests are similar to those in development length tests (compare Fig. 9.5 and Fig. 9.19b). The

Fig. 9.19 *Spliced bars*

(a) Lap splice

(b) Splitting failure for lap splice

discontinuity and stress concentrations at each end of a tension lap splice tend to induce early cracking and splitting from both ends. Evaluation of results of both development length and splice tests shows that the same parameters influence the length requirements in both cases and, in fact, the same expression (Eq. 9.4) can be used to determine both development and splice lengths (Ref. 9.2). The maximum reinforcement stress that is allowed in design is the yield strength. To avoid sudden failure at the splice, a full strength lap splice should develop more than the yield strength of the reinforcement. This will ensure yielding of bars and thereby a ductile behaviour of the member.

The length of splice required depends on the computed stress in the bar at the splice location and the proportion of the total number of bars that are spliced in the same region. As far as possible, splices should be located away from points of maximum stress in the reinforcement and should be staggered. The Code requirements for splice length are based on the development length, ℓ_d, the length being increased depending on the severity of the stress condition. However, the comparisons in Ref. 9.2 indicate that the Code requirements are very conservative for lap splices in regions of high stress.

9.15.3 CODE REQUIREMENTS FOR LAP SPLICES

CSA A23.3-M84 (12.14) permits lap splices for No. 35 and smaller size bars only except as provided in clauses 12.16.2 and 15.8.2.5. For larger bars, splice length requirements have not been established. Lap splices are generally made with the bar in contact, and in the case of noncontact splices the transverse bar spacing is limited to a maximum of one-fifth the splice length or 150 mm, whichever is less.

(a) Tension Splices of Deformed Bars

Tension lap splices are classified as A, B, or C, with corresponding minimum specified lap lengths. The classification as A, B, and C depends on the design stress in the bar in the splice region and the fraction of bars spliced within a lap length, as detailed in Table 9.3 wherein the corresponding minimum required lap lengths are also indicated. In this table, ℓ_d is the development length for a tensile stress of f_y computed as given in Section 9.9.1 (a) and (b), but the minimum limit of 300 mm on ℓ_d need not be applied. However, the lap length computed from Table 9.3 is subject to a minimum of 300 mm.

Table 9.3 Tension Lap Splice Classification and Specified Minimum Lap Lengths, ℓ_s

$\left(\dfrac{A_{s,\ provided}}{A_{s,\ required}}\right)$	Maximum Percent of A_s Spliced Within Required Lap Length		
	50	75	100
$\geqslant 2.0$	Class A $\ell_s = 1.0\ \ell_d$	Class A $\ell_s = 1.0\ \ell_d$	Class B $\ell_s = 1.3\ \ell_d$
< 2.0	Class B $\ell_s = 1.3\ \ell_d$	Class C $\ell_s = 1.7\ \ell_d$	Class C $\ell_s = 1.7\ \ell_d$

ℓ_s shall not be taken less than 300 mm.

(b) Compression Splices of Deformed Bars

The minimum lap length is specified as the development length in compression, but not less than $0.073\ f_y\ d_b$ for $f_y \leqslant 400$ MPa and $(0.133\ f_y - 24)\ d_b$ for $f_y > 400$ MPa, or 300 mm. For $f'_c < 20$ MPa, the lap length is increased by one-third.

When the splice is enclosed throughout its length by minimum ties or spirals, reduced lap lengths are permitted (CSA A23.3-M84(12.16)), subject to a minimum requirement of 300 mm.

In vertical bars required for compression only, end bearing splices are also permitted (Code clause 12.16.6).

(c) Other Cases

The Code gives specifications for the splicing details for bundled bars, deformed wire, welded wire fabric (both plane wire and deformed wire) and columns (see

CSA A23.3-M84 Section 12). For pairs of U-stirrups turned over to form a closed unit, the lap length required is 1.7 ℓ_d. However, in members at least 450 mm deep, such splices with $A_b f_y$ not more than 40 kN per leg may be considered effective if the legs extend the full available depth of the member.

REFERENCES

9.1 ACI Committee 408, *Bond Stress — The State of the Art*, J. ACI, Vol. 63, Oct. 1966, pp. 1161–1190.

9.2 Orangun, C.O., Jirsa, J.O., and Breen J.E., *A Reevaluation of Test Data on Development Length and Splices*, J. ACI, Vol. 74, March 1977, pp. 114–122.

9.3 Kemp, E.L., and Wilhelm, W.J., *Investigation of the Parameters Influencing Bond Cracking*, J. ACI, Vol. 76, Jan. 1979, pp. 47–71.

9.4 Jimenez, R., White, R.N., and Gergely, P., *Bond and Dowel Capacities of Reinforced Concrete*, J. ACI, Vol. 76, Jan. 1979, pp. 73–92.

9.5 Soretz, S., and Holzenbein, H., *Influence of Rib Dimensions of Reinforcing Bars on Bond and Bendability*, J. ACI, Vol. 76, Jan. 1979, pp. 111–125.

9.6 Ferguson, P.M., *Reinforced Concrete Fundamentals*, 4th ed., John Wiley and Sons, New York, 1979.

9.7 ACI Committee 318, Building Code Requirements for Reinforced Concrete (ACI Standard 318-63), American Concrete Institute, Detroit, Michigan, 1963.

9.8 Untrauer, R.E., and Warren, G.E., *Stress Development of Tension Steel in Beams*, J. ACI, Vol. 74, Aug. 1977, pp. 368–372.

9.9 Ferguson, Phil, M., *Small Bar Spacing or Cover — A Bond Problem for the Designer*, J. ACI, Vol. 74, Sept. 1977, pp. 435–439.

9.10 ACI Committee 318, Commentary on Building Code Requirements for Reinforced Concrete (ACI 318-83), 1st printing, American Concrete Institute, Detroit, Nov. 1983, 155 pp.

9.11 Marques, J.L.G., and Jirsa, J.O., *A Study of Hooked Bar Anchorages in Beam-Column Joints*, J. ACI, Vol. 72, May 1975, pp. 198–209.

CHAPTER 10

CONTINUITY IN REINFORCED CONCRETE CONSTRUCTION

10.1 GENERAL

Design of flexural members for a given bending moment and shear force was explained and illustrated in Chapters 5 to 8. Further, the detailing of reinforcement in members with known variations of bending moment was presented in Chapter 9. Most of the examples in these chapters dealt with simple beams in which the bending moments and shear forces were determined from statics; and where a continuous beam example was used, the moment variation was assumed to be known.

Reinforced concrete structures are usually cast-in-place, with all the structural elements such as slabs, beams, girders, columns and foundations cast monolithically. Such a continuous structure is statically indeterminate, and before the individual member dimensions and reinforcement can be determined by the methods described in Chapters 5 to 9, the distribution of design moments and shear forces along the individual members has to be determined. This is a problem of structural analysis. However, the analysis of a complete three-dimensional space frame subjected to various combinations of loading and different patterns of loading is very complex and hence simplifying approximations which are permitted by Codes are usually made. The basic approximation involves the subdivision of the structure into smaller, two-dimensional subframes (floors or roof, walls and plane frames; see Fig. 2.3), which in most instances can be further simplified into continuous beams or partial frames for the analysis of *gravity load* effects (see Section 10.2).

Even when a detailed analysis of the complete frame is envisaged, as for an unusual structure or structures of major consequence, the approximate analysis methods provide a handy tool to (1) proportion members in the preliminary design stage to obtain relative member stiffnesses needed for the detailed analysis, and (2) to provide a rough check on the results of the detailed analysis.

The maximum effects (moment, shear, etc.) on members of continuous structures due to the design loads (factored loads) are usually determined by elastic analysis (CSA A23.3-M84 (8.3)). It may be observed that there is an apparent inconsistency between the analysis and design procedures as the former uses elastic theory while the latter is based on inelastic section behaviour. However, the procedure leads to

Fig. 10.1   *Building frame approximated as series of plane frames*

Floor plan

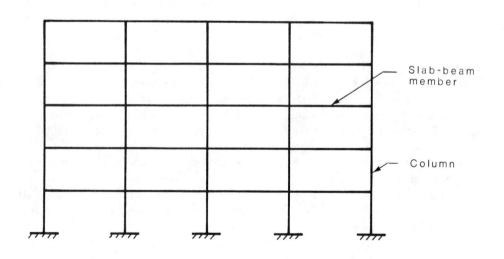

Longitudinal plane frame

safe and satisfactory designs and the procedure is justified if a linear moment-curvature $(M - \psi)$ relationship is assumed up to a moment of $M = M_r$. Since the design moments at various critical sections are determined for different loading patterns (see Section 10.2), for any one loading pattern most sections will have $M < M_r$ and a few sections will just reach $M = M_r$, so that the entire structure is within the linear range of the $M - \psi$ relationship. Under-reinforced sections, as permitted by the Code, have a nearly linear $M - \psi$ relationship up to $M = M_n$ (see Fig. 4.7a), and the inelastic analysis procedures such as "limit design" and "yield-line theory" for slabs usually assume such a linear $M - \psi$ variation up to $M = M_n$. The Code recognizes some inelastic response by permitting a limited redistribution of the moments in continuous flexural members calculated by the elastic theory (CSA A23.3-M84(8.4), see Section 10.7).

For a first order elastic analysis, where the effects of deflection and axial deformation are not included, the gravity load effects and lateral load (wind or earthquake) effects can be determined separately and superimposed. As is shown in Secton 10.2, the gravity load effects can often be determined by considering one floor at a time, together with the columns framing into that floor, and analyzing this partial frame. For lateral loads, however, each bent, consisting of the columns in *all* storeys in the bent together with the horizontal floor members (beam and/or slabs) connecting these columns is analyzed as a plane frame, with the horizontal loads generally applied at the joints (Fig. 2.3). Again, for buildings of moderate height, approximate methods of analysis may be used to determine wind effects on frame members.

In order to simplify the analysis, a three-dimensional framed structure is treated as a series of independent parallel plane-frames (bents) along the column lines in the longitudinal and transverse directions (Fig. 10.1). For gravity load effects, these plane frames can be further simplified into continuous beams or partial frames and analyzed with the loads so placed as to produce the worst effects at the design section considered. The critical loading patterns and the simplifications applicable for continuous beams and plane frames under vertical loads are considered next.

10.2 LOADING PATTERNS FOR CONTINUOUS BEAMS AND PLANE FRAMES

While the dead loads, by their nature, act throughout the structure at all times, the live load may act on any or all spans. In order to determine the maximum (and minimum) effects of the design load at any section in a member of a continuous beam or frame, it is first necessary to identify the spans to be loaded with live loads so as to create the worst effects. This is conveniently done by sketching, qualitatively, the shape of the influence line for the effect in question, using the Müller-Breslau principle (Ref. 10.1). The principle is that the influence line for any action (moment, shear force, reaction, etc.) at a section is given by the deflected shape of the structure resulting from a unit displacement *corresponding* to that action. Influence lines for the moments and shear forces at two sections in a continuous beam, and for the moments at two sections in a plane frame are shown in Figs. 10.2 and 10.3, respectively.

In continuous beams, the dead load which acts on all spans normally produces

Fig. 10.2 Influence lines and load (gravity) patterns for continuous beams

positive moment (that is, moment causing tension at the bottom fibres of the beam) in midspan regions and negative moments at support sections. Regarding the positioning of live loads, from the shape of the influence lines in Fig. 10.2 it can be concluded that:

1. the maximum positive moment in a span occurs when live load is placed on that span and every alternate span (Fig. 10.2 b(ii)),
2. the minimum positive moment in a span occurs when live load is placed on alternate spans, with the span under consideration unloaded (Fig. 10.2 b(iii)),
3. the maximum negative moment at the support section occurs when live load

is placed on the two neighbouring spans and then every alternate span (Fig. 10.2 c(ii)),
4. the influence on the desired effect at a section, of loads on spans far removed from the section under consideration, is relatively small.

Indeed, the alternating loading pattern shown in Fig. 10.2b gives the maximum positive moments, in *all* spans with live loads and the minimum positive moments in *all* spans with dead load only. With certain span proportions and live load/dead load ratios, the "minimum positive" span moment may be negative. For instance, a lightly loaded short span, flanked on either side by heavily loaded long spans, may have a net negative moment at midspan under this loading condition. Such a beam is subjected to positive or negative moment at midspan depending on the loading pattern and has to be reinforced with tension reinforcement placed at the top as well as at the bottom in the midspan regions. Minimum negative moments at support sections are not usually determined as there is seldom any chance of reversal of moment at this section under gravity loading. (Reversal of moment at support sections could, however, occur under lateral loading or settlement of supports.)

The fourth conclusion above indicates that when the spans and loadings do not vary a great deal, there is no great loss of accuracy if the influence of spans more than two supports away from the section under consideration are ignored. Thus in Fig. 10.2, the positive moment in span *BC* can be determined with reasonable accuracy by considering the portion *ABCD* only, and assuming that the beam is fixed at support *D* (beyond which the beam is, in fact, continuous). Similarly the negative moment at support *C* may be obtained by analyzing the part *ABCDE* of Fig. 10.2 c(ii), with support *E* assumed as fixed.

Figures 10.3a and b show the influence lines and loading patterns for the maximum positive moment at midspan, and the maximum negative moment near the right end, for the interior beam *BC* of a plane frame, respectively. The maximum positive moment at midspan of *BC* is obtained for the checkerboard pattern of live load placement, with the total load (dead + live) applied on the span in question and on alternate spans, and the dead load only on the remaining spans (Fig. 10.3a). It can also be seen that this loading pattern produces the maximum positive span moments in all spans with total load and minimum positive span moments in all spans with dead load only.

In Fig. 10.3b, the influence line is partially positive and partially negative along the beams above and below beam *BC*. Correctly, these two beams should receive live load only partially. However, for simplicity it is common practice in the analysis of building frames to apply distributed loads over their full length and the associated error is usually small. Furthermore, the ordinates of the influence lines decrease as one moves away from the spans under consideration, and particularly so as one moves on to floors at other levels. Hence in analyzing for gravity load effects on flexural members on a floor, the effects of loads on other floors may be neglected. Further, in usual building frames of regular shapes with no pronounced asymmetry in the geometry and loading, the sidesway caused by gravity loads is small and can be neglected. With these approximations, the analysis for gravity load effects on floor members may be made with reasonable accuracy by considering one floor at a time together with the columns framing into that floor, with the far ends of these

Fig. 10.3 *Gravity load patterns for plane frame*

(a) Loading for $+M_{F,max}$

(b) Loading for $(-M_G)_{max}$

columns assumed as fixed. This limited frame can be analyzed with the loads placed so as to produce the worst effect at critical sections. Thus the analysis for the cases in Fig. 10.3a and b may be simplified to the analysis of the corresponding partial frames shown in Fig. 10.3c and d, respectively.

For a column, the loading pattern producing the maximum moment and that producing the maximum axial load are different. Hence the maximum axial load and maximum moment do not occur simultaneously. The loading pattern for maximum moment in columns, shown in Fig. 10.4a for column CC', can once again be determined by sketching the influence line. It should be noted, however, that although the loading in Fig. 10.4a gives the maximum moments at the ends of column CC', the column bends in reverse curvature. In a relatively long column, a checkerboard pattern of loading causing single curvature bending of the column may be more

Fig. 10.3 *Continued*

(c) Partial frame and loading
for + $M_{F,max}$

(d) Partial frame and loading
for $(-M_G)_{max}$

critical than the pattern shown in Fig. 10.4a. Once again the moment at end C of column CC' in Fig. 10.4a is approximately given by the moment obtained from the analysis of the corresponding partial frame, shown in Fig. 10.4b. In analyzing this partial frame, the total column moment at a support such as C is distributed between the two columns above and below C in proportion to their stiffnesses. Somewhat greater column moments will be obtained, particularly in the exterior columns (column EE' in Fig. 10.4), if the influence of loading on the adjacent floors is also taken into account; however, even for these columns the partial frame analysis is generally satisfactory except for highly irregular spans or loadings. In general, column moments are much more sensitive to changes in the assumption and cannot be determined with the same degree of accuracy as beam moments.

10.3 APPROXIMATIONS PERMITTED BY CODE FOR FRAME ANALYSIS

CSA A23.3-M84 (8.3) permits the use of approximate methods of frame analysis for buildings of usual types of construction, spans and story heights. Specific approximations suggested are summarized below.

Fig. 10.4 *Gravity load pattern for maximum column moment*

(a) Loading pattern for column CC′

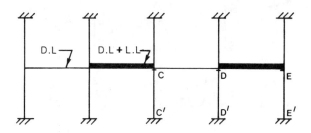

(b) Partial frame for maximum M$_{cc'}$

10.3.1 FLOOR MEMBERS

In accordance with the general frame behaviour discussed in Section 10.2, CSA A23.3-M84 (8.6–8.9) allows the analysis and design for vertical loading of floor (or roof) members of continuous construction by considering live load applied only to the floor (roof) under consideration and the far ends of the columns assumed as fixed. The loading patterns to be considered are (1) factored dead load on all spans plus factored live load on two adjacent spans (see Fig. 10.3d), giving the design negative moment at supports, and (2) factored dead load on all spans with factored live load on alternate spans (Fig. 10.3c), giving the maximum and minimum positive span moments on spans with and without live loads, respectively. In determining the bending moment at a support, the beam member may be assumed as fixed at any support two panels distant, provided the beam continues beyond that point (CSA A23.3-M84 (13.7.2.5). In analyzing for maximum and minimum positive moments in spans, the far ends of adjacent spans may similarly be considered as fixed.

10.3.2 COLUMNS

For columns (CSA A23.3-M84(8.8)), the loading pattern with factored live load on all floors above, which results in the maximum axial force (subject to the reduction allowed when the total tributary floor area supported by the column is large), and factored live load on a single (usually the larger and heavier loaded) adjacent span of the floor under consideration, which produces the maximum moment that can accompany the maximum axial force, is the most general combination for gravity loading. Further, because the interaction between axial strength and flexural strength of columns is non-linear (see Chapter 15), it is also necessary in column design to consider the loading condition which produces the maximum ratio of bending moment to axial force. This condition generally occurs under the checkerboard pattern of loading. In computing gravity load moments in columns by analyzing partial frames, the far ends of columns monolithic with the structure may be considered fixed as shown in Fig. 10.4b.

10.3.3 MOMENT COEFFICIENTS

For the design of continuous beams and one-way slabs with two or more approximately equal spans (the longer of two adjacent spans not exceeding the shorter by more than 20 percent) and with loads uniformly distributed, where the factored live load does not exceed three times the factored dead load, CSA A23.3-M84 (8.3.3) specifies approximate moment and shear values which may be used in lieu of more accurate analysis. These values are given in Table 10.1 and are generally conservative. It should be noted that the load patterns that produce the critical values for column moments (Section 10.3.2) are different from those for maximum negative moments in beams and hence the moment values in Table 10.1 should not be used to evaluate column moments.

10.3.4 MOMENT ENVELOPES FOR USE WITH CSA MOMENT COEFFICIENTS

To facilitate the detailing of bending or cut off of flexural reinforcement for members designed using the CSA moment coefficients, approximate moment envelopes and inflection points corresponding to these moment coefficients may be obtained based on the following considerations:

1. In the exterior span, the maximum positive moment in the span and the maximum negative moment at the exterior support occur simultaneously under the same loading pattern (see Fig. 10.5a).
2. In interior spans, when the span moment is maximum, the negative moments at the two supports are assumed equal (see Fig. 10.5b).
3. The specified positive moment in spans is assumed to be at the midspan section. Although the actual maximum positive moment may be slightly greater than the midspan moment, the latter is considered as the design moment. (An

Table 10.1 Moments and Shears in Continuous Beams Using CSA Coefficients (CSA A23.3-M84 (8.3.3)

(a) POSITIVE MOMENTS:

   (i)   End spans

      (1) if discontinuous end is unrestrained $\frac{1}{11}w_f\ell_n^2$

      (2) if discontinuous end is integral with support $\frac{1}{14}w_f\ell_n^2$

   (ii) Interior spans: $\frac{1}{16}w_f\ell_n^2$

(b) NEGATIVE MOMENTS

   (i)   At exterior face of first interior support:

      (1) two spans $\frac{1}{9}w_f\ell_n^2$

      (2) more than two spans $\frac{1}{10}w_f\ell_n^2$

   (ii) At other faces of interior supports $\frac{1}{11}w_f\ell_n^2$

   (iii) At faces of all supports for (1) slabs with spans not exceeding 3 m; and (2) beams where ratio of sum of column stiffnesses to beam stiffnesses exceeds eight at each end of the span $\frac{1}{12}w_f\ell_n^2$

   (iv) At interior faces of exterior supports for members built integrally with their supports

      (1) where the support is a spandrel beam or girder $\frac{1}{24}w_f\ell_n^2$

      (2) where the support is a column $\frac{1}{16}w_f\ell_n^2$

(c) SHEAR

   (i)   in end members at face of first interior support $1.15\,\dfrac{w_f\ell_n}{2}$

   (ii) at face of all other supports $\dfrac{w_f\ell_n}{2}$

Notes: w_f = factored load per unit length of beam

       ℓ_n = clear span for positive moment, negative moment at an exterior support, or shear, and the average of adjacent clear spans for negative moment at interior supports.

alternative procedure is to take the maximum span moment as the specified value.)

4. The loading patterns for the maximum negative moments at each end of a given span are different. However, to obtain the negative moment envelopes it is conservatively assumed that these maximum negative moments occur simultaneously. Two typical cases are shown in Fig. 10.5c.

Based on these approximations, the moment diagrams and the points of inflection may be computed using statics. The moment diagrams so obtained for the various moment coefficients are shown non-dimensionally in Fig. 10.6, which may be used

Fig. 10.5 *Moment diagrams for CSA coefficients (typical cases)*

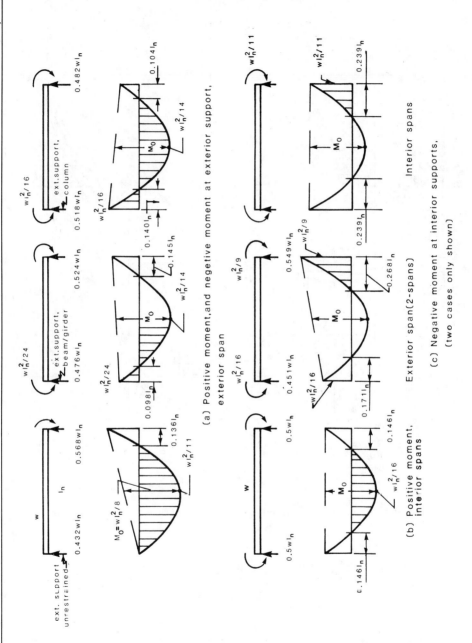

(a) Positive moment, and negative moment at exterior support, exterior span

(b) Positive moment, interior spans

(c) Negative moment at interior supports, (two cases only shown)
Interior spans

Exterior span (2-spans)

Fig. 10.6   *Non-dimensionalized moment diagrams for CSA moment coefficients*

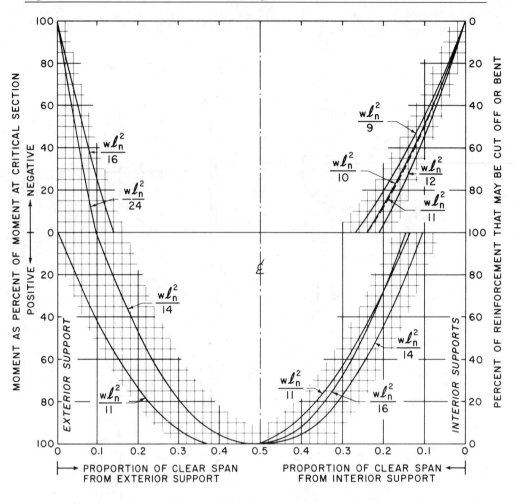

to determine locations where tension reinforcement is no longer required to resist flexure.

10.4   ANALYSIS PROCEDURES

There are several procedures for analysis of statically indeterminate structures (Refs. 10.1–10.3), and the most common involve the use of the slope–deflection equations, moment distribution or matrix methods. For detailed analysis of large frames, the matrix method using computers is the most powerful. In the absence of computer facilities, and particularly for the analysis of partial frames as detailed in Section 10.2, the method of moment distribution is most convenient and widely used. Ref.

10.4 shows that for building frames analyzed for gravity loads by considering each floor with the far ends of the columns as fixed, two-cycles of moment distribution are adequate to yield results of sufficient accuracy. Convenient tabular arrangements for computing maximum and minimum positive span moments, negative moments at beam ends, and column moments are also given in Ref. 10.4 and 10.5.

10.5 STIFFNESS OF MEMBERS

Regardless of the procedure used for the analysis of indeterminate frames, the stiffnesses of the members must be established first. For a prismatic member of length L, $4EI/L$ and GJ/L give the flexural and torsional stiffnesses, where I and J are the moment of inertia and the torsional inertia (cross-sectional property analogous to the polar moment of inertia of a circular section) of the member cross section. The modulus of elasticity $E = E_c$ is given by Eq. 1.1, and the shear modulus G may be taken as $E_c/2 \ (1 + \mu) \simeq E_c/2$, (assuming Poisson's ratio, μ, for concrete as zero). To compute I and J, the cross-sectional dimensions which are not initially known must be assumed or estimated based on preliminary design. The relative stiffnesses and the analysis may have to be back-checked and perhaps revised if the final dimensions selected are markedly different from the preliminary values.

Moment of inertia, I, of a reinforced concrete section is usually computed on the basis of either the gross concrete section (neglecting reinforcement) or the cracked section transformed to concrete. Ideally, the effective member stiffness values used in the frame analysis should reflect the degree of cracking, inelastic action and the amount of reinforcement along the member; however, it is difficult to accurately account for all such influences. CSA A23.3-M84 (8.6) allows the use of any reasonable and "consistent" assumptions for computing the flexural and torsional stiffnesses. For braced frames, only relative values of stiffnesses are required and in such situations, the assumption used to compute I does not significantly affect the analytical results as long as the same assumption is used for all members. The most common procedure is to use the moment of inertia of the gross concrete section to compute the stiffness. An alternative procedure, which better reflects the higher degree of flexural cracking in beams relative to columns, is to use one-half of the gross section I of the beam stem for beams and the gross I for columns (Ref. 10.6).

The precracking torsional stiffness of a rectangular beam is only about 10 to 20 percent of its flexural stiffness. Torsional cracking reduces the torsional stiffness to about 10 to 20 percent of its precracking value, so that the torsional stiffness becomes almost negligible compared to the corresponding flexural stiffness. Hence in indeterminate structures where torsion arises because the member must twist to maintain compatibility, it is reasonable to neglect torsional stiffness in the analysis of frames. Thus the torsional stiffness of beams transverse to the frame being analyzed is usually neglected except in two-way slab floor systems (Ref. 10.6).

Another question that arises in computing the stiffness of beams is the effect of flanges, and the flange width to be considered as contributing to the stiffness. In a one-way floor system such as shown in Fig. 10.7a, for the analysis of a plane frame along the columns in the east-west direction the floor loading to be considered is

Fig. 10.7 *Definition of slab-beam member for analysis of bent in east-west direction*

(a)

(b)

Section AA

Section BB

that placed over the width b. It would appear that the slab-beam member whose flexural stiffness is to be considered in the analysis of the frame is the T-beam with a flange width of b. However, with large values of b, the assumption that the entire flange width available is fully effective over the entire span may be questionable, particularly with the flanges being in the tension zone in regions of negative moment. It may be recalled, however, that when flanges are in tension, the Code (10.6.6) does provide for the distribution of main tension reinforcement over the "effective flange width" used in the design of sections. Hence it would be appropriate to use the same effective flange width to compute the moment of inertia as is used in design computations according to CSA A23.3-M84 (8.10), and the authors prefer this procedure.

An alternative procedure that has been suggested in Ref. 10.4 is to use twice the moment of inertia of the gross web section ($I = 2 \times b_w h^3 / 12$). This closely corresponds to an effective flange width of six times the web width for T-sections with a flange thickness to overall depth ratio (h_f/h) of 0.2 to 0.4. One advantage of this procedure is that it eliminates the need to compute the moment of inertia of the T-shaped section.

Yet another common situation where the assumptions regarding the moment of inertia adopted for the floor members vary amongst designers is the one-way slab-beam-girder floor system shown in Fig. 10.7b. One procedure is to treat interior beams marked $B1$ as a continuous T-beam supported on girders and carrying the gravity load on its tributary area of width b'. In the analysis of this continuous beam, the I of the effective T-section of the beam $B1$ is used for computing flexural stiffness. The torsional restraint offered by the girders may be neglected and the girders assumed to offer only vertical supports, except that some allowance is made for the restraint (and resulting negative moment) at the exterior support. A separate analysis is made for the beams $B2$ along the column line, considering it as a continuous T-beam supported and restrained by the columns. This procedure will usually result in slightly different designs for beams $B1$ and $B2$. An alternative procedure is to make all the beams within a panel width b tributary to a bent identical, and take their combined stiffness as the beam-member stiffness in the analysis of the frame along the column line. This procedure will be more appropriate when the girders are torsionally very stiff. In the case of two-way slab systems with or without beams along column lines (Figs. 2.5, 2.6, and 2.9b), CSA A23.3-M84 (13.7) specifies the use of the gross moment of inertia of the full panel width to compute the stiffness of the slab–beam member for frame analysis.

Sometimes beams are provided with haunches in order to increase their depth at supports in continuous spans where the negative moment and shear force are greater than in the span. Similarly, drop panels are frequently used around columns in flat slabs (see Chapter 14). In such cases the effects of the varying depth must be included in the analysis. Stiffness coefficients and fixed-end moment coefficients for haunched members are given in Ref. 10.7.

10.6 USE OF CENTRE-TO-CENTRE SPAN AND MOMENT AT SUPPORT FACE

In conventional analysis of frames, the members are represented by their centrelines and the resulting member lengths used (see Fig. 10.8a,c). CSA A23.3-M84 (8.7.2) also specifies the use of centre-to-centre distances for span lengths in the determi-

Fig. 10.8 *Moment correction to account for support width in continuous beams*

nation of the moment diagram. The use of centreline dimensions implies the following approximations:

1. The support reaction is concentrated at the support centrelines instead of being distributed in some fashion over the width of support (width of column).
2. The beam is prismatic right up to the support centreline and the beam stiffness is unaffected by the increased effective beam section between the column face and column centreline where the beam merges into the column, (see portion *BC* in Fig. 10.8a,b).
3. The beam centreline can deflect a column face (see point *B'* in Fig. 10.8c) rather than such movement being restricted by the column.

The effect of the first assumption on the moment diagram is usually insignificant and can be ignored (Ref. 10.4) The neglect of the increased stiffness and restraint of the beam within the column width results in a slight over-estimation of the positive moment at the midspan and a corresponding under-estimation of the negative moment at the support. These effects can be reasonably accounted for by assuming that the moment of inertia of the beam section is infinite over the width of the column (Fig. 10.8b). Such an adjustment will result in an upward shift of the theoretical moment diagram in Fig. 10.8d by $Vb/6$, where V is the computed beam

Fig. 10.9 *Column moments from frame analysis*

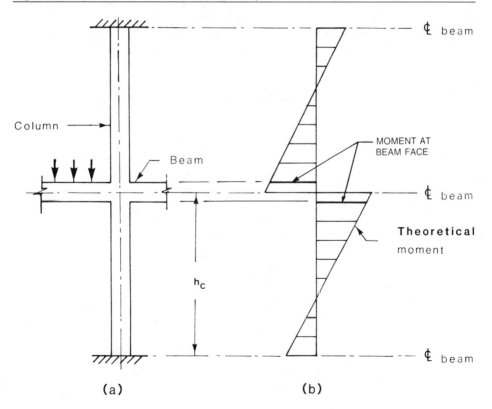

(a)　　　　　　　(b)

shear at the column centreline and b is the width of the column (Ref. 10.4). Since the effective beam depth is greatly increased once it merges into the column, the critical section of the beam for negative moment is generally at the column face. The Code (8.7.3) also permits beams and girders to be designed for the moments at the faces of supports. As can be seen in Fig. 10.8d, the adjusted design moment at the face of the support is given by $M_c - Vb/3$, where M_c is the theoretically computed moment at the support centreline. As far as the midspan section is considered, it is satisfactory and conservative to design for the theoretical moment, ignoring the correction $Vb/6$ which is usually small. It should be noted here that for the analysis of frames with two-way slab systems, CSA A23.3-M84 (13.7) specifies that account be taken of the enhanced moment of inertia of both slab–beam members and columns at their intersections (Chapter 14). When such allowance is made in the analysis, the shift of the moment diagrams indicated in Fig. 10.8d need not be made, and the design moment at the face of the support is obtained directly as $\simeq M_c - Vb/2$.

The conditions described above and in Fig. 10.8 with reference to a beam built into the supporting member also exist in a column, as shown in Fig. 10.9. The general practice for the analysis is to take the length of the column as equal to the centre-to-centre distance or storey height and to ignore the increase in moment of inertia of the column section within the beam depth (except for two-way floor systems dealt with in Chapter 14). With the lower moment gradient that usually exists in columns, the adjustments to be made to the centreline moment to obtain the column moment at the beam face is small (see Fig. 10.9b). Hence the column moment may be taken as that at the centreline of the beam as obtained from the analysis.

EXAMPLE 10.1

Centreline dimensions of a partial frame, consisting of one floor (slab-beam member) and the columns immediately above and below the floors, are given in Fig. 10.10a. The slab is one-way, spanning the frames spaced at 4.5 m (Fig. 10.10b), and has been designed as 160 mm thick. The specified live load on the floor is 5.44 kN/m² and there is a dead load of 0.5 kN/m² due to floor finish, ceiling, etc. The column size is estimated as 400 mm × 400 mm. Beams are provided only along the column lines in the longitudinal directions as shown in Fig. 10.10b. Use $f'_c = 30$ MPa and $f_y = 400$ MPa.

Analyze the frame for gravity loads and determine the moment envelope.

SOLUTION

The load per metre length of the beam is that coming from a width of 4.5 m (equal to the sum of one-half panel width on either side of the frame).

$$\text{Dead load from slab} = 0.16 \times 4.5 \times \frac{2400 \times 9.81}{1000} = 17.0 \text{ kN/m}$$

Dead load of beam stem, assuming overall beam size of 300 mm × 450 mm,
$$= 0.3 \times 0.29 \times 2.4 \times 9.81 = 2.05 \text{ kN/m}$$

Fig. 10.10 *Example 10.1*

(a) Frame

(b) Floor plan

Dead load due to floor finish, etc. = 0.5 × 4.5 = 2.25 kN/m
Live load from slab = 5.44 × 4.5 = 24.5 kN/m
Factored dead load = 1.25(17.0 + 2.05 + 2.25) = 26.7 kN/m
Factored live load = 1.5 × 24.5 = 36.7 kN/m

To estimate beam size, the maximum negative moment in the beam may be taken approximately as:

$$\frac{w_f \ell_n^2}{10} = \frac{(26.7 + 36.7)(6.0 - 0.4)^2}{10} = 199 \text{ kN·m}$$

and maximum shear as:

$$V_f = \frac{w_f \ell_n}{2} = \frac{(26.7 + 36.7)(6.0 - 0.4)}{2} = 178 \text{ kN}$$

For negative moment, the T-beam acts as a rectangular beam. Assuming a reinforcement ratio of about $0.65\rho_{\max}$, from Table 4.1, $K_u = 4.23$.

Required $bd^2 = \dfrac{M_u}{K_u} = \dfrac{199 \times 10^6}{4.23} = 47.3 \times 10^6 \text{ mm}^3$

Taking $b = 300$ mm, required $d = \sqrt{47.3 \times 10^6/300} = 397$ mm.
An overall depth of $h = 450$ mm initially assumed should be adequate for moment although the reinforcement ratio may be slightly higher than $0.65\rho_{max}$. The shear resistance, V_c, with this depth ($d \simeq 450 - 65 = 385$) is:

$V_c = 0.2\, \lambda\phi_c\sqrt{f_c'}\, b_w d = 0.2 \times 1.0 \times 0.6 \times \sqrt{30} \times 300 \times 385 \times 10^{-3} = 75.9 \text{ kN}$
$V_s = V_f - V_c = 178 - 75.9 = 102 \text{ kN}$
$\quad < V_{s,\text{lim}} = 0.8\, \lambda\phi_c\sqrt{f_c'}\, b_w d = 304 \text{ kN}$

Depth is adequate for shear

Minimum depth for deflection control (Table 5.2) is, for end spans, $5000/18.5 = 270$ mm, and for interior spans, $6000/21 = 286$ mm. The preliminary design of 300×450 mm for the beam section is OK.

Member Stiffness

Stiffnesses will be computed based on gross concrete area. For the slab–beam member, the moment of inertia of the effective T-section used in design (Section 5.9) will be used.

For columns,

$$I_c = 400 \times 400^3/12 = 2.13 \times 10^9 \text{ mm}^4$$

$$K_c = \frac{4E_cI_c}{h_c} = \frac{4E_c \times 2.13 \times 10^9}{3500} = 2.44\, E_c \times 10^6$$

For the T-beam, the effective width of flange is the least of:

$$
\begin{array}{lll}
0.25 \text{ span} & = 1250 \text{ mm for exterior span} \left.\right\} & \\
 & = 1500 \text{ mm for interior spans} \left.\right\} & \text{controls} \\
24\, h_f + b_w & = 24 \times 160 + 300 = 4140 \text{ mm} & \\
\text{spacing of beams} & = 4500 \text{ mm} &
\end{array}
$$

For exterior span, the section is shown in Fig. 10.11a

$$\bar{y} = \frac{950 \times 160 \times 80 + 300 \times 450 \times 225}{950 \times 160 + 300 \times 450} = 148 \text{ mm}$$

$$I_{b1} = \frac{1250 \times 148^3}{3} + \frac{950 \times 12^3}{3} + \frac{300 \times 302^3}{3}$$

$$= 4.11 \times 10^9 \text{ mm}^4$$

$$K_{b1} = \frac{4EI_{b1}}{L_1} = \frac{4E_c \times 4.11 \times 10^9}{5000} = 3.28\, E_c \times 10^6$$

(Note that if I_b is taken as twice I of the stem of the Tee, $I_b = 2 \times 300 \times 450^3/12 = 4.56 \times 10^9 \text{ mm}^4$. This is close enough to the I_{b1}, computed by detailed calculation

Fig. 10.11 *Effective section for slab–beam member*

(a) Beam AB   (b) Beam BC

(4.11×10^9), and is sufficiently accurate for relative stiffness computations).
Similarily, for interior beams *BC* and *CD* (Fig. 10.11b);

$$I_{b2} = 4.35 \times 10^9 \text{ mm}^4$$
$$K_{b2} = 2.90 \, E_c \times 10^6$$

The torsional stiffness of the edge beams at *A* and *E* will be ignored.

The frame analysis will be done by the method of moment distribution. The distribution factors at each joint for the beam members are:

$$\text{Joint } A\text{–}D_{AB} = \frac{K_{b1}}{\Sigma K} = \frac{3.28}{2 \times 2.44 + 3.28} = 0.402$$

$$\text{Joint } B\text{–}D_{BA} = \frac{K_{b1}}{\Sigma K} = \frac{3.28}{2 \times 2.44 + 3.28 + 2.90} = 0.297$$

$$D_{BC} = \frac{2.90}{2 \times 2.44 + 3.28 + 2.90} = 0.262$$

$$\text{Joint } C\text{–}D_{CB} = D_{CD} = \frac{2.90}{2 \times 2.44 + 2 \times 2.90} = 0.272$$

The fixed-end moments ($w_f L^2/12$) and simple span moments ($w_f L^2/8$) needed for moment calculations are tabulated in Table 10.2.

The load patterns for design moments at various locations are different as shown in Fig. 10.12. Instead of analyzing the frame for every loading pattern individually, all required cases can be obtained by appropriately combining the moment diagrams for (1) dead load only on all spans, and (2) live load only on each individual span, in turn. To get the latter, only two analyses need be done in this example, namely live load in spans *AB* and *BC* in turn. Because of symmetry, moment diagrams for live load in spans *CD* and *DE* will be the mirror image (end to end) of those due to loads in *BC* and *AB*, respectively. (Separation of dead load effects and live load effects may be

Fig. 10.12 *Loading patterns for critical moments in beam*

Table 10.2

	Exterior span		Interior span	
	FEM (kN·m)	S.S. moment (kN·m)	FEM (kN·m)	S.S. Moment (kN·m)
Dead load only (26.7 kN/m)	55.6	83.4	80.1	120
Live load only (36.7 kN/m)	76.5	115	110	165
Total load (63.4 kN/m)	132	198	190	285

1. Max. positive in AB, and CD; min. in BC, and DE

2. Max. positive in BC, and DE, min. positive in AB, and CD

(identical to case 1 reversed end to end)

3. Max. negative at B

4. Max. negative at C

required if gravity load effects have to be combined with lateral load effects.) Thus in all, only three moment distributions are required, namely:
1. dead load on all spans
2. live load on span *AB* only
3. live load on span *BC* only

These three distributions are set up in Table 10.3 at *a*, *b*, and *c*, respectively. (In these tables the distribution is set up in the conventional pattern for four cycles.)

In practice, the part of the frame beyond the support second removed from the section under consideration can be ignored, as previously explained, and the distribution may be limited to two cycles. Detailed explanation of such a set up is presented in Ref. 10.4. In Table 10.3, moments on only the beam members are presented with clockwise moment taken as positive. For any loading case, the total column moment at a joint can be obtained as the net unbalanced beam moment at the support, and the individual column moments can be obtained by distributing the total column moment between the columns immediately above and below the floor in proportion to their relative stiffnesses.

Combining the Distributions

The loading patterns for critical moments at the design sections in the beams are shown in Fig. 10.12. The moments for loading pattern No. 1 in Fig. 10.12 can be obtained by superposition of the moments for dead load on all spans (Table 10.3a), live load on span *AB* (Table 10.3b) and live load on span *CD* (Table 10.3c reversed end-to-end). This is done in Table 10.4 at *a*. This loading case gives the maximum span moment in *AB* and the minimum span moment in *BC*; as well as the maximum moment in span *CD* and the minimum moment in span *DE*. The latter two values, because of symmetry, are also equal to the maximum span moment in *BC* and the minimum span moment in *AB*. Thus from the single combination in Table 10.4a, the maximum as well as the minimum span moments can be obtained for both spans *AB* and *CD*. The complete moment diagram for all the spans obtained from Table 10.4a is plotted in Fig. 10.13.

Knowing the end moments in each span, the end shears and points of inflection in all spans may also be computed from statics. These computations are evident from the freebody diagram of each span, also given in Fig. 10.13. The moment diagram in Fig. 10.13 is plotted following the usual convention that sagging moment is positive and hogging moment is negative, and moments are drawn on the side on which each produces tension. It should be noted that when the end moments in a span are unequal as they usually are, the peak positive span moment will not be at midspan. The location of peak moment (zero shear) and the magnitude of the peak span moment can be computed from the freebody diagram for each span. However, for regular spans and loadings (and particularly in interior spans) the location of peak span moment is not far removed from midspan, and the peak moment is only slightly greater than the midspan moment. In such cases, the midspan moment may be taken as the design moment for the span. For instance, with the moment diagram in Fig. 10.13, for span *AB*, the moment is maximum (that is, shear is zero) at a distance from *A* of $150/63.4 = 2.37$ m compared to 2.5 m to midspan. The maximum moment at this section is 89.8 kN·m which is only 1.8 percent in excess of the moment at midspan. The points of influence are needed to detail bar extensions. Once again, for regular spans and loadings, exact calculations for location of inflection points are seldom needed, and bar extensions can be detailed in terms of span proportions (see Section 5.7).

Table 10.3   Moment Distributions

	A	B		C		D		E
Distribution Factor	0.402	0.297	0.262	0.272	0.272	0.262	0.297	0.402
C.O.F.	½	½		½		½		½
(a) DL on All Spans								
FEM	−55.6	55.6	−80.1	80.1	−80.1	80.1	−55.6	55.6
Bal.	22.4	7.3	6.4	—	—	6.4	−7.3	−22.4
C.O.	3.6	11.2	−2.9	3.2	−3.2	2.9	−11.2	−3.6
Bal.	−1.4	−3.3	—	—	—	—	3.3	1.4
C.O.	−1.7	−0.7	0.2	−1.5	1.5	−0.2	0.7	1.7
Bal.	0.7	0.2	—	—	—	—	−0.2	−0.7
C.O.	0.1	0.3	−0.1	0.1	−0.1	0.1	−0.3	−0.1
Bal.	—	−0.1	—	—	—	—	0.1	—
Total	−31.9	70.5	−76.5	81.9	−81.9	76.5	−70.5	31.9
(b) LL on Span AB								
FEM	−76.5	76.5						
Bal.	30.8	−22.7	−20.0					
C.O.	−11.4	15.4		−10.0				
Bal.	4.6	−4.6	−4.0	2.7	2.7			
C.O.	−2.3	2.3	1.4	−2.0		1.4		
Bal.	0.9	−1.1	−1.0	0.5	0.5	−0.4	−0.4	
C.O.	−0.5	0.5	0.2	−0.5	−0.2	0.3		−0.2
Bal.	0.2	−0.2	−0.2	0.2	0.2	−0.1	−0.1	0.1
Total	−54.2	66.1	−23.5	−9.1	3.2	1.2	−0.5	−0.1
(c) LL on Span BC								
FEM			−110	110				
Bal.		32.7	28.8	−30	−30			
C.O.	16.3		−15.0	14.4		−15.0		
Bal.	−6.6	4.5	3.9	−3.9	−3.9	3.9	4.5	
C.O.	2.2	−3.3	−2.0	2.0	2.0	−2.0		2.2
Bal.	−0.9	1.6	1.4	−1.1	−1.1	0.5	0.6	−0.9
C.O.	0.8	−0.5	−0.5	0.7	0.3	−0.5	−0.5	0.3
Bal.	−0.3	0.3	0.3	−0.3	−0.3	0.3	0.3	−0.1
Total	11.5	35.4	−93.2	91.8	−33.0	−12.9	5.0	1.5

Table 10.4

(a) Loading 1 of Fig. 10.12

	A	AB	B	BC	C	CD	D	DE	E
w, kN/m		63.4		26.7		63.4		26.7	
$wL^2/8$		198		120		285		83.4	
DL (Table 10.3a)	−31.9		70.5 / −76.5		81.9 / −81.9		76.5 / −70.5		31.9
LL on AB (T. 10.3b)	−54.2		66.1 / −23.5		−9.1 / 3.2		1.2 / −0.5		−0.1
LL on CD (T. 10.3c)	−1.5		−5.0 / 12.9		33.0 / −91.8		93.2 / −35.4		−11.5
Total	−87.6	88.2	132 / −87.1	23.5	106 / −171	114	171 / −106	20.3	20.3

(b) Loading 3 of Fig. 10.12

	A	AB	B	BC	C	CD	D	DE	E
$w =$		63.4		63.4		26.7		63.4	
$wL^2/8 =$		198		285		120		198	
DL (Table 10.3a)	−31.9		70.5 / −76.5		81.9 / −81.9		76.5 / −70.5		31.9
LL in AB (T. 10.3b)	−54.2		66.1 / −23.5		−9.1 / 3.2		1.2 / −0.5		−0.1
LL in BC (T. 10.3c)	11.5		35.4 / −93.2		91.8 / −33.0		−12.9 / 5.0		1.5
LL in DE (T. 10.3b)	0.1		0.5 / −1.2		−3.2 / 9.1		23.5 / −66.1		54.2
Total	−74.5	74	173 / −194	108	161 / −103	24	88.3 / −132	88	87.5

(c) Loading 4 of Fig. 10.12

	A	AB	B	BC	C	CD	D	DE	E
$w =$		26.7		63.4		63.4		26.7	
$wL^2/8 =$		83.4		285		285		83.4	
DL (T. 10.3a)	−31.9		70.5 / −76.5		81.9 / −81.9		76.5 / −70.5		31.9
LL in BC (T. 10.3c)	11.5		35.4 / −93.2		91.8 / −33.0		−12.9 / 5.0		1.5
LL in CD (T. 10.3c)	−1.5		−5.0 / 12.9		33.0 / −91.8		93.2 / −35.4		−11.5
Total	−21.9	22	101 / −157	103	207 / −207	103	157 / −101	22	21.9

Fig. 10.13 *Moment diagram for load pattern 1 of Fig. 10.12*

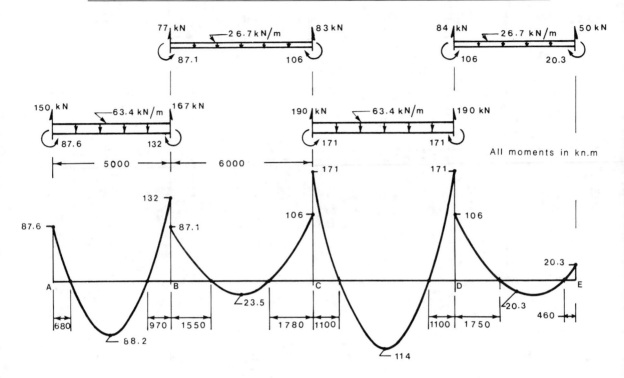

Fig. 10.14 *Maximum and minimum span moments*

All moments in kN.m

Fig. 10.15 *Moment diagrams for maximum negative moments*

(a) Maximum negative moment at B

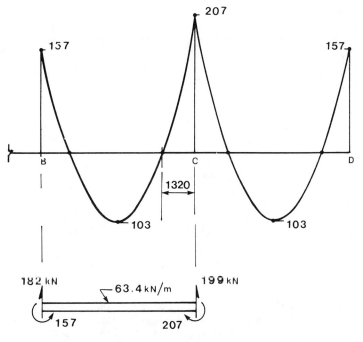

(b) Maximum negative moment at C

Combining the two halves of Fig. 10.13, the combined maximum and minimum span moment diagrams for spans *AB* and *BC* are drawn in Fig. 10.14. The shaded part of the moment diagram in Fig. 10.14 indicates the envelope for positive moment.

Following the procedure described above, the bending moments for loading cases 3 and 4 of Fig. 10.12 (giving the maximum negative moments at supports *B* and *C*, respectively) can be obtained by combining appropriately the results of the three moment distributions in Table 10.3. These combinations are also set up in Table 10.4 at *b* and *c*. The resulting moment diagrams are shown in Fig. 10.15, where only the spans on either side of the support considered are shown, as these loading cases have no relevance beyond the respective negative moment regions. For the exterior support *A*, the loading pattern for the peak negative moment is the same as at 1 in Fig. 10.12, for which the moment diagram is shown in Fig. 10.13. Combining these peak negative moment diagrams with the peak positive moment diagram in Fig. 10.14, the complete moment envelope shown in Fig. 10.16 can be obtained.

Fig. 10.16 *Moment envelopes for spans AB and BC*

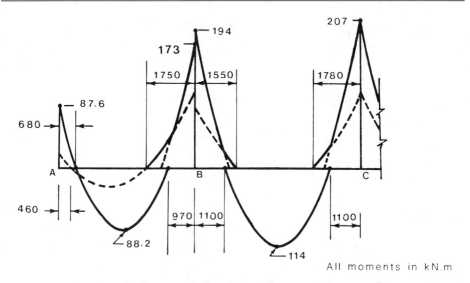

All moments in kN.m

EXAMPLE 10.2

For the problem in Example 10.1, determine the design moments, both positive and negative, for the spans *AB* and *BC* and compare with the approximate values recommended by the Code.

SOLUTION

The theoretical moment envelope, using centreline dimensions, was determined in Example 10.1 and is shown in Fig. 10.16. Adjustment to this diagram must be made to account for the increased beam stiffness within the column width, and the design negative moment must be determined at the column face. As explained in Section 10.6, these adjustments will result, approximately, in reductions in the span moment by $Vb/6$ and in the support moment at centreline by $Vb/3$; where V is the shear force at the support for the relevant loading case, and b is the width of column. The reduction in span moment is usually ignored so that the final design positive moments are:

$$\text{for span } AB, = 88.2 \text{ kN} \cdot \text{m}$$
$$\text{for span } BC, = 114 \text{ kN} \cdot \text{m}$$

Fig. 10.17 *Moments at column face*

For negative moments at support B, the centreline moments and shears were presented in Fig. 10.15a. Making the adjustments of $Vb/3$, as shown in Fig. 10.17,

$$M_{BA} = 173 - \frac{178 \times 0.40}{3} = 149 \text{ kN} \cdot \text{m}$$

$$M_{BC} = 194 - \frac{196 \times 0.4}{3} = 168 \text{ kN} \cdot \text{m}$$

Similar computations for the peak negative moments at A and C give:

$$M_{AB} = 87.6 - \frac{150 \times 0.4}{3} = 67.6 \text{ kN} \cdot \text{m}$$

$$M_{CB} = 207 - \frac{199 \times 0.4}{3} = 180 \text{ kN} \cdot \text{m}$$

The final design moments are presented in Fig. 10.18, for midspan sections and face-of-support sections.

The continuous beam in Example 10.1 satisfies the limitations in CSA A23.3-M84 (8.3.3) and hence the moment values specified therein and summarized in Table 10.1 may be used for design. These values are computed below:

For span *AB*, $\qquad\qquad \ell_n = 5 - 0.4 = 4.6 \text{ m}$

and for span *BC*, $\qquad \ell_n = 5.6 \text{ m}$

$$w_f = 63.4 \text{ kN/m}^2$$

Fig. 10.18 *Comparison of design moments by analysis and CSA coefficients*

(a) Analysis	67.6	88.2	149	168	114	180
(b) CSA cl. 8.3.3	83.8	95.8	165	150	124	181
Ratio $\frac{\text{(a)}}{\text{(b)}}$	0.81	0.92	0.90	1.12	0.92	0.99

All moments in kN·m

Positive span moments are:

(i) for span *AB* (discontinuous end *A* integral with support),

$$= \frac{1}{14} w_f \ell_n^2 = \frac{1}{14} \times 63.4 \times 4.6^2$$

$$= 95.8 \text{ kN} \cdot \text{m}$$

(ii) for span *BC*,

$$= \frac{1}{16} w_f \ell_n^2 = \frac{1}{16} \times 63.4 \times 5.6^2$$

$$= 124 \text{ kN} \cdot \text{m}$$

Negative moments are:

(i) at interior face of exterior support (member built integrally with supporting column),

$$M_{AB} = \frac{1}{16}\, w_f \ell_n^2 = \frac{1}{16} \times 63.4 \times 4.6^2 = 83.8 \text{ kN} \cdot \text{m}$$

(ii) at exterior face of first interior support (more than two spans),

$$M_{BA} = \frac{1}{10}\, w_f \ell_n^2 = \frac{1}{10} \times 63.4 \times \left(\frac{4.6 \times 5.6}{2}\right)^2$$

$$= 165 \text{ kN} \cdot \text{m}$$

(iii) at other faces of interior supports:

$$M_{BC} = \frac{1}{11}\, w_f \ell_n^2 = \frac{1}{11} \times 63.4 \left(\frac{4.6 + 5.6}{2}\right)^2$$

$$= 150 \text{ kN} \cdot \text{m}$$

$$M_{CB} = \frac{1}{11}\, w_f \ell_n^2 = \frac{1}{11} \times 63.4 \left(\frac{5.6 + 5.6}{2}\right)^2$$

$$= 181 \text{ kN} \cdot \text{m}$$

These values obtained by the Code equations are also presented in Fig. 10.18. The Code coefficients are, in general, conservative. This is so because they make allowances for several influences such as live load to dead load ratios of up to three and differences between all adjacent spans of up to 20 percent.

--- **EXAMPLE 10.3** ---

For the frame in Example 10.1, determine the design moments due to gravity loads for the columns framing into *A* and *B* at the floor level considered.

SOLUTION

The load pattern for the maximum bending moment in the column consists of live load placed on a single adjacent span and every alternate span. For the column at *A*, this loading pattern is identical to that shown as loading pattern No. 1. in Fig. 10.12 for which the moments were determined in Table 10.4a. The total column moment at *A* is equal to the unbalanced beam moment − 87.6 kN·m. Distributing this between the

Fig. 10.19 Column moments, Example 10.3

columns immediately above and below the floor, in proportion to their relative stiffnesses (the column stiffnesses are equal in this Example), at A,

$$M_{COL,\ ABOVE} = M_{COL,\ BELOW} = \tfrac{1}{2}\,(87.6) = 43.8\ \text{kN} \cdot \text{m}$$

At support B, the live load must again be placed on a single adjacent span. Loadings 1 and 2 in Fig. 10.12 correspond to live load being placed on the span to the left or right of B, respectively. From the moment values given in Table 10.4a, the larger unbalanced moment at B occurs when the load is placed to the right of B (loading 2 of Fig. 10.12), moments for which can be obtained as the end-for-end reversed values in Table 10.4a.

Total column moment at $B = 171 - 106 = 65\ \text{kN} \cdot \text{m}$.

Moments on columns above and below B are each equal to $\frac{1}{2}(65) = 32.5$ kN·m. The column moments are indicated in Fig. 10.19. Note that no adjustment has been made for the beam depth and the increased column stiffness within the beam depth.

10.7 INELASTIC ANALYSIS AND MOMENT REDISTRIBUTION

10.7.1 LIMIT ANALYSIS

Reinforced concrete frames may be analysed by inelastic theory (Refs. 10.8, 10.9). The procedure generally applied to reinforced concrete is termed *Limit Analysis* and is very similar to the *Plastic Analysis* procedure applied to steel structures. Limit analysis is based on the moment curvature (M–ψ) relationship of under-reinforced sections.

Fig. 10.20 *Idealized moment-curvature relation for under-reinforced beams*

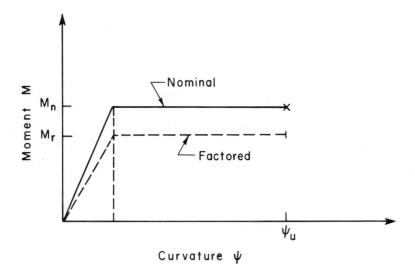

The M–ψ relationship for under-reinforced sections with steel bars having a well defined yield plateau (Fig. 4.7a) can be idealized as a bilinear elastic–plastic relation as shown in Fig. 10.20. This idealization implies that in a section subjected to flexure, the moment increases linearly with curvature until the applied moment reaches the value M_n. The latter can be taken equal to the nominal ultimate strength of the section developed in Eq. 4.19, or using Eq. 4.38 with ϕ_s and ϕ_c taken equal to unity, as:

$$M_n = \rho f_y \left(1 - \frac{\rho f_y}{1.7 f_c'} \right) b d^2$$

On further straining, the moment at the section cannot increase; however, the section "yields" and the curvature will continue to increase at constant moment. The section is then said to have developed a "plastic hinge." The section fails when the curvature reaches its ultimate value ψ_u. If the material resistance factors, ϕ_s and ϕ_c, are also taken into account, the factored moment resistance is M_r (given by Eq. 4.38) instead of M_n, the M–ψ relationship can be idealized as shown by the broken lines in Fig. 10.20, and the plastic hinge will form when the moment reaches the value M_r.

Consider an indeterminate structure such as the fixed-ended beam of uniform strength shown in Fig. 10.21a, subject to an increasing uniform load of w per unit length. As long as the maximum moment in the beam is less than the factored resistance of the section, M_r, the beam behaves elastically. When the maximum moment, which is at the support sections A and B and equal to $wL^2/12$, reaches the value M_r at a load of, say, w_1, the limit of elastic behaviour is reached and these sections are about to develop plastic hinges. The moment diagram at this load is as shown in Fig. 10.21b, and gives:

$$w_1 \frac{L^2}{12} = M_r$$

$$w_1 = \frac{12\,M_r}{L^2}$$

Any further loading will not increase the support moment, but will increase the curvature at these sections so that the beam now behaves like a pin-ended beam with constant end moments equal to M_r (Fig. 10.21c). Additional loading causes an increase in span moment only, as indicated by the broken lines in Fig. 10.21d. This increase can continue until the midspan moment also reaches M_r, provided of course that the curvature at the support does not exceed the ultimate value ψ_u. At this load, say, w_2, a plastic hinge will form at midspan as well. The bending moment diagram will now have the distribution shown by the solid line in Fig. 10.21d. The beam, now connected by three plastic hinges, behaves like a hinged "mechanism" which can deform with no further increase in load (Fig. 10.21e). It is also evident from the moment diagram in Fig. 10.21d that the load cannot increase above w_2 without an increase in moment at the supports or midspan, which is not possible. Thus for the fixed ended beam in Fig. 10.21a, a "collapse mechanism" is formed as the third plastic hinge develops, and the corresponding collapse load (or ultimate load) is obtained from Fig. 10.21d as:

$$2\,M_r = \frac{w_2 L^2}{8}$$

$$w_2 = \frac{16 M_r}{L^2}$$

In this example the collapse load $w_2 = 16\,M_r/L^2$ is one-third greater than the load at the end of elastic behaviour, $w_1 = 12\,M_r/L^2$. This simple example illustrates

Fig. 10.21 *Inelastic behaviour of fixed-ended beam*

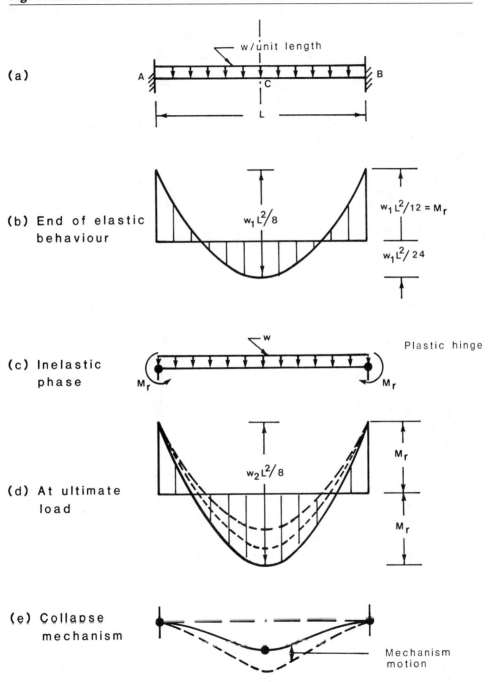

(a)

w/unit length

A B

C

L

(b) End of elastic behaviour

$w_1L^2/8$

$w_1L^2/12 = M_r$

$w_1L^2/24$

(c) Inelastic phase

w

Plastic hinge

M_r M_r

(d) At ultimate load

$w_2L^2/8$

M_r

M_r

(e) Collapse mechanism

Mechanism motion

Fig. 10.22 *Design based on elastic and limit analyses*

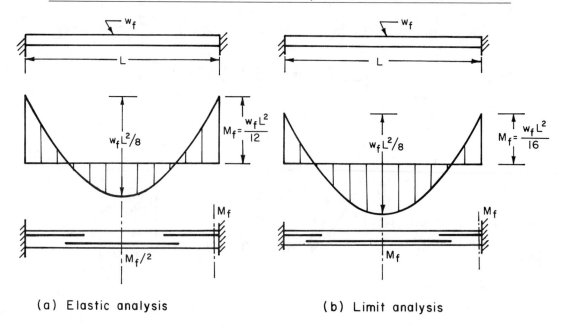

(a) Elastic analysis   **(b) Limit analysis**

that in an indeterminate structure, the attainment of the ultimate strength at one section does not necessarily mean that the maximum strength of the structure is reached. On the contrary, with additional loading, such highly stressed sections will yield locally (develop plastic hinges) which could lead to a *redistribution* of moments to lower stressed sections resulting in substantially increased capacity.

Referring to factored moment, a fixed ended beam with a uniform load w_f analyzed by elastic theory (Fig. 10.22a) will have to be designed for a moment $M_f = w_f L^2/12$. The maximum moment at midspan will be only $w_f L^2/24$. If the beam has uniform strength throughout, this will mean that the strength at the midspan section is not fully utilized. The same beam analyzed by limit analysis procedure with full moment redistribution developed, will have a design moment of $M_f = w_f L^2/16$ (Fig. 10.22b), which is only 75 percent of the design moment based on elastic analysis. Thus economy could result from recognizing inelastic behaviour. In steel structures, prismatic members of uniform strength are, in fact, mostly used, and substantial savings result by adopting plastic analysis and design. However, in reinforced concrete, even when a prismatic member is used, it is possible to design the amount of flexural reinforcement to suit the actual moment at any section and hence the savings may not be so significant. Thus the positive moment reinforcement for the beam in Fig. 10.22a could be designed for the moment $w_f L^2/24$, in which case both elastic analysis and limit analysis will give identical results. Of course, in continuous structures the situation is somewhat different in that the loading conditions for critical design moments at support sections and at midspan are usually different. Hence for a specific loading pattern causing maximum moment at a chosen section (such as a support), there will be spare capacity elsewhere in the structure (such as at midspan), and advantage could be taken of the economy resulting from possible inelastic action and consequent moment redistribution. Despite this, the major advantage to the designer of recognizing moment redistribution is that it enables him or her to slightly "adjust" the moment diagram, reducing (or increasing) the moment at some sections, and correspondingly increasing (or decreasing) the moment elsewhere leading to the design of an economic structure of more balanced proportions.

One other consideration that has somewhat limited the use of limit analysis is the restricted rotation capacity of hinging regions in reinforced concrete members. Because of the large inelastic strain which steel can sustain, plastic hinges in steel structures can usually develop the rotations necessary to effect full redistribution of moment leading to the formation of a collapse mechanism. However, the maximum compressive strain which concrete can sustain and hence the maximum inelastic rotation that a plastic hinge in a reinforced concrete member can develop before failure is limited. Therefore, full redistribution cannot be taken for granted, and in using limit analysis it is necessary to evaluate the rotation capacities of hinging regions and to ensure that these are adequate to meet the respective rotation requirements implied in the analysis.

10.7.2 REDISTRIBUTION OF MOMENTS PERMITTED BY CODE

Recognizing the inelastic behaviour of under-reinforced flexural members CSA A23.3-M84(8.4) permits a limited amount (up to 20 percent) of redistribution of negative moments in continuous beams. Accordingly, the negative moments at supports computed by elastic theory for any loading pattern may be increased or decreased by up to

$$(30 - 50 \; c/d) \leq 20 \text{ percent}$$

and the negative moments so modified used for calculation of the span moment. Here c/d is the neutral axis depth ratio. Sections having low percentages of tension reinforcement and/or containing compression reinforcement have low c/d ratios and can effect larger redistributions. Under-reinforced continuous beams possess adequate rotation capacity to effect the limited moment redistribution permitted above.

──────────────── *EXAMPLE 10.4* ────────────────

Assuming that the reinforcement ratios are such that a 15 percent adjustment of support moments is permitted for all spans, determine a more economic and balanced distribution of design moments for the continuous beam analyzed in Example 10.1.

SOLUTION

The moment envelope obtained by elastic analysis is given in Fig. 10.16. For flexure, the overall beam size is controlled by the maximum negative moment at supports where the continuous T-beam acts as a beam of rectangular section. Hence it may be desirable to reduce the maximum negative moment, which is 207 kN·m at support C for the loading case shown in Fig. 10.15b. The corresponding moment diagram obtained by elastic analysis is shown in Fig. 10.23a for span BC. Reducing the support moment by the full 15 percent brings it down to 207 × 0.85 × 176 kN·m. In order to cause the least increase in the span moment, the negative moment at B in span BC for this loading is increased to 176 kN·m (an increase of only 12 percent) so that both support sections will now have the same modified design moment. These two adjustments cause the span moment to increase to 285 − ½(176 + 176) = 109 kN·m which is still less than the maximum span moment of 114 kN·m in the analytical moment envelope (Fig. 10.16). The modified moment diagram is shown by the broken lines in Fig. 10.23a.

The moment diagram giving the maximum negative moment at B obtained in Example 10.1 is shown in Fig. 10.23b. Having seen that the lowest adjusted negative moment at C is 176 kN·m, and decided that both supports B and C will be designed for the same adjusted moment, once again the negative moments at B and C in Fig. 10.23b are adjusted to 176 kN·m each, corresponding to a decrease of 9.3 percent at B and an increase of 9.3 percent at C. The modified diagram is also shown in Fig. 10.23b and is identical to the modified diagram in Fig. 10.23a.

The elastic analysis distribution giving the maximum span moment for BC is presented in Fig. 10.23c. In order to reduce the span moment, the support moments are adjusted upward to the previously established value of 176 kN·m, an increase of 3 percent. The adjusted span moment is again 109 kN·m, as shown in Fig. 10.23c

Considering span AB and the moment diagram for maximum negative moment at B (Fig. 10.23c) there is little to be gained by adjusting any of the moments as the support section B is conveniently reinforced for the moment of 176 kN·m by continuing the bars designed for the negative moment at the left end of span BC. The moment diagram giving the maximum negative moment at support A and maximum span moment in AB is shown in Fig. 10.23e. Since there is reserve capacity at support B, this moment may be adjusted upwards by 15 percent to a value of 132 × 1.15 = 152 kN·m. This results in a reduction of span moment to 78.3 kN·m.

The moment at all critical sections obtained from elastic analysis is presented along with the corresponding moments obtained by allowing moment redistribution as permitted by the Code in Table 10.5 for comparison. It should be noted that the redistribution of moments may be effected in other ways as well, depending on specific requirements.

Table 10.5   Moment Redistribution as Permitted by Code

	A			B			C
		Span *AB*			Span *BC*		
	− ve	+ ve	− ve	− ve	+ ve	− ve	
Moment from elastic analysis	87.6	88.2	173	194	114	207	
Adjusted moments	87.6	78.3	173	176	109	176	

All moments in kN·m

Fig. 10.23 *Redistribution of moments, Example 10.4*

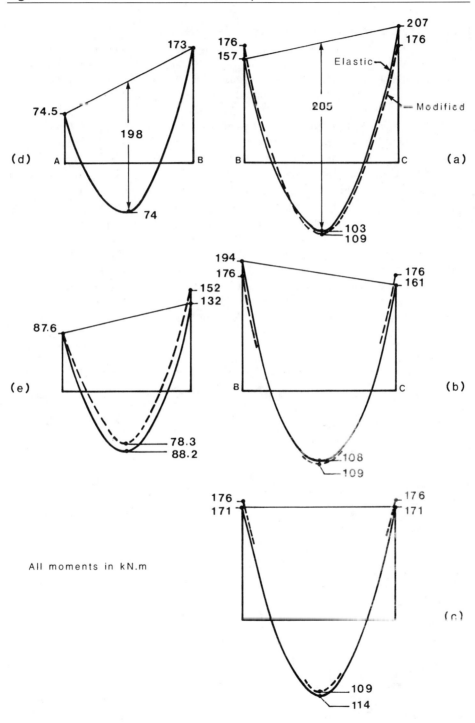

All moments in kN.m

10.8 PRELIMINARY DESIGN

Before a structure can be analyzed for load effects and its members designed, it is necessary to estimate cross-sectional dimensions in order to assess the dead loads due to self weight and to compute the moment of inertia of sections and member stiffness properties needed for the analysis. This may be done by a preliminary design, based on member loads estimated using tributary areas supported by each member, and approximate moment factors, although with experience the member sizes can be estimated without any such calculations. For preliminary designs, the floor areas tributary to beams, girders and columns may be taken approximately as shown in Fig. 10.24. These areas are based on the assumption that if a two-way slab is divided into four areas by lines midway between lines of support as shown in Fig. 10.24a, the load on each area is transferred to the adjacent support. One-way slabs are assumed to transfer the loads only to the longer (supported) sides. Girders have a tributary area equal to the floor area directly supported by them, if any, plus one-half of the sum of the tributary areas of beams supported by them. For columns, the tributary area per floor is one-half the sum of the tributary area of the floor members framing into it. The moments and shears may be estimated using the Code coefficients (see Secton 10.3.3); or alternatively the moment may be taken equal to the fixed-end moment for the span and loading, and the shears as those under simple support conditions. For continuous beams, cross-sectional di-

Fig. 10.24 *Tributary areas for beams, girders and columns*

(a) Two-way slab

(b) Two-way slab with beams

(c) Two-way slab with beams and girders

(d) One-way slab with beams and girders

mensions are governed by the moment and shears at the support sections. For columns, the tributary area is the sum of the tributary areas for the column from all levels above the storey under consideration. Sizes of interior columns are primarily determined by the axial loads on them since the unbalanced moments are not large for regular column spacings. The exterior columns with the floor extending on one side only are generally subjected to greater unbalanced moments and their size, determined on the basis of axial load, must be increased to allow for these moments, particularly in the upper storeys where the axial loads are relatively low compared to the moments.

PROBLEMS

---

1. The centreline dimensions of a plane frame are as shown in Fig. 10.25a, and the dimensions and section properties of the partial frame used for the second floor design in Fig. 10.25b. The load on the slab-beam member consists of a dead load (including self weight of slabs and beams) of 19.8 kN/m and a live load of 16.8 kN/m. Using elastic analysis, determine the moment envelope for the slab-beam member.

2. For the floor member in the partial frame in Problem 1, determine the design

Fig. 10.25 *Problems 1, 2, and 3*

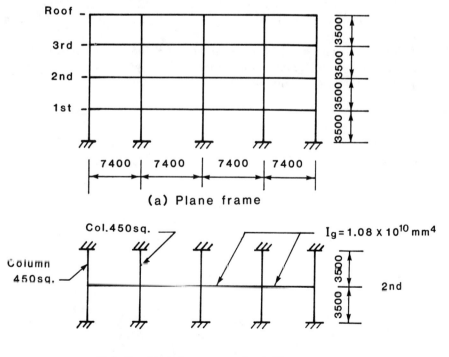

(a) Plane frame

(b) Partial frame – 2nd floor

moments at all critical sections and compare with the moments obtained by the Code coefficients.

3. Assuming that the reinforcement ratio will be so limited as to allow a moment redistribution of 10 percent, determine optimum redistributed design moments at all critical sections for the floor member in Problem 1.

REFERENCES

10.1 Hsieh, Y.Y., *Elementary Theory of Structures*, Prentice-Hall, Inc., New Jersey, 1970, 226 pp.

10.2 Wang, C.K., *Statically Indeterminate Structures*, McGraw Hill, New York, 1953.

10.3 Weaver, W. Jr., and Gere, J.M., *Matrix Analysis of Framed Structures*, 2nd Ed., Van Nostrand Co., New York, 1980, 492 pp.

10.4 *Continuity in Concrete Building Frames*, 4th Ed., Portland Cement Association, Chicago, Illinois, 1959, 56 pp.

10.5 *Frames and Continuous Structures — Analysis by Moment Distribution*, Concrete Information IS 210.01D, Portland Cement Association, Skokie, Illinois, 1980.

10.6 ACI Committee 318, *Commentary on Building Code Requirements for Reinforced Concrete* (ACI 318–83), American Concrete Institute, Detroit, Michigan, November 1983, pp. 33–36.

10.7 *Handbook of Frame Constants*, Engineering Bulletin, Portland Cement Association, Skokie, Illinois, available through Canadian Portland Cement Association, Ottawa, Ont., 1958, 33 pp.

10.8 ACI-ASCE Committee 428, *Limit Design of Reinforced Concrete Beams and Frames — Addendum*, (supplement to bibliographies published in ACI Journal in Nov. 1961 and Dec. 1962), J. ACI, Vol. 60, Oct. 1963, pp. 1471–1473.

10.9 Cohn, M.Z., *Limit-Design Solutions for Concrete Structures*, ASCE J. Struct. Div., Vol. 83, ST1, Feb. 1967, pp. 37–57.

CHAPTER 11

SERVICEABILITY LIMITS

11.1 INTRODUCTION

The need for considering two classes of limit states, namely *ultimate limit states* (concerned with safety) and *serviceability limit states* (concerned with satisfactory performance under intended use and occupancy) was explained in Chapter 3. Reinforced concrete structures and structural components are designed to have sufficient strength and stability to withstand the effects of factored loads, thereby satisfying the safety requirements. The design for serviceability limits is made at specified (service) loads (generally, the design is first made for strength and the serviceability limits are checked). The two major serviceability requirements of reinforced concrete structures in general relate to limiting deflections and crack control (CSA A23.3-M84(9.1)).

11.2 DEFLECTIONS

Accurate prediction of deflection of reinforced concrete members is difficult because of several factors such as: (1) uncertainties in the flexural stiffness EI of the member, as influenced by the varying degrees of tensile cracking of the concrete, varying amounts of flexural reinforcement, and variations in the modulus of elasticity and modulus of rupture of concrete; (2) curvature and deflection induced by differential shrinkage and creep, and the uncertainties about the time-dependent and environmental effects that influence these; (3) inelastic flexural behaviour of members; and (4) the inherently high degree of variability of measured deflections even under controlled laboratory conditions. Furthermore, the deflection that may be considered as allowable in a given situation is dependent on several aspects (such as aesthetic and sensory acceptability, effect on attached structural and non-structural elements, possible interference with expected performance (funtional), etc.) particular to that situation, and may often be somewhat arbitrary (Ref. 11.1). Thus approximations and simplifications are essential in deflection computations. At the same time, because of the random nature of deflection and its high variability, the apparent accuracy of elaborate and sophisticated calculation procedures is often illusory. Results of deflection computations must be regarded as representative values used for the purpose of comparison with the empirically set limiting values.

Deflection of reinforced concrete members is considered in two parts; namely, the *immediate* (or short-time) deflection occurring on application of the load, and the additional *long-time deflection* resulting mostly from shrinkage and creep under sustained loading. The immediate deflection due to applied dead load is often not a controlling factor as it can be compensated for by cambering the members. The long-time deflection is generally of more concern to the designer. The discussion that follows deals with deflection under static loading only.

11.3   DEFLECTIONS BY ELASTIC THEORY

Structural response at service load levels is nearly elastic and hence short-time deflection calculations for serviceability limits are generally based on elastic theory. Expressions for the elastic deflections of homogeneous beams for any loading and support conditions can be derived using standard methods of structural analysis; formulas for several standard cases are presented in handbooks (Refs. 11.2–11.4). These are usually in the form:

$$\Delta = k \frac{WL^4}{EI} = k_1 \frac{ML^2}{EI} \tag{11.1}$$

where Δ is typically the midspan deflection; W is the total load on the span; L is the span; EI is the flexural rigidity of a reference section; M is the maximum moment; and k and k_1 are constants which depend on the load distribution, conditions of end restraint and variation of EI (if any). For example, $k = 5/384$ for a uniformly loaded simple beam of uniform section. When a beam or loading is nonsymmetrical about the midspan, the maximum deflection is not at midspan. However, the midspan deflection of beams supported at both ends is very close to the maximum value (within about 3.5 percent for a span fixed at one end and simply supported at the other with distributed loading) so that for serviceability checks, the former value is generally used. A standard case frequently encountered in design is that of a continuous beam of uniform section with uniform loading w per unit length (Fig. 11.1). Since the support and midspan moments in such a case are generally known from the analysis made for design of the section, it would be convenient to express the midspan deflection of such a beam in terms of these moments. As shown in Fig. 11.1, this may be done by superimposing the deflections at midspan due to the three component loadings (computed in this case conveniently using the conjugate beam method). It may be recalled that (Ref. 11.2) the conjugate beam is subjected to a loading equal to the M/EI diagram of the real beam. The moment and shear in the conjugate beam at a section correspond to the deflection and slope of the real beam at the same section.

Assuming the flexural rigidity EI to be uniform along the span, and with the notations in Fig. 11.1, the midspan deflection, Δ_m, is obtained as:

$$\Delta_m = \frac{5}{48} \frac{M_o L^2}{EI} - \frac{M_1 L^2}{16\ EI} - \frac{M_2 L^2}{16\ EI}$$

$$= \frac{5L^2}{48\ EI} \left[M_o - (3/5)(M_1 + M_2) \right]$$

Fig. 11.1 *Midspan deflection — uniformly loaded continuous beam*

Beam and loading

Moment diagram

Conjugate beams
and loadings

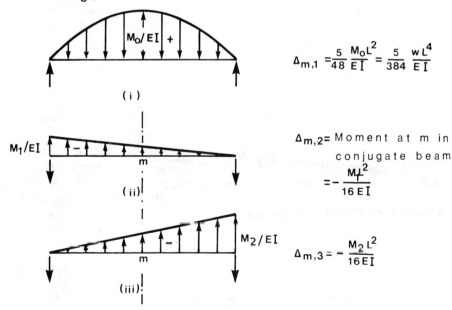

$$\Delta_{m,1} = \frac{5}{48} \frac{M_0 L^2}{EI} = \frac{5}{384} \frac{w L^4}{EI}$$

$$\Delta_{m,2} = \text{Moment at m in}$$
$$\text{conjugate beam}$$
$$= -\frac{M_1 L^2}{16 EI}$$

$$\Delta_{m,3} = -\frac{M_2 L^2}{16 EI}$$

Substituting $M_m = M_o - (1/2)(M_1 + M_2)$, and eliminating M_o,

$$\Delta_m = \frac{5L^2}{48\ EI}\left[M_m - (1/10)(M_1 + M_2)\right] \tag{11.2}$$

Alternatively, eliminating $(M_1 + M_2)$,

$$\Delta_m = K \frac{5}{48} \frac{M_m L^2}{EI} \tag{11.3}$$

where $\qquad K = 1.20 - 0.20\ M_o/M_m \tag{11.3a}$

Values of K for several cases are tabulated in Ref. 11.4. Similar expressions can also be worked out for concentrated loadings (see Ref. 11.5). For the moment patterns associated with the Code moment coefficients, the values of K can be computed using Eq. 11.3a. For a few typical cases, K is given in Table 11.1.

11.4 IMMEDIATE DEFLECTIONS

The principal member property that influences short-time deflection is the value of the flexural rigidity, EI. The modulus of elasticity of concrete, $E = E_c$, depends on factors such as concrete quality, age, stress level, and rate or duration of applied stress. However, for short-time loading to service load levels, the approximate re-lation in Eq. 1.1 gives a satisfactory value for E_c.

The moment of inertia, I, of a section of a beam is influenced by the steel per-centages as well as the extent of flexural cracking, and hence is not a constant along the span even for a beam of uniform cross-section. Flexural cracking will depend on the applied bending moment at a section and the modulus of rupture of the concrete. During the first time loading of a reinforced concrete beam, the portions of a beam where the applied moment is less than the cracking moment ($M < M_{cr}$, see Example 4.1 for the computation of M_{cr}) remain uncracked and hence have the moment of inertia of the gross transformed section (including reinforcement). Where the moment exceeds M_{cr}, the concrete in tension fails at the outer fibres and develops flexural cracks at random spacings. However, the concrete close to the neutral axis and in between the cracks still carries some tension and contributes to the effective stiffness of the beam. The tensile stress in the steel in between cracks may be as low as 60 percent of the stress at the cracks due to the tension contribution of concrete — the so-called "tension stiffening" effect. Thus, even in the portion of the beam with flexural cracking, the effective stiffness will be greater than that corresponding to a fully cracked transformed section.

11.4.1 EFFECTIVE MOMENT OF INERTIA

The typical load-deflection behaviour of a reinforced concrete beam under short-time loading is shown in Fig. 11.2. The initial slope of the curve corresponds closely

Table 11.1   Coefficient K in Eq. 11.3

Loading case	K
	1.00
	0.60
	0.80
	0.925*
Col./beam	0.85*
	0.80*

*Using Code moment coefficients

to the gross transformed section moment of inertia. As the load increases and cracking spreads, the slope of the curve decreases. It is apparent that in this region, the effective moment of inertia, I_e, within the service load range, will have a value in between the gross section moment of inertia, I_g, and the cracked transformed section moment of inertia, I_{cr}, and that the value of I_e will depend on the load level.

For use in the computation of immediate deflection of simple spans, CSA A23.3-M84 (9.5.2.3) recommends Eq. 11.4, developed by Branson in 1973 (see Ref. 11.5 for a discussion on this), giving an average effective moment of inertia I_e as:

$$I_e = \left(\frac{M_{cr}}{M_a}\right)^3 I_g + \left[1 - \left(\frac{M_{cr}}{M_a}\right)^3\right] I_{cr} \le I_g \qquad (11.4)$$

where M_{cr} = $f_r I_g / y_t$ = cracking moment   (11.4a)

M_a = maximum moment in member at the load stage at which deflection is being computed

I_g –   moment of inertia of gross concrete section about centroidal axis, neglecting the reinforcement

I_{cr} = moment of inertia of cracked section transformed to concrete

f_r = modulus of rupture of concrete, = $0.6 \lambda \sqrt{f_c'}$

y_t = distance from centroidal axis of gross section to extreme fibre in tension

λ = factor to account for low density concrete

Fig. 11.2 *Typical load-deflection curve for reinforced concrete beams*

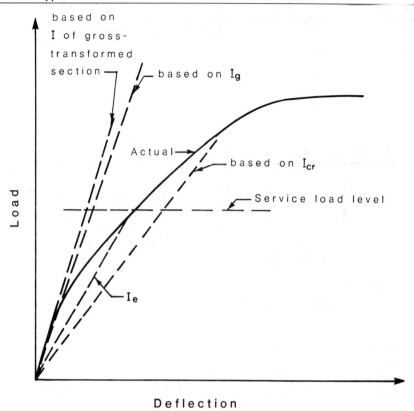

Equation 11.4 is relatively simple and gives the correct limiting values of $I_e = I_g$ for $M_a \leq M_{cr}$ and $I_e \simeq I_{cr}$ for $M_a >> M_{cr}$, where the beam is severely cracked. When $M_a/M_{cr} \geq$ about 3, $I_e \simeq I_{cr}$. Although the Code recommends neglecting the reinforcement in computing I_g, the gross transformed section (including reinforcement) moment of inertia may be more appropriate in place of I_g in Eq. 11.4, particularly for beams with high reinforcement ratios and structural low density concrete (high modular ratio) (Refs. 11.5, 11.6).

Comparison with test results indicates reasonable agreement (within the range of about ± 20 percent) between the deflections measured and those computed using Eq. 11.4 (Ref. 11.6). It may be noted that there are a number of other methods (such as the Yu and Winter method (Ref. 11.7)) to estimate the effective I, which result in deflection predictions with the same degree of accuracy as obtained by the Branson formula (see Ref. 11.6). A review and comparison of several methods proposed for computing short-time deflections is presented in Ref. 11.5.

Based on the effective stiffness, I_e, given by Eq. 11.4, the moment (or load)-deflection curve of a beam under short-time loading has the form shown in Fig. 11.3. The value of I_e depends on the magnitude of the moment. Thus, for different

Fig. 11.3 *Moment-deflection curve for short-term loading with effective stiffness I_e (Eq. 11.4)*

load (and hence moment) levels, such as dead load alone or dead load plus live load, the respective deflections (Δ_D and Δ_{D+L}) should be computed with the I_e values given by Eq. 11.4 for the corresponding total moment levels (M_D and M_{D+L}) as indicated in Fig. 11.3. The incremental deflection, such as the part Δ_L due to live load, is obtained as the difference between the deflections computed with and without the live load, that is, $\Delta_L = \Delta_{D+L} - \Delta_D$.

11.4.2 AVERAGE EFFECTIVE MOMENT OF INERTIA FOR CONTINUOUS SPANS

Equation 11.4 was developed for use as a single average I_e for simply supported beams or for the region between inflection points of continuous beams. The negative moment regions of a continuous beam may have an effective stiffness, I_e, different from that of the positive moment region. The variation in I_e may be due to changes in cross-sectional dimensions, changes in reinforcement ratios, or differences in the influence of cracking (for example, in continuous T-beams, the value of I_{cr} may be substantially different under positive and negative moments). To account approximately for these variations, for continuous spans the Code suggests the use of the average of the I_e obtained from Eq. 11.4 for the critical positive and negative moment sections. On this basis, for a beam continuous at one end,

$$\text{average } I_e = 0.75 \times (I_{e,\,m}) + 0.25 \times (I_{e,\,\text{cont.end}}) \qquad (11.5a)$$

and, for a beam continuous at both ends,

$$\text{average } I_e = \tfrac{1}{2}[I_{e,\,m} + \tfrac{1}{2}(I_{e,\,1} + I_{e,\,2})] \qquad (11.5b)$$

where $I_{e,\,m}$ is the value of I_e at the critical positive moment section (taken as the midspan section), and $I_{e,\,1}$ and $I_{e,\,2}$ are values of I_e at the critical negative moment sections (continuous ends).

Typically in a continuous (interior) span, the loading pattern producing the maximum deflection (and maximum positive moment) is different from the loading pattern that produces the maximum negative moment at supports. However, the degree of cracking in the negative moment regions of the beam will depend on the maximum negative moments there, as obtained from the moment envelope. To account for the influence of more widespread cracking in the negative moment region which could have occurred under possible larger negative moments, Branson recommends (Ref. 11.4, 11.5) the use of the moment envelope (for M_a in Eq. 11.4) to evaluate the negative and positive moment values for I_e. However, once the average I_e for the beam is so determined, the deflection is calculated using the loading pattern and corresponding moment diagram which produces the maximum deflection. In general, in continuous spans, the negative moment I_e values ($I_{e,\,1}$ and $I_{e,\,2}$) have a much smaller effect on the deflections than the positive moment I_e value ($I_{e,\,m}$). This can be seen from the conjugate beam method, where the span moment in the conjugate beam (which gives the deflection of the real beam) is more influenced by the loading on it ($= M/EI$) in the midspan regions than the loading near the ends. In fact for continuous spans with Neg.I_e/Pos.$I_e < 2$, and with moment patterns in between those for a simple span and for a fixed-ended span, the use of midspan I_e alone ($= I_{e,\,m}$) gives reasonable accuracy (in a limited comparison with measured deflections of continuous prismatic members, this procedure was seen to give generally better results than the simple averaging procedure suggested in the Code, see Refs. 11.8, 11.9). For spans with larger variation in flexural rigidity, or when the negative moment at either end is relatively larger, use of a weighted average I_e (such as in Eq. 11.6) is preferable (Refs. 11.5, 11.8, 11.9). For cantilever beams, I_e is appropriately based on the support section.

For beams with both ends continuous,

$$\text{Weighted average } I_e = 0.7\,I_{e,\,m} + 0.15\,(I_{e,\,1} + I_{e,\,2}) \qquad (11.6a)$$

For beams with one end continuous,

$$\text{Weighted average } I_e = 0.85\,I_{e,\,m} + 0.15\,I_{e,\,\text{cont.end}} \qquad (11.6b)$$

11.4.3 CRACKED SECTION PROPERTIES

For the general case of a doubly reinforced rectangular section, the cracked transformed section is shown in Fig. 11.4. Under elastic behaviour, the neutral axis passes through the centroid of the section. Equating moment of areas about the centroid,

$$b\frac{(kd)^2}{2} + A'_s\,(n-1)(kd - d') = nA_s\,(d - kd)$$

Fig. 11.4 *Cracked transformed section properties*

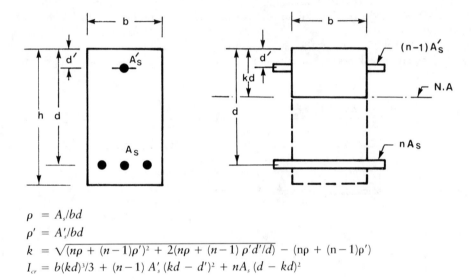

$\rho = A_s/bd$

$\rho' = A'_s/bd$

$k = \sqrt{(n\rho + (n-1)\rho')^2 + 2(n\rho + (n-1)\,\rho'd'/d)} - (n\rho + (n-1)\rho')$

$I_{cr} = b(kd)^3/3 + (n-1)\,A'_s\,(kd - d')^2 + nA_s\,(d - kd)^2$

Substituting $\rho = A_s/bd$ and $\rho' = A'_s/bd$, and solving for k,

$$k = \sqrt{\left[n\rho + (n-1)\rho'\right]^2 + 2\left[n\rho + (n-1)\rho'\frac{d'}{d}\right]} - [n\rho + (n-1)\rho'] \quad (11.7)$$

$$I_{cr} = \frac{b(kd)^3}{3} + (n-1)A'_s(kd - d')^2 + nA_s\,(d - kd)^2 \quad (11.8)$$

When there is no compression reinforcement,

$$k = \sqrt{(n\rho)^2 + 2n\rho} - n\rho \quad (11.9)$$

$$I_{cr} = \frac{b(kd)^3}{3} + nA_s\,(d - kd)^2 \quad (11.10)$$

EXAMPLE 11.1

For the T-beam designed in Example 5.7, compute the immediate deflection due to dead load and live load. (Normal density concrete, $f'_c = 20$ MPa, $f_y = 400$ MPa.)

SOLUTION

1. Loads and moments
    From Example 5.7,

$$\text{Dead load} = 13 + 2.68 = 15.68 \text{ kN/m}$$
$$\text{Specified live load} = 29.7 \text{ kN/m}$$
$$\text{Dead load moment, } M_D = 15.68 \times 8^2/8 = 125 \text{ kN·m}$$
$$\text{Moment under combined dead plus live load, } M_{D+L} = (15.68 + 29.7)8^2/8$$
$$= 363 \text{ kN·m}$$
$$E_c = 5000 \sqrt{f'_c} = 5000 \sqrt{20} = 22\,400 \text{ MPa}$$

The effective T-beam section is shown in Fig. 11.5. The cross-section properties are:

Fig. 11.5 *Example 11.1*

(a) Gross section

(b) Cracked-transformed section

2. Gross section, I_g

$$\text{Depth to centroid, } y_c = \frac{(3180 - 300) \times 120 \times 60 + 300 \times 500 \times 250}{(3180 - 300) \times 120 + 300 \times 500}$$

$$= 118 \text{ mm}$$

$$I_g = \frac{3180 \times 118^3}{3} + \frac{(3180 - 300) \times 2^3}{3} + \frac{300 \times (500 - 118)^3}{3}$$

$$= 7.32 \times 10^9 \text{ mm}^4$$

3. Cracked transformed section, I_{cr}

$$\text{The modular ratio, } n = \frac{E_s}{E_c} = \frac{200\,000}{22\,400} = 9.0$$

The depth of neutral axis, kd, assuming it to be within the flange, is obtained by equating moments of area,

$$3180 \frac{(kd)^2}{2} = 9 \times 4200\,(400 - kd)$$

Solving, $\qquad kd = 86.4 \text{ mm} \ (<h_f, \text{OK})$

$$I_{cr} = \frac{3180 \times 86.4^3}{3} + 9 \times 4200(400 - 86.4)^2$$

$$= 4.40 \times 10^9 \text{ mm}^4$$

4. The cracking moment, M_{cr}

$$f_r = 0.6\lambda \sqrt{f_c'}$$

$$= 2.68 \text{ MPa}$$

$$M_{cr} = f_r I_g / y_t$$

$$= \frac{2.68 \times 7.32 \times 10^9}{(500 - 118)} \times 10^{-6} = 51.4 \text{ kN·m}$$

5. Effective stiffness, I_e
   For dead load only, $(I_e)_D$, using Eq. 11.4, and with $M_a - M_D - 125 \text{ kN·m} > M_{cr}$

$$(I_e)_D = \left(\frac{51.4}{125}\right)^3 \times 7.32 \times 10^9 + \left[1 - \left(\frac{51.4}{125}\right)^3 \right] \times 4.40 \times 10^9$$

$$= 4.6 \times 10^9 \text{ mm}^4$$

For dead plus live load, $(I_e)_{D+L}$ is

$$(I_e)_{D+L} = \left(\frac{51.4}{363}\right)^3 \times 7.32 \times 10^9 + \left[1 - \left(\frac{51.4}{363}\right)^3\right] \times 4.40 \times 10^9$$

$$= 4.41 \times 10^9 \text{ mm}^4$$

6. Immediate deflections
    Midspan deflection for simply supported uniformly loaded beam is (Eq. 11.3 and Table 11.1),

$$\Delta = \frac{5}{48}\frac{M_m L^2}{EI}$$

For dead load only,

$$\Delta_D = \frac{5}{48} \times \frac{(125 \times 10^6) \times 8000^2}{22\ 400 \times 4.6 \times 10^9} = 8.1 \text{ mm}$$

For dead plus live load,

$$\Delta_{D+L} = \frac{5}{48} \times \frac{363 \times 10^6 \times 8000^2}{22\ 400 \times 4.41 \times 10^9} = 24.5 \text{ mm}$$

Live load deflection is,

$$\Delta_L = 24.5 - 8.1 = 16.4 \text{ mm}$$

7. Allowable immediate deflection due to live load by CSA A23.3-M84 (9.5.2.6) (see Section 11.11) is, for floors not supporting or attached to elements likely to be damaged by large deflection,

$$= \frac{L}{360} = \frac{8000}{360} = 22.2 \text{ mm}$$

$$\Delta_L < \Delta_{L,\,allow} \qquad\qquad \text{OK}$$

EXAMPLE 11.2

Compute the immediate deflection due to live load of the exterior panel slab of the one-way floor slab designed in Example 5.5.

SOLUTION

The Code moment coefficients will be used. Considering a slab strip of 1 m width, the specified loads are (Example 5.5):

Dead load = 3.94 kN/m

Live load = 9.0 kN/m

$$E_c = 5000 \sqrt{f_c'} = 22\,400 \text{ MPa, and } n = 9.0$$

$$I_g = 1000 \times 120^3/12 = 144 \times 10^6 \text{ mm}^4$$

$$f_r = 0.6 \sqrt{20} = 2.68 \text{ MPa}$$

$$M_{cr} = \frac{2.68 \times 144 \times 10^6}{60} \times 10^{-6} = 6.43 \text{ kN·m}$$

1. Effective moments of inertia

   I_e for midspan section (I_m) and the two support sections $(I_{e,1}$ and $I_{e,2})$ will be computed for the respective maximum moments.

   (a) Midspan section (Fig. 11.6b)

   From Example 5.5 (Table 5.6),

   $$d = 94.4 \text{ mm, } A_s = 417 \text{ mm}^2, \rho = A_s/bd = 0.00442$$

   $$k = \sqrt{2n\rho + (n\rho)^2} - n\rho = 0.245 \qquad\qquad \text{(Eq. 11.9)}$$

   $$kd = 23.1 \text{ mm}$$

   $$I_{cr} = \frac{1000 \times 23.1^3}{3} + 9 \times 417(94.4 - 23.1)^2 = 23.2 \times 10^6 \text{ mm}^4$$

   $$M_{m,D} = wL_n^2/14 = 3.94 \times 3^2/14 = 2.53 \text{ kN·m}$$

   $$M_{m,D+L} = (3.94 + 9) \times 3^2/14 = 8.32 \text{ kN·m}$$

   Since $M_{m,D} < M_{cr}$,

   $$(I_{e,m})_D = I_g = 144 \times 10^6 \text{ mm}^4$$

   $$(I_{e,m})_{D+L} = \left(\frac{6.43}{8.32}\right) \times 144 \times 10^6 + \left[1 - \left(\frac{6.43}{8.32}\right)^3\right] \times 23.2 \times 10^6$$

   $$= 79 \times 10^6 \text{ mm}^4 \qquad\qquad \text{(Eq. 11.4)}$$

Fig. 11.6 *Example 11.2*

(a)

(b) Section CC   (c) Section BB

(b) Exterior support

$$M_D = wL_n^2/14 = 3.94 \times 3^2/24 = 1.48 \text{ kN} \cdot \text{m}$$
$$M_{D+L} = (3.94 + 9) \times 3^2/24 = 4.85 \text{ kN} \cdot \text{m}$$

Since under both loadings, the moment is less than the cracking moment, M_{cr}, the section is uncracked and there is no need to compute I_{cr} at this section.
$I_{e,1} = I_g = 144 \times 10^6 \text{ mm}^4$, for both dead load only and for dead plus live load.

(c) Interior support (Fig. 11.6c)

$$A_s = 500 \text{ mm}^2, \rho = 0.0053$$

The bottom reinforcement from midspan, continued into the support, acts as compression reinforcement and has area $A_s' = 208 \text{ mm}^2$ ($\rho' = 0.0022$). The neutral axis depth may be computed using Eq. 11.7 as:

$$kd = 25 \text{ mm, and}$$
$$I_{cr} = 26.9 \times 10^6 \text{ mm}^4$$
$$M_D = wL_n^2/12 = 3.94 \times 3^2/12 = 2.96 \text{ kN} \cdot \text{m}$$
$$M_{D+L} = (3.94 + 9) \times 3^2/12 = 9.71 \text{ kN} \cdot \text{m}$$

Since $M_D < M_{cr}$, for dead load condition,

$$(I_{e,2}) = I_g = 144 \times 10^6 \text{ mm}^4$$

For dead load plus live load, using Eq. 11.4,

$$(I_{e,2})_{D+L} = \left(\frac{6.43}{9.71}\right)^3 \times 144 \times 10^6 + \left[1 - \left(\frac{6.43}{9.71}\right)^3\right] \times 26.9 \times 10^6$$
$$= 60.9 \times 10^6 \text{ mm}^4$$

2. Average I_e for span (Eq. 11.5b)
   Since the slab is integral with supporting beams, for I Average, it will be considered as continuous at both ends. For dead load only, all critical sections have $I_e = I_g$ so that,

$$\text{Average } (I_e)_D = I_g = 144 \times 10^6 \text{ mm}^4$$

For dead load plus live load,

$$\text{Average } (I_e)_{D+L} = \frac{1}{2}\left[79 + \frac{1}{2}(144 + 60.9)\right] \times 10^6$$
$$= 90.7 \times 10^6 \text{ mm}^4$$

Calculation of deflection:

$$\Delta = K\frac{5}{48}\frac{M_m L^2}{EI_e}$$

$$K = 1.20 - 0.20 M_o/M_m \qquad \text{(Eq. 11.3a)}$$

Loading for maximum deflection is same as for maximum positive moment, for which $M_m = wL_n^2/14$

$$K = 1.20 - 0.20(wL^2/8)/(wL^2/14)$$

$$= 0.85$$

$$\Delta_D = 0.85 \times \frac{5}{48} \times \frac{M_m L^2}{E(I_e)_D}$$

$$= 0.85 \times \frac{5}{48} \times \frac{2.53 \times 10^6 \times 3000^2}{22\,400 \times 144 \times 10^6} = 0.625 \text{ mm}$$

$$\Delta_{D+L} = 0.85 \times \frac{5}{48} \times \frac{8.32 \times 10^6 \times 3000^2}{22\,400 \times 90.7 \times 10^6} = 3.26 \text{ mm}$$

$$\Delta_L = 3.26 - 0.625 = 2.64 \text{ mm}$$

The maximum allowable immediate deflection due to live load for floors not supporting or attached to nonstructural elements likely to be damaged by large deflection is (CSA A23.3-M84 — Table 9-2):

$$\Delta_{all} = L/360 = 3000/360 = 8.33 \text{ mm}$$
$$\Delta_L < \Delta_{all} \qquad\qquad\qquad \text{OK}$$

[Note that the reinforcement ratio of 0.00442 at midspan is only 0.20 ρ_b (Table 5.3). Hence it could have been anticipated that deflection will not be critical; see Table 5.3.]

11.5 TIME-DEPENDENT DEFLECTION

Under sustained loads, the deflection increases with time, due principally to the effects of creep and shrinkage, although additional factors such as formation of new cracks, widening of earlier cracks and effects of repeated load cycles also contribute to these increases. The additional longtime deflection may be as large as two to three times the immediate deflection. The factors which influence creep and shrinkage were described in Sections 1.11 and 1.12. Thus the major parameters that influence the long-time deflection are the stress in concrete, amount of tensile and compressive reinforcement, size of member, curing conditions, temperature, relative humidity, age of concrete at time of loading, duration of loading, and conditions of restraint.

11.6 DEFLECTION DUE TO SHRINKAGE

In an unrestrained reinforced concrete member, shrinkage causes shortening of the member which is resisted by the reinforcement thereby developing compressive stress in the steel and tensile stress in the concrete which balance each other. When the reinforcement is symmetrically placed in the cross section, the shrinkage does not cause any curvature of the member. However, when the reinforcement is not symmetric, as is usually the case in flexural members, the shortening of the reinforced face is far less than that of the unreinforced (or lesser reinforced) face (see Fig. 11.7 a,b). This differential shrinkage causes a curvature of the member. The location of flexural reinforcement in beams is such that the shrinkage curvature has the same sign as the curvature due to flexure and hence magnifies the deflections due to applied loads. It is also obvious that the addition of compression reinforcement decreases the differential shrinkage strains and resulting deflections.

Shrinkage stresses are sustained stresses and shrinkage and creep are interdependent. The shrinkage curvature is also affected by the flexural cracking and hence the transverse load. However, as an approximation, the effects of shrinkage may be considered separate from those of creep and transverse loads.

11.7 SHRINKAGE CURVATURE

The two broad approaches used for the computation of shrinkage curvature are (1) the so called "equivalent tensile force" method, and (2) empirical methods. Detailed reviews of these methods are contained in Refs. 11.5 and 11.6. Because of its better agreement with test results, and its recommendation by ACI Committee 435 (Ref. 11.6), only the empirical method proposed by Miller as modified by Branson is briefly discussed below.

Fig. 11.7 *Shrinkage curvature*

(a)

(b)

(c)

For a singly reinforced section, Miller (Ref. 11.10) assumed a shrinkage strain distribution as shown in Fig. 11.7c, which gives the curvature as:

$$\phi_{sh} = \frac{\epsilon_{sh} - \epsilon_s}{d} = \frac{\epsilon_{sh}}{d}\left(1 - \frac{\epsilon_s}{\epsilon_{sh}}\right) \tag{11.11}$$

The fibres on the unreinforced face are assumed to have shrinkage strains equal to the free shrinkage strain of the concrete, ϵ_{sh}, ϵ_s in Eq. 11.11 is the strain in the steel resulting from shrinkage. The value of ϵ_s/ϵ_{sh} depends on the amount of reinforcement; and empirical values for this were suggested by Miller. As a modification and extension of Miller's method, Branson in 1963 suggested the following equations which are applicable to doubly reinforced beams as well.

For $(\rho - \rho') \leqslant 3$,

$$\phi_{sh} = 0.7\frac{\epsilon_{sh}}{h}(\rho - \rho')^{1/3}\left(\frac{\rho - \rho'}{\rho}\right)^{1/2} \tag{11.12}$$

For $(\rho - \rho') > 3$,

$$\phi_{sh} = \epsilon_{sh}/h \tag{11.13}$$

where ρ and ρ' are tension and compression reinforcement ratios expressed as percentages $(= 100\ A_s/bd$ and $100\ A_s'/bd)$ and h is the thickness of the member. In a comparison of several methods of computing shrinkage deflection, Eqs. 11.12 and 11.13 were seen to give the closest agreement with the test results (Refs. 11.5, 11.6). For use in these equations, the free shrinkage strain, ϵ_{sh}, obtained from test data incorporating effects of local aggregates is preferred. Methods for estimating the free shrinkage, both at time t, $(\epsilon_{sh})_t$, and at ultimate, $(\epsilon_{sh})_u$, were explained in Section 1.12. For situations where $(\epsilon_{sh})_u$ under local conditions is not known, ACI Committee 435 and Branson have suggested the use of $(\epsilon_{sh})_u = 400 \times 10^{-6}$ mm/mm for routine deflection calculations (Refs. 11.5, 11.9).

11.8 COMPUTATION OF DEFLECTION DUE TO SHRINKAGE

If the distribution of the shrinkage curvature, ϕ_{sh}, along the span is known, the resulting deflection may be computed by any of the classical procedures such as integration, moment-area, or conjugate beam methods (with ϕ_{sh} replacing the curvature due to moment, M/EI). It should be recognized that in restrained beams, the warping due to shrinkage may induce secondary moments, producing a curvature which should be accounted for in the deflection calculations. In general, the shrinkage deflections, Δ_{sh}, can be expressed as:

$$\Delta_{sh} = K_{sh}\,\phi_{sh}\,L^2 \tag{11.14}$$

where ϕ_{sh} in the shrinkage curvature, K_{sh} is a coefficient depending on the support conditions and the distribution of ϕ_{sh} along the span, and L is the span length. Assuming that $\rho - \rho'$, and hence ϕ_{sh}, has the same value along the span (including the positive and negative curvature regions in continuous beams), together with simplifying assumptions about the location of inflection points in continuous spans, the values given in Table 11.2 can be derived for K_{sh} (see Ref. 11.5, pp. 173–174 for derivations) for beams of uniform section.

In actual beams, the ratio $(\rho - \rho')$ may not be uniform. However, as in the case of deflections due to transverse loads, the positive curvature, ϕ_{sh}, in the midspan region has the dominant effect on deflection and hence it is satisfactory to use in Eq. 11.14 the ϕ_{sh}, based on $(\rho - \rho')$ for the midspan region for simple and continuous spans, and based on $\rho - \rho'$ for the support section for cantilever spans. For T-beams, Ref. 11.9 recommends the use of an average of ρ and ρ_w in Eqs. 11.12 and 11.13, where $\rho = 100\,A_s/bd$ and $\rho_w = 100\,A_s/b_w d$.

Table 11.2   Values of K_{sh}

K_{sh}	
0.5	cantilever beam
0.125	simply supported beam
0.0625	fixed ended beams
0.084	beams continuous at one end (in continuous beams with 2 spans only)
0.090	beams continuous at one end (in continuous beams with 3 or more spans)
0.065	beams continuous at both ends.

11.9 DEFLECTION DUE TO CREEP

As a result of creep under sustained stresses, the compressive strain in the concrete at the extreme fibre increases with time as shown in Fig. 11.8, while the strain in the tension steel increases only marginally (with the shifting of the neutral axis and consequent decrease in internal lever arm, the steel force and hence strain increases slightly to maintain equilibrium with applied loads, see Fig. 11.8c). The creep strain causes an increase in curvature, ϕ_{cp}, which, like the shrinkage curvature, causes additional deflection of the member.

Since $\phi_i = \epsilon_i/x_i$, and $\phi_{cp} = \epsilon_{cp}/x_{cp}$; and with $x_{cp} > x_i$, (see Fig. 11.8),

$$\frac{\phi_{cp}}{\phi_i} = k_r \frac{\epsilon_{cp}}{\epsilon_i} = k_r C_t \tag{11.15}$$

where ϕ_i is the initial elastic curvature, ϵ_i is the initial elastic strain, C_t is the creep coefficient (Section 1.11), and the coefficient k_r has a value less than unity. The

Fig. 11.8 *Creep strain and curvature*

(a) Section

(b) Strains

(c) Stresses

increase in deflection may be taken in proportion to the increase in curvature, so that,

$$\frac{\Delta_{cp}}{\Delta_i} = k_r C_t$$

$$\Delta_{cp} = k_r C_t \Delta_i \qquad (11.16)$$

where Δ_{cp} is the additional deflection due to creep and Δ_i is the immediate deflection due to the sustained part of the load. As in the case of shrinkage deflection, compression reinforcement will restrain creep strains and reduce curvature and deflection due to creep. For k_r in Eq. 11.16, ACI Committee 435 has recommended the expression given below, which includes the effect of compression reinforcement (Ref. 11.9).

$$k_r = 0.85/(1 + 50\,\rho')$$ (11.17)

where $\rho' = A'_s/bd$

Expressions for computing the creep coefficient C_t at age t were presented in Section 1.11. For average conditions, in the absence of more precise data, the ACI Committee recommended a value of 1.6 for the ultimate creep coefficient, C_u, for use in Eqs. 1.6 and 11.16.

11.10 LONG-TIME DEFLECTION BY CODE

Because of the large number of interacting parameters affecting long-time deflections, the uncertainty regarding these parameters, and the lower degree of reliability of deflection computations in general, under normal situations it is more practical as well as satisfactory to estimate the combined additional deflection due to all time-dependent effects with a single time-dependent factor applied to the initial deflection. Such a procedure is recommended in CSA A23.3-M84 (9.5.2.5). According to this, the additional long-time deflection $\Delta_t = (\Delta_{cp} + \Delta_{sh})$ may be obtained by multiplying the immediate deflection, Δ_i, due to the sustained load considered, by a factor $[s/(1 + 50\,\rho')]$.

$$\Delta_t = \Delta_{cp} + \Delta_{sh} = \Delta_i \times \frac{s}{1 + 50\,\rho'}$$ (11.18)

where $\rho' = A'_s/(bd)$ is the compression reinforcement ratio, taken as the value at midspan for simple and continuous spans and at the support for cantilevers, and s is a time-dependent factor for sustained loads, taken as:

2.0 for 5 years or more
1.4 for 12 months
1.2 for 6 months
1.0 for 3 months

Typically, Δ_i includes the immediate deflection due to dead load and due to the portion of live load that is sustained. The multiplication factor in Eq. 11.18 was developed by Branson, and Eq. 11.18 has been found to give good agreement with results of tests (Ref. 11.5, 11.9, 11.11). The factor accounts for the effect of compression reinforcement in reducing long-time deflection as well as for the duration of sustained loading.

The Code does permit the use of more detailed procedures such as computing Δ_{cp} and Δ_{sh} individually using Eqs. 11.14 and 11.16.

EXAMPLE 11.3

For the slab in Example 11.2, compute the long-time deflection and compare with the allowable values specified by the Code. Assume 20 percent of the live load to be sustained for over five years.

SOLUTION

1. By Code procedure
   From Example 11.2, the immediate deflections are:

   $$(\Delta_i)_D = 0.625 \text{ mm}, (\Delta_i)_L = 2.64 \text{ mm}$$

   Using Eq. 11.18, with $\rho' = 0$, 20 percent of live load sustained, and $s = 2.0$ for duration of sustained load of over five years,

   $$\Delta_t = \Delta_{cp} + \Delta_{sh} = \Delta_i \times \frac{s}{1 + 50 \rho'}$$

   $$= (0.625 + 0.2 \times 2.64) \times \frac{2.0}{1 + 0} = 2.31 \text{ mm}$$

   Sum of long time deflection due to sustained load and immediate deflection due to live load

   $$= \Delta_t + (\Delta_i)_L = 2.31 + 2.64 = 4.95 \text{ mm}$$

   Allowable limits in CSA A23.3-M84(9.5.2.6) are:

   (a) Floor attached to nonstructural elements not likely to be damaged by large deflections,

   $$= \frac{L}{240} = \frac{3000}{240} = 12.5 \text{ mm} > 4.95 \text{ mm} \qquad \text{OK}$$

   (b) If attached to nonstructural elements likely to be damaged by large deflection,

   $$= \frac{L}{480} = \frac{3000}{480} = 6.25 \text{ mm} > 4.95 \text{ mm} \qquad \text{OK}$$

   In both cases, the computed deflections are less than the maximum allowable deflection specified and, hence, the design is satisfactory.

2. By detailed computation
   Using Eqs. 11.14 and 11.16,

   $$\Delta_{sh} = k_{sh} \, \emptyset_{sh} L^2 \qquad (11.14)$$

   $$\Delta_{cp} = k_r \, C_t \, \Delta_i \qquad (11.16)$$

   (i) Shrinkage deflection Δ_{sh}
      ϕ_{sh} is given by Eqs. 11.12 and 11.13. For the slab, based on midspan section,

$\rho' = 0$, $\rho = 100\ A_s/bd = 0.442 < 3$, and Eq. 11.12 controls. Assuming average conditions and $(\epsilon_{sh})_u = 400 \times 10^{-6}$,

$$\phi_{sh} = 0.7 \frac{\epsilon_{sh}}{b} (\rho - \rho')^{1/3} \left(\frac{\rho - \rho'}{\rho}\right)^{1/2}$$

$$= 0.7 \times \frac{400 \times 10^{-6}}{120} (0.442)^{1/3} = 1.78 \times 10^{-6}\ \text{mm}^{-1}$$

For slab continuous at both ends, from Table 11.2 (since slab is integral with beams at both ends, it is considered continuous at both ends), $k_{sh} = 0.065$

$$\Delta_{sh} = 0.065 \times 1.78 \times 10^{-6} \times 3000^2$$
$$= 1.04\ \text{mm}$$

(ii) Creep deflection Δ_{cp}
k_r, using Eq. 11.17, with $\rho' = 0$
$$= \frac{0.85}{1 + 50\,\rho'} = 0.85$$

Ultimate C_t, for average conditions, $= 1.6$
$$\Delta_{cp} = k_r C_t \Delta_i$$
$$= 0.85 \times 1.6 \times (0.625 + 0.2 \times 2.64)$$
$$= 1.57\ \text{mm}$$

(iii) Total longtime deflection
$\Delta_t = \Delta_{sh} + \Delta_{cp} = 1.04 + 1.57 = 2.61$ mm, compared with 2.31 mm computed at 1.
Including immediate deflection due to live load, $(\Delta_i)_L + \Delta_t = 2.64 + 2.61 = 5.25$ mm.
This is within the allowable limits.

EXAMPLE 11.4

The dimensions of the exterior span of a continuous interior beam of a slab-beam girder floor are shown in Fig. 11.9a. The slab is 110 mm thick and the beams are spaced at 3.5 m centres. The specified loads consist of a dead load (including self weight) of 19.8 kN/m and a live load of 16.8 kN/m of which 50 percent is sustained. Compute the immediate and long-term deflections of the beam and compare with the Code requirements. Use $f_c' = 20$ MPa and $f_y = 300$ MPa.

SOLUTION

$$E_c = 5000 \sqrt{20} = 22\ 360\ \text{MPa}$$

$$n = E_s/E_c = 9$$

Fig. 11.9 *Example 11.4*

(a)

Gross section   mid-span section   Interior support
                                           (negative moment)

(b)

Effective flange width of T-section is

$$0.25 \text{ span} = 0.25 \times 7400 = 1850 \text{ mm}$$
$$24 h_f + b_w = 2940 \text{ mm}$$
$$\text{spacing} = 3500 \text{ mm}$$

The gross and cracked transformed sections of the beam at various critical sections are shown in Fig. 11.9b. The Code moment coefficients are used to compute maximum moments at critical sections. The computations for deflections are set up in Table 11.3. In computing the average moment of inertia for the continuous span, the beam is assumed (conservatively) as discontinuous at the exterior support in view of the relatively low negative moment at this section. Hence Eq. 11.5a is used in averaging I_e rather than Eq. 11.5b. Also note that at the interior support section, I_e is computed for the maximum negative moment at this section ($wL^2/10$), although the moment here corresponding to the load causing maximum deflection will be lesser than this maximum value.

Table 11.3 Design Example 11.4

Mom. Coeff.	1/24	1/14	1/10	Comments
I_g, 10^9 mm^4	10.8	10.8	10.8	
I_{cr}, 10^9 mm^4	2.27	3.27	3.08	
M_{cr}, kN·m	− 160	69.1	− 160	Eq. 11.4a
Dead load only				
M_{Dmax}, kN·m	39.8	68.3	95.6	
$M_{cr}/M_{D'max}$	>1	>1	>1	
$I_{e,D}$, 10^9 mm^4	10.8	10.8	10.8	
$(I_{e,D})_{av}$, mm^4	—	10.8×10^9	—	Eq. 11.5a
Dead load + Live load				
$M_{D+L, max}$, kN·m	73.7	126	177	
$M_{cr}/M_{D+L, max}$	>1	0.548	0.904	
$I_{e,D+L}$, 10^9 mm^4	10.8	4.51	8.78	
$(I_{e, D+L})_{av}$, 10^4 mm^4	—	5.58	—	Eq. 11.5a

Deflections

$$K = 1.2 - 0.2(wL^2/8)\,(wL^2/14) = 0.85 \qquad \text{Eq. 11.3a}$$

$$(\Delta_i)_D = 0.85 \times \frac{5}{48} \times \frac{68.31 \times 10^6 \times (6950)^2}{22\,360 \times 10.8 \times 10^9} = 1.21 \text{ mm}$$

$$(\Delta_i)_{D+L} = 0.85 \times \frac{5}{48} \times \frac{126 \times 10^6 \times (6950)^2}{22\,360 \times 5.58 \times 10^9} = 4.32 \text{ mm}$$

$(\Delta_i)_L = 4.32 - 1.21 = 3.11$ mm
$s/(1 + 50\,\rho') = 2.0$
$\Delta_{cp+sh} = 2(1.21 + 0.5 \times 3.11) = 5.53$ mm
$\Delta_{cp+sh} = (\Delta_i)_L = 5.53 + 3.11 \times 8.64$ mm
Check allowable limits:
$(\Delta_i)_L = 3.11$ mm $< L/360 = 19.3$ mm   OK
$\Delta_{cp+sh} + (\Delta_i)_L = 8.64 < L/480 = 14.5$ mm   OK
At both conditions, deflections are within limits.

EXAMPLE 11.5

Compute the deflection due to creep and shrinkage, over a period of three years for the beam in Example 11.4. Assume that the ultimate creep coefficient, $C_u = 2.35$, ultimate shrinkage strain, $(\epsilon_{sh})_u = 780 \times 10^{-6}$ mm/mm and that the relative humidity is 60 percent.

SOLUTION

1. Creep deflection, using Eq. 11.16 is:

$$\Delta_{cp} = k_r\, C_t\, \Delta_i \qquad\qquad \text{(Eq. 11.16)}$$

Considering the midspan section properties only, which has the dominant effect on deflections,

$$k_r = 0.85/(1 + 50\rho') \qquad\qquad \text{(Eq. 11.17)}$$
$$= 0.85, \text{ as } \rho' = 0$$

$$C_t = \left(\frac{t^{0.6}}{10 + t^{0.6}}\right) C_u \times \text{(Correction Factors } CF) \qquad \text{(Eq. 1.6)}$$

CF for humidity is (Eq. 1.8):

$$1.27 - 0.0067 \times (H = 60) = 0.868$$

$$C_t = \frac{(3 \times 365)^{0.6}}{10 + (3 \times 365)^{0.6}} \times 2.35 \times 0.868 = 1.77$$

From Example 11.4, immediate deflection due to sustained part of load is:

$$\Delta_i = 1.21 + 0.5 \times 3.11 = 2.77 \text{ mm}$$
$$\Delta_{cp} = k_r\, C_t\, \Delta_i$$
$$= 0.85 \times 1.77 \times 2.77 = 4.17 \text{ mm}$$

2. Shrinkage deflection

$$\Delta_{sh} = K_{sh}\phi_{sh}L^2 \qquad\qquad \text{(Eq. 11.14)}$$

Shrinkage strain is given by (Eq. 1.9):

$$(\epsilon_{sh})_t = \frac{t}{35 + t}(\epsilon_{sh})_u \times CF$$

CF for humidity is (Eq. 1.11):

$$(CF)_h = 1.40 - 0.01\, H$$
$$= 1.40 - 0.01 \times 60 = 0.8$$

$$(\epsilon_{sh})_{3 \text{ yrs}} = \frac{3 \times 365}{35 + 3 \times 365} \times 780 \times 10^{-6} \times 0.8$$
$$= 605 \times 10^{-6}$$

Shrinkage curvature is:

$$\phi_{sh} = 0.7 \frac{\epsilon_{sh}}{h} (\rho - \rho')^{1/3} \left(\frac{\rho - \rho'}{\rho} \right)^{1/2} \qquad \text{(Eq. 11.12)}$$

$$= 0.7 \frac{\epsilon_{sh}}{h} \rho^{1/3}$$

For T-beams

$$\rho = 100 \, A_s/bd$$
$$= 100 \times 1600/(1850 \times 533) = 0.162$$
$$\rho_w = 100 \, A_s/b_w d = 100 \times 1600/(300 \times 533) = 1.00$$

Average $\rho = 0.581$ percent

$$\phi_{sh} = 0.7 \times \frac{605 \times 10^{-6}}{600} \times (0.581)^{1/3} = 0.589 \times 10^{-6}$$

For beam continuous at one end (more than two spans):

$$K_{sh} = 0.09 \text{ (Table 11.2)}$$
$$\Delta_{sh} = 0.09 \times 0.589 \times 10^{-6} \times 6950^2 = 2.56 \text{ mm}$$
$$\Delta_{cp} + \Delta_{sh} = 4.17 + 2.56 = 6.73 \text{ mm}$$

11.11 DEFLECTION CONTROL IN CODE

Two methods are used in CSA A23.3-M84 for control of deflections of reinforced concrete members. The first method is to provide sufficient thickness for the member to have adequate flexural stiffness. Thus for beams and one-way slabs (solid or ribbed), not supporting or attached to other construction likely to be damaged by deflection, recommended minimum thicknesses are given in the Code (9.5.2.1) (see Table 5.2). For two-way construction, minimum thickness requirements are given in Clause 9.5.3. Only for these types of members, the recommended minimum thickness values may be used in lieu of deflection calculations. (A more extensive table including minimum thicknesses for beams and one-way slabs supporting or attached to nonstructural elements likely to be damaged by large deflections is presented in Ref. 11.9). The specified minimum thicknesses are established primarily from past experience with usual structures, and are not intended to be applied to special cases such as for members supporting heavy concentrated loads.

For members not meeting the minimum thickness requirements, or those supporting or attached to nonstructural elements likely to be damaged by large deflections, the deflections must be calculated and limited to the allowable values in the Code (9.5.2.6).

11.12 CONTROL OF CRACKING

Cracks in reinforced concrete structures may result from (1) volume changes (due to shrinkage, creep, thermal and chemical effects, etc.), (2) imposed loads and/or displacements (such as continuity effects, settlement of supports, etc.), and (3) flexural stress due to bending (Ref. 11.15). Cracks due to the first two causes could mostly be controlled by ensuring good quality concrete, and correct construction (control joints, temperature and shrinkage reinforcement, etc.). The discussion that follows deals with cracking due to flexural stresses.

Because of the low strength and failure strain of concrete in tension, some minute cracking of concrete in the tension side of flexural members is inevitable. This is particularly so when relatively higher strength reinforcing steels are used, in which case the steel stresses and strains at service load levels will also be high. From considerations of corrosion protection for reinforcement as well as aesthetics, it is preferable to have several well distributed fine hairline cracks than a few wide cracks. This is the basic objective of crack control procedures.

The cracking of concrete is of a random nature, and the width of cracks is subject to relatively wide scatter. The methods for crack width calculation only attempt to predict the most probable maximum crack width as observed in laboratory tests. These are further simplified to arrive at satisfactory reinforcing details meeting the objectives of crack control.

Experimental investigations of both tensile and flexural members have shown that crack width is proportional to tensile steel stress, f_s. The other important parameters that influence crack width are (1) the thickness of concrete cover, (2) the effective tension area of concrete surrounding the flexural tension reinforcement and having the same centroid as the reinforcement (A_e in Fig. 11.10), and (3) the number of bars, m (Ref. 11.12). Based on statistical analysis of extensive test data reported by several investigators, Gergely and Lutz developed Eq. 11.19 for the probable maximum crack width in a flexural member (Ref. 11.12).

$$w = C\beta f_s \sqrt[3]{d_c A} \qquad (11.19)$$

Fig. 11.10 *Parameters for crack width computation*

where, w = the maximum crack width; β = ratio of distances to the neutral axis from the extreme tension fibre to that from the centroid of the main reinforcement (= h_2/h_1 in Fig. 11.10); f_s = stress in steel at specified load calculated by elastic cracked section theory; d_c = thickness of concrete cover measured from the extreme tension fibre to the centre of the bar located closest thereto; A = effective tension area of concrete surrounding the main tension reinforcement and having the same centroid as that reinforcement, divided by the number of bars, = A_e/m; m = number of bars, taken as A_s divided by area of largest bar used when bars are of different sizes; and C = a numerical constant determined from statistical analysis of experimental data. C has the inverse unit of stress and was determined in Ref. 11.12 as 76×10^{-6} in²/kip with f_s in ksi. In SI units, the corresponding value of C is 11×10^{-6} mm²/N with stress, f_s, in MPa (= N/mm²). Eq. 11.19 suggests that the maximum crack width will be less with several moderately spaced small size bars well distributed over the zone of maximum tension than with a few bars of a larger diameter giving the same steel area. With a view to emphasizing the selection of reinforcing details for crack control, Eq. 11.19 can be rearranged as:

$$z = f_s \sqrt[3]{d_c A} \times 10^{-3} = \frac{w}{C\beta} \times 10^{-3} \qquad (11.20)$$

For general design use, β is assumed to have an approximate value of 1.2. The Code further assumes, on the basis of experience with existing structures, limiting crack widths of 0.40 mm and 0.33 mm for interior and exterior exposure, respectively. With these approximations, the right hand side of Eq. 11.20 reduces to:

$$\text{for interior exposure,} \quad \frac{0.40}{11 \times 10^{-6} \times 1.2} \times 10^{-3} = 30 \text{ kN/mm}$$

$$\text{for exterior exposure,} \quad \frac{0.33}{11 \times 10^{-6} \times 1.2} \times 10^{-3} = 25 \text{ kN/mm}$$

On this basis, CSA A23.3-M84(10.6.4) limits the value of $z = f_s \sqrt[3]{d_c A} \times 10^{-3}$ to a maximum of 30 kN/mm for interior exposure and to 25 kN/mm for exterior exposure. In computing z, f_s is the tensile steel stress at specified loads computed by elastic cracked section theory (see Section 4.6). However, the Code permits the use of $f_s = 0.6 f_y$, (f_y = specified yield stress) in lieu of a detailed calculation. In most practical situations, this is a conservative assumption as the actual steel stress at specified load levels will be below this value. The check on z is needed only when the specified yield strength for tension reinforcement exceeds 300 MPa.

Equation 11.19 has been shown to be applicable to one-way slabs as well (Ref. 11.13). However, the average value of β for slabs will be larger than for beams, and Ref. 11.14 suggests $\beta = 1.35$ for slabs. On this basis the maximum values of z for one-way slabs are:

$$\text{for interior exposure,} \quad z = \frac{1.2}{1.35} \times 30 = 26.7 \text{ kN/mm}$$

for exterior exposure, $z = \dfrac{1.2}{1.35} \times 25 = 22.2$ kN/mm

The limiting values for z given above are for normal conditions and do not apply to structures subject to very aggressive exposure or designed to be watertight (Code 10.6.5)). Recommended maximum permissible crack widths for such conditions are given in Ref. 11.15.

In T-beams where the flange is in tension, too wide a spacing of the reinforcement may result in formation of wide cracks in the slab near the web. To safeguard against this the Code (10.6.6) limits the distribution of negative moment reinforcement to a width equal to the lesser of the effective width or 1/10 span. In the latter case (span/10 < b), the outer portions of the flange must be protected with additional reinforcement.

In relatively deep flexural members, controlling the crack width at the level of tension reinforcement by limiting the value of z may still leave a deep zone of tension in the web between the neutral axis and the reinforcement level unprotected and susceptible to wide cracking. Hence, for flexural members with depth of web in excess of 750 mm, the Code (10.6.7) requires the placement of longitudinal skin reinforcement near the vertical faces of the web in the tension zone.

EXAMPLE 11.6

The actual placement details of flexural reinforcement at midspan of the beam in Example 11.4 are shown in Fig. 11.11. Assume the steel as Grade 400. Check the section against the crack control requirements of the Code.

SOLUTION

Distance of centroid of steel area from bottom face is:

$$d_s = \frac{2 \times 500 \times (51.3 + 25.2/2) + 2 \times 300 \times (51.3 + 19.5/2)}{1600}$$

$$= 63 \text{ mm}$$

$d = 600 - 63 = 537$ mm
$n = 9$, and $\rho = A_s/bd = 0.0016$

Using Eqs. 11.9 and 11.10,
$kd = 84$ mm, $I_{cr} = 3.32 \times 10^9$ mm^4
Service load moment (from Example 11.4) = 126 kN·m

Steel stress, $f_s = n \dfrac{M y_s}{I_{cr}}$

$$= \frac{9 \times 126 \times 10^6 \times (537 - 84)}{3.32 \times 10^9} = 155 \text{ N/mm}^2$$

Fig. 11.11 *Example 11.6*

From Fig. 11.11,

$d_c = 51.3 + 25.2/2 = 63.9$ mm

$A_e = 2\, d_s b_w = 2 \times 63 \times 300 = 37\,800$ mm^2

$$\text{Number of bars, } m = \frac{A_s}{\text{area of largest bar}}$$

$$= \frac{1600}{500} = 3.2$$

$$A = \frac{A_e}{m} = \frac{37\,800}{3.2} = 11\,800 \text{ mm}^2$$

$$z = f_s \sqrt[3]{d_c A} \times 10^{-3}$$
$$= 155 \times \sqrt[3]{63.9 \times 11\,800} \times 10^{-3}$$
$$= 14.1 \text{ kN/mm} < 25 \qquad\qquad \text{OK}$$

PROBLEMS

---

1. A simply supported beam of rectangular section is subjected to the service loads shown in Fig. 11.12. For this beam,
   (a) Compute the long-term deflection due to the sustained dead load only,

Fig. 11.12 Problem 1

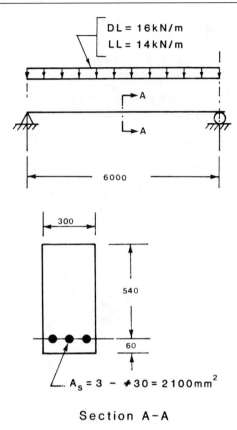

DL = 16 kN/m
LL = 14 kN/m

6000

300

540

60

$A_s = 3 - \#30 = 2100\,mm^2$

Section A-A

(b) Under full service loads, check whether the cracking is under control for "interior exposure" conditions. Take $f'_c = 25$ MPa, $f_y = 350$ MPa.

2. The end span of a continuous rectangular beam is shown in Fig. 11.13. Assuming that the negative moment at the first interior support is $wl^2/10$, compute the immediate deflection at midspan due to live load. Dead load = 3 kN/m, live load = 7 kN/m, $f'_c = 30$ MPa, and $f_y = 300$ MPa.

3. The dimensions, reinforcement details and service dead load moments using Code coefficients for a one-way slab are shown in Fig. 11.14. Compute the long time deflection due to sustained dead load only for the exterior span. Assume: Dead load, including weight of slab = 6 kN/m²
$$f'_c = 30 \text{ MPa}, f_y = 400 \text{ MPa}$$

4. For the beam in Problem 1 above (Fig. 11.12), check whether the crack width is within acceptable limits.

Fig. 11.13   *Problem 2*

REFERENCES

11.1   Subcommittee 1, ACI Committee 435, *Allowable Deflections*, J. of ACI, Vol. 65, No. 6, June 1968, pp. 433–444.

11.2   Hsieh, Y.Y., *Elementary Theory of Structures*, Prentice Hall, Inc., New Jersey, 1970, 432 pp.

11.3   Roark, R.J., and Young, W.C., *Formulas for Stress and Strain*, 5th Ed., McGraw Hill Book Co., New York, 1975.

11.4   *Concrete Design Handbook*, Canadian Portland Cement Association, Ottawa, Ont., Canada, 1985.

11.5   Branson, D.E., *Deformation of Concrete Structures*, McGraw-Hill, Inc., New York, 1977.

11.6   ACI Committee 435, *Deflections of Reinforced Concrete Flexural Members*, J. of ACI, Vol. 63, No. 6, June 1966, pp. 637–674.

11.7   Yu, W.W., and Winter, G., *Instantaneous and Long-time Deflections of*

Fig. 11.14 Problem 3

$$M_D = wL^2/24 \qquad wL^2/14 \qquad wL^2/10$$

$$= 3.42 \qquad = 5.87 \qquad = 8.22 \qquad \text{kN.m}$$

$$A_s = \quad 333 \qquad 500 \qquad 667 \qquad \text{mm}^2$$

Section   A–A   B–B   C–C

Reinforced Concrete Beams under Working Loads, J. of ACI, Vol. 57, No. 1, July 1960, pp. 29–50.

11.8 ACI Committee 435, *Deflections of Continuous Concrete Beams,* J. of ACI, Vol. 70, No. 12, Dec. 1973, pp. 781–787.

11.9 ACI Committee 435, *Proposed Revisions by Committee 435 to ACI Building Code and Commentary Provisions on Deflections,* J. of ACI, Vol. 75, No. 6, June 1978, pp. 229–238.

11.10 Miller, A.L., *Warping of Reinforced Concrete Due to Shrinkage*, J. of ACI, Vol. 54, No. 11, May 1958, pp. 939–950.

11.11 Branson, D.E., *Compression Steel Effect on Long-time Deflections*, J. of ACI, Vol. 68, No. 8, Aug. 1971, pp. 555–559.

11.12 Gergely, P. and Lutz, L.A., *Maximum Crack Width in Reinforced Concrete Flexural Members*, Causes, Mechanism, and Control of Cracking in Concrete, (SP–20), American Concrete Institute, Detroit, 1968, pp. 87–117.

11.13 Lloyd, J.P., Rejali, H.M., and Kesler, C.E., *Crack Control in One-Way Slabs Reinforced with Deformed Wire Fabric*, J. of ACI, Vol. 66, No. 5, May 1969, pp. 366–376.

11.14 *Commentary on Building Code Requirements for Reinforced Concrete (ACI 318–77)*, American Concrete Institute, Detroit, 3rd Printing, May 1979, p. 44.

11.15 ACI Committee 224, *Control of Cracking in Concrete Structures*, J. of ACI, Vol. 69, No. 12, Dec. 1972, pp. 717–753.

CHAPTER 12

ONE-WAY FLOOR SYSTEMS

12.1 ONE-WAY SLAB — BEAM AND GIRDER FLOOR

A plan of a typical one-way slab-beam girder floor is shown in Fig. 12.1a. The beam and girder layout subdivides the floor slab into long rectangular panels, the length of the panel being greater than twice its width. Such slabs transfer the loads on them primarily along the short span onto the supporting beams on their longer sides (see Section 2.3.2 and Fig. 2.4b). The slab can therefore be designed as a continuous one-way slab supported on the beams. A typical design strip of 1 m width is shown in Fig. 12.1b. Moments in continuous one-way slabs are usually determined by means of approximate coefficients such as those presented in Section 10.3.3. More extensive tables of moment coefficients, covering a large range of parameters, are presented in the Appendix to Ref. 12.1. These include the cases of a single concentrated load at midspan, two equal loads at third-points and three equal loads at quarter-points, as well as illustrative examples.

The load on the floor slab is in turn carried by the transverse beams. Each beam supports the load from approximately one-half width of the slab panel on either side of it. Thus an interior beam such as B1–B2–B1 carries the load from a tributary slab area of width, b, as shown in Fig. 12.1a, while the exterior beam B3–B4–B3 carries the load from a slab area of width, $b/2$. These transverse beams are continuous T-beams (L-beams for the exterior beam), with those along column lines framing into the columns, and the others framing into the longitudinal girders (Fig. 12.1 c, d). Two alternative procedures for analyzing these beams for gravity loads were discussed in Section 10.5 (Fig. 10.7b). In one-way slab systems, usually each transverse beam, with its tributary floor slab is analyzed individually as a continuous T-beam (rather than lumping together all the transverse beams within a panel width, B, as an equivalent single slab-beam member framing into the columns in the transverse bent, and analyzing the resulting partial frame). This procedure will be followed in the design example to follow in this chapter. It should be noted that the transverse beams framing into the columns (column line beams) and those framing into the girders will have different degrees of restraint at their supports. In usual situations involving continuous beams of approximately equal spans with uniform loads, the moments may be determined using approximate coefficients as in the case of continuous slabs (Section 10.3.3). Additional tables giving coefficients for three different

Fig. 12.1 *One-way slab-beam and girder floor*

concentrated loadings are contained in Ref. 12.1. For regular spans and loadings and, in particular, where moment coefficients are used to determine moments, there is little difference between the design of beams supported by columns and those supported by girders.

The longitudinal girders in Fig. 12.1e, such as G1–G2–G1, are continuous over the columns and are loaded mostly through the transverse beams. Interior girders have beams framing in from both sides, whereas exterior girders have beams on only one side. The floor loads transmitted through the beams act as concentrated loads on the girders.

The Code moment coefficients, which are for uniform loads, are not applicable to design of girders. Girders are usually analyzed as a partial frame following the procedure illustrated in Example 10.1. However, where appropriate, the approximate moment coefficients in Ref. 12.1 may be used for determining the moments in girders also.

12.2 DESIGN EXAMPLE

The design of a one-way slab-beam girder floor is illustrated in this Section with the design of a typical floor shown in Fig. 12.2. The specified live load is 4.8 kN/m². In addition to the weight of the floor itself, allowance is to be made for a dead load of 2.0 kN/m² obtained as follows:

suspended ceiling $= 0.25$ kN/m², mechanical and electrical fixtures $= 0.50$ kN/m², floor finish $= 0.25$ kN/m², and partitions $= 1.0$ kN/m². Concrete of strength $f'_c = 20$ MPa and steel reinforcement of Grade 300 ($f_y = 300$ MPa) are specified. Assume that concrete has a mass of 24 kN/m³.

12.2.1 DESIGN OF SLABS

The slab panels have length greater than twice the width and hence can be designed as one-way slabs. Typical slab panels are identified in Fig. 12.2 as $S1$ (exterior spans) and $S2$ (interior spans).

Suggested minimum thickness of slab for deflection control (CSA) A23.3-M84 Table 9-1; see Chapter 5, Table 5.2) with the modification factor for f_y other than 400 MPa is:

$$\text{For exterior spans,} \quad \frac{\ell}{24}\left(0.4 + \frac{f_y}{670}\right) = \frac{3075}{24}\left(0.4 + \frac{300}{670}\right) = 109 \text{ mm}$$

$$\text{For interior spans,} \quad \frac{\ell}{28} \times 0.848 = \frac{3500}{28} \times 0.848 = 106 \text{ mm}$$

A slab thickness of 110 mm will be used.

Fig. 12.2 *Floor framing scheme*

(i) Plan

(ii) Section AA (enlarged)

(iii) Section BB (enlarged)

FACTORED LOADS

Live load = 4.8 × 1.5 = 7.20 kN/m²
Dead load, including slab weight = (2 + 0.11 × 24) × 1.25 = 5.80 kN/m²
Total load, w_f = (7.20 + 5.8) kN/m² = 13.0 kN/m²

CLEAR SPANS AND DESIGN MOMENTS

Assuming a width of stem for the transverse beams of 300 mm, clear span for S1 = 3075 − 300 = 2775 mm, and for S2, = 3500 − 300 = 3200 mm. The Code moment coefficients (Table 10.1) are applicable for the continuous slab and will be used to compute the moments. Considering a continuous strip of slab 1 m wide, the moment coefficients and design moments for critical sections are set forth in lines 1 and 3 of Table 12.1.

DESIGN OF SECTIONS

The absolute maximum moment in the slab is at an interior support and equal to 12.1 kN·m. The required slab depth will be based on this moment. Selecting a reinforcement ratio $\rho = 0.5\rho_{max}$, from Table 4.1, $\rho = 0.0113$.
For this ρ, $K_r = 2.48$ MPa.

$$\text{Required } d = \sqrt{M_f/K_r b} = \sqrt{\frac{12.1 \times 10^6}{2.48 \times 1000}} = 69.9 \text{ mm}$$

An overall thickness of 110 mm is adequate. Assuming No. 15 bars and a clear cover of 20 mm, the depth actually provided is:

$$d = 110 - 20 - 16/2 = 82 \text{ mm}$$

$$\text{Actual } K_r = \frac{12.1 \times 10^6}{1000 \times 82^2} = 1.80 \text{ MPa}$$

For $K_r = 1.80$, using Table 5.4 or solving Eq. 5.3, $\rho = 0.00782$

Required $A_s = 0.00782 \times 1000 \times 82 = 642$ mm²

Similar calculations for areas of reinforcement required at other critical sections are set forth in Table 12.1.

CHECK SHEAR

The maximum factored shear forces are:

$$\text{in span S1, } 1.15 \frac{w_f \ell_n}{2} = 1.15 \times \frac{13.0 \times 2.775}{2} - 20.7 \text{ kN}$$

$$\text{and in span S2, } \frac{w_f \ell_n}{2} = 13.0 \times 3.2/2 = 20.8 \text{ kN}$$

Table 12.1 Design of Slab

	SPAN S1			SPAN S2		
	Exterior support	Midspan	First interior support	First interior support	Midspan	All other supports
1. Moment of coefficients C	$-\dfrac{1}{24}$	$\dfrac{1}{14}$	$-\dfrac{1}{10}$	$-\dfrac{1}{11}$	$\dfrac{1}{16}$	$-\dfrac{1}{11}$
2. Length ℓ_n, m	2.775	2.775	2.988	2.988	3.20	3.20
3. Factored moment $M_f = C_{uf}\ell_n^2$, kN·m	-4.17	7.15	-11.6	-10.6	8.32	-12.1
4. $K_r = \dfrac{M_f \times 10^6}{1000\,d^2}$, MPa	0.620	1.06	1.73	1.58	1.24	1.80
5. $100\,\rho$, (Eq. 5.3, or Table 5.4)	0.251 →	0.440	0.748 →	0.677 ←	0.520	0.782 →
6. ρ_{min}		0.002			0.002	
7. $A_s = \rho \times 1000 \times 82$, mm²	206 →	361	614 →	555 ←	426	642 →
8. Maximum, spacing, mm		330			330	
9. Bar selection (No. and spacing in mm)	No. 10 @ 330	No. 10 straight @ 600 + No. 15 bent @ 600	No. 15 @ 300, alternate bent on each side		No. 10 st. @ 600 + No. 15 bt @ 600	No. 15 @ 300 alternate bent on each side
10. A_s provided, mm²	303	500	667	667	500	667

11. Sketch

Factored shear resistance provided by concrete, V_c, is:

$$V_c = 0.2\, \lambda \phi_c \sqrt{f_c'}\, b_w d$$
$$= 0.2 \times 1.0 \times 0.6 \times \sqrt{20} \times 1000 \times 82 \times 10^{-3}$$
$$= 44.0 \text{ kN}$$
$$V_f = 20.8 \text{ kN} < V_c$$

and hence no shear reinforcement is needed in slab.

Shrinkage and Temperature Reinforcement

Since the principal reinforcement is provided only along the short direction (one-way slab), secondary reinforcement normal to the principal reinforcement must be provided to resist shrinkage and temperature stresses. The minimum area for such reinforcement is (CSA A23.3-M84(7.8).

$$A_s = 0.002 \times 1000 \times 110 = 220 \text{ mm}^2$$

Maximum spacing of such reinforcement is:

$$5h = 550 \text{ mm or } 500 \text{ mm}$$

No. 10 bars at 450 mm will be adequate to provide this area.

12.2.2 DESIGN OF BEAMS B1–B2–B1

The Code moment coefficients will be used for the design of transverse beams. All interior beams will be provided with the same cross section although there is some difference between the restraints at supports for beams supported by girders and for beams framing into columns. The girders will be assumed to have a width of 450 mm.

Beam depth: the minimum depth for deflection control is

$$\frac{\ell}{24} \times 0.848 = \frac{7400}{24} \times 0.848 = 262 \text{ mm}$$

For dead load calculation, an overall depth of 600 mm will be assumed. The beams are spaced at 3.5 m so that each interior transverse beam has a tributary floor area of width 3.5 m.

Live load = $7.20 \times 3.5 = 25.2$ kN/m
Dead load from floor = $5.8 \times 3.5 = 20.3$ kN/m
Weight of beam stem = $0.3\,(0.6 - .11) \times 24 \times 1.25 = 4.41$ kN/m
Total beam load = 49.9 kN/m

The beam spans and loads satisfy the limitations in CSA A23.3-M84 (8.3.3) and the approximate coefficients may be used to compute the design moments and shears at all critical sections in the beam. The coefficients, moments and shears for beams

B1 and B2 are given in Fig. 12.3. The negative moment at the interior face of an exterior support is taken as $w_f \ell_n^2/16$, applicable for a column support, and this is also used for all beams supported on girders, so that the same design can be used for all interior beams.

Design of Sections

The clear span $\ell_n = 7.4 - 0.45 = 6.95$ m. The overall beam size is controlled by the maximum moment which is at the interior support of B1 and equal to -241 kN·m. The beam effectively acts as a rectangular section under the negative moment. Assuming a reinforcement ratio of $\rho = 0.6\rho_{max} = 0.0136$, from Table 4.1, $K_r = 2.88$

$$\text{Required } d = \sqrt{\frac{241 \times 10^6}{2.88 \times 300}} = 528 \text{ mm}$$

Keeping the overall depth at 600 mm, and assuming No. 30 bars, No. 10 stirrups and 40 mm clear cover,

$$d = 600 - (40 + 11.3 + 29.9/2) = 533 \text{ mm}$$

The adequacy of this depth is now checked for shear.

Shear design will be done with the Simplified Method in clause 11.3 of the Code. Factored shear resistance provided by concrete is:

$$V_c = 0.2\lambda\phi_c \sqrt{f_c'}b_w d = 0.2 \times 1.0 \times 0.6 \times \sqrt{20} \times 300 \times 533 \times 10^{-3}$$
$$= 85.8 \text{ kN}$$

Required shear resistance, V_s, to be provided by shear reinforcement is:

$$V_s = V_f - V_c = 199 - 85.8 = 113.2 \text{ kN}$$

Upper limit on V_s, (cl. 11.3.6.6) is:

$$V_{s,\lim} = 0.8\lambda\phi_c \sqrt{f_c'} \, b_w d = 4 \times 85.8 = 343 \text{ kN} > V_s \text{ required} \qquad \text{OK}$$

The depth of 600 mm is adequate to carry shear with web reinforcement.

To compute flexural reinforcement at the exterior face of the first interior support where the bending moment is maximum,

$$K_r = \frac{241 \times 10^6}{300 \times 533^2} = 2.83 \text{ MPa}$$

$$\rho = 0.0133 \text{ (Table 5.4)}$$

$$A_s = 0.0133 \times 300 \times 533 = 2123 \text{ mm}^2$$

This reinforcement ratio is about 0.59 ρ_{max} and is satisfactory.

Fig. 12.3 Beam B1–B2–B1

		B1			B2	
1. Moment coefficient, C	−1/16	1/14	−1/10		−1/11	1/16
2. Moment = C × 49.9 × 6.95², (kN·m)	−151	172	−241		−219	151
3. Shear = $\dfrac{49.9 \times 6.95}{2}$, 1.15 × $\dfrac{w\ell_n}{2}$ (kN)	173	172	199		173	151
4. A_s, required, (mm²)	1227	1286	2123		1892	1128
5. Reinforcement selected and	1–No. 25 st.	2–No. 20 st.	2–No. 20 st.			2–No. 20 st.
	2–No. 25 bt.	2–No. 25 bt.	2–No. 25 bt.			2–No. 20 bt.
			2–No. 20 bt.			
area, A_s (mm²)	1500	1660	2200			1200
6. Factored resistance, M_r, kN·m	−181	215	−250		−250	162

At exterior support of B1,

$$K_r = \frac{151 \times 10^6}{300 \times 533^2} = 1.77 \text{ MPa}$$

$$\rho = 0.00768$$

$$A_s = 1227 \text{ mm}^2$$

At support of B2,

$$K_r = \frac{219 \times 10^6}{300 \times 533^2} = 2.57 \text{ MPa}$$

$$\rho = 0.0118$$

$$A_s = 1892 \text{ mm}^2$$

At midspan of B1 and B2, the beams act as T-beams. Effective flange width is,

$$0.25 \text{ span} = 1850 \text{ mm} \qquad \text{(least)}$$
$$24 \, h_f + b_w = 24 \times 110 + 300 = 2940 \text{ mm}$$
$$\text{spacing} = 3500 \text{ mm}$$
Hence, $$b = 1850 \text{ mm}$$

Since the same beam acting as a rectangular section at the support under a larger moment of 241 kN·m is under-reinforced, the T-section with the lower span moment will surely be under-reinforced and hence there is no need to check this.

For beam B1 at midspan:
The reinforcement area is determined using the alternative procedure explained in Example 5.7,

$$d - h_f/2 = 533 - 110/2 = 478 \text{ mm}$$
$$0.9d = 0.9 \times 533 = 480 \text{ mm}$$
$$z \approx 480 \text{ mm}$$

$$\text{Trial } A_s = \frac{172 \times 10^6}{0.85 \times 300 \times 480} = 1405 \text{ mm}^2$$

$$a = \frac{1405 \times 0.85 \times 300}{0.85 \times 0.6 \times 20 \times 1850} = 19.0 \text{ mm} < h_f$$

$$z = d - a/2 = 533 - 19.0/2 = 524 \text{ mm}$$

$$\text{Revised } A_s = \frac{172 \times 10^6}{0.85 \times 300 \times 524} = 1287 \text{ mm}^2$$

(Detailed calculation gives $A_s = 1286$ mm².)

For beam B2 at midspan:

$$M_f = 151 \text{ kN·m}$$

$$A_s \simeq 1287 \times \frac{151}{172} = 1130 \text{ mm}^2$$

$$a = \frac{1130 \times 0.85 \times 300}{0.85 \times 0.6 \times 20 \times 1850} = 15.3 \text{ mm}$$

$$z = 533 - 15.3/2 = 525 \text{ mm}$$

$$A_s = \frac{151 \times 10^6}{0.85 \times 300 \times 525} = 1128 \text{ mm}^2$$

Reinforcement is provided as shown in Fig. 12.3. The locations where individual bars are no longer required for flexure may be determined using Fig. 10.6. This, together with the Code requirements in Fig. 5.14, determine the bar bending details. In arriving at the bend locations in Fig. 12.3, it has been assumed that for satisfying bar extension requirements, the bent bar is effective up to the section where it crosses the mid-depth of the beam. The bar cut-off and bending locations in Fig. 12.3 meet all the requirements for development (see Example 9.4). Note that with the No. 25 bars the effective depth will be 536 mm compared to the 533 mm used in the computations. This difference will hardly change the results of the calculations and, further, is on the safe side. Part of the tension reinforcement (two No. 20 bars) at the top over the support section will be placed in the flanges of the T-section. The factored flexural strengths, M_r of all critical sections are also given in line 6 of Fig. 12.3 (see Example 9.4 for calculations).

Design of Shear Reinforcement

For Beam B1 the variation of shear force based on the Code coefficients is as shown in Fig. 12.4.

At distance d from face of first interior support,

$$V_f = \frac{199 \, (3996 - 533)}{3996} = 172 \text{ kN}$$

Factored shear resistance provided by concrete, V_c, is 85.5 kN. Shear resistance to be provided by shear reinforcement is:

$$V_s = V_f - V_c = 172 - 85.8 = 86.2 \text{ kN}$$

Fig. 12.4 *Shear force in beams*

V_s is less than $0.4\lambda\phi_c\sqrt{f'_c}\ b_wd = 171$ kN, and hence the maximum spacing in clause 11.3.8.1 need not be halved.

Using No. 10 vertical U stirrups, the required spacing is:

$$s = \frac{\phi_s A_v f_y d}{V_s} = \frac{0.85 \times 200 \times 300 \times 533}{86.2 \times 10^3} = 315 \text{ mm}$$

Spacing required for providing the minimum shear reinforcement is:

$$s = \frac{A_v f_v}{0.35\ b_w} = \frac{200 \times 300}{0.35 \times 300} = 571 \text{ mm}$$

$$s_{\text{max}} = 0.5d = 266 \text{ mm} \qquad\qquad\qquad \text{(controls)}$$

For both beams B1 and B2, the required spacing, being greater than the maximum permitted, the latter controls. For both beams starting from each end, provide the first stirrup at 130 mm from the face of the support followed by 12 stirrups at 265 mm spacing placed towards midspan.

12.2.3 DESIGN OF GIRDER

The interior girders G1–G2–G1 have beams framing in from both sides. The girder moments and shears will be determined by an elastic analysis of the partial frame consisting of the girders and the columns immediately above and below. The girder cross-sectional dimensions are needed for the elastic analysis and this is first estimated using a preliminary design.

Preliminary Design of Girder Section

The loads on the girder are the dead and live loads from the floor slab and floor beams transmitted by the latter as point loads, and the distributed loading acting directly on the girder including its own weight.

The concentrated live load transmitted by the interior beams from a tributary area of 3.5 m × 7.4 m is:

$$7.20 \times 3.5 \times 7.4 = 186 \text{ kN}$$

(This includes the live load on the part of the floor directly over the girder. This part of the live load is in fact distributed over the length of the girder; however, there is little error in combining it with the load from the rest of the floor transmitted through the beams as concentrated loads.)

Concentrated dead load from the beams is:

$$(20.3 + 4.41) \times 6.95 = 172 \text{ kN}$$

Distributed dead load due to partition, floor finish, etc., acting directly on the girder (width = 450 mm) is:

$$1.25 \times 2 \times 0.45 = 1.13 \text{ kN/m length}$$

Weight of girder, assuming a total depth of 1 m,

$$= 1.25 \times 0.45 \times 1 \times 24 = 13.5 \text{ kN/m}$$

Total distributed dead load $= 1.13 + 13.5 = 14.6 \text{ kN/m}$

The loads and span dimensions are shown in Fig. 12.5a. For the preliminary design, the maximum negative moment in the girder may be taken as 80 percent of the simple span moment and the maximum shear as 1.15 times the simple span shear. These factors give the same ratios as those used in the Code coefficients for distributed loading ($w_f \ell_n^2/10 = 0.8 \times w_f \ell_n^2/8$). On this basis, for the longer span G2,

$$M_f \approx 0.8 \left[(186 + 172) \times 3.275 + 14.6 \times 10.05^2/8 \right] = 1122 \text{ kN·m}$$
$$V_f \approx 1.15 \left(186 + 172 + 14.6 \times 10.05/2 \right) = 496 \text{ kN}$$

With a reinforcement ratio of about $0.60\rho_{max}$, the depth required is:

$$d = \sqrt{\frac{1122 \times 10^6}{2.88 \times 450}} = 930 \text{ mm}$$

Fig. 12.5 *Analysis of girder*

(a) Girder loading

**(b) Partial frame for
analysis of girder**

Girder section

With this depth, shear resistance provided by concrete, V_c, is:

$$V_c = 0.2 \times 1.0 \times 0.6 \times \sqrt{20} \times 450 \times 930 \times 10^{-3} = 225 \text{ kN}$$

Shear resistance required to be provided by reinforcement is:

$$V_s = 496 - 225 = 271 \text{ kN}$$

Upper limit of V_s for section dimensions is:

$$0.8\lambda\phi_c\sqrt{f_c'}\, b_w d = 900 \text{ kN} > 271 \text{ kN}$$

The dimensions are adequate for shear with shear reinforcement.

Since the reinforcement ratio assumed is still relatively low, and the bending moment over most of the girder length will be lower than the value considered above, an overall girder depth of 900 mm will be assumed for use in detailed analysis.

Detailed Elastic Analysis

The partial frame dimensions are shown in Fig. 12.5b. The analysis will be made using the moment distribution method. The girder is assumed to have a T-section with an effective flange width of $0.25 \times$ span $= 2500$ mm. The member stiffnesses are:

1. columns:
$$K_c = \frac{4 E \times 450^4}{12} \times \frac{1}{3500} = 3.91 E \times 10^6$$

2. for girder (T-section):

$$I = 5.02 \times 10^{10} \text{ mm}^4$$

$$\text{for G1}, K_{G1} = \frac{4 E \times 5.02 \times 10^{10}}{10\,000} = 20.1 E \times 10^6$$

$$\text{for G2}, K_{G2} = \frac{4 E \times 5.02 \times 10^6}{10\,500} = 19.1 E \times 10^6$$

The distribution factors to the girders are (Fig. 12.5b):
at exterior support,

$$\text{to G1}, = \frac{20.1}{2 \times 3.91 + 20.1} = 0.720$$

at interior support,

$$\text{to G1}, = \frac{20.1}{2 \times 3.91 + 20.1 + 19.1} = 0.427$$

$$\text{and to G2}, = \frac{19.1}{2 \times 3.91 + 20.1 + 19.1} = 0.406$$

The critical loading conditions together with the corresponding simple span moment diagrams are shown in Fig. 12.6. The results of the moment distribution are presented in Fig. 12.7 (moments) and Fig. 12.8 (shears).

Some economy may be achieved by redistributing the moments obtained by the elastic analysis, as permitted by CSA A23.3-M84(8.4); however this is not attempted in this example. The complete moment envelope for girders G1–G2 is presented in Fig. 12.9.

Design of Sections

1. Section B (Fig. 12.9)

The maximum design moment for the girders is at the exterior face of the first interior column. The moment here being negative, the girder acts as a beam of

Fig. 12.6 Load patterns and simple span moments

(a) Loading case I

(b) Loading case 2

(c) Loading case 3

rectangular section. The moment at the column face is (see Section 10.6) given by $(M_{CL} - Vb/3)$, and is:

$$1130 - \frac{494 \times 0.45}{3} = 1056 \text{ kN} \cdot \text{m}$$

Since the moment is much lower at all other sections, a slightly higher reinforcement ratio of about $0.75 \, \rho_{max}$ will be used at this section.

For $\rho = 0.75 \, \rho_{max}$, $K_r = 3.41$

$$\text{Required } d = \sqrt{\frac{1056 \times 10^6}{3.41 \times 450}} = 830 \text{ mm}$$

Fig. 12.7 Theoretical moment diagrams

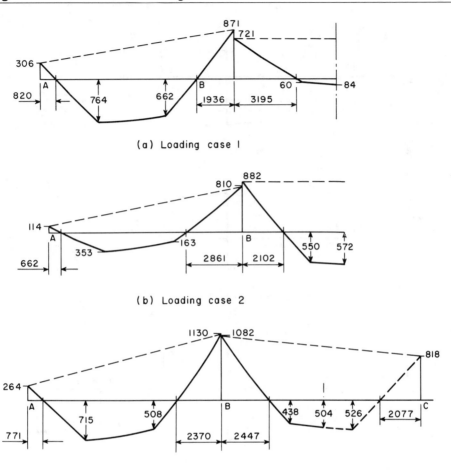

(a) Loading case 1

(b) Loading case 2

(c) Loading case 3

All moments in kN.m

Provide an overall depth of 900 mm. The negative reinforcement in the girder will be placed with a clear cover of 80 mm so that it will clear the reinforcement of the beams framing into it. Assuming No. 35 bars,

$$d = 900 - 80 - \frac{35.7}{2} = 802 \text{ mm}$$

$$K_r = \frac{1056 \times 10^6}{450 \times 802^2} = 3.65 \text{ MPa}$$

$$\rho = 0.0187$$

$$A_s = 0.0187 \times 450 \times 802$$

$$= 6750 \text{ mm}^2$$

Fig. 12.8 *Shear force distribution*

(a) Loading case 1

(b) Loading case 2

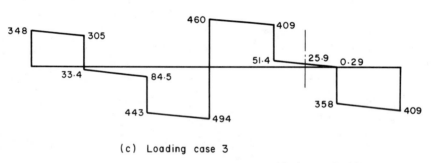

(c) Loading case 3

All shears in kN

2. Exterior support, A:

Design moment $= 306 - \dfrac{379 \times 0.45}{3} = 249$ kN·m

$$K_r = \frac{249 \times 10^6}{450 \times 802^2} = 0.861 \text{ MPa}$$

$$\rho = 0.00353$$

$$\rho_{\min} = 1.4/f_y = 0.00467$$

$$A_s = 1690 \text{ mm}^2$$

Fig. 12.9 *Girder design*

(a) Moment envelope

(b) Reinforcement details

(c) Cross sections

3. Span *AB*:

$M = 764 \text{ kN·m}$

Girder has a T-Section, with $b = 2500$ mm. With 40 mm clear cover, No. 15 stirrups and No. 30 bars, $d = 900 - (40 + 16 + 29.9/2) = 829$ mm.

$$\text{Approximate lever arm} \left(d - \frac{h_f}{2}, \text{ or } 0.9d\right) = 774 \text{ mm}$$

$$A_s = \frac{764 \times 10^6}{0.85 \times 300 \times 774} - 3870 \text{ mm}^2$$

With this A_s,

$$a = \frac{3870 \times 0.85 \times 300}{0.85 \times 0.6 \times 20 \times 2500} = 38.7 \text{ mm}$$

$$d - \frac{a}{2} = 829 - 38.7/2 = 810 \text{ mm}$$

$$\text{Revised } A_s = \frac{764 \times 10^6}{0.85 \times 300 \times 810} = 3700 \text{ mm}^2$$

4. Span *BC*:

As in the case of span *AB*, the reinforcement area is obtained as 2750 mm². The flexural reinforcement details are shown in Fig. 12.9. The bar cut-off and bend details satisfy the Code requirements presented in Fig. 5.14.

Shear Reinforcement

Maximum shear force at distance *d* from the face of support (802 + 450/2 = 1027 mm from centre of support) is:

$$V_f = 494 - 14.6 \times 1.027 = 479 \text{ kN}$$
$$V_c = 0.2 \times 1.0 \times 0.6 \times \sqrt{20} \times 450 \times 802 \times 10^{-3} = 194 \text{ kN}$$
$$V_s = 479 - 194 = 285 \text{ kN}, < 0.4\lambda\phi_c\sqrt{f_c'}b_w d$$

Hence normal spacing will suffice.

Using No. 15–U stirrups, the required spacing is,

$$s = \frac{0.85 \times 400 \times 300 \times 802}{285 \times 10^3} = 287 \text{ mm}$$
$$s_{\text{max}} = d/2 = 401 \text{ mm}$$

In the outer thirds of both girder spans, the shear is relatively high. Hence in this region a uniform spacing of 280 mm may be provided. The first stirrup is placed at 140 mm from the face of the support followed by 12 spaces at 280 mm. In the remaining portion of the middle third of each span, nominal shear reinforcement of No. 10–U stirrups may be provided at a spacing of about 800 mm. (In this region $V_f < V_c/2$ and no shear reinforcement is required to carry shear stress.) Note that in designing for shear, the shear strength contribution of the bent bars has been neglected.

Additional Considerations

Since the height of beam is in excess of 750 mm, skin reinforcement conforming to CSA A23.3-M84(10.6.7) must be placed near the faces of the member over the half of the height nearest the principal reinforcement. Three No. 10 bars on each face will be adequate for this. The Code (10.6.6) requires some tension reinforcement to be distributed over the effective flange width when the flanges are in tension. In this case the main reinforcement in the slab is placed parallel to the girder and this will be adequate to meet the above requirement. Furthermore, as required in clause 8.10.5, transverse reinforcement is needed at the top of the girder flange. The shrinkage and temperature reinforcement provided in the slab in a direction normal to its principal reinforcement will usually be adequate to meet this requirement.

12.3 ONE-WAY JOIST FLOOR

12.3.1 GENERAL

A one-way concrete joist floor (also known as ribbed slab) consists of a series of closely spaced parallel joists with an integrally cast thin concrete slab on top, and framing into girders at both ends (Fig. 12.10). Each joist together with its share of the top slab acts as a continuous T-beam spanning in one direction between the girders. Joist floors are usually formed using standard removable steel forms ("pans") placed between the joist ribs. Because much of the ineffective concrete in the tension zone is eliminated, joist floor construction can provide increased effective depth and stiffness for the floor member without increasing its dead weight. Standard size reusable forms make this type of construction economical for large spans (in excess of about 5 m) where a solid one-way slab becomes uneconomical on account of its large self-weight.

Standard pan form dimensions include widths of 500 mm and 750 mm and depths from 150 mm to 500 mm (Ref. 12.2). Pans with tapered ends are also available for use where larger joist width is desired near the end supports due to higher shear or flexural (negative moment) requirements.

12.3.2 CODE REQUIREMENTS

The Code limitations on cross-sectional dimensions of standard concrete joist floor construction are given in CSA A23.3-M84 (8.11) and summarized in Fig. 12.11. The top slab spanning the joists must have flexural reinforcement at right angles to the joists, sufficient to carry applied slab loads. However, under normal joist spacings, the minimum reinforcement specified for shrinkage and temperature will provide adequate flexural strength for the slab. Because of the close spacing and long span of the joists, and the possibility of distribution of local overloads, a 10 percent greater concrete shear stress is permitted for joists. Also, there is no minimum shear reinforcement requirement for joists (CSA A23.3-M84 (11.2.5.1b).

12.3.3 DESIGN

Just as a one-way slab is designed considering a typical 1 m wide strip as a continuous beam, a one-way joist-floor is designed by considering a typical single joist together with its flange (of width equal to the centre to centre spacing of the joists) as a continuous T-beam (Fig. 12.10b). In the design of this T-beam, the width of stem, b_w, is taken as the least width of the joist rib, at the bottom of the joist, neglecting the slight increase in width at higher levels due to the taper of the sides of the forms. As for regular T-beams, for joists the minimum flexural reinforcement ratio, $1.4/f_y$, is based on the width of stem, b_w. Although web reinforcement may be used to increase the shear capacity of joists, it is usually avoided (limit $V_f \leq V_c$). The shear strength can be increased by using deeper joists or by widening the joists near the support using tapered forms.

Fig. 12.10 *One-way concrete joist floor construction*

(a) Plan

(b) Section AA (enlarged)

(c) Section CC (enlarged)

Fig. 12.11 *Code limitations for one-way joist construction*

In the design of the girder, the load transmitted by the joists is considered as a distributed loading. The girder spanning the columns is also designed as a continuous T-beam whose section has the shape shown in Fig. 12.10c.

Design examples and load tables for standard one- and two-way joist construction are presented (in imperial units) in Ref. 12.3.

PROBLEMS

1. For the floor system shown in Fig. 12.2, design the exterior beam B3–B4–B3.
2. For the floor system shown in Fig. 12.2, design the exterior girder.

REFERENCES

12.1 "Reinforced Concrete Design Handbook," 3rd ed., ACI Publication SP-3, American Concrete Institute, Detroit, Michigan, 1965, pp. 255–271.

12.2 "Code of Standard Practice — Joist Construction," CRSI Manual of Standard Practice, Concrete Reinforcing Steel Institute, Chicago, Illinois, 1977.

12.3 "CRSI Handbook 1980," 45h ed., Concrete Reinforcing Steel Institute, Chicago, Illinois, 1980.

CHAPTER 13

TWO-WAY SLABS
SUPPORTED ON WALLS
OR STIFF BEAMS

13.1 ONE- AND TWO-WAY ACTION OF SLABS

Slabs which deform essentially into a cylindrical surface can be designed as one-way slabs or beam-strips spanning in the direction of the curvature. Such a behaviour truly exists in rectangular slabs supported only on the two opposite sides by unyielding supports and carrying a load that is uniformly distributed along the direction parallel to the supports (Fig. 2.4a, d). When an elastic, homogeneous and isotropic slab is bent into a cylindrical surface, secondary moments in the direction parallel to the supports are developed due to the Poisson effect (Ref. 13.1). However, in reinforced concrete slabs, the magnitude of this secondary (longitudinal) moment is relatively small and it is taken care of by the temperature and shrinkage reinforcement provided along this direction at right angles to the principal reinforcement.

When the support and/or loading conditions are different from those mentioned above, the slab bends into a surface with curvature in both the longitudinal and transverse directions, and the slab develops two-way action with primary moments in both directions. Principal flexural reinforcement is then needed in both directions. Typical elastic theory distributions of longitudinal and transverse moments in uniformly loaded square and rectangular plates, simply supported on all four sides, are shown in Fig. 13.1. For the square panel, the maximum moments in both directions are equal. In addition to the bending moments, the slab elements are also subject to twisting moments which have their maximum values at the corners of the panel. These twisting moments can be resolved into principal moments (bending) M_1 and M_2 at the corners. It can be seen that at slab corners, the moment M_1 is along the diagonal and negative, causing tension on the top fibres of the slab along the diagonal; and the moment M_2 is positive and at right angles to the diagonal, causing tension in this direction on the bottom fibres of the slab. The potential crack patterns resulting from the twisting moments are shown in Fig. 13.1c.

As the ratio long span/short span increases, the curvature and moments along the long direction progressively decrease and more and more of the slab load is transferred to the two longer supports by flexure in the short direction (Fig. 13.1b). Thus, in general, edge-supported rectangular reinforced concrete slabs with ratios of long to short span of two or more can also be designed as a one-way slab spanning in

Fig. 13.1 *Moment variation in elastic homogenous slab supported on all four sides*

(c) Potential cracking

(d) Corner reinforcement

(a)

(b) Long rectangular slab

$$\frac{B}{A} = 2$$

$$\Upsilon = 0.3$$

the short direction, with the principal reinforcement placed at right angles to the long support. The small moment in the long direction as well as the lengthwise distribution of normal non-uniform loads can usually be taken care of by the shrinkage and temperature reinforcement in the long direction. In addition, a minimum amount of negative reinforcement must be provided perpendicular to the short edge and some special reinforcement is also needed at the corners of panels to carry the twisting moment effects (Fig. 13.1d).

Slab panels which deform with significant curvature in both directions must be designed as two-way slabs with principal reinforcement placed in both directions. This category includes rectangular slabs (generally with long to short span ratios of less than two) supported by walls or by beams framing into columns (Figs. 2.4c, 2.9b); all slabs supported directly on columns (flat plates and flat slabs — Figs. 2.5 and 2.6); and waffle slabs (Fig. 2.8).

13.2 BACKGROUND OF TWO-WAY SLAB DESIGN

Prior to the 1973 edition of CSA A23.3, rectangular two-way floor systems supported on all sides by walls or relatively stiff beams (that is, essentially non-deflecting edge supports) were designed using moment coefficients based on semi-elastic analysis of homogeneous isotropic slabs. This procedure assumed that all load on the slab panel is transmitted to the surrounding walls or beams rather than directly to the columns. On the other hand, flat plates and flat slabs, which transmit the load directly to the columns, develop significant frame action and were designed on the basis of elastic analysis (frame analysis) considering the interaction between the slab and the columns. The 1973 and 1977 editions of CSA A23.3 provided for the design of all two-way slab systems (with or without beams along the edges of the panel) on the basis of the same general principles, considering it as part of a frame consisting of a row of columns or supports and the supported slab-beam members. These procedures, namely, the "Direct Design Method" and the "Equivalent Frame Method," are more complex and are described in Chapter 14. They are essentially meant for two-way slab systems, with or without beams, and supported by columns at the corners of the panels (Figs. 2.5, 2.6, 2.8, 2.9b).

For rectangular slabs supported on four sides by walls or stiff beams, CSA A23.3-M84 (13.3.1.2) has, in Appendix E of the Code, reintroduced one of the earlier methods of using moment coefficients for determining slab thickness, moments and shears in slabs and loads on supporting beams or walls. This method of using moment (and shear) coefficients for edge-supported slabs is simpler although slightly less economical. It is still widely used for such cases, and is particularly useful in situations where there is little or no interaction between the floor systems and supporting columns, (that is, no frame action is developed), for example, two-way slab systems supported entirely on masonry walls. Similarly, in simple cases such as the design of a one or two-panel slab supported on walls or beams, the elaborate and complex procedure of Chapter 14 may not often be justified and the simpler pre–1973 procedure may be more appropriate. The use of this simpler design procedure for edge-supported slabs is detailed in this chapter.

13.3 DESIGN OF TWO-WAY SLABS SUPPORTED ON WALLS AND STIFF BEAMS

Only rectangular uniformly loaded slabs supported on all four sides by walls or relatively stiff beams (in general, slabs with essentially non-deflecting edge-supports) are considered in this category. In such cases, the design of the slab part and that of the supporting elements (walls and beams) can be separated. The former is done using moment coefficients for edge-supported slabs based on elastic analysis, considerations of inelastic redistribution of moments, and simplifying approximations. In the case of wall supports, this completes the floor design part. When the edge support consists of a beam between columns, the beam is designed for the share of slab load that is transmitted directly to it, considering frame action, if any.

Three alternative methods for computing design moments in two-way slabs at critical sections are presented in Ref. 13.2. All three methods allow for the effects of continuity at supports and the ratio of short span to long span. Subject to their respective specified limitations, all three methods lead to satisfactory designs. One of these, based on the developments by Marcus, and identified as "Method 3" in Ref. 13.2, with minor modifications, is the method recommended in CSA A23.3-84 (App.E), and is described below. In order to adopt this method, a supporting beam

Fig. 13.2 *Column and middle strips, and variation of moments for edge-supported two-way slab*

Variation of short span positive moment, $M_{a,pos}$, along 11

may be considered stiff, if the ratio $b_w h_w^3/(\ell_n h_s^3)$ is not less than 2.0, where b_w, h_w and ℓ_n are respectively the width, overall depth and clear span of the beam, and h_s is the overall depth of the slab.

The two-way slab panel is divided in both directions into a middle strip, one-half panel in width and symmetrical about the panel centreline, and two column strips, each one-quarter panel in width and occupying the two quarter-panel areas outside the middle strip, as shown in Fig. 13.2. The bending moments (per unit width of slab) in the middle strips are given by:

$$M_A = C_A wA^2 \tag{13.1}$$

and

$$M_B = C_B wB^2 \tag{13.2}$$

where M_A, M_B = moment in the middle strip along the short span (A) and long span (B), respectively. These are moments per unit width of slab.

C_A, C_B = moment coefficients given in Tables 13.1–13.3

w = applicable factored uniform load per unit area; total (dead + live) load for negative moments and shears, but separated into dead load and live load for positive moments

A, B = length of clear span in short and long directions, respectively

Values of C_A and C_B for positive and negative moments for various edge conditions and A/B ratios are given in Tables 13.1–13.3 CSA A23.3-M84 (App. E) also gives identical coefficients in Tables E-1 and E-2. The support conditions considered in these tables include only combinations of essentially fixed (or continuous) and simply supported (supports with negligible torsional resistance) edge conditions. Similarly, the tables cover a range of A/B ratios from 1.0 to 0.5. When this ratio is less than 0.5, the slab should be designed as a one-way slab spanning the short span; however, negative moment reinforcement as required for a ratio of 0.5 must be provided along the short edge (reinforcement near the top face and perpendicular to the short edge).

The negative moment coefficients are given in Table 13.1 and these are applicable to the respective continuous (or fixed) edges, indicated by cross-hatching in the table. To allow for the slight restraint that may be provided by the torsional rigidity of the supporting beam or wall, all discontinuous (simply supported) edges must be designed for a negative moment equal to three-quarters of the positive moment in the same direction (CSA A23.3-M84-App. E). For negative moment calculations with Table 13.1, w is the total factored uniform load (dead plus live). The critical sections for negative moment calculations are along the edges of the panel at the faces of the supports. At supports over which the slab is continuous, if the negative moment on one side is less than 80 percent of that on the other side, the difference is distributed between the two sides in proportion to the relative stiffnesses of the slabs.

The coefficients for positive moments are given in Table 13.2 for dead load and in Table 13.3 for live load. Dead and live loads are treated separately here to allow

Table 13.1   Coefficients for Negative Moments in Slabs*

$$\left.\begin{array}{l} M_{A\ neg} = C_{A\ neg} \times w \times A^2 \\ M_{B\ neg} = C_{B\ neg} \times w \times B^2 \end{array}\right\} \text{ where } w = \text{total uniform dead plus live load}$$

Ratio $m = \dfrac{A}{B}$		Case 1	Case 2	Case 3	Case 4	Case 5	Case 6	Case 7	Case 8	Case 9
1.00	$C_{A\ neg}$		0.045		0.050	0.075	0.071		0.033	0.061
	$C_{B\ neg}$		0.045	0.076	0.050			0.071	0.061	0.033
0.95	$C_{A\ neg}$		0.050		0.055	0.079	0.075		0.038	0.065
	$C_{B\ neg}$		0.041	0.072	0.045			0.067	0.056	0.029
0.90	$C_{A\ neg}$		0.055		0.060	0.080	0.079		0.043	0.068
	$C_{B\ neg}$		0.037	0.070	0.040			0.062	0.052	0.025
0.85	$C_{A\ neg}$		0.060		0.066	0.082	0.083		0.049	0.072
	$C_{B\ neg}$		0.031	0.065	0.034			0.057	0.046	0.021
0.80	$C_{A\ neg}$		0.065		0.071	0.083	0.086		0.055	0.075
	$C_{B\ neg}$		0.027	0.061	0.029			0.051	0.041	0.017
0.75	$C_{A\ neg}$		0.069		0.076	0.085	0.088		0.061	0.078
	$C_{B\ neg}$		0.022	0.056	0.024			0.044	0.036	0.014
0.70	$C_{A\ neg}$		0.074		0.081	0.086	0.091		0.068	0.081
	$C_{B\ neg}$		0.017	0.050	0.019			0.038	0.029	0.011
0.65	$C_{A\ neg}$		0.077		0.085	0.087	0.093		0.074	0.083
	$C_{B\ neg}$		0.014	0.043	0.015			0.031	0.024	0.008
0.60	$C_{A\ neg}$		0.081		0.089	0.088	0.095		0.080	0.085
	$C_{B\ neg}$		0.010	0.035	0.011			0.024	0.018	0.006
0,55	$C_{A\ neg}$		0.084		0.092	0.089	0.096		0.085	0.086
	$C_{B\ neg}$		0.007	0.028	0.008			0.019	0.014	0.005
0.50	$C_{A\ neg}$		0.086		0.094	0.090	0.097		0.089	0.088
	$C_{B\ neg}$		0.006	0.022	0.006			0.014	0.010	0.003

*A cross-hatched edge indicates that the slab continues across or is fixed at the support; an unmarked edge indicates a support at which torsional resistance is negligible.

Table 13.2   Coefficients for Dead Load Positive Moments in Slabs*

$$\left.\begin{array}{l} M_{A\,pos\,DL} = C_{A\,DL} \times w \times A^2 \\ M_{B\,pos\,DL} = C_{B\,DL} \times w \times B^2 \end{array}\right\} \text{ where } w = \text{total uniform dead load}$$

Ratio $m = \dfrac{A}{B}$		Case 1	Case 2	Case 3	Case 4	Case 5	Case 6	Case 7	Case 8	Case 9
1.00	$C_{A\,DL}$	0.036	0.018	0.018	0.027	0.027	0.033	0.027	0.020	0.023
	$C_{B\,DL}$	0.036	0.018	0.027	0.027	0.018	0.027	0.033	0.023	0.020
0.95	$C_{A\,DL}$	0.040	0.020	0.021	0.030	0.028	0.036	0.031	0.022	0.024
	$C_{B\,DL}$	0.033	0.016	0.025	0.024	0.015	0.024	0.031	0.021	0.017
0.90	$C_{A\,DL}$	0.045	0.022	0.025	0.033	0.029	0.039	0.035	0.025	0.026
	$C_{B\,DL}$	0.029	0.014	0.024	0.022	0.013	0.021	0.028	0.019	0.015
0.85	$C_{A\,DL}$	0.050	0.024	0.029	0.036	0.031	0.042	0.040	0.029	0.028
	$C_{B\,DL}$	0.026	0.012	0.022	0.019	0.011	0.017	0.025	0.017	0.013
0.80	$C_{A\,DL}$	0.056	0.026	0.034	0.039	0.032	0.045	0.045	0.032	0.029
	$C_{B\,DL}$	0.023	0.011	0.020	0.016	0.009	0.015	0.022	0.015	0.010
0.75	$C_{A\,DL}$	0.061	0.028	0.040	0.043	0.033	0.048	0.051	0.036	0.031
	$C_{B\,DL}$	0.019	0.009	0.018	0.013	0.007	0.012	0.020	0.013	0.007
0.70	$C_{A\,DL}$	0.068	0.030	0.046	0.046	0.035	0.051	0.058	0.040	0.033
	$C_{B\,DL}$	0.016	0.007	0.016	0.011	0.005	0.009	0.017	0.011	0.006
0.65	$C_{A\,DL}$	0.074	0.032	0.054	0.050	0.036	0.054	0.065	0.044	0.034
	$C_{B\,DL}$	0.013	0.006	0.014	0.009	0.004	0.007	0.014	0.009	0.005
0.60	$C_{A\,DL}$	0.081	0.034	0.062	0.053	0.037	0.056	0.073	0.048	0.036
	$C_{B\,DL}$	0.010	0.004	0.011	0.007	0.003	0.006	0.012	0.007	0.004
0.55	$C_{A\,DL}$	0.088	0.035	0.071	0.056	0.038	0.058	0.081	0.052	0.037
	$C_{B\,DL}$	0.008	0.003	0.009	0.005	0.002	0.004	0.009	0.005	0.003
0.50	$C_{A\,DL}$	0.095	0.037	0.080	0.059	0.039	0.061	0.089	0.056	0.038
	$C_{B\,DL}$	0.006	0.002	0.007	0.004	0.001	0.003	0.007	0.004	0.002

*A cross-hatched edge indicates that the slab continues across or is fixed at the support; an unmarked edge indicates a support at which torsional resistance is negligible.

Table 13.3 Coefficients for Live Load Positive Moments in Slabs*

$$\left.\begin{array}{l} M_{A\ pos\ LL} = C_{A\ LL} \times w \times A^2 \\ M_{B\ pos\ LL} = C_{B\ LL} \times w \times B^2 \end{array}\right\} \text{ where } w = \text{total uniform dead load}$$

$m = \dfrac{A}{B}$ Ratio		Case 1	Case 2	Case 3	Case 4	Case 5	Case 6	Case 7	Case 8	Case 9
1.00	$C_{A\ LL}$	0.036	0.027	0.027	0.032	0.032	0.035	0.032	0.028	0.030
	$C_{B\ LL}$	0.036	0.027	0.032	0.032	0.027	0.032	0.035	0.030	0.028
0.95	$C_{A\ LL}$	0.040	0.030	0.031	0.035	0.034	0.038	0.036	0.031	0.032
	$C_{B\ LL}$	0.033	0.025	0.029	0.029	0.024	0.029	0.032	0.027	0.025
0.90	$C_{A\ LL}$	0.045	0.034	0.035	0.039	0.037	0.042	0.040	0.035	0.036
	$C_{B\ LL}$	0.029	0.022	0.027	0.026	0.021	0.025	0.029	0.024	0.022
0.85	$C_{A\ LL}$	0.050	0.037	0.040	0.043	0.041	0.046	0.045	0.040	0.039
	$C_{B\ LL}$	0.026	0.019	0.024	0.023	0.019	0.022	0.026	0.022	0.020
0.80	$C_{A\ LL}$	0.056	0.041	0.045	0.048	0.044	0.051	0.051	0.044	0.042
	$C_{B\ LL}$	0.023	0.017	0.022	0.020	0.016	0.019	0.023	0.019	0.017
0.75	$C_{A\ LL}$	0.061	0.045	0.051	0.052	0.047	0.055	0.056	0.049	0.046
	$C_{B\ LL}$	0.019	0.014	0.019	0.016	0.013	0.016	0.020	0.016	0.013
0.70	$C_{A\ LL}$	0.068	0.049	0.057	0.057	0.051	0.060	0.063	0.054	0.050
	$C_{B\ LL}$	0.016	0.012	0.016	0.014	0.011	0.013	0.017	0.014	0.011
0.65	$C_{A\ LL}$	0.074	0.053	0.064	0.062	0.055	0.064	0.070	0.059	0.054
	$C_{B\ LL}$	0.013	0.010	0.014	0.011	0.009	0.010	0.014	0.011	0.009
0.60	$C_{A\ LL}$	0.081	0.058	0.071	0.067	0.059	0.068	0.077	0.065	0.059
	$C_{B\ LL}$	0.010	0.007	0.011	0.009	0.007	0.008	0.011	0.009	0.007
0.55	$C_{A\ LL}$	0.088	0.062	0.080	0.072	0.063	0.073	0.085	0.070	0.063
	$C_{B\ LL}$	0.008	0.006	0.009	0.007	0.005	0.006	0.009	0.007	0.006
0.50	$C_{A\ LL}$	0.095	0.066	0.088	0.077	0.067	0.078	0.092	0.076	0.067
	$C_{B\ LL}$	0.006	0.004	0.007	0.005	0.004	0.005	0.007	0.005	0.004

*A cross-hatched edge indicates that the slab continues across or is fixed at the support; an unmarked edge indicates a support at which torsional resistance is negligible.

Table 13.4 **Ratio of Load *w* in *A* and *B* Directions for Shear in Slab and Load on Supports***

Ratio $m = \dfrac{A}{B}$		Case 1	Case 2	Case 3	Case 4	Case 5	Case 6	Case 7	Case 8	Case 9
1.00	W_A	0.50	0.50	0.17	0.50	0.83	0.71	0.29	0.33	0.67
	W_B	0.50	0.50	0.83	0.50	0.17	0.29	0.71	0.67	0.33
0.95	W_A	0.55	0.55	0.20	0.55	0.86	0.75	0.33	0.38	0.71
	W_B	0.45	0.45	0.80	0.45	0.14	0.25	0.67	0.62	0.29
0.90	W_A	0.60	0.60	0.23	0.60	0.88	0.79	0.38	0.43	0.75
	W_B	0.40	0.40	0.77	0.40	0.12	0.21	0.62	0.57	0.25
0.85	W_A	0.66	0.66	0.28	0.66	0.90	0.83	0.43	0.49	0.79
	W_B	0.34	0.34	0.72	0.34	0.10	0.17	0.57	0.51	0.21
0.80	W_A	0.71	0.71	0.33	0.71	0.92	0.86	0.49	0.55	0.83
	W_B	0.29	0.29	0.67	0.29	0.08	0.14	0.51	0.45	0.17
0.75	W_A	0.76	0.76	0.39	0.76	0.94	0.88	0.56	0.61	0.86
	W_B	0.24	0.24	0.61	0.24	0.06	0.12	0.44	0.39	0.14
0.70	W_A	0.81	0.81	0.45	0.81	0.95	0.91	0.62	0.68	0.89
	W_B	0.19	0.19	0.55	0.19	0.05	0.09	0.38	0.32	0.11
0.65	W_A	0.85	0.85	0.53	0.85	0.96	0.93	0.69	0.74	0.92
	W_B	0.15	0.15	0.47	0.15	0.04	0.07	0.31	0.26	0.08
0.60	W_A	0.89	0.89	0.61	0.89	0.97	0.95	0.76	0.80	0.94
	W_B	0.11	0.11	0.39	0.11	0.03	0.05	0.24	0.20	0.006
0.55	W_A	0.92	0.92	0.69	0.92	0.98	0.96	0.81	0.85	0.95
	W_B	0.08	0.08	0.31	0.08	0.02	0.04	0.19	0.15	0.05
0.50	W_A	0.94	0.94	0.76	0.94	0.99	0.97	0.86	0.89	0.97
	W_B	0.06	0.06	0.24	0.06	0.01	0.03	0.14	0.11	0.03

*A cross-hatched edge indicates that the slab continues across or is fixed at the support; an unmarked edge indicates a support at which torsional resistance is negligible.

for their different patterns of placement to produce the maximum positive moment. The critical section for positive moment is along the centreline of the panel.

The middle strips are designed for the full intensity of negative and positive moments computed using the Tables 13.1 to 13.3, assuming the moment variation to be uniform across the width of the strips. The moments (both positive and negative) in the column strip are gradually reduced from the full value M_A and M_B at the edge of the middle strip to one-third of these values at the edge of the panel, as shown in Fig. 13.2.

For computing the shear stresses in the slab, Ref. 13.2 recommends that the load, w, on the slab be assumed distributed in the "A" and "B" directions in the proportions given in Table 13.4. The same proportional distribution is assumed to compute the load on the supporting beams (or walls), provided that the load on the beam along the short direction is taken as not less than that of a slab area bounded by 45° lines from the corners. However, CSA A23.3-M84 (App. E) does not contain the equivalent of Table 13.4. Instead, the Code specifies that the loads on the supporting beams may be assumed as the load within the tributary areas of the panel bounded by the intersection of 45° lines from the corners and the median line of the panel parallel to the long sides. Further, the bending moments in the supporting beams may be determined using the equivalent uniform load given by Eqs. 13.3 and 13.4. The slab shears may be computed on the assumption of the above distribution of the load to the supports.

Equivalent uniform loads on beams are:

(i) for beam along short span: $\qquad \dfrac{w_f A}{3}$ $\hfill$ (13.3)

(ii) for beam along long span: $\qquad \dfrac{w_f A}{3} \times \dfrac{(3 - m^2)}{2}$ $\hfill$ (13.4)

where $m = A/B$

The effects of twisting moments described in Section 13.1 and illustrated in Fig. 13.1c are significant at exterior corners. Hence special reinforcement should be provided at exterior corners in both the bottom and top of the slab, details for which are presented in CSA A23.3-M84 (13.4.6). Accordingly this reinforcement should be provided for a distance in each direction from the corner equal to one-fifth the longer span. Both the bottom and top reinforcement should be sufficient to resist a moment equal to the maximum positive moment per metre of width in the slab. The top reinforcement should be placed in a single band in the direction of the diagonal from the corner, and the bottom reinforcement in a direction at right angles to the diagonal (see Fig. 13.1d). Alternatively, in either the top or bottom of the slab, the reinforcement may be placed in two bands parallel to the sides of the slab, with each band having an area sufficient to resist the moment mentioned above.

Additional requirements to be met by two-way slab reinforcement in general are given in CSA A23.3-M84 (13.4). The spacing of bars at critical sections in solid two-way slabs should not exceed the lesser of two times the slab thickness and 500 mm. All positive reinforcement perpendicular to a discontinuous edge should be extended to the edge of the slab and embedded for at least 150 mm into spandrel beams, walls, or columns. Similarly all negative reinforcement perpendicular to a discontinuous edge must be anchored into spandrel beams, walls, or columns, to be developed at the face of the support. The area of reinforcement must be adequate to resist the bending moments at all critical sections but should not be less than the minimum specified as shrinkage and temperature reinforcement (Code 7.8).

CSA A23.3-M84 (App. E and cl. 9.5.3) requires that the slab have adequate thickness for deflection control (see Section 14.3); however, for slabs designed by this method the thickness should in no case be less than (a) 100 mm, (b) the perimeter of the slab divided by 140, in the case of slabs discontinuous on one or more edges, or (c) the perimeter of the slab divided by 160, in the case of slabs continuous on all edges.

--- **EXAMPLE 13.1** ---

The plan of a floor consisting of two-way slabs supported on and cast integrally with stiff beams is shown in Fig. 13.3. The specified dead load, in addition to the slab weight, is 1.5 kN/m² and the specified live load is 4.8 kN/m². Design the corner panel. Assume $f'_c = 20$ MPa, $f_y = 400$ MPa, and the beams to be 300 mm wide.

SOLUTION

1. Minimum Thickness for Deflection Control
   Clear spans are, $7 - 0.3 = 6.7$ m; and $6 - 0.3 = 5.7$ m. For design using the provisions of CSA A23.3-M84 (App. E), the minimum thickness should be determined by cl. 9.5.3, but in no case should it be less than (this being a corner panel with two discontinuous edges):

$$h = \frac{\text{perimeter of slab}}{140} = \frac{2(6700 + 5700)}{140} = 177 \text{ mm}$$

Minimum thickness using A23.3-M84 Equations 9-7, 9-8 and 9-9 (see Section 12.3) are:

$\alpha_m \geqslant 2$ for stiff beams
β = ratio of long to short clear spans
   $= 6.7/5.7 = 1.18$
β_s = ratio of continuous edge to perimeter of slab
   $= 0.5$ (corner panel)
ℓ_n = clear span in long direction = 6700 mm

Fig. 13.3 Example 13.1

(a)

(b) Moment coefficients, middle strips

Code (Eq. 9-7) gives:

$$h = \frac{\ell_n (800 + f_y/1.5)}{36\,000 + 5000\,\beta \left[\alpha_m - 0.5\,(1 - \beta_s)\left(1 + \dfrac{1}{\beta}\right)\right]}$$

$$= \frac{6700(800 + 400/1.5)}{36\,000 + 5000 \times 1.18\,[2 - 0.5(1 - 0.5)\,(1 + 1/1.18)]}$$

$$= 158 \text{ mm}$$

The thickness must be not less than (Code (Eq. 9-8)):

$$h = \frac{\ell_n (800 + f_y/1.5)}{36\,000 + 5000\,\beta\,(1 + \beta_s)}$$

$$= \frac{6700\,(800 + 400/1.5)}{36\,000 + 5000 \times 1.18\,(1 + 0.5)} = 159 \text{ mm}$$

The thickness need not be more than (Code (Eq. 9-9)):

$$h = \frac{\ell_n\,(800 + f_y/1.5)}{36\,000}$$

$$= \frac{6700(800 + 400/1.5)}{36\,000} = 198 \text{ mm}$$

The controlling minimum thickness is perimeter/140 = 177 mm. In this case a thickness of 180 mm is selected. (A smaller thickness may be used provided deflections are computed and shown to be within allowable limits).

$$\text{Weight of slab} = 0.18 \times \frac{2400}{1000} \times 9.81 = 4.24 \text{ kN/m}^2$$

Factored loads are:

Dead load $= 1.25\,(1.5 + 4.24)$ $= 7.17$ kN/m²

Live load $= 1.5 \times 4.8$ $= 7.20$ kN/m²

Total $= 14.4$ kN/m²

Ratio, short to long span $= m = 5.7/6.7 = 0.85$

2. Design Moments in Middle Strip
    The moment coefficients are given in Table 13.1 for negative moments at continuous ends and in Tables 13.2 and 13.3 for positive moments at midspan (for dead and live load respectively). These coefficients are shown in Fig. 13.3b at the appropriate locations. At continuous edges, the negative moment coefficients for the neighbouring panels are also shown to facilitate redistribution of the unbalanced value if greater than 20 percent. In this case such a situation exists at the continuous edge 2–3, with the coefficients being 0.034 and 0.046. Since the span lengths and loading are the same, the coefficients may be distributed instead of the actual

moment values. The slab thickness being the same for both slabs, the difference is distributed equally between the two sides, to yield a moment coefficient of (0.034 + 0.046)/2 = 0.04. (Note that the difference in the restraint conditions at the far edges has little effect on the relative stiffnesses of the slabs at the near edges and hence the average is used).

At the continuous edge 3–4 (Fig. 13.3b), the negative moment coefficients on the two sides do not differ by more than 20 percent. The procedure described in Section 13.3 does not specify a distribution of the difference in the moments for such cases. However, normally it is the same negative reinforcement that resists the moment on either side of the beam along edge 3–4 and hence it has to be designed for the larger of the moments from either side, that is in this case for a moment with a coefficient $C_{A, neg}$ = 0.072.

Alternatively, the reinforcement may be designed for the moment obtained by distributing the difference in proportion to the relative stiffnesses (i.e., $C_{A, neg}$ = (0.066 + 0.072)/2 = 0.069) as for the edge 2–3; and this latter procedure is followed in this example. It is emphasized that slabs are highly indeterminate structures and are normally very much under-reinforced so that some amount of inelastic moment redistribution can be depended upon. CSA A23.3-M84 (13.6.7) also permits a modification of the moment by up to 10 percent provided there is a reserve capacity elsewhere to accommodate the total static moment. Since the positive and negative reinforcements are designed for different loading patterns, for a given loading condition generally there is a reserve strength elsewhere to accommodate the total static design moment.

The factored moments at various critical sections are computed below. Subscripts A and B refer to short and long spans, respectively.

(i) Negative moments at continuous edges:
$M_{A, neg}$ = 0.069 × 14.4 × 5.7^2 = 32.3 kN·m
$M_{B, neg}$ = 0.04 × 14.4 × 6.7^2 = 25.9 kN·m

(ii) Positive moments at midspan:
$M_{A, pos, DL}$ = 0.036 × 7.17 × 5.7^2 = 8.39 kN·m
$M_{A, pos, LL}$ = 0.043 × 7.2 × 5.7^2 = 10.1 kN·m

$M_{A, pos, total}$ = 18.5 kN·m

$M_{B, pos, DL}$ = 0.019 × 7.17 × 6.7^2 = 6.12 kN·m
$M_{B, pos, LL}$ = 0.023 × 7.2 × 6.7^2 = 7.43 kN·m

$M_{B, pos, total}$ = 13.6 kN·m

(iii) Negative moment at discontinuous edges, equal to 3/4 of positive moment at midspan:
$M_{A, neg}$ = 0.75 × 18.5 = 13.9 kN·m
$M_{B, neg}$ = 0.75 × 13.6 = 10.2 kN·m

3. Design of Sections
The largest moment anywhere in the slab is the negative moment at the continuous edge = 32.3 kN·m. Selecting 20 mm clear cover, and No. 10 bars, and placing the reinforcement parallel to the short side as the outer layer (to provide the greater effective depth in this direction to resist the relatively larger moment along the short span),

Figure 13-4 *Effective depths for short and long spans*

Section of slab parallel to
long side

$$d_A = 180 - 20 - 11.3/2 = 154 \text{ mm (Fig. 13.4)}$$
and
$$d_B = 180 - 20 - 11.3 - 11.3/2 = 143 \text{ mm}$$

(i) At continuous end of short span, middle strip

$$M_{A, neg} = 32.3 \text{ kN} \cdot \text{m}$$

$$\text{Required } K_r = \frac{M_f}{bd^2} = \frac{32.3 \times 10^6}{1000 \times 154^2}$$

$$= 1.36 \text{ MPa}$$

Referring to Table 5.4, this gives a reinforcement ratio of $\rho = 0.00431$, which is within limits and in the desirable range for slabs.

$$\text{Area of steel } A_s = 0.00431 \times 1000 \times 154$$
$$= 664 \text{ mm}^2$$

No. 10 bars at spacing of 150 mm gives

$$A_s = 667 \text{ mm}^2, \text{ as required}$$

Alternatively, in the entire middle strip, which has a width of 6700/2 = 3350 mm, a total of 23 bars can be distributed giving an area per metre width of 2300/3.35 = 687 mm² (compared to 664 mm² required). This may be made up of, to the extent available, bent up bars to a maximum of one-half the positive reinforcement from the midspan regions, on both sides of the support. The balance may be provided by additional straight bars.

(ii) The computations for the reinforcement at all other critical sections in middle strips, following the same procedure as illustrated above, are set forth in Table 13.5.

(iii) Reinforcement in column strip.
Across the width of each column strip, the moment per unit width varies linearly from the full value (of the corresponding moment in the middle strip)

Table 13.5 Example 13.1

		Short Span			Long Span		
		Cont. end Neg. mom.	Midspan Pos. mom.	Discont. end Neg. mom.	Cont. end Neg. mom.	Midspan Pos. mom.	Discont. end Pos. mom.
Middle Strip:							
1. Factored moment, M_f	kN·m	32.3	18.5	13.9	25.9	13.6	10.2
2. Effective depth, d	mm	154	154	154	143	143	143
3. $K_r = M_f/(1000\,d^2)$	MPa	1.36	0.780	0.586	1.27	0.665	0.499
4. 100 ρ, (Tab e 5.1)		0.431	0.239	—	0.401	0.202	—
5. 100 ρ_{max}		—	—	0.200	—	0.200	0.200
6. $A_s = \rho \times 1000\,d$, per m	mm²	664	368	—	573	290	—
7. $A_{s\,min} = 0.002 \times 1000\,b$	mm²	—	—	360	—	360	360
8. Spacing of No. 10 bars	mm	151	271	277	175	277	277
9. Select spacing, ($\leq 2\,h$)	mm	150	270	270	170	270	270
10. No. of bars in middle strip		23	12	12	17	12	12
Column Strip:							
11. No. of bars required in each column strip = $^1/_3$ number in mid-strip)		8	4	4	6	4	4
12. Selected number and spacing (max. spacing = 320 mm)		8 at 209	5 at 320	5 at 320	6 at 240	4 at 320	4 at 320

at the inner edge to one-third of this value at the slab edge (Fig. 13.2). Hence the average column strip moment is two-thirds the corresponding value in the middle strip. The reinforcement required to resist this moment can be computed as for the middle strip. However, it is sufficiently accurate and slightly conservative to determine the reinforcement in proportion to the moments from the values already determined for the middle strip. Thus, each column strip has one-half the width of the middle strip and two-thirds its moment. Hence, the total number of bars required will be roughly one-third the number of bars required in the middle strip. For example, each column strip in the short direction will need, at the continuous end, one-third of 23 or approximately 8 bars. On this basis, the number of bars required for moment in each column strip at all critical sections is given in row 11 of Table 13.5. These numbers may have to be increased to meet maximum spacing requirements. The bars in the column strip may be placed at spacings which gradually increase from the inner edge towards the supported edge. Alternatively, column strip reinforcement may be placed at uniform spacing. The spacing given in row 12 of Table 13.5 follows the latter pattern.

4. Reinforcement Detailing
    To detail bar cut-off and bend points, the inflection points may be assumed to be one-sixth of the span length from the face of supports over which the slab is continuous. At exterior spans, all positive reinforcement perpendicular to the discontinuous edge should extend to the edge of the slab and have embedment (straight or hooked) of at least 150 mm into spandrel beams or walls. The bar development requirements are the same as for beams. Special care has to be exercised in anchoring the negative reinforcement at a discontinuous end into the supporting beam or wall so as to develop the bar stress at the critical section at the face of the support. Typical bar bending details are shown in Fig. 13.5a.

5. Shear in Slab
    Using the coefficients in Table 13.4, the larger fraction of load transferred, in the short direction, to the beams on the longer side is 0.66.
    On a 1 m wide slab strip in the short direction, the shear force is:

$$V_f = \frac{w_a A}{2} = \frac{0.66 \times 14.4 \times 5.7}{2} = 27.1 \text{ kN}$$

$$V_c = 0.2 \lambda \phi_c \sqrt{f'_c}\, b_w d = 0.2 \times 1 \times 0.6 \times \sqrt{20} \times 1000 \times 154 \times 10^{-3}$$

$$= 82.7 \text{ kN}$$

$V_f < V_c$, no shear reinforcement needed.   OK

Alternately, if the load distribution to the support is assumed as in CSA A23.3-M84 (E 3.1), the maximum shear force per metre on a slab strip in short direction at midspan is,

$$V_f = 0.5\, w_f A = 0.5 \times 14.4 \times 5.7 = 41 \text{ kN}$$

Again, $V_f < V_c$,   OK

Fig. 13.5 *Reinforcement details, example 13.1*

(a) Suggested reinforcement details

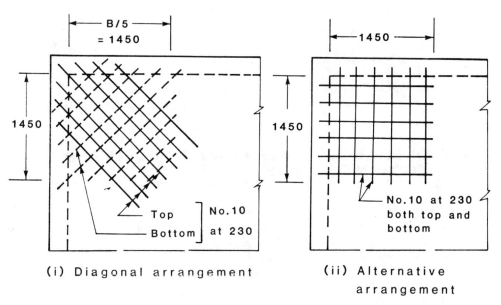

(i) Diagonal arrangement   (ii) Alternative arrangement

(b) Details of corner reinforcement,
exterior corners

6. Load Transmitted to Beams
   If the coefficients in Table 13.4 are used, the load per unit length on a long beam from the slab panel is:

$$\frac{0.66 \times 14.4 \times 5.7}{2} = 27.1 \text{ kN/m}$$

$$\text{Load on short beam} = \frac{0.34 \times 14.4 \times 6.7}{2} = 16.4 \text{ kN/m}$$

$$\text{Minimum} = \frac{wA}{3} = \frac{14.4 \times 5.7}{3} = 27.4 \text{ kN/m} > 16.4 \text{ kN/m}$$

Loading on shorter beam from slab panel = 27.4 kN/m

Alternatively, using Eqs. 13.3 and 13.4 specified in CSA A23.3, load on longer beam is:

$$\frac{w_f A}{3} \times \frac{(3 - m^2)}{2} = \frac{14.4 \times 5.7}{3} \times \frac{(3 - 0.85^2)}{2} = 31.2 \text{ kN/m}$$

and load on shorter beam is:

$$\frac{w_f A}{3} = \frac{14.4 \times 5.7}{3} = 27.4 \text{ kN/m}$$

These loads may be used as equivalent uniform load for computing bending moments on the respective beams.

7. Corner Reinforcement

The corner reinforcement required at exterior corners is equal to the maximum positive moment reinforcement, that is No. 10 bars at 270 mm (Table 13.5). These bars are provided for a length of 6700/5 = 1340 mm as shown in Fig. 13.5b.

13.4 DIFFERENCE BETWEEN WALL SUPPORTS AND COLUMN SUPPORTS

It was explained in Section 13.3 that, in general, a load w placed on a two-way slab is transmitted partly (w_A) along the short span to the longer edge supports, and partly (w_B) along the long span to the shorter edge supports. In wall supported panels, the portions w_A and w_B of the load are transmitted by the respective walls vertically direct to their foundations (or other supports) as shown in Fig. 13.6a. In contrast, when the edge supports consist of beams between columns, as in Fig. 13.6b, the portion, w_A, of the load transmitted in the North–South (N–S) direction by the slab is in turn transmitted in the E–W direction to the supporting columns by the beams along the E–W direction. Thus, considering the combined action of both slab and E–W direction beams, the full load (part w_B by slab and part w_A through the beams) is transmitted in the E–W direction, as is evident also from the equilibrium considerations of the section of floor shown in Fig. 13.6b (ii). The same considerations also show that the slab together with the beams in the N–S direction (Fig. 13.6b(iii)) transmit the full slab load along the N–S direction as well, to the supporting columns.

In general, in column supported slabs, with or without beams along column lines, 100 percent of the slab load has to be transmitted by the floor system in both

Fig. 13.6 *Load transfer in wall-supported and column-supported slabs*

(a) Wall supported

b(i)

b(ii) Load transfer in
E–W direction

b(iii) Load transfer in
N–S direction

directions towards the columns. For the case of the stiff beams considered in this chapter, the analysis and design of the slab part and the beam part are considered separately. When the beams are relatively flexible (or shallow), and when there are no beams, the analysis of the floor system (including beams, if any) is integrated and the frame action developed between the floor system and supporting columns is taken into consideration. This procedure is described in the next chapter.

―――――――――――――― **PROBLEMS** ――――――――

1. A garage, 6 m × 4 m, is to be provided with a single panel slab roof. The slab is to be supported on walls on three sides and on a monolithically cast beam on the entrance side (4 m). The service live load on the roof is 2 kN/m². Use f'_c = 30 MPa and f_y = 300 MPa. Design the roof.
2. For the floor in Example 13.1, (Fig. 13.3) design the reinforcement details for an interior panel. Assume slab thickness to be 180 mm.

REFERENCES

13.1 Timoshenko, S., and Woinosky-Krieger, S., *Theory of Plates and Shells*, McGraw-Hill Book Co., New York, 1959, 580 pp.
13.2 *Building Code Requirements for Reinforced Concrete* (ACI 318-63), American Concrete Institute, Detroit, 1963.

CHAPTER 14

DESIGN OF TWO-WAY
SLAB SYSTEMS

14.1 INTRODUCTION

This chapter deals with the general case of slab systems supported on columns (or other non-deflecting supports), with or without beams along column lines, and includes flat plates, flat slabs, waffle slabs, and solid two-way slabs with beams along column lines. It was shown in Section 13.4 that in two-way slabs supported on columns, 100 percent of the floor loads have to be carried in both the longitudinal and transverse directions to the supporting columns. The mechanisms of load transfer from the slab to the columns include flexure, shear and torsion. To obtain the load effects on the elements of the floor system and its supporting members using an elastic analysis, the structure may be considered as a series of bents, each consisting of a row of columns (or other supports) and the portion of the floor system tributary to it, which can be taken as the part of the floor member bounded by the panel centrelines on either side of the columns. Such bents must be taken both longitudinally and transversely in the building, to assure load transfer in both directions, as indicated in Fig. 14.1a. One such isolated bent is shown in Fig. 14.1b. The horizontal member consists of the slab, together with drop panels or beams along column lines, if any.

It can be seen that the part of the structure in Fig. 14.1b is very similar to the plane frame considered in Fig. 10.3a and b, the primary difference being that the horizontal member in the former is a relatively wide slab-beam member. In the elastic analysis of the plane frame in Fig. 10.3, it was shown in Section 10.2 that several approximations can be made subject to certain limitations. Similar approximations can also be made in the present case. For example, in the analysis for gravity load effects, instead of considering the entire frame, a partial frame consisting of each floor (or roof) may be analyzed separately together with the columns immediately above and below, the columns being assumed fixed at their far ends (Fig. 14.1c). Such a procedure is described in the "Equivalent Frame Method" presented in Section 14.9 (compare with the simplification in Fig. 10.3c and d). When frame geometry and loadings meet certain limitations, the positive and negative factored moments at critical sections of the slab-beam member may be calculated using moment coefficients applied to the simple span moments, M_o. This is analogous to the use of the CSA moment coefficients in the analysis of continuous beams

Fig. 14.1 *Equivalent frame*

(a) Definition of equivalent frame

(b) E–W int. equivalent frame (full)

(c) E–W int. equivalent frame (partial)

described in Section 10.3.3. A somewhat similar procedure for two-way slabs, termed "Direct Design Method" is given in Section 14.5.

In the plane frames considered in Chapters 10 and 12, the horizontal member is a beam which can be proportioned for the full design moment obtained from the

Fig. 14.2 *Moment variations in a two-way slab panel*

analysis. In contrast, the horizontal member in the frame in Fig. 14.1c is a very wide beam (slab with or without beam along the column line) which is supported generally on a limited width. Hence the outer portions of this member are less stiff than the part along the support lines, and the distribution of the moment along the width of the member is not uniform. Considering one panel of the frame, the variation of the moment in the floor member along the span is as shown in Fig. 14.2b. Here

M_{ab} represents the moment in the slab-beam member over its full width AB.

This moment is distributed over the width AB non-uniformly as shown (qualitatively) by the solid curve in Fig. 14.2c. The actual variation along AB depends on several factors such as the span ratio, relative stiffness of beams along column lines (if any), presence of drop panels, torsional stiffness of transverse beams (if any), etc., and is highly indeterminate; and hence approximations are desired. The procedure generally adopted is to divide the slab panel into column strips (along the column lines) and middle strips (along the panel centrelines) and then to apportion the moment between these strips, the distribution of the moment within the width of each strip being assumed uniform. This is shown by the broken lines in Fig. 14.2c and also in Fig. 14.2d. When beams are provided along the column line, the beam portion is relatively stiffer than the slab and hence the column strip moment itself has to be further apportioned between the beam part and the slab part of the column strip (Fig. 14.2e).

14.2 DESIGN PROCEDURES IN CSA CODE

CSA A23.3-M84 (13.3) permits the design of two-way slab systems by any procedure that satisfies the conditions of equilibrium and geometrical compatibility and the requirements of strength and serviceability. Apart from the specific method for design of rectangular slabs supported on four sides by walls or stiff beams described in Chapter 13, two specific methods which satisfy these requirements, namely the Direct Design Method and the Equivalent Frame Method, are described in detail in the Code for design of two-way slab systems under gravity loads. However, other methods such as yield-line analysis and finite element analysis are acceptable provided they satisfy all the above-mentioned requirements.

The two methods specified in the Code are applicable for slab systems reinforced for flexure in more than one direction, such as flat plates, flat slabs, waffle slabs, and two-way slabs with beams along column lines. These methods are based on the analyses of results of extensive series of tests, comparison with theoretical results obtained using the theory of flexure for plates, and design practices used successfully in the past. Some of the background material used in the development of the design procedures recommended in the Code is presented in Refs. 14.1–14.3. Although the fundamental principles behind these design methods are quite general, the detailed design rules for the two methods in Section 13 of the Code are applicable, in the case of slabs with beams, only when the beams are located along the edges of the panel and when the beams are supported by columns (or other essentially non-deflecting supports) at the corners of the panel. The methods are not applicable to slabs reinforced for flexure in one direction only (one-way slabs). Both methods are somewhat complicated and require the determination of numerous parameters.

In both the Direct Design Method and the Equivalent Frame Method, a slab panel is divided into column strips and middle strips (Fig. 14.3a). A column strip is defined as a design strip along the column line and having a width equal to the lesser of $0.25\,\ell_1$ or $0.25\,\ell_2$, on each side of the column centreline, and includes within this

Fig. 14.3 *Column strip and middle strip*

(a)

(b)

width any drop panel or beam along the column line if used. Here, ℓ_1 is the length of span in the direction moments are to be determined and ℓ_2 is the length of span transverse to ℓ_1, both measured centre to centre of the supports. (In general, subscript 1 identifies parameters in the direction moments are being determined, and subscript 2 those in the transverse direction.) The middle strip is a design strip bounded by two column strips. In considering an equivalent frame along a column line, the slab width, ℓ_2, consists of two half middle strips flanking the column strip (Fig. 14.3b). When monolithic (or fully composite) beams are used along column lines, the effective beam sections are usually T-shaped, and are considered to include a portion of the slab on each side of the beam extending a distance equal to the projection of the beam above or below the slab, whichever is greater, but not greater than four times the slab thickness, as shown in Fig. 14.4. In cases where the beam stem is

Fig. 14.4 *Definition of beam section*

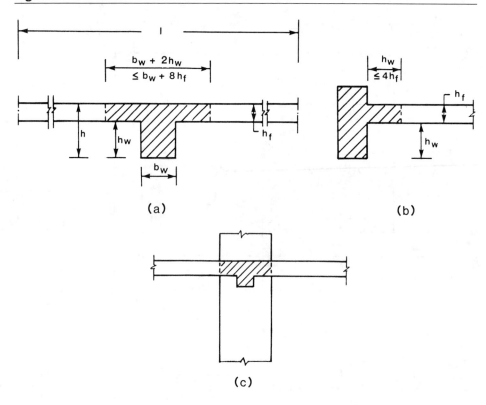

(a)   (b)

(c)

short, the T-beam may be assumed to have a width equal to that of the support (Fig. 14.4c).

The two methods differ essentially in the manner of determining the longitudinal distribution of bending moments in the slab-beam member between the negative and positive moment sections (Fig. 14.2b). While the Direct Design Method uses moment coefficients to determine the longitudinal distribution, the Equivalent Frame Method uses an elastic partial frame analysis. The procedure for the lateral distribution of the factored moments between column strip and middle strip (and to the beam element along the column line, if present) is the same for both design methods. Both methods require the values of several relative stiffness parameters in order to obtain the longitudinal and transverse distributions of factored moments in the design strips. Hence, it is necessary to assume, initially, the gross section dimensions of the floor system and the columns (or at least the relative stiffness values). If, on the basis of the analysis and subsequent design of sections, these dimensions have to be modified so as to affect the relative stiffness values significantly, the analysis and design have to be revised. The two design procedures are described in detail in Sections 14.5 and 14.9.

14.3 MINIMUM THICKNESS OF TWO-WAY SLABS FOR DEFLECTION CONTROL

Computation of deflections of two-way slab systems is quite complex. For the type of floor systems considered in this Chapter, CSA A23.3-M84 (9.5.3) recommends empirical equations (Eqs. 14.1 to 14.3 below) for the minimum overall thickness of slab necessary for the control of deflections. If these minimum thickness requirements are satisfied, deflections need not be computed. These equations are also applicable for two-way slabs supported on stiff beams, considered in Chapter 13. Equations 14.1 to 14.3 were established based on results of extensive tests on floor slabs, and have been supported by experience with this type of construction.

The required minimum thickness, h, of two-way floor slabs (or other two-way floor construction), having a ratio of long to short span not exceeding two, and having beams between all supports, with the relative stiffness of beams in two perpendicular directions, $\alpha_1 \ell_2^2/(\alpha_2 \ell_1^2)$, being not less than 0.2 nor greater than 5, is:

$$h = \frac{\ell_n(800 + f_y/1.5)}{36\,000 + 5000\,\beta\left[\alpha_m - 0.5(1 - \beta_s)\left(1 + \dfrac{1}{\beta}\right)\right]} \tag{14.1}$$

but not less than:

$$h = \frac{\ell_n(800 + f_y/1.5)}{36\,000 + 5000\,\beta(1 + \beta_s)} \tag{14.2}$$

The thickness need not exceed:

$$h = \frac{\ell_n(800 + f_y/1.5)}{36\,000} \tag{14.3}$$

where ℓ_1 = length of span in the direction that moments are being determined, centre-to-centre of supports
ℓ_2 = length of span transverse to ℓ_1, centre-to-centre of supports
ℓ_n = length of clear span in long direction
f_y = specified yield strength, in MPa
α_m = average value of α for all beams on edges of the panel being considered
α = the beam stiffness parameter defined as the ratio of flexural stiffness of beam section to the flexural stiffness of a width of slab bounded laterally by the centreline of the adjacent panel, if any, on each side of the beam (see below)
α_1, α_2 = α in directions of ℓ_1 and ℓ_2, respectively
β = ratio of clear span in long to short direction
β_s = ratio of length of continuous edges to total perimeter of a slab panel

The beam stiffness parameter α is used extensively in the two design methods. Referring to Fig. 14.4a, $\alpha = E_{cb}I_b/E_{cs}I_s$; where I_b is the centroidal moment of inertia of the gross beam section shown; I_s is the centroidal moment of inertia of the gross section of the slab, $= \ell\,h^3/12$; and E_{cb} and E_{cs} are the modulus of elasticity for the concrete in the beam and the slab, respectively (usually the same). The parameter α can have values ranging from zero (for the case of no beams) to infinity (for the case of rigid beams or wall supports).

When the floor slabs are of uniform thickness and panel size, the critical panel to be checked for minimum thickness requirement is the corner panel with two discontinuous edges ($\beta_s = 0.5$). For a panel with a discontinuous edge, the beam at this edge must have $\alpha \geqslant 0.80$, failing which the minimum thickness calculated above for that panel should be increased by 10 percent.

Further to the limitations given above, the thickness must be not less than:
(a)   for slabs having beams on all edges with a value of $\alpha_m \geqslant 2.0$ .... 90 mm
(b)   for slabs with a value of $\alpha_m < 2.0$ .... 120 mm

Equations 14.1 to 14.3 are plotted in Fig. 14.5 for the cases of $f_y = 300$ MPa and $\beta_s = 0.5$ and 1.0 (corner and interior panels) and for two values of $\beta = 1.0$ and 2.0. For other values of f_y, the value of ℓ_n/h obtained from Fig. 14.5 has to be multiplied by the ratio $1000/(800 + f_y/1.5)$, that is by 0.968 for $f_y = 350$ MPa, and 0.938 for $f_y = 400$ MPa.

For two-way slabs without beams (or with beams only at discontinuous edges) and having a ratio of long to short span not exceeding 2, the required minimum thickness is governed by Eq. 14.3. Again, at a discontinuous edge, an edge beam

Fig. 14.5 CSA A23.3-M84 (9.5.3.2) minimum thickness for two way slabs with beams between all supports, for $f_y = 300$ MPa [for other f_y values, multiply (ℓ_n/h) by $1000/(800 + f_y/1.5)$]

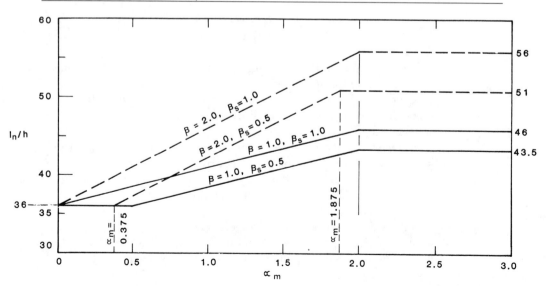

with a value of $\alpha \geq 0.8$ must be provided, failing which the minimum thickness required by Eq. 14.3 must be increased by at least 10 percent in the panel with the discontinuous edge. On the other hand, for slabs provided with drop panels satisfying Code clause 9.5.3.3.2 (that is, minimum drop panel size of one-sixth the centre-to-centre span length in each direction and minimum drop projection below the slab of $h/4$), the thickness required by Eq. 14.3 may be reduced by 10 percent. However, the slab thickness must not be less than:

(a) for panels with drop panels conforming to cl. 9.5.3.3.2 .... 100 mm
(b) all other panels .... 120 mm

These minimum thickness requirements for slabs without beams between interior supports are summarized in Table 14.1.

Table 14.1 Minimum Thickness for Two-Way Slabs Without Beams Between Interior Supports

	Without Drop Panels*			With Drop Panels**		
	Exterior Panel		Interior Panel	Exterior Panel		Interior Panel
	No Edge Beam	With Edge Beam***		No Edge Beam	With Edge Beam***	
$f_y = 300$ MPa	$\ell_n/32.7$	$\ell_n/36.0$	$\ell_n/36.0$	$\ell_n/36.4$	$\ell_n/40.0$	$\ell_n/40.0$
$f_y = 400$ MPa	$\ell_n/30.7$	$\ell_n/33.8$	$\ell_n/33.7$	$\ell_n/34.1$	$\ell_n/37.5$	$\ell_n/37.5$

  * Thickness must be not less than 120 mm.
 ** With drop panels conforming to Code clause 9.5.3.3.2. Thickness must be not less than 100 mm.
*** Edge beams must have $\alpha \geq 0.80$.

Note: With the permission of CSA, this material is reproduced from CSA Standard CAN 3-A23.3-M84, *Design of Concrete Structures for Buildings*, which is copyrighted by the Canadian Standards Association and copies of which may be purchased from the Association, 178 Rexdale Boulevard, Rexdale, Ontario M9W 1R3.

14.4 TRANSFER OF SHEAR AND MOMENTS TO COLUMNS

Lateral loads, as well as unbalanced gravity loads, cause transfer of moments between the slab system and supporting columns and walls built integrally with the slab. This moment is computed by the frame analysis when the Equivalent Frame Method is used. When the Direct Design Method is used, the unbalanced moment at an interior support due to unbalanced gravity loads is obtained using an empirical equation, and the moment at the exterior support is taken as equal to the nominal moment resistance, M_n, provided in the column strip (cl. 13.6.3.6). The total unbalanced moment at a support must be resisted by the columns above and below the slab in proportion to their stiffnesses (Fig. 14.6a).

In slabs without beams along the column line, the transfer of the moment from the slab to the columns requires special consideration. Based on data in Ref. 14.4, A23.3-M84(13.3.3) specifies that a portion, M_{fb}, of the total unbalanced moment, M_f, be considered as transferred to the columns by flexure; and the balance, M_{fv}, through shear and torsion, as given in Eqs. 14.4–14.6, and shown in Fig. 14.6.

Fig. 14.6 *Transfer of unbalanced moment to columns*

$M_f = M_2 - M_1$

$M_2 > M_1$

(a)

Unbalanced slab moment transferred by flexure,

$M_{fb} = \gamma_f M_f$

(b)

$M_{fv} = (1 - \gamma_f) M_f$

(c)

$$M_{fb} = \gamma_f M_f \tag{14.4}$$

$$M_{fv} = (1 - \gamma_f) M_f \tag{14.5}$$

$$\gamma_f = \cfrac{1}{1 + \cfrac{2}{3}\sqrt{\cfrac{c_1 + d}{c_2 + d}}} \tag{14.6}$$

where M_{fb} = portion of moment transferred by flexure
  M_{fv} = portion of moment transferred by shear
  M_f = total unbalanced moment transferred to columns
  c_1, c_2 = the size of equivalent rectangular column, capital or bracket, measured in the direction moments are being determined, and transverse to it, respectively

For square and round columns, $\gamma_f = 0.6$. Based on test results and experience, the width of slab considered effective in resisting the moment M_{fb} is taken as the width between lines a distance $1.5\,h$ on either side of the column or column capital (Fig. 14.6b), and hence this strip should have adequate reinforcement to resist this moment. The section considered for moment transfer by eccentricity of shear stress is at a distance $d/2$ from the periphery of the column or column capital (Fig. 14.6c). The shear stresses introduced because of the moment transfer must be added to the shear stresses due to the vertical support reaction. Shear stresses are considered in Section 14.6. The detailing of the reinforcement for the transfer of moments from slab to the columns, particularly at the exterior column where the unbalanced moment is usually the largest, is critical to both the performance and the safety of flat slabs without edge beams.

14.5 DIRECT DESIGN METHOD

14.5.1 LIMITATIONS

The Direct Design Method is a simplified procedure where the negative and positive design moments at critical sections in the slab-beam member, and the unbalanced moments transferred to columns, are computed using empirical moment coefficients. In order to ensure that the moments obtained by this procedure are not significantly different from those obtained by an elastic analysis, the Direct Design Method is restricted to slab systems meeting the following limitations (Code (13.6)):

1. there must be at least three continuous spans in each direction;
2. the panels must be rectangular, with the ratio of the longer to shorter span, centre-to-centre of supports, within a panel not exceeding 2.0;
3. the successive span lengths, centre-to-centre of supports, in each direction must not differ by more than one-third of the longer span;
4. columns are not offset more than 10 percent of the span (in the direction of offset) from either axis between centrelines of successive columns;

5. the factored live load must not exceed three times the factored dead load (the loads are assumed to be gravity loads, uniformly distributed);
6. if a panel is supported by beams on all sides, the relative stiffness of the beams in the two perpendicular directions, $\alpha_1 \ell_2^2 / \alpha_2 \ell_1^2$, must not be less than 0.2 nor greater than 5.0. Subscripts 1 and 2 refer to the direction moments are being determined and the direction transverse to it, respectively;
7. moment redistribution as permitted in CSA A23.3-M84(8.4) is not applicable.

14.5.2 NEGATIVE AND POSITIVE FACTORED MOMENTS IN SLAB-BEAM MEMBER

(a) Total Factored Static Moment
In any given span ℓ_1, the *total factored moment* for the span, M_o, (same as the simple span moment) is given by Eq. 14.7. From considerations of equilibrium and safety (see Fig. 14.2b), the absolute sum of the positive and average negative design moments must not be less than M_o.

$$M_o = w_f \ell_2 \, \ell_n^2 / 8 \qquad (14.7)$$

where w_f = factored load per unit area of the slab
   ℓ_1 = length of span in the direction moments are being determined
   ℓ_2 = length of span transverse to ℓ_1, same as the full width of the slab-beam member of the equivalent frame, and equal to the average of transverse spans on either side of the column line when these spans are different
   ℓ_n = clear span in direction moments are determined, measured face-to-face of columns, capitals, brackets, or walls, but not less than $0.65 \, \ell_1$. Circular supports are treated as equivalent square supports having the same area

(b) Longitudinal Distribution of Factored Moments
The factored static moment is distributed between critical negative moment sections (taken at the face of rectangular supports) and positive moment sections (at or near midspan) as shown in Fig. 14.7. For interior spans, the distribution is as follows:

1. negative factored moment, $M_{neg} = 0.65 \, M_o$ (14.8)
2. positive factored moment, $M_{pos} = 0.35 \, M_o$ (14.9)

In the end span, the total factored static moment, M_o, is distributed according to the factors given in the Table in Fig. 14.7c.

At interior supports such as B in Fig. 14.7, the negative moment sections must be designed to resist the larger of the two negative factored moments determined for the spans on either side of the support, unless an analysis is made to distribute the unbalanced moment in accordance with the stiffnesses of the elements framing into the support (Code cl. 13.6.3.4). At exterior supports, edge beams or edges of

Fig. 14.7 *Longitudinal distribution of factored static moment — Direct Design Method*

(a) Slab supported on walls

(b) Slab without beams between interior supports

Case→	(1)	(2)	(3)	(4)	(5)
	Exterior edge unrestrained	Slab with beams between all supports	Slab without beams between interior supports		Exterior edge fully restrained
			Without edge beam	With edge beam	
Interior negative factored moment	0.75	0.70	0.70	0.70	0.65
Positive factored moment	0.68	0.59	0.53	0.50	0.35
Exterior negative factored moment	0	0.16	0.26	0.30	0.65

(c) Moment factors for end span

slabs acting as torsional members (see below) must be designed to resist, in torsion, their share of the exterior negative factored moments. For determining the torsion in the edge beam it may be assumed conservatively that the negative factored moment at the exterior support is uniformly distributed over the width of the slab-beam member, ℓ_2, with the torsion being zero at the panel centrelines, as shown in Fig. 14.8a. Alternatively, the distribution of torsion on the edge beam may be taken to be the same as the transverse distribution of the negative moment in the slab-beam member (see Section 14.5.3 below).

Fig. 14.8 *Torsional member and restraint at exterior edge*

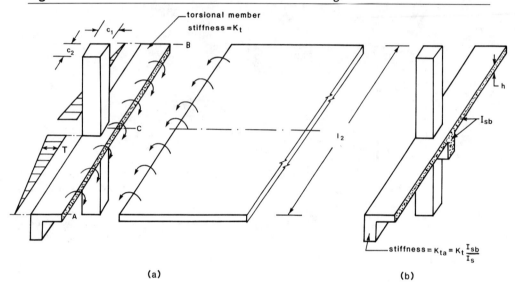

(a) (b)

(c) Torsional Member and Stiffness

The torsional member (Figs. 14.8 and 14.24) is assumed to have a constant cross section throughout its length, consisting of the larger of (Code (13.7.5)):

1. a portion of the slab having width equal to that of the column, bracket, or capital measured in the direction of ℓ_1;
2. the portion of the slab as in (1) above plus that part of the transverse beam above and below the slab;
3. the transverse beam section as defined in Fig. 14.4.

These three definitions for the section of the torsional member are illustrated in Fig. 14.9. The torsional stiffness, K_t, of the torsional member, needed for the Equivalent Frame Method (for computing the equivalent column stiffness, K_{ec} — see Section 14.9), may be computed using the approximate expression:

$$K_t = \Sigma \frac{9E_{cs}C}{\ell_2\left(1 - \dfrac{c_2}{\ell_2}\right)^3}$$ (14.10)

Fig. 14.9 *Definition of torsional member*

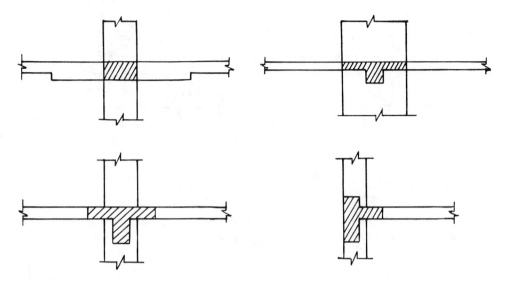

where E_{cs} = modulus of elasticity of concrete in slab

c_2 = size of rectangular or equivalent rectangular column, capital, or bracket, measured transverse to the direction in which moments are being determined

C = cross-sectional property of the torsional member having the same relationship to the torsional rigidity as does the polar moment of inertia for circular cross sections

The summation in Eq. 14.10 applies to the case where there is a torsional member on each side of the column extending to the centrelines of the adjacent panels. For an exterior equivalent frame, the attached torsional member exists on one side of the column only. For the torsional member cross section, which may be T- or L-shaped, the computation of an exact value for C is very difficult. As an approximation, C is computed by subdividing the cross section into separate rectangular parts and summing up the C-values for each of the component rectangles using the approximate expression for C for a rectangular section. On this basis,

$$C = \Sigma\left(1 - 0.63\frac{x}{y}\right)\frac{x^3 y}{3} \tag{14.11}$$

where x and y are the shorter and longer dimensions of a rectangular part (Fig. 14.10). Since this procedure will always give a value for C less than the correct value, the subdivision of the section into rectangles must be such as to maximize C. This is done by subdividing so as to minimize the length of the common boundaries. When beams frame into the column in the direction moments are being determined, as shown in Fig. 14.8b, the value of K_t computed using Eq. 14.10 leads to K_{ec} values which are too low. In such cases, the value of K_t given by Eq. 14.10 must be multiplied

Fig. 14.10 *Computation of torsional property C*

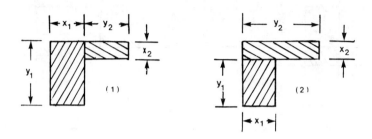

Use larger C computed for (1) or (2)

by the ratio of the moment of inertia of the slab with such beam to the moment of inertia of the slab without such beam, as follows:

$$K_{ta} = K_t \frac{I_{sb}}{I_s}$$ (14.12)

where K_{ta} = increased torsional stiffness due to beam in ℓ_1 direction

I_s = moment of inertia of the slab having the full width ℓ_2, excluding the portion of the beam stem extending above and below the slab, = $\ell_2 h^3/12$ (Fig. 14.8b)

I_{sb} = moment of inertia of the slab-beam member of width ℓ_2, including the portion of the beam stem extending above and below the slab (full T-section in Fig. 14.8b)

(d) Further Simplifications for Direct Design Method

The use of the Direct Design Method is limited to fairly simple and regular floor systems by the requirements described in Section 14.5.1. Hence certain simplifications may be made in the computations of the various member stiffnesses for use in this method. Thus, in computing the stiffness of the slab-beam member, the increase in the moment of inertia of slab-beams between column centreline and column face may be neglected, and a uniform cross section between column centrelines assumed. Similarly, the increase in column stiffness due to column capitals, drop panels, or the surrounding slab may be neglected and a uniform column cross-section assumed between the centre of floor and roof. The increase to K_t specified in Eq. 14.12 may also be neglected in the Direct Design Method. With these simplifications, and assuming the slab-beam member has the same section between faces of supports, the stiffnesses may be computed as:

$$K_c = 4E_{cc} I_c/\ell_c$$ (14.13)

$$K_s = 4E_{cs} I_s/\ell$$ (14.14)

$$K_b = 4E_{cb}\, I_b/\ell \qquad \qquad \cdot (14.15)$$

where E_{cc}, E_{cs}, E_{cb} = modulus elasticity of concrete in the member considered

I_c, I_s, I_b = moment of inertia of the cross section of the member considered

ℓ_c = height of column centre-to-centre of floor and roof

ℓ = span, centre-to-centre of support, in the direction stiffness is desired (ℓ_1 or ℓ_2)

Note that the above stiffness parameters are required in the Equivalent Frame Method also; however, for use in that method, the effect of variations in moment of inertia along the axis of columns and between face and centreline of supports of slab-beams must be taken into account.

(e) Modifications of Factored Moments

The negative and positive factored moments obtained by the preceding procedure (Fig. 14.7) may be modified by up to 10 percent provided the absolute sum of the positive and average negative bending moments in the span is not less than M_o given by Eq. 14.7 (Code (13.6.7)).

14.5.3 FACTORED MOMENTS AND SHEARS IN COLUMN STRIPS, BEAMS AND MIDDLE STRIPS

Having obtained the factored negative and positive moments at all critical sections in the slab-beam member, the next step is to divide these between the column strips, beams, and middle strips at each critical section. The CSA A23.3-M84(13.6) rules for apportioning the moments to the individual design strips is based on the studies reported in Ref. 14.3 and successful past practice. The transverse distribution at a section depends upon the span ratio ℓ_2/ℓ_1; the ratio, α_1, of the flexural stiffnesses (along the ℓ_1 direction) of the beam section and the slab section of width ℓ_2; and for the exterior support, on the ratio of torsional stiffness of the edge beam section to the flexural stiffness of a slab of width ℓ_2 and span ℓ_1, expressed by the parameter β_t. The parameters α_1 and β_t can be expressed as:

$$\alpha_1 = \frac{E_{cb}\, I_{b,1}}{E_{cs}\, I_{s,1}} \qquad \qquad (14.16)$$

$$\beta_t = \frac{E_{cb}\, C}{2\, E_{cs}\, I_{s,1}} \qquad \qquad (14.17)$$

where E_{cb}, E_{cs} = elastic modulus of concrete in the beam and slab, respectively

$I_{b,1}$ = centroidal moment of inertia of the cross section of the beam along the ℓ_1 direction

$I_{s,1}$ = moment of inertia of the slab in the ℓ_1 direction, $= \ell_2 h^3/12$

C = torsional property of the edge beam section, computed by Eq. 14.11

The transverse distribution of the negative moment at the exterior support is dependent on the relative torsional stiffness of the edge beam, expressed by the parameter β_t. Unless the torsional stiffness of the edge beam is high relative to the flexural stiffness of the slab it supports, all of the negative factored moment is assigned to the column strip and none to the middle strip.

The transverse distribution of moments described below is applicable to both Direct Design Method and Equivalent Frame Method.

(a) Factored Moments in Column Strips and Beams

For specific values of the parameters ℓ_2/ℓ_1, $\alpha_1 \ell_2/\ell_1$, and β_t, Table 14.2 presents the percentage of the total negative and positive factored moments assigned to the column strip. For values of the parameters in between those given in Table 14.2, linear interpolation must be made.

Where walls are used as supports along column lines, they can be considered flexurally as very stiff beams, with an $\alpha_1 \ell_2/\ell_1$ value greater than one. CSA A23.3-M84 (13.6.4.3) specifies that where the support is a column or wall extending transversely for a distance equal to or greater than $0.75 \, \ell_2$, the negative moment can be considered to be uniformly distributed over the width ℓ_2. If the slab is simply supported on the wall, the moment at the exterior support will, of course, be zero (see Fig. 14.7a).

When a beam is used along the column line in the ℓ_1 direction, the beam must be designed to resist 85 percent of the column strip moment provided $\alpha_1 \ell_2/\ell_1 \geq 1.0$. For values of $\alpha_1 \ell_2/\ell_1$ between one and zero, the proportion of column strip

Table 14.2 Percentages of Factored Moments at Critical Sections Assigned to Column Strip (CSA A23.3-M84(13.6))

	Ratio ℓ_2/ℓ_1		
	0.5	1.0	2.0
(1) Interior negative moment:			
$\quad \alpha_1 \ell_2/\ell_1 = 0$	75	75	75
$\quad \alpha_1 \ell_2/\ell_1 \geq 1.0$	90	75	45
(2) Exterior negative moment:			
$\quad \alpha_1 \ell_2/\ell_1 = 0, \begin{cases} \beta_t = 0 \\ \beta_t \geq 2.5 \end{cases}$	100 75	100 75	100 75
$\quad \alpha_1 \ell_2/\ell_1 \geq 1.0, \begin{cases} \beta_t = 0 \\ \beta_t \geq 2.5 \end{cases}$	100 90	100 75	100 45
(3) Positive moment:			
$\quad \alpha_1 \ell_2/\ell_1 = 0$	60	60	60
$\quad \alpha_1 \ell_2/\ell_1 \geq 1.0$	90	75	45

Note: With the permission of CSA, this material is reproduced from CSA Standard CAN3-A23.3-M84, *Design of Concrete Structures for Buildings*, which is copyrighted by the Canadian Standards Association and copies of which may be purchased from the Association, 178 Rexdale Boulevard, Rexdale, Ontario M9W 1R3.

moment to be resisted by the beam must be obtained by linear interpolation between 85 and zero percent. The beam must also resist moments caused by loads applied directly on it and not considered in the slab design. The portion of the column strip moment not resisted by the beam must be resisted by the slab portion of the column strip, outside the beam section.

(b) Factored Moments in Middle Strips

At all critical sections, the portion of the negative and positive factored moments not resisted by the column strip is assigned proportionately to the two half middle strips on either side of the column strip. Each full middle strip in a panel has moments assigned to its two halves from the equivalent frames on either side, and is designed for the total moment, as shown in Fig. 14.11 for middle strip No. 2. The middle strip adjacent to and parallel with an edge supported by a wall must be designed for twice the moment assigned to its interior half portion forming part of the equivalent frame along the first row of interior supports (see middle strip No. 1 in Fig. 14.11).

Fig. 14.11 *Factored moments in middle strips*

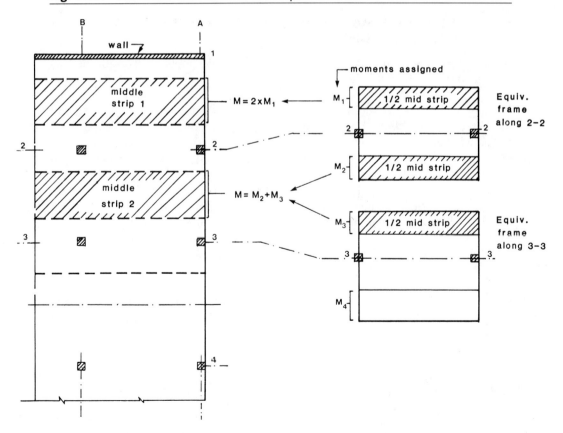

(c) Torsion in Spandrel Beam

It is evident from Fig. 14.8 that the exterior negative moment in a slab is transferred to the columns, in part, through the edge beam (or spandrel beam) acting as a torsional member. The spandrel beam must be designed for this torsion (see also Section 14.5.2b). Similar action, although to a lesser degree, develops in the transverse torsional member along interior supports.

(d) Shear Forces in Beams and Slab

For beams with stiffness such that $\alpha_1 \ell_2/\ell_1 \geqslant 1.0$ the Code specifies that the factored shear be computed as that caused by the loads in tributary areas bounded by 45° lines drawn from the corners of the panels and the centreline of the panels parallel to the long sides, as shown in Fig. 14.12. When $\alpha_1 \ell_2/\ell_1 < 1.0$, the beam shear is obtained by linear interpolation, assuming that for $\alpha_1 \ell_2/\ell_1 = 0$, the beam carries no load. In such cases, the beams framing into the column transmit only part of the shear to the column, and the balance of the shear in the panel is assumed to be transmitted by the slab directly to the columns. The latter part causes shear stresses in the slab around the column as in the case of a flat slab (see Section 14.6.2). In addition to the shear due to slab loads, beams must also resist shears due to factored loads applied directly on the beams.

14.5.4 MOMENTS IN SUPPORTING MEMBER AND EFFECTS OF PATTERN LOADINGS

As described in Section 14.4, columns and walls built monolithically with the slab must be designed to resist the unbalanced moments transferred from the slab. The moment transferred to the exterior column is the same as the negative factored moment at the exterior support, computed using the factors given in Fig. 14.7c in the Direct Design Method and obtained by the frame analysis in the Equivalent Frame Method. The negative factored moments at faces of interior supports computed by Eq. 14.8 and the factors in Fig. 14.7c correspond to the action of full factored live load plus dead load, whereas the maximum unbalanced moment at the support would occur under pattern loadings. The Code specifies that unless a general analysis under such loading is made for the Direct Design Method, the factored unbalanced gravity load moment at an interior support may be computed using Eq. 14.18. The total unbalanced moment must be resisted by the supporting member above and below the slab in proportion to their stiffnesses.

$$M_f = 0.07\left[(w_{df} + 0.5\ w_{\ell f})\ell_2\ell_n^2 - w_{df}'\ \ell_2'(\ell_n')^2 \right] \times 10^{-3} \qquad (14.18)$$

where w_{df} is the factored dead load per unit area; $w_{\ell f}$ is the factored live load per unit area; w_{df}', ℓ_2', and ℓ_n' refer to the shorter of the two spans on either side of the support, assumed to carry factored dead load only; and w_{df}, $w_{\ell f}$, ℓ_2 and ℓ_n refer to the longer of the two spans, assumed to carry factored dead load plus one-half the factored live load.

At the exterior support, the torsional member (edge beam or edge portion of the slab (see Fig. 14.8)) must be designed to resist, in torsion, its share of the exterior negative factored moment.

The moments at critical sections computed in Section 14.5.2b (Eqs. 12.8, 12.9 and Fig. 14.7c) correspond to the application of full factored load (dead plus live loads) on all spans. For slab systems meeting the limitations in Section 14.5.1, the maximum factored negative moment at a support is not much different from the value computed above, because the loading pattern for this case also has full factored loads on both spans adjoining the support considered. However, the positive moment in the slab increases under pattern loading. In order to limit such increases to less than 33 percent, when the ratio of unfactored dead load to live load, β_a, is less than two, the Code requires that the relative flexural stiffness of the columns, α_c, be not less than the minimum values, α_{min}, specified in Table 14.3. If the column stiffnesses do not satisfy these minimum requirements, the positive factored moments in the panels supported by such columns must be increased by a factor δ_s:

$$\delta_s = 1 + \frac{2 - \beta_a}{4 + \beta_a}\left(1 - \frac{\alpha_c}{\alpha_{min}}\right) \tag{14.19}$$

where β_a = ratio of dead load per unit area to live load per unit area (in each case without load factors)

     α_c = ratio of flexural stiffness of the columns above and below the slab to the combined flexural stiffness of the slabs and beams at a joint taken in the direction of the span for which moments are being determined, = $\Sigma K_c / \Sigma (K_s + K_b)$

   α_{min} = the minimum required values of α_c given in Table 14.3

Design of a two-way slab system by the Direct Design Method is illustrated in Example 14.1.

14.6 SHEAR IN TWO WAY SLABS

14.6.1 ONE-WAY ACTION (BEAM SHEAR)

In two-way slabs supported on walls or relatively stiff beams along column lines, a portion of the slab load is transmitted in each of the two directions. The shear strength of the slab can be checked by considering typical slab strips, one metre wide, taken in each of the two directions and loaded with the appropriate share of the load transmitted in the respective directions. In this computation, the slab strip in each direction is considered as a beam one metre wide, and the procedure for shear design is identical to that for beams discussed in Chapter 6. This process was illustrated in Example 13.1. This kind of shear design considering beam-type action may be termed one-way shear or beam shear. The critical section for shear in one-way or beam shear is taken at a distance equal to the effective depth, d, away from the face of the support. In beam-supported slabs, in addition to designing the slab

Table 14.3   Minimum required α_{min}

β_a	Ratio ℓ_2/ℓ_1	Relative beam stiffness, α				
		0	0.5	1.0	2.0	4.0
2.0	0.5–2.0	0	0	0	0	0
1.0	0.5	0.6	0	0	0	0
	0.8	0.7	0	0	0	0
	1.0	0.7	0.1	0	0	0
	1.25	0.8	0.4	0	0	0
	2.0	1.2	0.5	0.2	0	0
0.5	0.5	1.3	0.3	0	0	0
	0.8	1.5	0.5	0.2	0	0
	1.0	1.6	0.6	0.2	0	0
	1.25	1.9	1.0	0.5	0	0
	2.0	4.9	1.6	0.8	0.3	0
0.33	0.5	1.8	0.5	0.1	0	0
	0.8	2.0	0.9	0.3	0	0
	1.0	2.3	0.9	0.4	0	0
	1.25	2.8	1.5	0.8	0.2	0
	2.0	13.0	2.6	1.2	0.5	0.3

to resist the shear force in it, the beam must also be designed to resist the shear force caused by the loads in the area tributary to it (as shown in Fig. 14.12).

Slabs directly supported on columns must also be checked for one-way shear, as though it were a wide beam supported on and spanning between the columns. Considering the slab member as an equivalent frame, the critical section for one-way shear is at a distance, d, from the face of the support, (section 1–1 in Fig. 14.13a). The shear stress may be computed for the full slab width, ℓ_2, or for a typical strip of slab, one metre wide, shown shaded in Fig. 14.13a. Considering the one metre wide strip and assuming the shear to be zero at midspan, the factored shear, V_f is given by

$$V_f = w_f \times 1 \times \left(\frac{\ell_n}{2} - d\right)$$

In the case of a slab with drop panels (Fig. 14.13b) a second section where one-way shear stress may be critical is at section 2–2, distance d_2 from the edge of the drop panel, where d_2 is the effective depth of slab outside the drop.

Since shear reinforcement is not generally used in slabs, the factored shear force, V_f, in one-way action must not exceed the factored shear resistance of concrete, V_c.

Fig. 14.12 *Tributary areas for beam shear*

For one-way action, the value for V_c is given by Eq. 6.11. Slabs are generally safe in one-way shear, and where necessary, the shear stress is controlled by increasing the slab thickness.

14.6.2 SHEAR IN TWO-WAY ACTION (PUNCHING SHEAR)

In a relatively thin flat slab supported on columns (or the case of a large concentrated load applied on a small slab area, such as in footings supporting a column) and subjected to bending in two directions, shear failure may occur by punching through of the loaded area along a truncated cone or pyramid, with the surface sloping out in all directions from the perimeter of the loaded area, as shown in Fig. 14.14a. The shear strength in the vicinity of the slab-to-column connections, especially when moments are transferred, is of primary importance in the satisfactory performance of flat plate and flat slab construction.

Based on an evaluation of extensive research studies, Ref. 14.5 has shown that the critical section governing the ultimate shear strength in two-way action of slabs and footings is along the perimeter of the loaded area. Further, for square columns or loaded areas and for practical design situations, the ultimate shear stress at this section is a function of $\sqrt{f_c'}$ and the ratio of the side dimensions of the loaded area to the effective depth of slab. The effect of the latter parameter can be accounted

Fig. 14.13 Critical sections for one-way shear (wide-beam action)

shear =o at ℄

(a)

$V_c = 0.2 \lambda \phi_c \sqrt{f_c'} / bd$

normally stirrup are not used

Ghali

(b)

Fig. 14.14 *Shear stresses in slabs*

(a) Punching shear failure

(b) Assumed critical section

for by assuming the shear area to be that at a vertical section located at a distance $d/2$ beyond the edge of the loaded area (Fig. 14.14b), in which case the ultimate shear stress will depend only on $\sqrt{f_c'}$. On this basis, the permissible shear stress in slabs without shear reinforcement for two-way action, computed at an assumed critical section located at $d/2$ from the periphery of the loaded area (for square or nearly square loaded areas, see Fig. 14.14b) and with the use of a ϕ-factor of 0.85, was determined as $0.33 \sqrt{f_c'}$ MPa and this was specified as the upper limit for the shear stress in two-way action in CSA A23.3-M77. To partly offset the decrease in the shear resistance due to the introduction of a material resistance factor of $\phi_c = 0.6$, the factor 0.33 above has been increased to 0.40 in CSA A23.3-M84. With this, the factored shear resistance of concrete in two-way action, V_c, is:

$$V_c = 0.4 \lambda \phi_c \sqrt{f_c'} \, b_o d$$

where, b_o = the perimeter of a critical section assumed located at distance $d/2$ from the periphery of the loaded area, = $2(c_1 + c_2 + 2d)$ for the column in Fig. 14.14b.

Tests have shown (Ref. 14.6) that the shear resistance, V_c, decreases with increasing rectangularity of the loaded area. To account for this reduction in shear strength also, CSA A23.3-M84 specifies that the factored shear resistance, V_r, of slabs or footings for two-way action be taken as not greater than the factored shear resistance, V_c, computed by Eq. 14.20, unless shear reinforcement is provided.

$$V_c = (1 + 2/\beta_c)\, 0.2\, \lambda\, \phi_c\sqrt{f_c'}\, b_o d$$
$$\leqslant 0.4\, \lambda\, \phi_c\sqrt{f_c'}\, b_o d \qquad\qquad (14.20)$$

where β_c = ratio of the long side to the short side of the concentrated load or reaction area.

Equation 14.20 gives $V_c = 0.4\, \lambda\, \phi_c\sqrt{f_c'}\, b_o d$ for $\beta_c \leqslant 2.0$, and $V_c = 0.2\, \lambda\, \phi_c\sqrt{f_c'}\, b_o d$ for a wall type support ($\beta_c \to \infty$) where the slab is loaded like a wide beam.

In general, the factored shear force, V_f, causing punching shear may be computed as the net upward column reaction (or other concentrated load) less the downward load within the area of slab enclosed by the perimeter of the critical section. When a general frame analysis is made, as for the Equivalent Frame Method, the column reaction is obtained from the analysis. In the Direct Design Method, and for preliminary design purposes, V_f may be computed as the total design load acting on the shaded area shown in Fig. 14.15. The factored shear resistance, V_r for two-way

Fig. 14.15 *Critical sections and loading for punching shear*

(a)

(b) Punching around drop

action computed using Eq. 14.20 must be $\geq V_f$, unless shear reinforcement is provided. For computing V_c, the critical section is considered as the section perpendicular
to the plane of the slab and located so that its periphery is a minimum and approaches
no closer than $d/2$ to the periphery of the concentrated load or reaction area. An
illustration of this for a non-rectangular loaded area and the associated value of β_c
is shown in Fig. 14.16 (Ref. 14.7).

Fig. 14.16 *Critical section and β_c for non-rectangular loaded area (Ref. 14.7)*

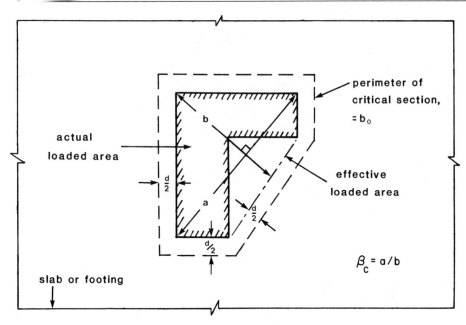

Shear reinforcement may consist of properly anchored bars or wires (Fig. 14.17a).
Nailhead type shear reinforcement, consisting of vertical rods anchored at each end
by plates having an area of at least ten times the cross-sectional area of the rods,
may also be used. When shear reinforcement is provided, the factored shear resistance, $V_r = V_c + V_s$, is limited to:

$$V_r = V_c + V_s \leq 0.6 \lambda \, \phi_c \sqrt{f'_c} \, b_o d \qquad (14.21)$$

where

$$V_c \leq 0.2 \lambda \, \phi_c \sqrt{f'_c} \, b_o d$$

When the factored shear force, V_f, exceeds $V_c = 0.2 \lambda \, \phi_c \sqrt{f'_c} \, b_o d$, the reinforcement must be adequate to resist the excess shear, $V_s = V_f - 0.2 \lambda \, \phi_c \sqrt{f'_c} \, b_o d$. The
reinforcement may be computed using the procedures of Chapter 6, (Eqs. 6.18,
6.19). Alternatively, reinforcement may be made up of shearhead reinforcement

consisting of steel I-sections or channel shapes embedded within the slab and designed in accordance with Ref. 14.8. Such shearhead reinforcement effectively moves the critical section farther out from the column as shown in Fig. 14.17b.

Fig. 14.17 *Shear reinforcement at slab-column connection*

(a) **Bars as shear reinforcement**

(b) **Embedded steel shapes as shearhead reinforcement**

Where openings in the slab are located near (at a distance of less than ten times the slab thickness from a concentrated load or reaction area, or within column strips in flat slabs) concentrated loads or reactions, the portion of the periphery of the critical section which is enclosed by radial projections of the openings to the centroid of the loaded area must be considered ineffective in computing the shear stress. Some examples of the effective portions of critical sections for slabs with openings are illustrated in Fig. 14.18.

14.6.3 SHEAR INCLUDING TRANSFER OF MOMENTS

The vertical shear stress, v, in the slab at the critical section around a column (see Figure 14.14b) transferring a design shear force, V_f, is given by Eq. 14.22 as:

$$v = \frac{V_f}{A_c} \tag{14.22}$$

where A_c is the area of concrete resisting shear at the critical section. As described in Section 14.4, a portion, M_{fv} of the total unbalanced moment transferred between

Fig. 14.18 *Effective perimeter for slabs with openings (Ref. 14.7)*

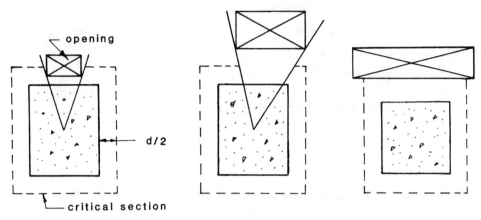

slab and column is considered transferred by the eccentricity of the shear about the centroid of the critical section (see Fig. 14.6c). Combining the effects of both V_f and M_{fv}, and assuming the shear stress to vary linearly about the centroid of the critical section, the shear stress distribution at the critical section is as illustrated in Fig. 14.19. The maximum shear stress is given by:

$$v_{AB} = \frac{V_f}{A_c} + \frac{M_{fv}}{J_c}c_{AB}$$ (14.23)

where $\quad A_c =$ area of concrete at the critical section $= b_o d$

$\quad\quad\quad J_c =$ property of the critical section analogous to the polar moment of inertia

$\quad\quad c_{AB} =$ distance from the centroidal axis, *cc*, of critical section to the face *AB* of the critical section, respectively

Expressions for J_c are also given in Fig. 14.19 for the two cases shown there. Equations for J_c for several other cases are presented in Ref. 14.9. The Code requires that the combined shear stress, v, computed by Eq. 14.23 be limited to a maximum of v_r given below (Eqs. 14.24, 14.25):

(i) for members without shear reinforcement,

$$v_r = \frac{V_c}{b_o d}$$ (14.24)

where V_c is given by Eq. 14.20;

(ii) for members with shear reinforcement,

$$v_r = \frac{V_r}{b_o d}$$ (14.25)

where V_r is given by Eq. 14.21.

Fig. 14.19 *Combined shear due to punching and transfer of moments*

$$J_c = \frac{(c_1+d)\,d^3}{6} + \frac{(c_1+d)^3\,d}{6} + \frac{(c_1+d)^2\,(c_2+d)\,d}{2}$$

$$c_{AB} = (c_1+d)/2$$

(a) Interior column

$$J_c = \left[\left(c_1+\frac{d}{2}\right)d^3 + \left(c_1+\frac{d}{2}\right)^3 d\right]\bigg/6 + (c_2+d)\,d\,c_{AB}^2$$

$$+\, 2\left(c_1+\frac{d}{2}\right)d\left[\left(c_1+\frac{d}{2}\right)\bigg/2 - c_{AB}\right]^2$$

$$c_{AB} = \left(c_1+\frac{d}{2}\right)^2\bigg/(2\,c_1 + c_2 + 2\,d)$$

(b) Edge column

14.7 DESIGN EXAMPLE FOR DIRECT DESIGN METHOD

EXAMPLE 14.1

The plan and cross section of a two-way slab floor, with beams along column lines, are shown in Fig. 14.20. The service live load is 5 kN/m² and there is a superimposed dead load of 1.0 kN/m². Material properties are $f'_c = 30$ MPa and $f_y = 400$ MPa. The member sizes shown are all based on preliminary estimates. Determine the design moments in the various strips in the E–W direction for an edge and an interior panel. Use the Direct Design Method.

Fig. 14.20 *Example 14.1*

Plan

Section XX

SOLUTION

The moments in panels $S1$ and $S2$ in Fig. 14.20 will be determined. This is done by considering the equivalent frame along column line 2–2, isolated and shown in Fig. 14.21a.

1. Check whether the limitations for the use of the Direct Design Method (CSA A23.3-M84(13.6.1)) are applicable.
   (i) There are three or more spans in each direction.
   (ii) Ratio longer to shorter span $= 7.5/6.0 = 1.25 < 2.0$
   (iii) Difference in successive span lengths $= 0 < \frac{1}{3}$ longer span
   (iv) No offset columns
   (v) Dead load, assuming a slab thickness of 180 mm, is:
   $1 + 0.18 \times 2400 \times 9.8 \times 10^{-3} = 5.23$ kN/m^2
   Live load $= 5$ kN/m^2
   Ratio, factored live load/factored dead load $= (1.5 \times 5)/(1.25 \times 5.23)$
   $= 1.15 < 3$
   Also, all loads are gravity loads and uniformly distributed over the panels.
   (vi) Relative stiffness of the beams:
   Same concrete is used for slab, beams and columns, so that E_c is same for all elements.
   $\alpha_1 = I_{b1}/I_{s1}$
   $\alpha_2 = I_{b2}/I_{s2}$
   Since the beam stem dimensions are the same along all edges of the panel, to check this limitation, it may be assumed $I_{b1} \simeq I_{b2}$, so that,

$$\alpha_1/\alpha_2 = I_{s2}/I_{s1} = \ell_1/\ell_2$$

$$\text{Ratio,} \frac{\alpha_1 \ell_2^2}{\alpha_2 \ell_1^2} \simeq \frac{\ell_1}{\ell_2} \times \frac{\ell_2^2}{\ell_1^2} = \frac{\ell_2}{\ell_1} = \frac{7.5}{6} = 1.25 < 5.0$$

All limitations are satisfied and the Direct Design Method is applicable.

Fig. 14.21 *Equivalent frame, example 14.1*

(a) Frame considered

Fig. 14.21 *Continued*

$$I_b = 7.48 \times 10^9 \ mm^4$$
$$I_s = 1.56 \times 10^9 \ mm^4$$

(b) Beam and slab sections for α along edge AD

$$I_b = 8.75 \times 10^9 \ mm^4$$
$$I_s = 3.65 \times 10^9 \ mm^4$$

(c) Beam, and slab sections for α = α₁ along AB

$$I_b = 8.75 \times 10^9 \ mm^4$$
$$I_s = 2.92 \times 10^9 \ mm^4$$

(d) Beam and slab sections for α along BC

$$I_{sb} = 15.4 \times 10^0 \ mm^4$$
$$I_{slab} = 3.65 \times 10^9 \ mm^4$$

(e) Slab–beam member along AB

2. Determine slab thickness for deflection control.

   The equations for minimum thickness, Eqs. 14.1–14.3, applicable for two-way slabs having beams between all supports, are in terms of the beam stiffness parameter, $\alpha = E_{cb} I_b / E_{cs} I_s = I_b / I_s$. The thickness requirement will be evaluated for the edge panel *S1*.

   The beam and corresponding slab section for computing α for the edges of the panel are shown in Fig. 14.21 *b–d*, wherein the centroidal moments of inertia of the gross sections are also given.

   For edge *AD*,
   $$I_b = 7.48 \times 10^9 \text{ mm}^4$$
   $$I_s = 3200 \times 180^3/12 = 1.56 \times 10^9 \text{ mm}^4$$
   $$\alpha_2 = 7.48/1.56 = 4.79$$

   For edges *AB, CD*,
   $$I_b = 8.75 \times 10^9 \text{ mm}^4$$
   $$I_s = 7500 \times 180^3/12 = 3.65 \times 10^9 \text{ mm}^4$$
   $$\alpha_1 = 8.75/3.65 = 2.40$$

   For edge *BC*,
   $$I_b = 8.75 \times 10^9 \text{ mm}^4$$
   $$I_s = 6000 \times 180^3/12 = 2.92 \times 10^9 \text{ mm}^4$$
   $$\alpha_2 = 8.75/2.92 = 3.00$$

   Average $\alpha = \alpha_m = (4.79 + 2 \times 2.4 + 3.0)/4 = 3.15$

   $$\beta_s = \frac{\text{continuous edge}}{\text{total perimeter}} = \frac{19.5}{27} = 0.722$$

   $$\beta = \frac{\text{larger clear span}}{\text{shorter clear span}} = \frac{7100}{5600} = 1.27$$

   Minimum thickness required is (Eq. 14.1):

   $$h = \frac{7100 \,(800 + 400/1.5)}{36\,000 + 5000 \times 1.27\,[3.15 - 0.5(1 - 0.722)(1 + 1/1.27)]}$$
   $$= 139 \text{ mm}$$

   But h must be not less than (Eq. 14.2):

   $$h = \frac{7100 \,(800 + 400/1.5)}{36\,000 + 5000 \times 1.27\,(1 + 0.722)} = 161 \text{ mm}$$

   h need not be more than (Eq. 14.3):

   $$h = \frac{7100 \,(800 + 400/1.5)}{36\,000} = 210 \text{ mm}$$

   A slab thickness of 180 mm is adequate and will be used in the design calculations.

3. Factored static moment
   Factored loads are:

Live load on slab = $1.5 \times 5 = 7.50$ kN/m²
Dead load on slab, including slab weight,
$$= 1.25 \times 5.23 = 6.54 \text{ kN/m}^2$$
Weight of beam stem $= 1.25 (0.4 \times 0.37 \times 2400 \times 9.8 \times 10^{-3})$
$$= 4.35 \text{ kN/m}$$
Factored static moment in the E–W direction in an interior equivalent frame is:

$$M_o = \frac{1}{8}[7.5 \ (7.50 + 6.54) + 4.35] \times 5.6^2 = 430 \text{ kN} \cdot \text{m}$$

Note that the weight of the beam stem has been included in the calculation of the static design moment. The Code (13.6.5.3) specifies that moments caused by loads applied on the beam and not considered in the slab design be determined directly and the beam proportioned to resist such moments also. Thus, the weight of the beam stem could be considered separately as direct loading on the beam. However, when such loading is not significant compared to the floor loading, as is usually the case with the self-weight of the stem, it is convenient and satisfactory to consider the former along with the slab loading as is done in this example.

4. Longitudinal distribution of factored moments
   In the interior span, the moment is distributed as follows (Eqs. 14.8–14.9 and Fig. 14.7):
   Negative factored moment at face of support = $0.65 \times 430 = 280$ kN·m
   Positive factored moment at midspan $= 0.35 \times 430 = 151$ kN·m
   In the exterior span, the distribution is as given in Fig. 14.7c for Case (2) — slab with beams between all supports.
   Negative factored moment at exterior support,

   $$= 0.16 \times 430 = 68.8 \text{ kN} \cdot \text{m}$$

   Negative factored moment at exterior face of first interior support,

   $$= 0.70 \times 430 = 301 \text{ kN} \cdot \text{m}$$

   Positive factored moment

   $$= 0.59 \times 430 = 254 \text{ kN} \cdot \text{m}$$

   Check for effects of pattern loadings:
   Since, β_a = dead load/live load = $5.23/5 < 2.0$, the positive factored moments must be increased by the factor in Eq. 14.19 unless the minimum column stiffness requirements in Table 14.3 are met.
   For the frame here, $\beta_a \approx 1.0$, $\ell_2/\ell_1 = 1.25$ and $\alpha = \alpha_1 = 2.4$, and Table 14.3 gives $\alpha_{min} = 0$. Actual $\alpha_c > 0$ and hence the column stiffness is adequate, and the effect of pattern loadings can be neglected.
   CSA A23.3 M84(13.6.7) permits a 10 percent modification of the factored moments, but this is not applied in this example.
   The longitudinal distribution of moments is shown in Fig. 14.22a.

Fig. 14.22 *Design moments in various elements in E–W direction*

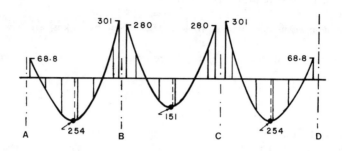

(a)   Positive and negative design moments

(b)   Column and middle strip moments

(c)   Beam and slab design moments in column strip

5.   Transverse distribution of moments to design strips.

At each critical section, part of the design moment is assigned to the column strip and the balance to the two half middle strips. The part assigned to the column strip is further apportioned between the beam element and remaining slab element within the column strip.

The parameters determining the transverse distribution are:

$$\ell_2/\ell_1 \,=\, 7.5/6 \,=\, 1.25,$$
$$\alpha_1 \,\ell_2/\ell_1 \,=\, 2.4 \times 7.5/6.0 \,=\, 3.0$$

and

$$\beta_t = \frac{E_{cb}C}{2E_{cs}I_s} = \frac{C}{2I_s}, \text{ for the edge beam}$$

The torsional member along the exterior edge (edge beam) has the cross section shown in Fig. 14.21b. The torsional constant (Eq. 14.11) is:

$$C = \left(1 - 0.63 \times \frac{400}{550}\right) \times \frac{400^3 \times 550}{3} + \left(1 - 0.63 \times \frac{180}{370}\right)$$
$$\times \frac{180^3 \times 370}{3}$$
$$= 6.86 \times 10^9 \text{ mm}^4$$

The section of slab for computing I_s in the calculations of β_t is shown in Fig. 14.21e, and

$$I_s = \frac{7500 \times 180^3}{12} = 3.65 \times 10^9 \text{ mm}^4$$

Hence,

$$\beta_t = C/2I_s = \frac{6.86 \times 10^9}{2 \times 3.65 \times 10^9} = 0.94$$

For these values, the percentage distribution to the various elements, determined from Table 14.2, using linear interpolation, are given below. Since $\alpha_1 \ell_2/\ell_1 > 1.0$, the beam is assumed to carry 85 percent of the column strip moments.

SECTION	Percent assigned to	
	Column strip	Middle strip
(i) Interior negative design moments	67.5	32.5
(ii) Exterior negative design moments	87.8	12.2
(iii) Positive design moments	67.5	32.5

The actual factored moments in the column strip and the middle strip are shown in Fig. 14.22b. The subdivision of the column strip moment to its beam part and slab part is shown in Fig. 14.22c. The width of the column strip is $6/4 = 1.5$ m on each side of column, and the middle strip is $7.5 - 2 \times 1.5 = 4.5$ m wide (two half strips 2.25 m wide each) as shown in Fig. 14.22b.

6. Slab reinforcement
    The design of flexural reinforcement in the E–W direction for the slab panels S1 and S2 is set up in Table 14.4. The effective depth of the beam, assuming 40 mm cover, No. 10 bar stirrups and No. 20 bars as reinforcement is 489 mm. For the larger moment at the interior support, $K_r = 1.81$ which corresponds to a reinforcement ratio of close to 0.25 ρ_{max} as seen from Table 4.1, thus ensuring that the beam is under-reinforced throughout.
    For the slab panel, the E–W direction being the shorter span, the reinforcement

in this direction is placed as the outer layer of slab reinforcement. Assuming 20 mm cover and No. 10 bar reinforcement, the effective depth of the slab in the E–W direction is 154 mm. The largest moment per metre width in the slab is 21.7 kN·m in the middle strip at the first interior support. If a reinforcement ratio of 0.4 ρ_{max} is selected, from Table 4.1 K_r = 2.80 MPa, and the required effective depth is:

$$d = \sqrt{\frac{21.7 \times 10^6}{2.80 \times 1000}} = 88.0 \text{ mm}$$

The effective depth provided, 154 mm, is more than this and the slab will be under-reinforced throughout. The slab thickness could have been reduced to 160 mm which is the minimum thickness required for deflection control. However, the design here is continued with the assumed thickness of 180 mm.

The minimum reinforcement in the slab, for shrinkage and temperature stresses, is 0.002 bh = 0.002 × 1000 × 180 = 360 mm² per metre width. Where the area of reinforcement computed for flexure is less than this minimum, the latter value is provided. The maximum bar spacing is limited to $2d$ = 308 mm. Where necessary, the number of bars must be increased to meet this maximum spacing requirement (not critical in the design in Table 14.4). The negative moment section at the interior support is designed to resist the larger of the two interior negative design moments determined for the spans on either side.

Details of flexural reinforcement for slabs are described in Section 14.10.

7. Transfer of moments to the columns

   In this example, since beams are provided along column lines, transfer of unbalanced moment to the supporting columns is not critical. For instance, the largest unbalanced moment is at the exterior support and equal to 68.8 kN·m. Of this, 51.3 kN·m or 75 percent (which is in fact 85 percent of the 87.8 percent moment in the column strip) is directly transferred by the beam. The requirement for moment transfer through flexure (Eq. 14.4) is only 60 percent. (Note that the Code requirement in clause 13.3.3.3 is concerned primarily with slab systems without beams.)

   The exterior negative moment in the slab part is 68.8 − 51.3 = 17.5 kN·m (that is 25 percent of total), and this is transferred to the column mostly through the spandrel beam in torsion. Design for torsion was considered in Chapter 7. In this example the torque being very low, the torsional stresses may be neglected. Similar stresses at interior supports will be even less.

8. Shear in slab and beams

   In slabs with beams along the edges of the panels, the slab shear is essentially that associated with one-way action. The assumed distribution of loads for shear strength consideration is as shown in Fig. 14.23. For a one metre wide strip of slab, the factored shear force at the critical section at a distance d from the face of the beam is:

   $$V_f = (7.5 + 6.54) \times \left(\frac{5.6}{2} - 0.154\right) = 37.2 \text{ kN}$$

   $$V_c = 0.2 \lambda \phi_c \sqrt{f'_c} b_w d$$

   $$= 0.2 \times 1.0 \times 0.6 \times \sqrt{30} \times 1000 \times 154 \times 10^{-3} = 101 \text{ kN}$$

   $$V_f < V_c, \qquad\qquad\qquad\qquad\qquad\qquad\qquad\qquad\qquad \text{OK}$$

Table 14.4 Design of Slab Reinforcement in E–W Direction for Panels S1 and S2

	Exterior Panel S1			Int. Panel S2	
	Ext. support	Mid-span	Int. support	Support	Mid-span
(a) Column Strip					
(1) Beam design:					
Moment, M_f (kN·m)	51.3	145	173	161	86.7
$K_r = M_f \times 10^6/(400 \times 489^2)$					
(MPa)	0.536	1.52	1.81	‡	0.906
ρ, from Table 5.4	0.00160	0.00472	0.00568	—	0.00275
$\rho_{min} = 1.4/f_y$	0.0035*	—	—	—	0.0035*
$A_s = \rho \times 400 \times 489$ (mm²)	685	923	1112	1112	685
Number of No. 20 bars	2 + 1 No. 10	3	4	4	2 + 1 No. 10
(2) Slab design, 2 strips, 1.3 m each:					
Moment (kN·m)	4.5	12.8	15.2	14.2	7.65
Moment/m width, M_f (kN·m/m)	3.46	9.85	11.7	—	5.88
$K_r = M_f \times 10^6/(1000 \times 154^2)$					
(MPa)			0.493		
ρ, from Table 5.4			0.00147		
$A_s = \rho \times 1000 \times 154$ (mm²)			227		
$A_{s\,min} = 0.002 \times 1000 \times 180$ (mm²)	360*	360*	360*	360*	360*
A_s per half column strip of 1.3 m (mm²)	468	468	468	468	468
Number of No. 10 bars in 1.3 m width	5	5	5	5	5
Spacing (≤ 309 mm) (mm)	260	260	260	260	260
(b) Middle Strip (full width = $2 \times 2.25 = 4.5$m)					
Moment in 4.5 m (kN·m)	8.4	82.6	97.8	91	49
Moment/m width, M_f (kN·m/m)	1.87	18.4	21.7	—	10.9
$K_r = M_f \times 10^6/(1000 \times 154^2)$					
(MPa)	—	0.776	0.915	—	—
ρ, from Table 5.4	—	0.00235	0.00278	—	—
$A_s = \rho \times 1000 \times 154$ (mm²)	—	361	428	428	—
$A_{s\,min}$ (mm²)	360*	—	—	—	360*
Area for full 4.5 m width (mm²)	1620	1620	1926	1926	1620
Number of No. 10 bars	16	16	21	21	16
Spacing (≤ 308 mm) (mm)	281	281	214	214	281

* Minimum reinforcement requirement controls.

‡ Provide same reinforcement as at section to the left of support.

Fig. 14.23 *Shear in slab and beams*

Since all beams in this example have $\alpha_1 \ell_2 / \ell_1$ values greater than 1.0, they are proportioned to resist the full shear from the loads on the respective tributary areas, and no part of this need be assigned to the slab to be resisted in two-way action (Code cl. 13.6.8). The design of the beam for shear follows the procedures described and illustrated in Chapter 6.

14.8 DESIGN AIDS

The discussions in preceding sections, and the design example above show that the design of two-way slabs involves the computation of numerous stiffness parameters and coefficients. This is also true of the Equivalent Frame Method discussed in the next section. To assist in these computations several design aids in the form of charts and tables have been prepared and these are presented in Ref. 14.9.

14.9 EQUIVALENT FRAME METHOD

A brief description of the background of the Equivalent Frame Method, with references to detailed descriptions, is presented in Ref. 14.2. The method (Code (13.7)) simplifies the analysis of a three-dimensional building by subdividing it into a series of two-dimensional frame bents or *equivalent frames* centred on column lines in both longitudinal and transverse directions, as shown in Fig. 14.1. However, unlike the Direct Design Method, the longitudinal distribution of the moments in the slab-beam member is determined using an elastic analysis of this equivalent plane frame. The analysis may be done considering an entire plane frame (Fig. 14.1b) or, for vertical loading, each floor may be analyzed separately considering the corresponding partial frame shown in Fig. 14.1c. Consistent with the usual approximations made in the analysis of continuous frames (see Chapter 10) it may further be assumed in determining the moment at a given support or span that the slab-beam is fixed at any support two panels distant provided the slab continues beyond that point.

The load transfer system in the equivalent frame involves three major parts, namely, (1) the horizontal slab-beam member, (2) the vertical supporting elements (columns or walls), and (3) a transverse torsional member at the column supports which assists in the transfer of moments between the slab-beam member and the vertical supporting members. These three distinct parts are identified in Fig. 14.24a. In order to account for the increased flexibility of the slab-to-column correction resulting from the torsional flexibility of the torsional member, in the gravity load analysis of the frame, at all joints, the real columns above and below and the transverse torsional members are replaced by equivalent columns of stiffness, K_{ec}. With this modification at all supports, the frame that is in fact analyzed elastically is a plane single storey multiple bay portal frame shown in Fig. 14.24b. In computing the stiffness parameters of the members, for the analysis of this frame, the effects of the variation in the moment of inertia of the slab-beams and columns along their axes must be taken into account. These effects are described below.

(a) Slab-Beam Member
The slab-beam member is bounded laterally by the centreline of the panel on each side of the column line. For an exterior frame, the slab-beam extends laterally from the edge to the centreline of the adjacent panel. Thus the slab-beam includes drop panels for flat slabs, and beams along column lines for two-way slabs, if used. The moment of inertia of the slab-beam between the faces of supports (columns or column capitals) is computed on the basis of the gross cross-sectional area of concrete. From the face of the support to the centreline of the support, the moment of inertia of the slab-beam is taken as the moment of inertia at the face of the support divided by $(1 - c_2/\ell_2)^2$, where c_2 and ℓ_2 are the column width and slab width, respectively, measured transverse to the direction in which moments are being determined (Fig. 14.24a). The magnification factor $1/(1 - c_2/\ell_2)^2$ allows for the increased cross-section of the slab-beam within the bounds of the supporting column cast mono-lithically with the slab. Variation of the moment of inertia along the span of the slab-beam for a typical case is shown in Fig. 14.25b. The stiffness factors, carry-over factors and fixed end moments of the slab-beam are dependent on the variation of the moment of inertia along the span. These factors for slab-beams with common geometric and loading configurations are presented in Ref. 14.9. Factors for two typical cases are reproduced in Tables 14.5 and 14.6.

Fig. 14.24 *Equivalent frame method*

(a) Elements of equivalent frame

(b) Equivalent frame to be analyzed

(b) Equivalent Column

The restraint to the slab-beam member at the support sections is offered by the flexural stiffness of the column and the torsional stiffness of the transverse torsional member (see Figs. 14.8 and 14.24a). The actual column stiffness has to be modified to account for the torsional flexibility of the transverse torsional member. Referring

Fig. 14.25 *Variation of moments of inertia along member axis*

(a) Slab system

(b) Variation of I of slab

(c) Sections —A–A
  —B–B
  —C–C

(d) Variation of I of column

to Fig. 14.24a, the angle of twist of the torsional member, and hence the rotation of the slab at support section, progressively increases from a minimum (equal to the flexural rotation of the column) at the column support, C, to a maximum at the outer ends A and B. To allow for this increased flexibility of the connection, the actual columns above and below the slab and the transverse beam (or torsional member) are replaced by an *equivalent column* whose flexibility (inverse of the stiffness) is equal to the sum of the flexibilities of the actual columns and the torsional member. The stiffness, K_{ec}, of this equivalent column is thus obtained from:

$$\frac{1}{K_{ec}} = \frac{1}{\Sigma K_c} + \frac{1}{K_t} \qquad (14.26)$$

or

$$K_{ec} = \frac{(\Sigma K_c)\, K_t}{\Sigma K_c + K_t} \qquad (14.27)$$

Table 14.5 Moment Distribution Constants for Slab-Beam Elements

Moment Distribution Constants for Slab-Beam Elements

$$FEM_{NF} = \sum_{i=1}^{n} m_{NFi} \, W_i \ell_1^2$$

$$K_{NF} = k_{NF} \cdot E_{cs} \, I_s / \ell_1$$

C_{N1}/ℓ_1	C_{N2}/ℓ_2	Stiffness Factors k_{NF}	Carry Over Factors C_{NF}	Unif. Load Fixed end M. Coeff. (m_{NF})	Fixed end moment Coeff. (m_{NF}) for (b−a) = 0.2				
					a = 0.0	a = 0.2	a = 0.4	a = 0.6	a = 0.8
				$C_{F1} = C_{N1}$; $C_{F2} = C_{N2}$					
0.00	—	4.00	0.50	0.0833	0.0151	0.0287	0.0247	0.0127	0.00226
	0.00	4.00	0.50	0.0833	0.0151	0.0287	0.0247	0.0127	0.00226
	0.10	4.18	0.51	0.0847	0.0154	0.0293	0.0251	0.0126	0.00214
0.10	0.20	4.36	0.52	0.0860	0.0158	0.0300	0.0255	0.0126	0.00201
	0.30	4.53	0.54	0.0872	0.0161	0.0301	0.0259	0.0125	0.00188
	0.40	4.70	0.55	0.0882	0.0165	0.0314	0.0262	0.0124	0.00174
	0.00	4.00	0.50	0.0833	0.0151	0.0287	0.0247	0.0127	0.00226
	0.10	4.35	0.52	0.0857	0.0155	0.0299	0.0254	0.0127	0.00213
0.20	0.20	4.72	0.54	0.0880	0.0161	0.0311	0.0262	0.0126	0.00197
	0.30	5.11	0.56	0.0901	0.0166	0.0324	0.0269	0.0125	0.00178
	0.40	5.51	0.58	0.0921	0.0171	0.0336	0.0276	0.0123	0.00156
	0.00	4.00	0.50	0.0833	0.0151	0.0287	0.0247	0.0127	0.00226
	0.10	4.49	0.53	0.0863	0.0155	0.0301	0.0257	0.0128	0.00219
0.30	0.20	5.05	0.56	0.0893	0.0160	0.0317	0.0267	0.0128	0.00207
	0.30	5.69	0.59	0.0923	0.0165	0.0334	0.0278	0.0127	0.00190
	0.40	6.41	0.61	0.0951	0.0171	0.0352	0.0287	0.0124	0.00167
	0.00	4.00	0.50	0.0833	0.0151	0.0287	0.0247	0.0127	0.00226
	0.10	4.61	0.53	0.0866	0.0154	0.0302	0.0259	0.0129	0.00225
0.40	0.20	5.35	0.56	0.0901	0.0158	0.0318	0.0271	0.0131	0.00221
	0.30	6.25	0.60	0.0936	0.0162	0.0337	0.0284	0.0131	0.00211
	0.40	7.37	0.64	0.0971	0.0168	0.0359	0.0297	0.0128	0.00195
				$C_{F1} = 0.5C_{N1}$; $C_{F2} = 0.5C_{N2}$					
0.00	—	4.00	0.50	0.0833	0.0151	0.0287	0.0247	0.0127	0.0023
	0.00	4.00	0.50	0.0833	0.0151	0.0287	0.0247	0.0127	0.0023
	0.10	4.16	0.51	0.0857	0.0155	0.0296	0.0254	0.0130	0.0023
0.10	0.20	4.31	0.52	0.0879	0.0158	0.0304	0.0261	0.0133	0.0023
	0.30	4.45	0.54	0.0900	0.0162	0.0312	0.0267	0.0135	0.0023
	0.40	4.58	0.54	0.0918	0.0165	0.0319	0.0273	0.0138	0.0023
	0.00	4.00	0.50	0.0833	0.0151	0.0287	0.0247	0.0127	0.0023
	0.10	4.30	0.52	0.0872	0.0156	0.0301	0.0259	0.0132	0.0023
0.20	0.20	4.61	0.55	0.0912	0.0161	0.0317	0.0272	0.0138	0.0023
	0.30	4.92	0.57	0.0951	0.0167	0.0332	0.0285	0.0143	0.0024
	0.40	5.23	0.58	0.0989	0.0172	0.0347	0.0298	0.0148	0.0024
	0.00	4.00	0.50	0.0833	0.0151	0.0287	0.0247	0.0127	0.0023
	0.10	4.43	0.53	0.0881	0.0156	0.0305	0.0263	0.0134	0.0023
0.30	0.20	4.89	0.56	0.0932	0.0161	0.0324	0.0281	0.0142	0.0024
	0.30	5.40	0.59	0.0986	0.0167	0.0345	0.0300	0.0150	0.0024
	0.40	5.93	0.62	0.1042	0.0173	0.0367	0.0320	0.0158	0.0025
	0.00	4.00	0.50	0.0833	0.0151	0.0287	0.0247	0.0127	0.0023
	0.10	4.54	0.54	0.0884	0.0155	0.0305	0.0265	0.0135	0.0024
0.40	0.20	5.16	0.57	0.0941	0.0159	0.0326	0.0286	0.0145	0.0025
	0.30	5.87	0.61	0.1005	0.0165	0.0350	0.0310	0.0155	0.0025
	0.40	6.67	0.64	0.1076	0.0170	0.0377	0.0336	0.0166	0.0026
				$C_{F1} = 2C_{N1}$; $C_{F2} = 2C_{N2}$					
0.00	—	4.00	0.50	0.0833	0.0151	0.0287	0.0247	0.0127	0.0023
	0.00	4.00	0.50	0.0833	0.0150	0.0287	0.0247	0.0127	0.0023
0.10	0.10	4.27	0.51	0.0817	0.0153	0.0289	0.0241	0.0116	0.0018
	0.20	4.56	0.52	0.0798	0.0156	0.0290	0.0234	0.0103	0.0013
	0.00	4.00	0.50	0.0833	0.0151	0.0287	0.0247	0.0127	0.0023
0.20	0.10	4.49	0.51	0.0819	0.0154	0.0291	0.0240	0.0114	0.0019
	0.20	5.11	0.53	0.0789	0.0158	0.0293	0.0228	0.0096	0.0014

*Reproduced, with permission, from Ref. 14.9

Table 14.6 Moment Distribution Constants for Slab-Beam Elements, Drop Thickness = 0.50*h*

Moment Distribution Constants for Slab-Beam Elements
Drop Thickness = 0.50h

$$FEM_{NF} = \sum_{i=1}^{n} m_{NFi} \, W_i \ell_1^2$$

$$K_{NF} = k_{NF} \, E_{cs} \, I_s / \ell_1$$

C_{N1}/ℓ_1	C_{N2}/ℓ_2	Stiffness Factors k_{NF}	Carry Over Factors C_{NF}	Unif. Load Fixed end M. Coeff. (m_{NF})	Fixed end moment Coeff. (m_{NF}) for (b−a) = 0.2				
					a = 0.0	a = 0.2	a = 0.4	a = 0.6	a = 0.8
				$C_{F1} = C_{N1};\ C_{F2} = C_{N2}$					
0.00	—	5.84	0.59	0.0926	0.0164	0.0335	0.0279	0.0128	0.0020
0.10	0.00	5.84	0.59	0.0926	0.0164	0.0335	0.0279	0.0128	0.0020
	0.10	6.04	0.60	0.0936	0.0167	0.0341	0.0282	0.0126	0.0018
	0.20	6.24	0.61	0.0940	0.0170	0.0347	0.0285	0.0125	0.0017
	0.30	6.43	0.61	0.0952	0.0173	0.0353	0.0287	0.0123	0.0016
0.20	0.00	5.84	0.59	0.0926	0.0164	0.0335	0.0279	0.0128	0.0020
	0.10	6.22	0.61	0.0942	0.0168	0.0346	0.0285	0.0126	0.0018
	0.20	6.62	0.62	0.0957	0.0172	0.0356	0.0290	0.0123	0.0016
	0.30	7.01	0.64	0.0971	0.0177	0.0366	0.0294	0.0120	0.0014
0.30	0.00	5.84	0.59	0.0926	0.0164	0.0335	0.0279	0.0128	0.0020
	0.10	6.37	0.61	0.0947	0.0168	0.0348	0.0287	0.0126	0.0018
	0.20	6.95	0.63	0.0967	0.0172	0.0362	0.0294	0.0123	0.0016
	0.30	7.57	0.65	0.0986	0.0177	0.0375	0.0300	0.0119	0.0014
				$C_{F1} = 0.5C_{N1};\ C_{F2} = 0.5C_{N2}$					
0.00	—	5.84	0.59	0.0926	0.0164	0.0335	0.0279	0.0128	0.0020
0.10	0.00	5.84	0.59	0.0926	0.0164	0.0335	0.0279	0.0128	0.0020
	0.10	6.00	0.60	0.0945	0.0167	0.0343	0.0285	0.0130	0.0020
	0.20	6.16	0.60	0.0962	0.0170	0.0350	0.0291	0.0132	0.0020
0.20	0.00	5.84	0.59	0.0926	0.0164	0.0335	0.0279	0.0128	0.0020
	0.10	6.15	0.60	0.0957	0.0169	0.0348	0.0290	0.0131	0.0020
	0.20	6.47	0.62	0.0987	0.0173	0.0360	0.0300	0.0134	0.0020
				$C_{F1} = 2C_{N1};\ C_{F2} = 2C_{N2}$					
0.00	—	5.84	0.59	0.0926	0.0164	0.0335	0.0279	0.0128	0.0020
0.10	0.00	5.84	0.59	0.0926	0.0164	0.0335	0.0279	0.0128	0.0020
	0.10	6.17	0.60	0.0907	0.0166	0.0337	0.0273	0.0116	0.0015

Table 14.7 Stiffness and Carry-Over Factors for Columns

Stiffness and Carry-Over Factors for Columns

$$K_c = k \frac{EI_c}{H}$$

For values of k_{BA} and c_{BA} read (t_a/t_b) as (t_b/t_a)

*g can be approximated as $z/2$.

t_a/t_b	H/H_c	1.05	1.10	1.15	1.20	1.25	1.30	1.35	1.40	1.45	1.50
0.00	k_{AB}	4.20	4.40	4.60	4.80	5.00	5.20	5.40	5.60	5.80	6.00
	C_{AB}	0.57	0.65	0.73	0.80	0.87	0.95	1.03	1.10	1.17	1.25
0.2	k_{AB}	4.31	4.62	4.95	5.30	5.65	6.02	6.40	6.79	7.20	7.62
	C_{AB}	0.56	0.62	0.68	0.74	0.80	0.85	0.91	0.96	1.01	1.07
0.4	k_{AB}	4.38	4.79	5.22	5.67	6.15	6.65	7.18	7.74	8.32	8.94
	C_{AB}	0.55	0.60	0.65	0.70	0.74	0.79	0.83	0.87	0.91	0.94
0.6	k_{AB}	4.44	4.91	5.42	5.96	6.54	7.15	7.81	8.50	9.23	10.01
	C_{AB}	0.55	0.59	0.63	0.67	0.70	0.74	0.77	0.80	0.83	0.85
0.8	k_{AB}	4.49	5.01	5.58	6.19	6.85	7.56	8.31	9.12	9.98	10.89
	C_{AB}	0.54	0.58	0.61	0.64	0.67	0.70	0.72	0.75	0.77	0.79
1.0	k_{AB}	4.52	5.09	5.71	6.38	7.11	7.89	8.73	9.63	10.60	11.62
	C_{AB}	0.54	0.57	0.60	0.62	0.65	0.67	0.69	0.71	0.73	0.74
1.2	k_{AB}	4.55	5.16	5.82	6.54	7.32	8.17	9.08	10.07	11.12	12.25
	C_{AB}	0.53	0.56	0.59	0.61	0.63	0.65	0.66	0.68	0.69	0.70
1.4	k_{AB}	4.58	5.21	5.91	6.68	7.51	8.41	9.38	10.43	11.57	12.78
	C_{AB}	0.53	0.55	0.58	0.60	0.61	0.63	0.64	0.65	0.66	0.67
1.6	k_{AB}	4.60	5.26	5.99	6.79	7.66	8.61	9.64	10.75	11.95	13.24
	C_{AB}	0.53	0.55	0.57	0.59	0.60	0.61	0.62	0.63	0.64	0.65
1.8	k_{AB}	4.62	5.30	6.06	6.89	7.80	8.79	9.87	11.03	12.29	13.65
	C_{AB}	0.52	0.55	0.56	0.58	0.59	0.60	0.61	0.61	0.62	0.63
2.0	k_{AB}	4.63	5.34	6.12	6.98	7.92	8.94	10.06	11.27	12.59	14.00
	C_{AB}	0.52	0.54	0.56	0.57	0.58	0.59	0.59	0.60	0.60	0.61
2.2	k_{AB}	4.65	5.37	6.17	7.05	8.02	9.08	10.24	11.49	12.85	14.31
	C_{AB}	0.52	0.54	0.55	0.56	0.57	0.58	0.58	0.59	0.59	0.59
2.4	k_{AB}	4.66	5.40	6.22	7.12	8.11	9.20	10.39	11.68	13.08	14.60
	C_{AB}	0.52	0.53	0.55	0.56	0.56	0.57	0.57	0.58	0.58	0.58
2.6	k_{AB}	4.67	5.42	6.26	7.18	8.20	9.31	10.53	11.86	13.29	14.85
	C_{AB}	0.52	0.53	0.54	0.55	0.56	0.56	0.56	0.57	0.57	0.57
2.8	k_{AB}	4.68	5.44	6.29	7.23	8.27	9.41	10.66	12.01	13.48	15.07
	C_{AB}	0.52	0.53	0.54	0.55	0.55	0.55	0.56	0.56	0.56	0.56
3.0	k_{AB}	4.69	5.46	6.33	7.28	8.34	9.50	10.77	12.15	13.65	15.28
	C_{AB}	0.52	0.53	0.54	0.54	0.55	0.55	0.55	0.55	0.55	0.55
3.2	k_{AB}	4.70	5.48	6.36	7.33	8.40	9.58	10.87	12.28	13.81	15.47
	C_{AB}	0.52	0.53	0.53	0.54	0.54	0.54	0.54	0.54	0.54	0.54
3.4	k_{AB}	4.71	5.50	6.38	7.37	8.46	9.65	10.97	12.40	13.95	15.64
	C_{AB}	0.51	0.52	0.53	0.53	0.54	0.54	0.54	0.53	0.53	0.53
3.6	k_{AB}	4.71	5.51	6.41	7.41	8.51	9.72	11.05	12.51	14.09	15.80
	C_{AB}	0.51	0.52	0.53	0.53	0.53	0.53	0.53	0.53	0.53	0.52
3.8	k_{AB}	4.72	5.53	6.43	7.44	8.56	9.78	11.13	12.60	14.21	15.95
	C_{AB}	0.51	0.52	0.53	0.53	0.53	0.53	0.53	0.52	0.52	0.52
4.0	k_{AB}	4.72	5.54	6.45	7.47	8.60	9.84	11.21	12.70	14.32	16.08
	C_{AB}	0.51	0.52	0.52	0.53	0.53	0.52	0.52	0.52	0.52	0.51
4.2	k_{AB}	4.73	5.55	6.47	7.50	8.64	9.90	11.27	12.78	14.42	16.20
	C_{AB}	0.51	0.52	0.52	0.52	0.52	0.52	0.52	0.51	0.51	0.51
4.4	k_{AB}	4.73	5.56	6.49	7.53	8.68	9.95	11.34	12.86	14.52	16.32
	C_{AB}	0.51	0.52	0.52	0.52	0.52	0.52	0.51	0.51	0.51	0.50
4.6	k_{AB}	4.74	5.57	6.51	7.55	8.71	9.99	11.40	12.93	14.61	16.43
	C_{AB}	0.51	0.52	0.52	0.52	0.52	0.52	0.51	0.51	0.50	0.50
4.8	k_{AB}	4.74	5.58	6.53	7.58	8.75	10.03	11.45	13.00	14.69	16.53
	C_{AB}	0.51	0.52	0.52	0.52	0.52	0.51	0.51	0.50	0.50	0.49
5.0	k_{AB}	4.75	5.59	6.54	7.60	8.78	10.07	11.50	13.07	14.77	16.62
	C_{AB}	0.51	0.51	0.52	0.52	0.51	0.51	0.51	0.50	0.49	0.49
6.0	k_{AB}	4.76	5.63	6.60	7.69	8.90	10.24	11.72	13.33	15.10	17.02
	C_{AB}	0.51	0.51	0.51	0.51	0.50	0.50	0.49	0.49	0.48	0.47
7.0	k_{AB}	4.78	5.66	6.65	7.76	9.00	10.37	11.88	13.54	15.35	17.32
	C_{AB}	0.51	0.51	0.51	0.50	0.50	0.49	0.48	0.48	0.47	0.46
8.0	k_{AB}	4.78	5.68	6.69	7.82	9.07	10.47	12.01	13.70	15.54	17.56
	C_{AB}	0.51	0.50	0.50	0.50	0.49	0.49	0.48	0.47	0.46	0.45
9.0	k_{AB}	4.79	5.69	6.71	7.86	9.13	10.55	12.11	13.83	15.70	17.74
	C_{AB}	0.50	0.50	0.50	0.50	0.49	0.48	0.47	0.46	0.45	0.45
10.0	k_{AB}	4.80	5.71	6.74	7.89	9.18	10.61	12.19	13.93	15.83	17.90
	C_{AB}	0.50	0.50	0.50	0.49	0.48	0.48	0.47	0.46	0.45	0.44

*Reproduced, with permission, from Ref. 14.9

where $\Sigma K_c =$ sum of flexural stiffnesses of the columns above and below the slab
        $K_t =$ torsional stiffness of the torsional member

Eq. 14.26 can be seen to approach the correct limits. When the torsional member is infinitely stiff, $(K_t = \infty)$, the slab-beam member along the entire length AB undergoes the same rotation as that of the column, and in this case Eq. 14.26 rightly gives $K_{ec} = \Sigma K_c$. On the other hand if $\Sigma K_c = \infty$, $K_{ec} = K_t$. In this case, the column, being infinitely stiff, does not rotate, but the slab still undergoes the same rotation as that of the torsional member (except for the width c_2 directly over the column). For the case $K_c = 0$, as for instance with a simple masonry support along the length AB of the slab, the transverse beam becomes torsionally unrestrained by the supporting member and can rotate freely as a rigid body. Hence the slab-section AB is flexurally unrestrained and in this case Eq. 14.27 appropriately gives $K_{ec} = 0$. For the hypothetical case of $K_t = 0$ (transverse member with zero torsional stiffness), Eq. 14.27 again gives $K_{ec} = 0$. In this case also the slab is flexurally unrestrained along the length AB (except for the short stretch of width c_2 directly over the column). When the slab is supported on a monolithic reinforced concrete wall, the slab rotation will be nearly uniform along the length AB and equal to the rotation of the wall. In this case the flexural stiffness of the wall replaces ΣK_c, and the torsional flexibility $1/K_t$ of the wall may be assumed zero, so that $K_{ec} = \Sigma K_{wall}$.

It should be noted that the concept of the equivalent column stiffness, K_{ec}, as defined in Eq. 14.26 and explained in the preceding paragraph is applicable for gravity load analysis only. For lateral load analysis, the effect of the flexibility of the connection from column-to-slab will be to reduce the effective slab stiffness.

In computing the equivalent column stiffness, K_{ec}, at a support using Eq. 14.26, K_t and K_c are the stiffnesses of the torsional member and of the column. The definition of the torsional member and the calculation of its torsional stiffness, K_t, were detailed in Section 14.5.2(c).

The length of an actual column, ℓ_c, is measured centre-to-centre of floors. For computing the column stiffness, K_c, the moment of inertia of the column section outside the joint is based on the gross concrete section. The column section moment of inertia is assumed to be infinite for the portion of the column integral with the slab-beam, that is from the top to the bottom of the slab-beam at a joint. Variation of the column moment of inertia for the top half of a column with a capital is shown in Fig. 14.25d. Stiffness and carry-over factors of columns for the common configurations are presented in Ref. 14.9 and are reproduced in Table 14.7.

(c) Analysis Procedure

Having computed the stiffness properties of the individual members of the equivalent frame shown in Fig. 14.24b, the elastic analysis of this frame for determining the bending moments and shear forces in the members may be made using any of the conventional methods such as moment distribution, slope-deflection, and matrix analysis. In the analysis illustration given in Example 14.2, the moment distribution procedure is employed.

When specific loading patterns are shown, the frame is analyzed for those patterns. When the live load is variable, and the specified live load does not exceed three-quarters of the specified (unfactored) dead load, or the nature of live load is such that all panels will be loaded simultaneously, the Code recommends a single loading case of full factored loads (dead plus live) on all spans, for determining the maximum positive and maximum negative factored moments at all critical sectioins in all spans (Fig. 14.26a). For loading conditions other than the above, the maximum positive factored moment in a span is determined with three-quarters of full factored live load on that span and on alternate spans (plus factored dead load on all spans, see

Fig. 14.26 *Arrangement of live loads for equivalent frame analysis*

(a) Loading pattern for factored moment at all
critical sections (♦) for LL ≤ ¾ DL

(b) Loading pattern for maximum positive moment*
at spans BC and DC, LL > ¾ DL

(c) Loading pattern for maximum negative moment*
at support B, LL ≥ ¾ DL

*Note: factored moments shall not be taken less than those from loading
pattern at (a)

Fig. 14.26b). For maximum negative moment at a support, three-quarters of full factored live load is placed on the two adjacent spans only with factored dead loads on all spans (Fig. 14.26c). However, in no case must the design moments be taken as less than those occurring with full factored loads (dead plus live) on all spans. Use of three-quarters of full factored live load in the two pattern loadings given above is in recognition of the fact that the maximum negative and maximum positive live load moments in a span do not occur simultaneously, and that some inelastic redistribution of moments does occur in slabs which are usually very much under-reinforced.

The results of the frame analysis using centreline dimensions gives negative moments at the centrelines of the supports, from which the factored negative moments at critical sections must be determined, as shown in Fig. 14.27. The critical section of the slab-beam member for negative moment at interior supports is taken at the face of the support (column, capital, or bracket), but in no case at a distance greater than $0.175 \, \ell_1$ from the centre of the column. At exterior supports with brackets or capitals, the critical section is taken at a distance from the face of the column not greater than one-half the projection of the bracket or capital beyond the face of the column (Fig. 14.27).

To maintain consistency in design requirements, the Code (13.7.7.4) provides that in a slab system, which meets the limitations of the Direct Design Method given in Section 14.5.1, but is analyzed by the Equivalent Frame Method, the analytical moments may be reduced in such proportion that the numerical sum of the positive

Fig. 14.27 *Factored negative moments in slab-beam and column*

and average negative bending moments used in design (M_o, in Fig. 14.27b) need not exceed the value obtained from Eq. 14.7.

The moment in the equivalent column, obtained from the elastic analysis, must be distributed between the actual columns above and below the slab-beam, in proportion to their stiffness, and these moments used in their design (see Fig. 14.27c).

After determining the negative and positive factored moments at all critical sections for the slab-beam member, these factored moments must be distributed transversely at each critical section to the column strips, beams (if any), and middle strips. This transverse distribution procedure is the same as the one used in the Direct Design Method, and explained in Section 14.5.3.

A design example using the Equivalent Frame Method is given in Section 14.11.

14.10 SLAB REINFORCEMENT DETAILS

14.10.1 GENERAL

The area of reinforcement must be that required for the factored moments at the critical sections, but not less than the minimum specified for shrinkage and temperature stresses (Code (7.8)). The Code limits the spacing of bars at critical sections in solid slabs to a maximum of two times the slab thickness, but not to exceed 500 mm. This limitation is intended to ensure slab action, reduce cracking, and provide for the distribution of concentrated loads.

The flexural restraint along the exterior edge of the slab panel is subject to considerable variation and uncertainty. Depending on the torsional rigidity of the spandrel beam or slab edge, and its interaction with an exterior wall, if any, the flexural restraint at the slab edge may range from one of almost complete fixity to a condition of simple support. To provide for such uncertainties, special requirements are made in the Code for detailing flexural reinforcement in the exterior span. Thus, in these spans, all positive reinforcement perpendicular to the discontinuous edge is required to be extended to the slab edge and embedded for a length, straight or hooked, of at least 150 mm. Similarly all negative moment reinforcement perpendicular to the discontinuous edge must be anchored into the spandrel beam, wall, or column (or within the slab when no spandrel beam or wall is present), to provide development at the face of such support.

For slabs supported on beams along column lines having a value of α greater than 1.0, special corner reinforcement is required at exterior corners as described in Example 13.1.

Location of bar cut-off or bend points must be based on the moment envelopes and the requirements for development length and bar extensions described in Section 5.10. When the Equivalent Frame Method is used, the frame analysis gives the moment envelopes and locations of points of inflection.

14.10.2 SLABS WITHOUT BEAMS

In addition to the above requirements, for slabs without beams, the Code prescribes bend point locations and minimum extensions for reinforcement as shown in Fig. 14.28.

Fig. 14.28 *Minimum length of reinforcement — slabs without beams*

(a) Column strip

Notes:

(1) See clause 12.11.1 for reinforcement extension into supports.

(2) See clause 13.4.9 for bottom reinforcement requirements for structural integrity.

(With the permission of CSA, this material is reproduced from CSA standard CAN3-A23.3-M84, *Design of Concrete Structures for Buildings*, which is copyrighted by the Canadian Standards Association and copies of which may be purchased from the Association, 178 Rexdale Boulevard, Rexdale, Ontario M9W 1R3.)

Fig. 14.28   *Continued*

(b) **Middle strip**

Notes:

(1) See clause 12.11.1 for reinforcement extension into supports.

(2) See clause 13.4.9 for bottom reinforcement requirements for structural integrity.

(With the permission of CSA, this material is reproduced from CSA standard CAN3-A23.3-M84, *Design of Concrete Structures for Buildings*, which is copyrighted by the Canadian Standards Association and copies of which may be purchased from the Association, 178 Rexdale Boulevard, Rexdale, Ontario M9W 1R3.)

However, for slab systems designed to resist moments due to combined lateral and gravity loads, the length of reinforcement must be determined by analysis, but must not be less than those prescribed in Fig. 14.28.

14.10.3 MINIMUM BOTTOM REINFORCEMENT FOR STRUCTURAL INTEGRITY

Shear failure of flat slabs around a column type support may lead to the tearing out of the top reinforcement over the support section from the top surface of the slab (see Fig. 14.29a) resulting in a punch through at the support. This, in turn, may lead to a progressive type of collapse. In order to provide adequate post-punching-failure resistance and thereby prevent progressive collapse of the structure after a local failure, a minimum amount of properly anchored bottom reinforcing bars, made effectively continuous over the span must pass through the columns (or other support reaction areas) in each span direction (see Fig. 14.29b). The Code (13.4.9) prescribes Eq. 14.28 for the minimum area of reinforcement, A_{sb}, so required, and the area A_{sb} must be made up of at least two bars.

$$A_{sb} = \frac{0.5w_s\ell_2\ell_n}{\phi_s f_y} \times 10^{-3} \qquad (14.28)$$

where w_s is the total (dead plus live) specified load per unit area, but not less than twice the slab dead load per unit area. This reinforcement will be able to hang up the slab from the support even after local failure of the slab around the support (Fig. 14.29b). In order to make these bars effectively continuous, they may be lap spliced within the reaction area with a Class A tension lap splice (see Chapter 10), or spliced outside the reaction area with a lap length of $2\ell_d$. At discontinuous edges, this reinforcement must be anchored into the support by bends, hooks or otherwise, so as to develop yield stress at the face of the support.

In spans with beams containing shear reinforcement between supports, the above reinforcement is not required.

14.11 DESIGN EXAMPLE FOR EQUIVALENT FRAME METHOD

—————————— *EXAMPLE 14.2* ——————————

Figure 14.30 shows the floor plan and cross section of a flat slab floor for an office building. In addition to the self weight of the slab, there is a superimposed dead load of 1.70 kN/m², made up of: ceiling = 0.20 kN/m², mechanical and electrical fittings = 0.50 kN/m², and partitions = 1.00 kN/m². The specified live load is 2.4 kN/m². Assume concrete weighs 24 kN/m³, $f_c' = 30$ MPa, and $f_y = 400$ MPa.

Analyze one typical interior bay in the E–W direction, shown shaded in Fig. 14.30a, using the Equivalent Frame Method. Compute the reinforcement in the various design strips and check shear stresses.

Fig. 14.29 *Minimum bottom reinforcement required to pass through column*

(a) Tearing out of top reinforcement
after shear failure

(b) Bottom reinforcement for
hanging up slab

SOLUTION

Drop panels are used over all columns to reduce the amount of negative reinforcement necessary and to decrease the shear stresses around the column head.

1. Slab thickness for deflection control
    The floor slab is provided the same thickness in all panels. At discontinuous edges, edge beams with a stiffness ratio $\alpha \geq 0.8$ will be provided so that no increase in slab thickness is required for the exterior panels. Equation 14.3 and Table 14.1 give the minimum thickness specified by the Code for deflection control. In the equation,

    $$\ell_n = \text{larger clear span} = 8000 - 500 = 7500 \text{ mm}$$

    For slabs with drop panels meeting minimum Code requirements and with edge beams having $\alpha \geq 0.8$, and $f_y = 400$ MPa,

    $$h_{\min} = \frac{\ell_n}{37.5} = \frac{7500}{37.5} = 200 \text{ mm}$$

    A slab of thickness of 200 mm is selected.

2. Drop panels
    Minimum size of drop panels is specified in CSA A23.3-M84 (9.5.3.3.2 and 13.4.7). Minimum extension of drop in each direction from centre of support = ⅙ span = 1.13 m (E–W direction) and 1.33 m (N–S direction). Minimum thickness = $h/4$ = 50 mm. A drop panel extending 1.5 m in each direction from column centre (3 m × 3 m panel), and projecting 100 mm below the slab is selected (Fig. 14.31).

3. Edge beam
    As already indicated, an edge beam with $\alpha \geq 0.8$ is provided along discontinuous edges. Try a beam with rib dimensions of 450 mm depth and 250 mm width, as shown in Fig. 14.31.

    $$I_b = 2.71 \times 10^9 \text{ mm}^4$$

Fig. 14.30 *Example 14.2*

(a) Floor framing plan

(b) Section AA (enlarged)

Fig. 14.31 *Slab dimensions*

(a)

(b)

(c)

(d)

For a beam along longer edge,
$$I_s = 2.43 \times 10^9 \text{ mm}^4 \text{ and,}$$
$$\alpha_a = I_b/I_s = 2.71/2.43 = 1.12 > 0.8$$
For a beam along shorter edge of 6.8 m,
$$I_s = 4250 \times 200^3/12 = 2.83 \times 10^9 \text{ mm}^4$$
$$\alpha = 2.71/2.83 = 0.96 > 0.8$$

4. Effective depths
   Minimum cover for slab by CSA A23.1-M77 is 20 mm. However, minimum cover to give a two hour fire resistance rating (Ref. 14.10) is 25 mm. Reinforcement will be placed in the outer layer for the N–S direction bars in order to resist the larger moments in this direction, and in the inner layer for the E–W direction. Selecting the 25 mm cover and assuming No. 15 bars, the effective depths are (Fig. 14.32):

$$d_{\text{E–W (slab)}} = 200\text{–}25\text{–}16\text{–}16/2 = 151 \text{ mm}$$
$$d_{\text{E–W (drop)}} = 300\text{–}25\text{–}16\text{–}16/2 = 251 \text{ mm}$$
$$d_{\text{N–S (slab)}} = 200\text{–}25\text{–}16/2 \qquad = 167 \text{ mm}$$
$$d_{\text{N–S (drop)}} = 300\text{–}25\text{–}16/2 \qquad = 267 \text{ mm}$$

5. Factored loads
   Dead loads are:

Slab concrete = 0.2 × 24	=	4.8 kN/m²
Additional *DL*	=	1.7 kN/m²
Total *DL* on slab	=	6.5 kN/m²
Concrete in drop panel	=	0.1 × 24 = 2.4 kN/m²

Fig. 14.32 *Effective depths*

Slab section (N–S direction)   Drop panel section (N–S direction)

Live load		$= 2.4 \text{ kN/m}^2$

Factored loads are:

Dead load on slab $= 1.25 \times 6.5$	$= 8.13 \text{ kN/m}^2$
Live load on slab $= 1.5 \times 2.4$	$= \underline{3.60 \text{ kN/m}^2}$
Total on slab	$= 11.7 \text{ kN/m}^2$
Dead load on drop panel $= 1.25 \times 2.4$	$= 3.00 \text{ kN/m}^2$

6. Preliminary checking of shear

   To ascertain the overall adequacy of slab thickness, the shear strength is checked now. However, for shear in two-way action, a more detailed check, including shear due to moment transfer, is deferred until after the moments are determined from the frame analysis. In order to ensure safety when moment transfer effects are included, the punching shear stress due to vertical shear only, calculated below, must be kept well below the allowable value.

   a. One-way (beam) shear

      The critical sections are at distance d from column face, and also from the edge of the drop panel (Fig. 14.33).

      At $d = 251$ mm from the column face, considering (conservatively) a 1 m wide strip outside the drop, shown shaded in Fig. 14.32 (to be more exact, the full width of slab may be considered as a single wide beam, allowing for the larger load and depth of slab within the drop panel):

      $$V_f = 11.7 \times 1 \times 2.899 = 33.9 \text{ kN}$$

      $$V_c = 0.2 \times 1.0 \times 0.6 \times \sqrt{30} \times 1000 \times 151 \times 10^{-3} = 99.3 \text{ kN}$$

      $$V_s \ll V_c \qquad\qquad\qquad\qquad \text{OK}$$

   It is obvious that the shear stress will be within limits at the critical section, a distance d from the edge of the drop. Note that one-way shear is more critical along the longer (8 m) span, than it is in the N–S direction. The corresponding shear stress also will be within limits in this example.

Fig. 14.33 *Critical sections for shear (one-way and two-way action)*

b. Two-way action (punching shear)

Critical sections are at $d/2$ from the face of the columns and from the edge of the drop panel. At an interior column, at distance $d/2 = 251/2$ from the face of the column (Fig. 14.33):

$$V_f = 11.7 \times (6.8 \times 8 - 0.751^2) + 3.00 \times (3 \times 3 - 0.751^2)$$
$$= 655 \text{ kN}$$
$$b_o = 4 \times 0.751 \times 3.004 \text{ m}$$

Average shear stress,

$$\frac{V_f}{A_c} = \frac{655 \times 10^3}{3004 \times 251} = 0.869 \text{ MPa}$$

For slab without shear reinforcement, the shear stress due to the factored shear force calculated above, together with the shear stress due to moment transfer (yet to be computed), is limited to v_r,

where
$$v_r = \frac{V_c}{b_o d} = (1 + 2/\beta_c)\, 0.2\, \lambda\, \phi_c \sqrt{f_c'} \leq 0.4\, \lambda\, \phi_c \sqrt{f_c'}$$

$\beta_c = 1.0$ for square columns and hence,

$$v_r = 0.4 \times 1.0 \times 0.6 \times \sqrt{30} = 1.31 \text{ MPa}$$

Shear stress due to factored shear force = 0.869 MPa $\ll v_r$

At $d/2 = 151/2$ mm from edge of drop panel, (Fig. 14.33):

$$V_f = 11.7 \times (6.8 \times 8 - 3.151^2) = 520 \text{ kN}$$

$$\text{Stress } v = \frac{520 \times 10^3}{(4 \times 3151) \times 151} = 0.273 \text{ MPa} \ll v_r$$

Similar calculations show that punching shear stress is less critical at the exterior column. Furthermore, at the exterior column, part of the shear is transmitted by the edge beam by beam action.

7. Relative stiffness parameters of equivalent frame
   a. Column stiffnesses, K_c
      Referring to Fig. 14.34,

$$\left.\begin{aligned} H &= 3.3 \text{ m} \\ H_c &= 3.3 - (0.2 + 0.1) = 3.0 \text{ m} \end{aligned}\right\} \quad H/H_c = 3.3/3 = 1.1$$

$$\left.\begin{aligned} t_a &= 200/2 + 100 &&= 200 \text{ mm} \\ t_b &= 200/2 &&= 100 \text{ mm} \end{aligned}\right\} \quad \begin{aligned} t_a/t_b &= 200/100 = 2.0 \\ t_b/t_a &= 100/200 = 0.5 \end{aligned}$$

Fig. 14.34 *Column properties*

From Table 14.7 giving stiffness and carry-over factors for columns:

for $H/H_c = 1.1$,
$t_a/t_b = 2$
$\left.\right\}$
$K_{AB} = 5.34$
$C_{AB} = 0.54$

for $H/H_c = 1.1$
$t_b/t_a = 0.5$
$\left.\right\}$
$K_{BA} = 4.85$
$C_{BA} = 0.60$

$K_c = K\, EI_c/H$, $\qquad I_c = 500^4/12 = 5.21 \times 10^9 \text{ mm}^4$
$K_{C,\, below} = 5.34\, E_c \times 5.21 \times 10^9/3300 = 8.43\, E_c \times 10^6$
$K_{C,\, above} = 4.85\, E_c \times 5.21 \times 10^9/3300 = 7.66\, E_c \times 10^6$

b. Torsional member stiffness, K_t

Torsional member at the two outer columns (1) and (6) has the section shown in Fig. 14.31b.

$$C_{ext} = \left(1 - 0.63 \times \frac{150}{250}\right) \times \frac{150^3 \times 250}{3} + \left(1 - 0.63 \times \frac{300}{500}\right)$$
$$\times \frac{300^3 \times 500}{3}$$
$$= 2.97 \times 10^9 \text{ mm}^4$$

For interior columns, the torsional member is the portion of slab having a width equal to the column, that is, a section 500 mm wide and 300 mm deep (Fig. 14.31d).

$$C_{int} = \left(1 - 0.63 \times \frac{300}{500}\right) \times 300^3 \times 500/3 = 2.8 \times 10^9 \text{ mm}^4$$

For all columns, $c_2/\ell_2 = 500/8000 = 0.0625$
Using Eq. 14.10,

$$K_{t,\, ext} = \frac{2 \times 9\, E_c \times 2.97 \times 10^9}{8000(1 - 0.0625)^3} = 8.11\, E_c \times 10^6$$

$$K_{t,\, int} = \frac{2 \times 9\, E_c \times 2.8 \times 10^9}{8000(1 - 0.0625)^3} = 7.65\, E_c \times 10^6$$

c. Equivalent column stiffnesses, K_{ec}, (Eq. 14.27)

$$K_{ec,\, ext} = \frac{(8.43 + 7.66) \times 10^6\, E_c \times 8.11 \times 10^6\, E_c}{(8.43 + 7.66) \times 10^6\, E_c + 8.11 \times 10^6\, E_c} = 5.39 \times 10^6\, E_c$$

$$K_{ec,\, int} = \frac{(8.43 + 7.66) \times 10^6\, E_c \times 7.65 \times 10^6\, E_c}{(8.43 + 7.66) \times 10^6\, E_c + 7.65 \times 10^6\, E_c} = 5.18 \times 10^6\, E_c$$

d. Slab stiffnesses and fixed-end moment coefficients

Although the Table 14.6 is for a drop extension of $\ell_1/6 = 1.13$ m, this Table will be used in this example with drop extensions of 1.5 m. The slab-beam geometry is given in Fig. 14.31a and the loading in Fig. 14.35a.

Fig. 14.35 *Loading details for slab-beam*

$$C_{F1} = C_{N1}, C_{F2} = C_{N2}$$
$$C_{N1}/\ell_1 = 0.5/6.8 = 0.074$$
$$C_{N2}/\ell_2 = 0.5/8.0 = 0.063$$

Interpolating from Table 14.6 for these values,

Stiffness factor, $K_{NF} = 5.93$
Carry-over factor, $C_{NF} = 0.60$
FEM coefficient, $m_{NF1} = 0.0931$, for uniform load

For additional load due to drop panel, Table 14.6 gives moment coefficients for partial loadings over lengths of $0.2\ell_1 (b - a = 0.2)$. Weight/m of span due to drop $= 3.00 \times 3 = 9.00$ kN, acting for lengths of 1.5 m $(= 0.22 \ell_1)$ from column centrelines. For using Table 14.6, this load may be considered as $9.00 \times 0.22/0.20 = 9.90$ kN/m distributed over a length of $0.2 \ell_1$, as indicated in Fig. 14.35b.

For weight of drop at near end, $a = 0$, $b - a = 0.2$, and for $C_{N1}/\ell_1 = 0.074$ and $C_{N2}/\ell_2 = 0.063$, interpolating from Table 14.6, fixed-end moment coefficient is:

$$m_{NF, 2} = 0.0165$$

Similarly, for weight of drop at far end,

$$a = 0.8, b - a = 0.2$$

and

$$m_{NF, 3} = 0.00191$$

Moment of inertia of the slab section beyond drop, I_s is:

$$I_s = 8000 \times 200^3/12 = 5.33 \times 10^9 \text{ mm}^4$$
$$K_s = K_{NF}E_cI_s/\ell_1$$
$$= 5.93 \times E_c \times 5.33 \times 10^9/6800 = 4.65 \times 10^6 E_c$$

8. Equivalent frame analysis
   The analysis of the frame is done by the method of moment distribution in this example.
   The moment distribution factors for slab-beam members are:

at exterior support, $DF_{1 - 2} = DF_{6 - 5} = \dfrac{K_s}{\Sigma K} = \dfrac{4.65 \times 10^6 E_c}{4.65 \times 10^6 E_c + 5.39 \times 10^6 E_c}$

$$= 0.463$$

at interior supports, $DF = \dfrac{4.65 \times 10^6}{2 \times 4.65 \times 10^6 + 5.18 \times 10^6} = 0.321$

Since specified live load $(= 2.4$ kN/m²$)$ is less than three-quarters of dead load $(= \frac{3}{4} \times 6.5 = 4.88$ kN/m²$)$, the factored moments at all critical sections may be obtained from the single loading pattern of full factored live load plus dead load on all spans. The factored loads on the slab-beam member are (Fig. 14.35):

uniformly distributed load $\qquad$ = 11.7 × 8 = 93.6 kN/m

dead load due to weight of
drop assumed redistributed $\left.\right\}$ = 9.90 kN/m
over $0.2\ell_1$ of span

Slab fixed end moments are:

$$FEM_{NF} = \Sigma m_{NF}w\ell_1^2$$
$$= 0.0931 \times (11.7 \times 8) \times 6.8^2 + 0.0165$$
$$\times 9.90 \times 6.8^2 + 0.00191 \times 9.90 \times 6.8^2$$
$$= 411 \text{ kN·m}$$

The moment distribution is set up in Table 14.8, wherein only the slab moments are given.

The freebody diagram for a typical span is shown in Fig. 14.35c. The shear forces at each end of the span are given by:

taking moments about right end,

$$V_L \times 6.8 + M_L + M_R - 9.00 \times 1.5 \times (6.8 - 0.75) - (11.7 \times 8 \times 6.8^2)/2$$
$$- 9.00 \times 1.5^2/2 = 0$$

$$V_L = 331.7 - \frac{(M_L + M_R)}{6.8}$$

Similarly,
$$V_R = 331.7 + \frac{(M_L + M_R)}{6.8}$$

Location of zero shear (and maximum moment) is given by (Fig. 14.35d):

$$V_L = 9.00 \times 1.5 + 11.7 \times 8 \,(x)$$

$$x = \frac{V_L - 13.5}{93.6}$$

Maximum positive moment at distance (x) is given by:

$$+ M_{max} = M_L + V_L \,(x) - 9.00 \times 1.5 \times (x - 0.75) - 11.7 \times 8(x)^2/2$$

Values of V_L, V_R, x, and $+ M_{max}$ are also tabulated in Table 14.8 for all spans.

The critical sections for negative design moment are at the faces of the columns. The moment at face of columns may be computed as (Fig. 14.35e):

$$M_{CF} = M_L + V_L \times 0.25 - 103 \times 0.25^2/2$$

and similarly for the left face of the support at right. These values are also computed and given in Table 14.8.

The slab satisfies all the limitations for the Direct Design Method. In such situations Code clause 13.7.7.4 permits a proportional reduction of the analytical moments such that the sum of positive and average negative moments, $\bar{M}$, does

Table 14.8   Equivalent Frame Analysis (Moment Distribution)

	s1 (L)	s1 (mid)	s1 (R)	s2 (L)	s2 (mid)	s2 (R)	s3 (L)	s3 (mid)	s3 (R)	s4 (L)	s4 (mid)	s4 (R)	s5 (L)	s5 (mid)	s5 (R)
$K_s \times 10^6 E_c =$ / $K_{ec} \times 10^6 E_c =$	5.39	4.65		5.18	4.65		5.18	4.65		5.18	4.65		5.18	4.65	5.39
DF / COF	0.463	0.6	0.321	0.321	0.6	0.321	0.321	0.6	0.321	0.321	0.6	0.321	0.321	0.6	0.463
FEM / Bal	−411 / 190		411	−411		411	−411		411	−411		411	−411		411 / −190
C.O. / Bal.			114 / −37	−37								37	−114 / 37		
C.O. / Bal.	−22 / 10					−22 / 7	7		−7	22 / −7					22 / −10
C.O. / Bal.			6 / −3	4 / −3		1	−4 / 1		4 / −1	−1		−4 / 3	−6 / 3		
C.O.	−2			1		−2	−1		1	2		−1			2
M, Col., ℄ Σ (kN·m)	−234		491	−446		396	−407		407	−396		446	−491		234
V_L, V_R, (kN)	294		369	339		324	332		332	324		339	369		294
x, (m)		3.00			3.48			3.40			3.48			3.00	
$+ M_{max}$ (kN·m)		196			130			145			130			196	
$- M_{CF}$ (kN·m)	−164		402	−364		318	−327		327	−318		364	−402		164
$\bar{M}$, (kN·m)		479			471			472			471			479	
M_o, (kN·m)		471			471			471			471			471	
correction factor, $M_o/\bar{M}$		0.983			1.0			1.0			1.0			0.983	
Design Neg. mom. (kN·m)	161		395	364		318	327		327	318		364	395		161
Design Pos. Mom. (kN·m)		193			130			145			130			193	

not exceed M_o given by Eq. 14.7. Values of $\bar{M}$ for the analytical results are computed for all spans in Table 14.8. The static design moment, M_o, is given by (Fig. 14.35f):

$$M_o = (11.7 \times 8) \times 6.3^2/8 + 9.0 \times 1.25^2/2 = 471 \text{ kN} \cdot \text{m}$$

$\bar{M} > M_o$ in only the end spans, for which the correction factor $= M_o/\bar{M} = 471/479 = 0.983$. The correction factors and corrected moments are also shown in Table 14.8.

9. Transverse distribution of moments
   $\alpha_1 \ell_2/\ell_1 = 0$, since no beam along column line $\ell_2/\ell_1 = 8.0/6.8 = 1.18$
   a. Column strip share of moments
      From Table 14.2, interior negative design moment in column strip = 75 percent. At exterior support,

$$\beta_t = \frac{E_{cb} C}{2 E_{cs} I_s} = \frac{2.97 \times 10^9}{2 \times 5.33 \times 10^9} = 0.279$$

Interpolating from Table 14.2, for $\ell_2/\ell_1 = 1.18$ and $\beta_t = 0.279$, exterior negative factored moment in column strip = 97 percent.
Positive factored moment in column strip, from Table 14.2, is ($\ell_2/\ell_1 = 1.18$, $\alpha_1 \ell_2/\ell_1 = 0$): = 60 percent.

   b. Middle strip share

Interior negative factored moment = 100 − 75 = 25 percent
Exterior negative factored moment = 100 − 97 = 3 percent
Positive factored moment = 100 − 60 = 40 percent

Both column strip and middle strip moments are shown in Fig. 14.36.

10. Column moments
    Total unbalanced slab moment at exterior support, from Table 14.8, is 234 kN·m.
    For column above,
    stiffness $= 7.66 \times 10^6 E_c$, carry-over factor (COF) = 0.60
    For column below,
    stiffness $= 8.43 \times 10^6 E_c$, COF = 0.54

$$\text{Proportion of moment in column above} = \frac{7.66}{7.66 \times 8.43} = 0.476$$

Proportion of moment in column below = 1 − 0.476 − 0.524

Fig. 14.36 *Factored moments in column and middle strips*

Factored Moment	−161	193	−395	−364	130	−318	−327	145	kN·m
Column Strip	97	60	75	75	60	75	75	60	percent
Middle Strip	3	40	25	25	40	25	25	40	percent

At exterior support of slab,

In column above: M at base $= 0.476 \times 234 = 111$ kN·m

M at top $= 0.60 \times 111 = 66.8$ kN·m

In column below: M at top $= 0.524 \times 234 = 123$ kN·m

M at base $= 0.54 \times 123 = 66.2$ kN·m

At columns 2 and 5,

$$M = 491 - 446$$
$$= 45 \text{ kN·m}$$

In column above: M at base $= 0.476 \times 45 = 21.4$ kN·m

M at top $= 0.60 \times 21.4 = 12.9$ kN·m

In column below: M at top $= 0.524 \times 45 = 23.6$ kN·m

M at base $= 0.54 \times 23.6 = 12.7$ kN·m

At columns 3 and 4,

$$M, \text{from Table 14.8} = 407 - 396 = 11 \text{ kN·m}$$

Using Eq. 14.18,

$$\ell_1 = \ell_1' = 6.8 \text{ m}, \ell_n = 6.8 - 0.5 = 6.3 \text{ m}$$
$$w_d = w_d' = 8.13 \text{ kN/m}^2 \text{ (neglecting weight of drop)}$$
$$0.5 \, w_L = 3.60/2 = 1.80 \text{ kN/m}^2$$

$$M = 0.07[(8.13 + 1.80)8 \times 6.3^2 - 8.13 \times 8 \times 6.3^2] = 40.0 \text{ kN·m}$$
$$> 11 \text{ kN·m}$$

In column above: M at base $= 0.476 \times 40 = 19.0$ kN·m
M at top $= 0.60 \times 19.0 = 11.4$ kN·m
In column below: M at top $= 0.524 \times 40 = 21$ kN·m
M at base $= 0.54 \times 21.0 = 11.3$ kN·m

All column moments are shown in Fig. 14.37.

Fig. 14.37 Column moments

all moments in kN.m.

11. Flexural reinforcement

The flexural reinforcement at all critical sections is computed in Table 14.9 for column strip and Table 14.10 for middle strips.

The areas as well as spacings of reinforcement in column strips are both within the specified limits. Note that No. 20 bars will give slightly more convenient spacings at interior supports.

The Code (13.3.3.3) requires sufficient reinforcement to be concentrated over the column to resist the portion of unbalanced moment transferred by flexure. This

Table 14.9 Design of Reinforcement in Column Strip, E–W Direction

		1		2		3		℄	
Moment in col. strip, M_f;	kN·m	156	116	296	273	78	239	245	87
Width of strip or drop, b;	mm	3000	3400	3000	3000	3400	3000	3000	3400
Effective depth, d;	mm	251	151	251	251	151	251	251	151
$K_r = M_f \times 10^6/(bd^2)$;	MPa	0.825	1.50	1.57	—	1.01	—	1.30	1.12
ρ (Table 5.4);		0.0025	0.00466	0.00488	—	0.00307	—	0.00401	0.00342
$A_s = \rho bd$;	mm²	1883	2392	3675	—	1576	—	3020	1756
Number of No. 15 bars;	No.	10	12	19	—	9	—	16	9
Spacing;	mm	300	283	158	378	378	—	188	378

Table 14.10 Design of Reinforcement in Middle Strip, E–W Direction

		1		2		3		℄	
Middle strip moment, M_f;	kN·m	4.8	78	98.8	91	52	79.6	82	58
Width of strip, 8000 − col. strip b;	mm	5000	4600	5000	5000	4600	5000	5000	4600
Effective depth, d;	mm	151	151	151	151	151	151	151	151
$K_r = M_f \times 10^6/(bd^2)$;	MPa		0.744	0.867	—	0.496	—	0.719	0.553
$\rho =$		0.002*	0.00225	0.00263	—	0.002*	—	0.00217	0.002*
$A_s = \rho bd$;	mm²	1510	1560	1984	—	1390	—	1638	1390
Number of No. 15 bars;	No.	8	8	10	—	7	—	9	7
Spacing;	mm	625‡	575‡	500‡	—	657‡	—	556‡	657‡
Number of No. 15 bars to limit spacing < 2b = 400;	No.	12	12	12	12	12	12	12	12

* Minimum reinforcement controls
‡ Maximum allowable spacing controls

may be critical and should be checked only at the exterior support in this example as the unbalanced moments at all other supports are very small. Unbalanced moment at exterior support, from Table 14.8 = 234 kN·m.

Portion transferred by flexure (Eq. 14.4) is 60 percent for a square column,

$$= 0.6 \times 234 = 140 \text{ kN·m}$$

Width of slab considered effective in transferring this moment,

$$= c_2 + 3h = 500 + 3 \times 300 = 1400 \text{ mm}$$
$$K_r = M/bd^2 = 140 \times 10^6/(1400 \times 251^2) = 1.59 \text{ MPa}$$
$$\rho = 0.00495, \qquad A_s = 1739 \text{ mm}^2$$

Thus, at the exterior support 9-No. 15 bars must be distributed over a width of 1400 mm centred over the column. The remaining outer portions of the drop panel = (3000 − 1400)/2 = 800 mm may be provided with two bars each to limit the spacing to < 2h. The column strip negative moment reinforcement thus adds up to 9 + (2 × 2) = 13 bars compared to the 10 required for moment in Table 14.9.

In the middle strips, reinforcement at the exterior support and at midspan in interior spans is controlled by minimum area requirements. However, with No. 15 bars, the actual number of bars at all sections is controlled by the maximum allowable spacing limitation. Six bars are required in each of the two half middle strips.

12. Continuous bottom reinforcement required for structural integrity

$$\text{self weight of slab + specified dead load} = 6.5 \text{ kN/m}^2$$

$$\text{equivalent self weight of drop} = \frac{2.4 \times 3^2}{8 \times 6.3} = 0.4 \text{ kN/m}^2$$

$$\text{specified live load} = 2.4 \text{ kN/m}^2$$

$$\text{total specified load} = 9.3 \text{ kN/m}^2$$

$$w_s = 9.3 \text{ kN/m}^2, \text{ but not less than twice dead load}$$
$$= 2 \times (6.5 + 0.4) = 13.8 \text{ kN/m}^2$$

Along E–W direction, required area A_{sb} is:

$$A_{sb} = \frac{0.5 \, (w_s \times 10^{-3})}{\phi_s f_y} \ell_2 \ell_n$$

$$- \frac{0.5 \times 13.8 \times 10^{-3} \times 8000 \times 6300}{0.85 \times 400} = 1022 \text{ mm}^2$$

Five No. 15 bottom bars must be made effectively continuous and pass through the column within its width of 500 mm.

13. Checking shear stresses

The shear stresses in two-way action around the column-slab connection, including the shear due to transfer of moments must now be checked.

a. Interior column

The maximum column reaction and unbalanced slab moment is at columns 2 and 5. From Table 14.8, the vertical column reaction is obtained as the sum of the shears on either side, and is:

$$R = 369 + 339 = 708 \text{ kN}$$

For shear in two-way action, the average of the effective depths in E–W and N–S directions may be used for d.

$$d = (251 + 267)/2 = 259 \text{ mm}$$

For the critical section at $d/2$ from the column face (Fig. 14.38), the shear is

$$V_f = 708 - (11.7 + 3.00) \times 0.759^2 = 700 \text{ kN}$$

and perimeter $b_o = 4 \times 759 = 3036$ mm

Moment transferred by shear is 40 percent of unbalanced moment, and equal to:

$$M_{fv} = 0.4 \times (491 - 446) = 18 \text{ kN} \cdot \text{m}$$

For an interior column, the term J_c/c in Eq. 14.23 is given by (Ref. 14.9 and Fig. 14.19):

$$c_{AB} = (c_1 + d)/2 = 759/2 = 380 \text{ mm}$$

$$J_c = \frac{(c_1 + d)d^3}{6} + \frac{(c_1 + d)^3 d}{6} + \frac{(c_1 + d)^2(c_2 + d)d}{2}$$

$$= \frac{759 \times 259^3}{6} + \frac{759^3 \times 259}{6} + \frac{759^2 \times 759 \times 259}{2}$$

$$= 77.7 \times 10^9 \text{ mm}^4$$

$$\frac{J_c}{c_{AB}} = \frac{77.7 \times 10^9}{380} = 204 \times 10^6 \text{ mm}^3$$

Maximum shear stress is (Eq. 14.23):

$$v = \frac{700 \times 10^3}{3036 \times 259} + \frac{18 \times 10^6}{2.04 \times 10^8}$$

$$= 0.978 \text{ MPa} < v_r = 1.31 \text{ MPa} \qquad \textbf{OK}$$

b. Exterior column

Because of the edge beam, part of the shear will be transmitted by this beam to the column by beam action. However, to simplify the calculation, the shear

Fig. 14.38 *Critical sections for two-way shear*

stress is conservatively computed neglecting the contribution of the edge beam.

Referring to Table 14.8 and Fig. 14.38:
Column reaction = 294 kN
At critical section $d/2$ from column face,

$$V_f = 294 - (11.7 + 3.00) \times 0.759 \times 0.630$$
$$= 287 \text{ kN}$$
$$M_{fv} = 0.4 \times 234 = 93.6 \text{ kN·m}$$
$$b_o = 2 \times 630 + 759 = 2019 \text{ mm}$$

From Fig. 14.19,

$$C_{AB} = \frac{(c_1 + d/2)^2}{2c_1 + c_2 + 2d} = \frac{(500 + 259/2)^2}{2 \times 500 + 500 + 2 \times 259} = 196 \text{ mm}$$

$$J_c = \frac{(c_1 + d/2)d^3 + (c_1 + d/2)^3 d}{6} + (c_2 + d)d \, (c_{AB})^2$$

$$+ \; 2(c_1 + d/2)d[(c_1 + d/2)/2 - c_{AB}]^2$$

$$= \frac{630 \times 259^3 + 630^3 \times 259}{6} + 759 \times 259 \times 196^2 + 2$$

$$\times \; 630 \times 259 \, (630/2 - 196)^2$$

$$= 24.8 \times 10^9 \text{ mm}^4$$

$$\frac{J_c}{c_{AB}} = \frac{24.8 \times 10^9}{196} = 126 \times 10^6 \text{ mm}^3 \qquad\qquad \text{OK}$$

$$v = \frac{287 \times 10^3}{2019 \times 259} + \frac{93.6 \times 10^6}{1.26 \times 10^8}$$

$$= 1.29 \text{ MPa} < v_r = 1.31 \text{ MPa}$$

The two-way shear must also be checked at a critical section, a distance $d_s/2$ (half depth of slab) from the edge of the drop panels. The procedure is the same as shown above and the stresses at these sections are well within the limits in this example.

PROBLEMS

1. Redesign the interior equivalent frame in Example 14.2, assuming that beams 300 mm wide and 500 mm deep (overall) are provided along the column lines, and no drop panels are used.
2. For the floor designed in Example 14.2, design the exterior equivalent frame in the N–S direction, using the Equivalent Frame Method.
3. For the floor system in Example 14.2, using the Direct Design Method, determine the design moments in the various design strips of a N–S interior equivalent frame.

REFERENCES

14.1   Sozen, M.A. and Siess, C.P., *Investigation of Multiple-Panel Reinforced Concrete Floor Slabs: Design Methods — Their Evolution and Comparison*, ACI Journal, Proc. V. 60, No. 8, Aug. 1963, pp. 999–1028.

14.2   Corley, W.G., and Jirsa, J.O., *Equivalent Frame Analysis for Slab Design*, ACI Journal, Proc. V. 67, No. 11, Nov. 1970, pp. 875–884.

14.3 Gamble, W.L., *Moments in Beam Supported Slabs*, ACI Journal, Proc. V. 69, No. 3, Mar. 1972, pp. 149–157.

14.4 Hanson, N.W., and Hanson, J.M., *Shear and Moment Transfer Between Concrete Slabs and Columns*, Journal, PCA Research and Development Laboratories, V. 10, No. 1, Jan. 1968, pp. 2–16.

14.5 ACI-ASCE Committee 326, *Shear and Diagonal Tension*, ACI Journal, Proc. V. 59, No. 3, Mar. 1962, pp. 352–396.

14.6 ACI-ASCE Committee 426, *The Shear Strength of Reinforced Concrete Members*, Proc. ASCE, V. 100, No. ST 8, Aug. 1974, pp. 1543–1591.

14.7 *Commentary on Building Code Requirements for Reinforced Concrete* (ACI 318–83), American Concrete Institute, Detroit, Michigan, 1983.

14.8 *Building Code Requirement for Reinforced Concrete*, (ACI 318–83), American Concrete Institute, Detroit, Michigan, 1983, 111 pp.

14.9 *Concrete Design Handbook*, Canadian Portland Cement Association, Ottawa, Canada, 1985.

14.10 *Supplement to the National Building Code of Canada 1985* (Chapter 2, Fire Performance Rating), National Research Council, Ottawa, 1985.

CHAPTER 15

DESIGN OF COMPRESSION MEMBERS — SHORT COLUMNS

15.1 COMPRESSION MEMBERS

15.1.1 TYPES OF COLUMNS

In a building, columns are the vertical members carrying primarily compressive forces. CSA A23.3-M84 classifies a compression member with height in excess of three times its least lateral dimension as a column and refers to less slender members as pedestals. Several other structural elements such as inclined members, rigid frame members, truss members, as well as beams may also be subjected to substantial axial compressive forces. The design considerations explained in this chapter with reference to *columns* are applicable to compression members in general.

There are few compression members which are subjected to purely axial compressive forces. In general, compressive forces in a member exist in combination with bending. The bending moments may result from applied end moments, lateral loading on the member and/or eccentricity of applied compressive forces.

The three basic types of reinforced concrete columns are shown in Fig. 15.1. Tied columns are reinforced with main longitudinal bars enclosed within lateral *ties*. The spiral column has its longitudinal bars enclosed within closely spaced and continuously wound spiral reinforcement. A composite column is one which is reinforced longitudinally with structural steel shapes, pipe, or tubing, with or without longitudinal bars. This chapter is primarily concerned with the first two types as these are most common in reinforced concrete construction. The most common cross-sectional shapes for tied columns are the square and rectangle, although T-, L-, and other shapes are occasionally used, particularly for edge and corner columns in buildings. Spiral columns are usually circular in cross-section, the alternative shapes being square and octagonal sections.

15.1.2 GENERAL REQUIREMENTS ON REINFORCEMENT

(a) Longitudinal Bars

CSA A23.3-M84(10.9) limits the longitudinal reinforcement in reinforced concrete compression members to between 1 and 8 percent of the gross area of a section.

Fig. 15.1   *Types of reinforced concrete columns*

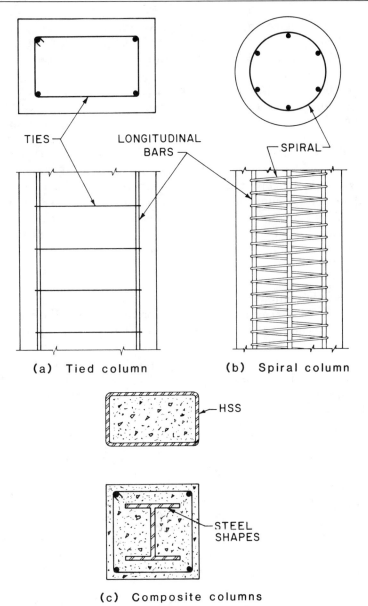

TIES   LONGITUDINAL
                  BARS   SPIRAL

(a) Tied column   (b) Spiral column

HSS

STEEL
SHAPES

(c) Composite columns

The lower limit of 1 percent is specified to ensure the use of a reinforced concrete (rather than plain concrete) column and to provide some flexural resistance even for members where computations show no applied moment. Furthermore, shrinkage and creep effects tend to transfer the load from the concrete to the longitudinal reinforcement. The minimum reinforcement ensures that these effects do not lead to the eventual yielding of the reinforcement under sustained service loads. Occasionally, considerations other than strength (architectural, for instance) may dictate the overall column size, resulting in an oversize column. In such a case, the minimum reinforcement percentage applied to the gross section may lead to excessive reinforcement. The Code therefore permits the factored resistance and reinforcement area to be based on a reduced effective gross area not less than one-half the actual total area.

The upper limit of 8 percent represents a practical maximum for longitudinal reinforcement from considerations of both economy and requirements for proper placing of concrete. The maximum steel ratio in the regions containing lap-splices must also not exceed 8 percent. Hence, where column bars are required to be lap-spliced, the longitudinal reinforcement ratio should usually be limited to a maximum of 4 percent. Use of reinforcement in excess of 4 percent in regions outside lap splices may also result in difficulties in placing and compacting concrete. For spirally reinforced columns, a minimum of six bars and for columns with rectangular or circular ties a minimum of four bars (one at each corner) are required. For other shapes (L, T, etc.) one bar must be provided at each corner. Additional detailing requirements for longitudinal reinforcement in columns are given in CSA A23.3-M84 (7.5).

(b) Lateral Reinforcement

All longitudinal reinforcement in a compression member must be enclosed within lateral reinforcement. The latter may consist of closed ties or spirals. Lateral reinforcement helps to: (1) hold the longitudinal bars in position during construction; (2) provide lateral restraint to the individual longitudinal bars in compression and prevent their premature buckling; (3) at close spacings, provide lateral confinement to the concrete; and (4) carry shear and torsion when reinforcing for such stresses are required.

(c) Ties

The Code (cl. 7.6.5) prescribes the requirements for ties. Reinforcement for ties must have a diameter of at least 30 percent of that of the largest longitudinal bar for bars up to No. 30, and must be at least No. 10 in size for larger size longitudinal bars and for bundled bars. The spacing of ties must not exceed the least of: (1) 16 times the diameter of the smallest longitudinal bar; (2) 48 times the tie diameter; (3) the least lateral dimension of the compression member; and (4) 300 mm when bundled bars are used. Additional ties are usually provided at each end of lap spliced bars, above and below end bearing splices, and immediately below sloping regions of offset bent longitudinal bars.

The arrangement of ties must be such as to enclose and laterally support every corner bar and every alternate longitudinal bar by the corner of a tie having an included angle of not more than 135°. Furthermore, no bar must be farther than 150 mm clear on either side from such a laterally supported bar. Several arrangements of ties to satisfy this requirement are presented in Refs. 15.1 and 15.2.

(d) Spirals

Requirements for spirals are presented in CSA A23.3-M84 (7.6.4). The minimum amount of spiral reinforcement prescribed and the basis for this will be described in Section 15.2 (Eq. 15.4). The diameter of the reinforcing steel used for the spiral must not be less than 6 mm. The pitch of the spiral must not exceed one-sixth of the diameter of the core (measured out-to-out of spirals), and the clear spacing between successive turns must be between 25 mm and 75 mm, inclusive.

15.1.3 SLENDERNESS EFFECTS

(a) Limiting Slenderness Ratio

Columns which are long and slender may fail due to instability at a load lower than the strength of a short column of identical cross-sectional dimensions. Furthermore, slender columns have larger lateral deflections under the applied primary moments, which create greater secondary moments due to the axial compression interacting with lateral deflection, the so-called $P - \Delta$ effect. These effects are discussed in more detail in Section 16.1. The ratio $k\ell_u/r$ is taken as the measure of the slenderness, where ℓ_u is the unsupported length of the member, k is the effective length factor, and r is the radius of gyration of the cross section. Relatively short and stocky columns with $k\ell_u/r$ ratios such that the reduction of strength due to slenderness effects is small enough to be neglected will be termed *short* columns. The strength of such a column is dependent on the cross-sectional dimensions and material strength. It is reported, based on surveys of existing practice, that over 90 percent of the columns in braced frames and over 40 percent of the columns in unbraced frames fall into this category (Ref. 15.3).

The Code (cl. 10.11.4) prescribes that for *compression members braced against lateral deflection*, the effects of slenderness may be neglected if:

$$k\ell_u/r < 34 - 12M_1/M_2 \qquad (15.1)$$

where M_1 and M_2 are the numerically smaller and larger factored end moments, respectively, obtained from a conventional (first order) frame analysis, and M_1/M_2 is positive if the member is bent in single curvature, and negative if bent in double curvature. The radius of gyration, r, may be taken as 0.3 times the overall depth of a section, h, for rectangular sections, and 0.25 times the diameter for circular sections (Code (10.11.13)). The slenderness limit in Eq. 15.1 corresponds to lengths for which the slender column strength is at least 95 percent of the cross-sectional strength (Refs. 15.3, 15.4).

For compression members *not* braced against lateral deflection, the Code requires the effects of slenderness, both in terms of member stability and lateral drift, to be taken into account irrespective of the value of the ratio $k\ell_u/r$ of the member (see Chapter 16). Similarly, for all compression members with $k\ell_u/r$ greater than 100, the design forces and moments must be determined from a "second order" analysis including the effects of the changing geometry caused by deflections, of the influence of axial loads and variable moment of inertia on member stiffnesses, and of the duration of loads.

(b) Effective Length Factor, k

The effective length, $k\ell$, of a column of length ℓ is the distance between the inflection points in the deflected shape of the column. Columns in actual structures are usually

Fig. 15.2 *Effective length of columns*

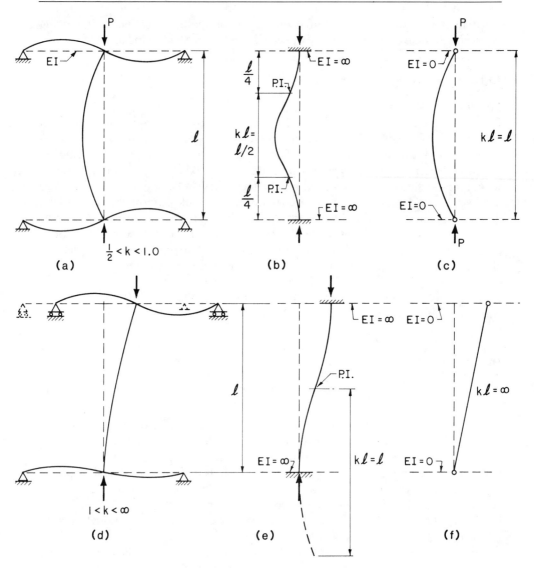

attached to beams or other framing members, which introduce some rotational restraint at the column ends. When relative transverse displacement between the upper and lower ends of columns is prevented, the frame is said to be braced against sidesway. A column in a braced frame is shown schematically in Fig. 15.2a. If the framing beam members have infinite flexural stiffness, the column ends cannot rotate and the column is effectively fixed at both ends, as shown in Fig. 15.2b. The effective length of the column in this case is $k\ell = \ell/2$, or $k = 0.5$. If, on the other hand, the beams have zero flexural stiffness, the column behaves as pinned at both ends

(Fig. 15.2c) for which $k\ell = \ell$. Thus columns in braced frames will have $0.5 < k < 1.0$, and it is conservative to take $k = 1.0$.

Columns in an unbraced frame can sway sideways, as shown in Fig. 15.2d. In this case, beams of infinite flexural stiffness will prevent end rotation and the deflected column shape will be as shown in Fig. 15.2e for which $k = 1.0$. At the other theoretical limit of zero flexural stiffness for the beams, the column reduces to one that is pinned at both ends and permitted to sway. In this case the column is unstable and $k = \infty$ (that is, the critical load reduces to zero). Hence braced frames have $1.0 < k < \infty$. Approximate methods for computing k factors for columns in frames are given in Section 16.2.

15.2 DESIGN OF SHORT COLUMNS

Design of short columns is considered in the rest of this chapter. Slenderness effects and design of slender columns are described in Chapter 16.

15.3 STRENGTH AND BEHAVIOUR UNDER AXIAL COMPRESSION

As already mentioned, compression members are usually subjected also to flexure. It may be observed that CSA A23.3-M84 also treats flexure and axial loads in the same section. Thus the assumptions regarding stress and strain distributions across the cross section of an axially loaded member are the same as those used in flexure and given in Section 4.8 (Code (10.2)). The theoretical factored resistance, P_{ro}, of the section under axial loading is a useful property and gives one of the limiting points in the strength behaviour of the column. Axial loading may be defined as loading that produces a uniform stress distribution across the cross section. For symmetrically reinforced sections (Fig. 15.3a), this would correspond to a load applied at the geometric centroid of the section. With the assumptions of stress and strain distributions given in Section 4.8, the ultimate strength of the section is reached when the axial strain reaches 0.003. The average factored concrete stress now may be assumed as $0.85\phi_c f'_c$ and the steel at this strain would have yielded. The strain and stress distributions are as shown in Fig. 15.3a, and the corresponding factored axial load resistance, P_{ro}, is given by:

$$P_{ro} = 0.85\phi_c f'_c(A_g - A_{st}) + \phi_s f_y A_{st} \qquad (15.2)$$

where P_{ro} = factored axial load resistance with zero moment
     A_g = gross area of column cross section
 and A_{st} = total area of longitudinal reinforcement

When the section is not symmetrically reinforced, as with the section shown in Fig. 15.3b, axial load conditions exist when the load P_{ro}, once again given by Eq. 15.2, coincides with the resultant of the internal stresses shown. The point through

Fig. 15.3   *Axial loading*

$$C_s = A_{st} \phi_s f_y / 2$$

$$C_c = 0.85 \phi_c f_c' (A_g - A_{st})$$

**(a)   Symmetrically
reinforced**

$$C_{s1} = A_{s1} \phi_s f_y, \quad C_{s2} = A_{s2} \phi_s f_y$$

$$C_c = 0.85 \phi_c f_c' (A_g - A_{s1} - A_{s2})$$

**(b)   Unsymmetrically
reinforced**

which P_{ro} acts is termed the *plastic centroid*, whose location can be obtained from Eq. 15.3. In eccentrically loaded members, the eccentricity, e, is measured from the plastic centroid so that $e = 0$ corresponds to an axially loaded member. For the symmetrically reinforced section in Fig. 15.3a the plastic centroid coincides with the geometric centroid.

$$\bar{d} = \frac{C_c x_c + C_{s1} d + C_{s2} d'}{C_c + C_{s1} + C_{s2}}$$

For composite members, the value of P_{ro} can similarly be obtained as:

$$P_{ro} = 0.85\phi_c f_c'(A_g - A_{st} - A_t) + \phi_s f_y A_{st} + \phi_a f_y A_t \qquad (15.3)$$

where A_t is the area of structural steel shape, pipe, or tubing in the composite section, and ϕ_a is the resistance factor for structural steel, $= 0.90$.

The Code (cl. 10.3.5) limits the maximum factored axial load resistance, $P_{r(max)}$, of compression members to $0.85\,P_{ro}$ for members with spiral reinforcement and to $0.80\,P_{ro}$ for members with tie reinforcement. This limitation is to allow for unforeseen factors and accidental eccentricities that may result from misalignment and other construction errors, and to recognize that concrete strength may be less than f_c' under sustained high loads. With this limitation, the maximum factored axial load resistances are given by Eqs. 15.4 and 15.5.

For columns with spiral reinforcement,

$$P_{r(max)} = 0.85\,[0.85\,\phi_c f_c'(A_g - A_{st}) + \phi_s f_y A_{st}] \qquad (15.4)$$

For columns with tie reinforcement,

$$P_{r(max)} = 0.80\,[0.85\,\phi_c f_c'(A_g - A_{st}) + \phi_s f_y A_{st}] \qquad (15.5)$$

If no allowance is made for the material resistance factors (i.e., $\phi_c = \phi_s = 1.0$), the *nominal axial ultimate strength* with zero moment, P_{no}, of a column is given by:

$$P_{no} = 0.85\,f_c'(A_g - A_{st}) + f_y A_{st} \qquad (15.6)$$

A tied column fails at a nominal load, P_{no}, given by Eq. 15.6. The failure is by crushing and shearing of the concrete and by buckling of the longitudinal bars, and is sudden and brittle. A typical load versus axial shortening diagram has the shape shown in Fig. 15.4.

A spirally reinforced column behaves in a manner identical to a tied column of the same section, up to a nominal load, P_{no}, given by Eq. 15.6. At this load, the outer shell of concrete covering the spiral spalls off; however, the concrete in the core, laterally confined by the closely spaced spirals, continues to carry load. Failure ultimately takes place when the spiral reinforcement yields. The load capacity beyond spalling depends on the amount of spiral reinforcement, as indicated by the load–axial shortening curves shown in Fig. 15.4. Based on extensive experimental results, it has been found (Ref. 15.5) that the strength contributed by the spiral is approximately given by $2f_{sy}A_{sp}$, where f_{sy} is the yield strength of spiral reinforcement, and A_{sp} is the volume of spiral reinforcement per unit length of column. However, this additional strength is not realized until the column has been subjected to a load and deformation sufficient to cause the spalling of the concrete shell. If the amount of spiral reinforcement is such that the load capacity contributed by it compensates the loss of strength due to spalling of the concrete shell, there will be no sudden deterioration of strength and the column will continue to carry the load at spalling,

Fig. 15.4   *Behaviour of axially loaded tied and spiral columns*

given by Eq. 15.6. To achieve such a behaviour, the amount of spiral reinforcement needed can be obtained by equating the strength of the concrete in the shell to the strength contributed by spiral, as follows:

$$0.85 \, f_c'(A_g - A_c) = 2 \, f_{sy} A_{sp}$$

or

$$\rho_s = \frac{A_{sp}}{A_c} = 0.425 \left(\frac{A_g}{A_c} - 1 \right) f_c'/f_{sy} \tag{15.7}$$

where A_c = total area of concrete core, measured out-to-out of the spirals
ρ_s = ratio of volume of spiral reinforcement to volume of the core, per unit length of the column

CSA A23.3-M84(10.9.3) requires a minimum spiral reinforcement of

$$\rho_s \geqslant 0.45 \left(\frac{A_g}{A_c} - 1 \right) \frac{f_c'}{f_{sy}} \tag{15.8}$$

With an amount of spiral reinforcment as given by Eq. 15.8, the strength of a spiral column will be as shown by the continuous line in Fig. 15.4. The column strength is maintained even after spalling of the concrete cover, and failure is gradual, preceded by considerable yielding. Thus the spiral column has substantial ductility and toughness. Spirals in excess of that given by Eq. 15.8 will result in an increase in column strength after spalling; however, such increases are not of much practical value in normal situations because of the accompanying large deformations and extensive cracking and spalling.

Thus with the minimum amount of spiral reinforcement provided, the *factored* axial load resistance, P_{ro}, of both tied and spirally reinforced columns under axial loading is given by Eq. 15.2. Their differences in behaviour are with respect to their ductility, toughness and mode of failure. The ductile failure mode of the spiral column accounts for its larger $P_{r(max)} = 0.85\,P_{ro}\,\phi$ factor ($\phi = 0.75$) compared to tied columns ($P_{r(max)} = 0.80\,P_{ro}$).

15.4 COMPRESSION WITH BENDING

15.4.1 GENERAL BEHAVIOUR

The behaviour of members under combined axial compression and bending or, alternatively, under eccentric compressive force is discussed in this section. For the sake of simplicity, a rectangular section, with reinforcement placed near opposite faces, is used initially. Nonetheless, the basic principles involved are quite general and applicable to other shapes and distributions of reinforcement. The assumptions are the same as those described in Section 4.8 with reference to flexure. The two general principles used are the compatability of strains and equilibrium of forces. The compressive stress distribution in concrete will be assumed to be that represented by the equivalent rectangular stress block described in Section 4.8. Combined axial load, P, and bending moment, M, can be represented by a statically equivalent eccentric load, P, acting at an eccentricity $e = M/P$ (Fig. 15.5). The distribution of strain across the section at failure load, for various values of e, is shown in Fig. 15.6a. As seen in the previous section, axial compression alone ($e = 0$) causes a

Fig. 15.5 *Compression plus bending (eccentric loading)*

Fig. 15.6 *Strength under combined compression and bending*

(a) Variation of strain

(b) P_r - M_r interaction diagram

uniform strain in the cross section (line 1 in Fig. 15.6a). With eccentric loading, the strain variation is non-uniform. At low eccentricities the entire cross section may still be in compression. As eccentricity increases, part of the section may be subjected to tensile strains and stresses. In the limit, when e becomes infinity (that is, $P = 0$), the section is under pure flexure. The strain and stress distributions corresponding to this have already been discussed in Chapters 4 and 5.

In between the limiting cases of $e = 0$ and $e = \infty$, there exists a certain eccentricity, $e = e_b$, for which the strain distribution is such that as the failure load is reached, the yielding of the steel on the tension side and the attainment of a compressive strain $\epsilon_c = 0.003$ at the extreme fibre in the concrete occur simultaneously. This is the balanced strain condition. The factored axial load resistance corresponding to this condition is the balanced axial load resistance, P_{rb}. For eccentricity $e > e_b$, failure will be initiated by yielding of the steel on the tension side (tension failure). For $e < e_b$, failure occurs by crushing of the concrete (compression failure). The combinations of factored axial load resistance and factored moment resistance corresponding to these different values of e can be represented by a curve such as shown in Fig. 15.6b. For any given e, the value of the corresponding factored resistances, P_r and $M_r = P_r e$, may be computed using the equilibrium and compatability conditions as illustrated below for three different cases.

15.4.2 BALANCED STRAIN CONDITIONS

The cross-sectional dimensions, the strain distribution and the factored internal forces for this case are shown in Fig. 15.7. From the known maximum strains in

Fig. 15.7 *Balanced strain condition*

concrete and steel, the depth to the neutral axis, x_b, and the depth of the equivalent rectangular concrete stress block, a_b, are directly calculated as:

$$x_b = \frac{0.003}{f_y/E_s + 0.003} d = \frac{600}{f_y + 600} d \qquad (15.9)$$

$$a = \beta_1 x_b$$

The strain and factored stress in the reinforcement on the compression side can then be computed as:

$$\epsilon'_s = \frac{0.003\,(x_b - d')}{x_b}$$

and

$$\phi_s f'_s = \phi_s E_s \epsilon'_s \leq \phi_s f_y$$

Knowing the stress distribution, the internal forces in concrete, the compression reinforcement and tension reinforcement are given, respectively, by:

$$\begin{aligned}
C_c &= 0.85\,\phi_c f'_c b a \\
C'_s &= A'_s\,(\phi_s f'_s - 0.85\phi_c f'_c) \\
&\simeq A'_s\,\phi_s f'_s \\
T &= A_s\,\phi_s f_y
\end{aligned} \qquad (15.10)$$

Since the external load P_{rb} and internal forces are in equilibrium, summing up the forces vertically,

$$P_{rb} = C_c + C'_s - T \qquad (15.11)$$

and equating moments about the plastic centroid

$$M_{rb} = P_{rb}\,e_b = C_c(\bar{d} - \frac{a}{2}) + C'_s(\bar{d} - d') + T(d - \bar{d}) \qquad (15.12)$$

Equations 15.11 and 15.12 give the values of P_{rb} and M_{rb} (or P_{rb} and e_b) corresponding to a balanced strain condition. A numerical example is given below.

———————————— **EXAMPLE 15.1** ————————————

For the column having the cross-sectional dimensions shown in Fig. 15.8 and loaded on the x–axis, compute the factored axial load resistance and factored moment resistance under balanced strain conditions. Take $f'_c = 30$ MPa and $f_y = 400$ MPa.

SOLUTION

Strain distribution for balanced conditions and corresponding internal forces are shown in Fig. 15.8. From these diagrams,

$$\epsilon_y = 400/200\ 000 = 0.002$$

$$\frac{x_b}{525} = \frac{0.003}{0.002 + 0.003}, \qquad x_b = 315 \text{ mm}$$

$$\alpha = \beta_1 x_b = 0.85 \times 315 = 268 \text{ mm}$$

$$\epsilon_s' = \frac{0.003\ (315 - 75)}{315} = 0.0023 > \epsilon_y, \qquad f_s' = f_y = 400 \text{ MPa}$$

$$C_c = 0.85 \times 0.6 \times 30 \times 350 \times 268 \times 10^{-3} = 1435 \text{ kN}$$

$$C_s' = 3000 \times (0.85 \times 400 - 0.85 \times 0.6 \times 30) \times 10^{-3} = 974 \text{ kN}$$

$$T = 3000 \times 0.85 \times 400 \times 10^{-3} = 1020 \text{ kN}$$

The load, P_{rb}, and eccentricity, e_b, are such that the internal forces are in equilibrium with P_{rb}.

$$P_{rb} = 1435 + 974 - 1020 = 1389 \text{ kN}$$

Fig. 15.8 *Example 15.1*

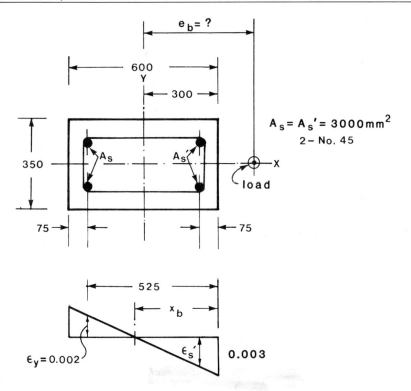

Taking moments about centroidal (plastic) axis

$$P_{rb}\, e_b = [1435 \times (300 - 268/2) + 974 \times (300 - 75)$$
$$+ 1020 \times (300 - 75)] \times 10^{-3}$$
$$= 687 \text{ kN} \cdot \text{m}$$
$$e_b = 687/1389 = 0.494 \text{ m}$$

15.4.3 COMPRESSION FAILURE

When $e < e_b$, the concrete strain reaches 0.003 before the steel on the tension side reaches yield strain. Depending on the value of e, the neutral axis depth, x, may even exceed h. If the neutral axis falls within the section, the strain distribution and the forces are as shown in Fig. 15.9a. Assume that the depth of neutral axis is x. All stresses and internal forces can be expressed in terms of x, and then the condition of equilibrium used to evaluate x and hence the load. The relevant equations are worked out below.

In terms of x,

$$\epsilon_s' = \frac{0.003\,(x - d')}{x}$$

$$f_s' = \epsilon_s'\, E_s \leq f_y$$

Similarly for the steel in tension,

$$\epsilon_s = \frac{0.003}{x}\,(d - x)$$

and

$$f_s = E_s\, \epsilon_s = E_s \left(\frac{0.003\,(d - x)}{x} \right)$$

The internal factored forces are:

$$\left.\begin{aligned}
C_c &= 0.85\ \phi_c f_c'\ b \beta_1\, x \\
C_s' &= A_s'(\phi_s f_s' - 0.85\ \phi_c f_c') \\
T &= A_s \phi_s f_s = A_s \phi_s E_s\, \frac{0.003\,(d - x)}{x}
\end{aligned}\right\} \qquad (15.13)$$

C_c, C_s' and T can all be expressed in terms of x. These forces together with P_r form a system in equilibrium. Taking moments about P_r,

Fig. 15.9 *Compression failure*

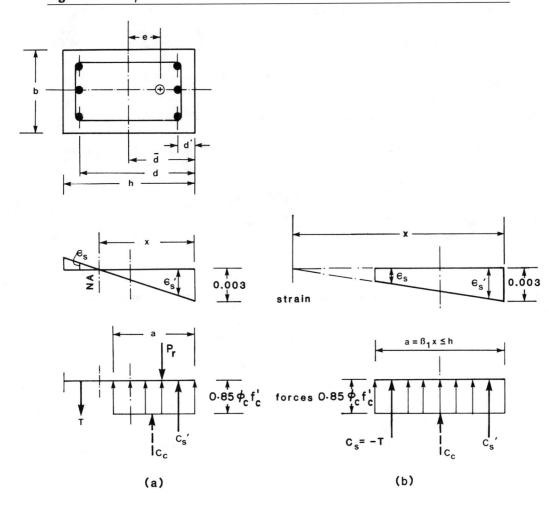

$$C_c (e - \bar{d} + \beta_1 x/2) - C_s'(\bar{d} - e - d') - T (e - d + d) = 0 \quad (15.14)$$

(Note that, in setting up the equilibrium equation, moments may be taken about any point. However, the point of application of P_r is a convenient point, as in this case P_r is eliminated from the equation).

Substituting in this equation, expressions for C_c, C_s' and T, a cubic equation in x will be obtained, which can be solved for x. Once x is determined, all internal forces can be computed using Eq. 15.13. The factored axial load resistance is given by:

$$P_r = C_c + C_s' - T$$

In substituting the expression for C_s' in Eq. 15.14, it is not known whether the compression steel has yielded or not. Hence, a suitable assumption may be made, such as $f'_s = f_y$, and the assumption backchecked after determining x.

Note that if $x > d$, the steel stress, f_s, is also compressive and hence the force T is compressive (in this case, to be more exact, $T = A_s (\phi_s f'_s - 0.85 \; \phi_c f'_c)$). If the value of x is such that the depth of stress block, $a = \beta_1 x$, exceeds h, then a in Eq. 15.13 has to be limited to h (see Fig. 15.9b).

Computation of the failure load for $e < e_b$ is illustrated with Example 15.2.

EXAMPLE 15.2

For the column analyzed in Example 15.1, compute the factored axial load resistance at an eccentricity $e = 90$ mm.

SOLUTION

Example 15.1 showed that the steel near the compression face yielded even at balanced conditions. Since, for $e < e_b$ the strain ϵ'_s will be even higher, it can be taken that

$$f'_s = f_y$$

If the depth to neutral axis is x, referring to the assumed strain distribution in Fig. 15.9a,

$$f_s = (200\,000) \frac{0.003}{x} (525 - x) = \frac{600 \, (525 - x)}{x}$$

For $f_c = 30$ MPa, $\beta_1 = 0.85$
The factored forces are:

$$C_c = (0.85)(0.6)(30)(350)(0.85x) \; 10^{-3} = 4.55 \; x \text{ kN}$$
$$C'_s = 3000 \, [0.85(400) - 0.85(0.6)(30)] \; 10^{-3} = 974 \text{ kN}$$
$$T = 3000 \, (0.85) \left(\frac{600(525 - x)}{x} \right) 10^{-3} = 1530 \left(\frac{525 - x}{x} \right) \text{ kN}$$

Taking moments about the resultant along P_r,

$$(4.55 \; x)(210 - 0.85x/2) + 974 \, (210 - 75) + 1530 \, (525 - x) \, (525 - 210)/x = 0$$

Simplifying,

$$x^3 - 494 \, x^2 + (181)10^3 x - (131)10^6 = 0$$

Solving,

$$x = 575 \text{ mm}$$

The actual strain and force distribution is as shown in Fig. 15.10. Note that $x > d$, and hence the steel, A_s, will be in compression, and T will have a negative value (that is, compressive). Substituting for x, the factored forces are:

Fig. 15.10 *Example 15.2*

$$C_c = 4.55x = 2616 \text{ kN}$$
$$T = 1530 \, (525 - 575)/575 = -133 \text{ kN (Comp.)}$$
$$P_r = C_s' + C_c - T = 974 + 2616 + 133 = 3723 \text{ kN}$$

The moment accompanying P_r is,

$$M_r = P_r \, e = 3723 \, (0.09) = 335 \text{ kN} \cdot \text{m}$$

15.4.4 TENSION FAILURE

For eccentricities greater than e_b, the steel on the tension side will yield first. This will be followed by a shifting of the neutral axis towards the compression side, an increase in the compressive strain in the extreme fibre, and ultimately failure of the concrete by crushing. At failure, the maximum compressive strain in the concrete is taken as 0.003. Once again, the depth of the neutral axis at failure, x, may be taken as the unknown, and a solution obtained using strain compatability and equilibrium conditions as in the case of compression failure. The procedure is illustrated in Example 15.3.

EXAMPLE 15.3

If the column in Example 15.1 is loaded at an eccentricity of 800 mm, compute the nominal strength.

SOLUTION

Since $e > e_b$, the steel on the tension side yields and $f_s = f_y = 400$ MPa. It may be initially assumed that the compression steel also yields. This can be checked after the neutral axis is located. The depth of the neutral axis is assumed as x.
   The factored forces are:

$$C_c = 0.85 \ \phi_c f'_c b \ \beta_1 x$$
$$= (0.85)(0.6)(30)(350)(0.85x) \ 10^{-3} = 4.55x \text{ kN}$$
$$C'_s = A'_s (\phi_s f_y - 0.85 \ \phi_c f'_c)$$
$$= 3000 \ (0.85(400) - 0.85(0.6)(30)) \ 10^{-3} = 974 \text{ kN}$$
$$T = A_s \phi_s f_y$$
$$= 3000(0.85)(400)10^{-3} = 1020 \text{ kN}$$

Taking moments about the line of action of P_r (Fig. 15.11),

$$4.55x(500 + 0.85x/2) + 974 \ (500 + 75) - 1020 \ (500 + 525) = 0$$

Solving,

$$x = 184 \text{ mm}$$

To verify whether the compression steel does in fact yield,

$$\epsilon'_s = \frac{0.003 \ (184 - 75)}{184} = 0.00178 < \epsilon_y$$

The compression steel does not yield. Hence, revise the calculation with,

$$f'_s = (200 \ 000) \frac{0.003 \ (x - 75)}{x} = \frac{600 \ (x - 75)}{x}$$

$$C'_s = A'_s \ (\phi_s f'_s - 085 \phi_c f'_c)$$

$$= 3000 \left(\frac{(0.85)600 \ (x - 75)}{x} - 0.85(0.6)(30) \right) 10^{-3}$$

$$= \frac{1530 \ (x - 75)}{x} - 45.9 \text{ kN}$$

Fig. 15.11 *Example 15.3*

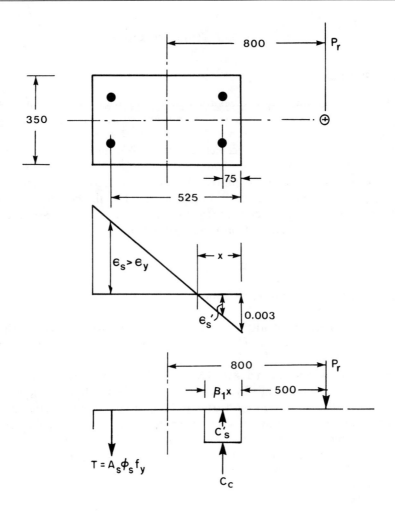

Using this correct expression for C'_s (in place of 974 kN) in the moment equation yields the cubic equation:

$$1.93\, x^3 + (2.28)\, 10^3 x^2 - (192)\, 10^3 x - (66.0)\, 10^6 = 0$$

Solving,

$$x = 198 \text{ mm}$$

Substituting for x,

$$C_c = 4.55\,(198) = 901 \text{ kN}$$

$$C'_s = \frac{1530\,(198 - 75)}{198} - 45.9 = 905 \text{ kN}$$

$$P_r = C_c + C'_s - T$$

$$= 901 + 905 - 1020 = 786 \text{ kN}$$

$$M_s = P_r e = 786 \,(0.80) = 629 \text{ kN} \cdot \text{m}$$

15.4.5 GENERAL CASE

The general case of analysis involves the determination of the factored axial load resistance of a column of known cross section, loaded at a given eccentricity.

The procedures described in Section 15.3.3 and 15.3.4 are quite general and can be used for sections with reinforcement distributed around the perimeter of the section at varying distances from the compression face (see Fig. 15.12). If x represents the depth to the neutral axis at failure, the strains, factored stresses, and factored forces in the reinforcement (C_s and T_s in Fig. 15.12c) and the factored compressive force in concrete, C_c, can all be expressed in terms of x, using the strain diagram and cross-sectional dimensions. Considering equilibrium of internal forces and applied load,

$$P_r = \Sigma(\text{internal forces})$$

and,   moment of $P_r = \Sigma(\text{moments of internal forces})$

These two equilibrium equations enable the determination of x and P_r. Verification of the yielding (or otherwise) of reinforcement and consequent revisions may be necessary.

The procedure was illustrated with Examples 15.2 and 15.3, wherein an exact solution was obtained directly by setting up and solving algebraic equations. A solution may also be obtained by a trial and adjustment method. Here a neutral axis depth is initially assumed and the corresponding factored internal forces computed. The correct depth to neutral axis is such that the resultant of the internal forces will act at the given eccentricity, e, and the magnitude of this resultant will give the factored axial load resistance, P_r. The neutral axis depth is adjusted until these conditions are satisfied.

15.5 STRENGTH INTERACTION DIAGRAMS

The discussion in the preceding section has shown that the factored axial load resistance, P_r, of a column depends on the eccentricity of loading. For the particular column cross section shown in Fig. 15.8, the combinations of factored axial load resistance, P_r, and factored moment resistance, M_r, for three values of $e = M_r/P_r$ were computed in Examples 15.1–15.3. For the same section, the factored axial load resistance under purely axial loading (zero eccentricity), P_{ro}, is given by Eq. 15.3 (see Fig. 15.13a).

$$P_{ro} = [0.85 \times 0.6 \times 30 \,(350 \times 600 - 6000) + 0.85 \times 400 \times 6000]10^{-3}$$
$$= 5160 \text{ kN}$$

Fig. 15.12 *General case*

(a)

(b) Strain

(c) Forces

At the other extreme, under pure bending, the strains and forces are as shown in Fig. 15.13b. In this case,

$$T = C_c + C_s'$$

$$3000 \times 0.85 \times 400 = 0.85 \times 0.6 \times 30 \times 350 \times 0.85c$$

$$+ 3000 \left[0.85 \times 200\,000 \times \frac{0.003\,(c - 75)}{c} \right.$$

$$\left. - 0.85 \times 0.6 \times 30 \right]$$

Fig. 15.13 *Limiting cases of ultimate strength*

(a) Axial loading   **(b) Pure bending**

Solving,
$$c = 110 \text{ mm}$$

Check: $\epsilon_s' = 0.003(c - 75)/c = 0.000955 < \epsilon_y$ OK

Taking moments about tension steel,

$$M_{ro} = (0.85 \times 0.6 \times 30 \times 350 \times 0.85 \times 110 \,(525 - 0.85 \times 110/2) + 3000$$

$$\left[0.85 \times 200\,000 \times \frac{0.003\,(110 - 75)}{110} - 0.85 \times 0.6 \times 30 \right] (525 - 75)) \times 10^{-6}$$

$$= 438 \text{ kN} \cdot \text{m}$$

The combinations of factored axial load and moment resistances for this section under five different eccentricities (ranging from $e = 0$ to $e = \infty$) are presented graphically (points A to E) in Fig. 15.14. The curve $ABCDE$ drawn through these points represents the full range of combinations of P_r and M_r, and is termed a factored axial load and moment resistances interaction diagram. Combinations of P and M that are plotted on this curve represent factored resistances, whereas all points within the curve represent combinations which are safe.

As already explained in Section 15.3, to allow for unforeseen factors, accidental eccentricities, etc., the Code limits the maximum axial load resistance of compression members to 80 percent of P_{ro} for tied columns and to 85 percent of P_{ro} for columns with spiral reinforcement (Eqs. 15.4 and 15.5). For the column shown in Fig. 15.14, this limiting factored resistance is $P_{r(max)} = 0.8 \times 5160 = 4130$ kN. This limitation is indicated by the cut off line $A'F$ in Fig. 15.14. The axial load limitation is roughly equivalent to specifying that all columns be designed for a minimum eccentricity of

Fig. 15.14 *Factored resistance interaction diagram*

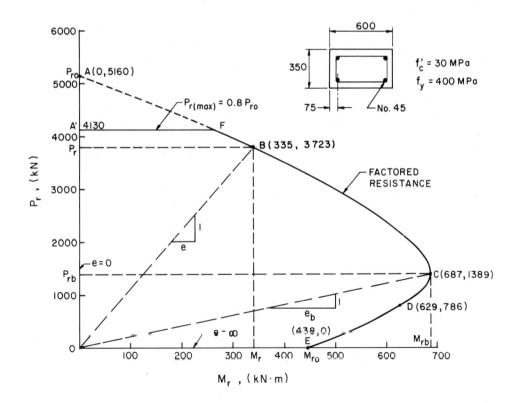

load of $e = 0.1 h$ for tied columns (point F in Fig. 15.14) and $e = 0.05h$ for spirally reinforced columns.

The complete factored resistance interaction diagram for the tied column shown in the inset in Fig. 15.14 is represented by the curve $A'FBCDE$ in Fig. 15.14 and is typical for common column sections. Points on the curve give factored resistances, P_r and M_r, corresponding to limiting strength conditions specified in CSA A23.3-M84. The design criteria for strength require that the factored load effects do not exceed the factored resistances. It is apparent that at very low eccentricities (segment $A'F$), the factored axial load resistance is independent of the value of e, as the limiting maximum resistance controls. The region FC of the curve represents compression failure, where an increase in axial load is accompanied by a decrease in the moment the section can carry. Columns with relatively high axial loads, such as those in the lower storeys of a high rise building, fall into this category. In the region CDE of the curve, the failure is initiated by yielding of the reinforcement on the tension side. In this zone, as the axial load decreases the moment that can be resisted simultaneously also decreases. Hence the loading condition giving the maximum moment in the column and the minimum axial load that accompanies this moment is of greater significance in this region. It should be noted that for columns CSA A23.3-M84(8.8.1) requires the consideration of the loading condition giving the maximum ratio of bending moment to axial load.

15.6 INTERACTION DIAGRAMS AS DESIGN AIDS

The factored resistance curve in Fig. 15.14, plotted in terms of P_r and M_r, is applicable for the section shown in the inset in Fig. 15.14. The factored resistance of this section, loaded at any given eccentricity, e, can be directly obtained from the graph by drawing the radial line OB (see Fig. 15.14) having a slope of $1/e$, and reading off the co-ordinates of the point B on the curve. However, by making the relationship independent of cross-sectional dimensions, as replotted in Fig. 15.15, the curve can be used for all sections geometrically similar to and having the same material properties as the one shown in the inset in Fig. 15.14. In Fig. 15.15, the curve is plotted in terms of parameters, P_r/A_s and $M_r(A_gh)$, both of which have the unit of MPa. The curve may also be plotted fully nondimensionalized, in terms of $P_r/(A_gf'_c)$ and $M_r/A_gh f'_c$.

Interaction relations such as that shown in Fig. 15.15, in tabular or graphical form, are presented in several publications (Refs. 15.6–15.8) as an aid to design (there are slight differences regarding the reference parameters chosen to represent the x and y axes). An extensive set of interaction diagrams, similar to that shown in Fig. 15.15, is presented in Ref. 15.6, from which the graphs shown in Figs. 15.16–15.22 have been reproduced. These graphs present the interaction between factored resistances, P_r and M_r, including the effect of material resistance factors ϕ_c and ϕ_s. The resistances of circular columns with ties and with spiral reinforcement (Figs.

Fig. 15.15 *Interaction diagram as a design aid*

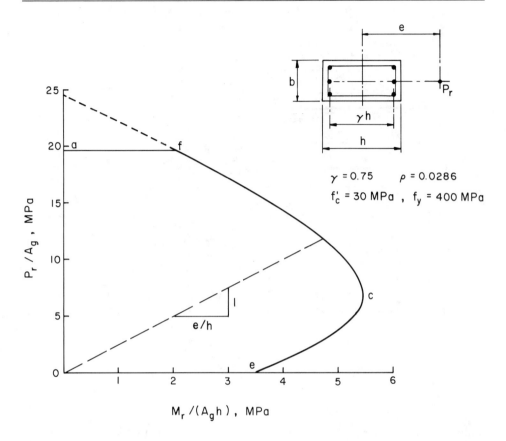

15.20–15.22) differ only at very small eccentricities where the cut-off line (*af* in Fig. 15.15) corresponding to the maximum limit for factored axial load resistance (Eqs. 15.4 and 15.5) controls. The cut-off line for a spirally reinforced column is indicated by the broken lines in Figs. 15.20–15.22. For values of γ in between those for which interaction diagrams are available, the factored resistances may be obtained by interpolation; however, it is conservative and generally satisfactory to use an inter-action diagram with γ smaller than the actual/expected value. The curves in Figs. 15.16–15.22 are for $f_y = 400$ MPa. These curves can be used for other values of f_y by multipliying the ρ_g obtained from these curves with the ration $400/f_y$ to obtain the ρ_g corresponding to f_y. The radial lines in the interaction diagram give values of the tensile stress in the steel layer closest to the tensile face of the column. Figures 15.17–15.19 are applicable for columns with equal reinforcement on the two end faces, and these should be used for the common case of rectangular columns with four bars located at the corners of the column. Analysis and design procedures with

Fig. 15.16 *Interaction diagrams for axial load and moment resistance for rectangular columns with an equal number of bars on all four faces*

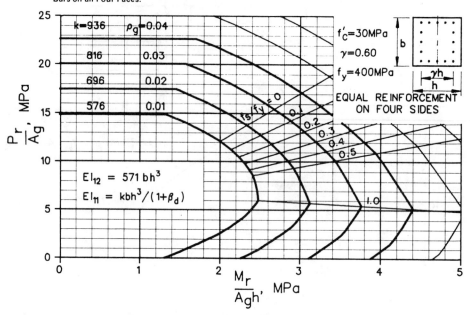

Interaction Diagrams for Axial Load and Moment Resistance for Rectangular Column with an Equal Number of Bars on all Four Faces.

Reproduced, with permission, from Ref. 15.6

the aid of these diagrams are illustrated in Examples 15.4–15.6.

The discussions in Section 15.4, have shown that the strength of a column section depends on several factors, such as cross-sectional dimensions, reinforcement ratio, disposition of reinforcement in the cross section, material strengths, and M/P ratio. It is apparent that for a design problem with given P_f, M_f, f_c' and f_y, there is no unique solution, and several alternative designs are possible. The design procedure involves an initial appropriate selection of some of these factors and then determining the remaining unknowns using the design aids.

15.7 ANALYSIS OF SECTIONS USING INTERACTION DIAGRAMS

Analysis problems involve the determination of the design strength of a given section loaded at a given eccentricity. A method of solution of such problems using basic principles was explained in Section 15.4 and illustrated in Examples 15.1–15.3.

Solutions to analysis problems can be obtained more easily using the interaction diagrams. The procedure is illustrated in Example 15.4 on page 538.

Fig. 15.17 *Interaction diagrams for axial load and moment resistance for rectangular columns with bars in the two end faces only*

Reproduced, with permission, from Ref. 15.6

Fig. 15.18 *Interaction diagrams for axial load and moment resistance for rectangular columns with bars in the two end faces only*

Reproduced, with permission, from Ref. 15.6

Fig. 15.19 *Interaction diagrams for axial load and moment resistance for rectangular columns with bars in the two end faces only*

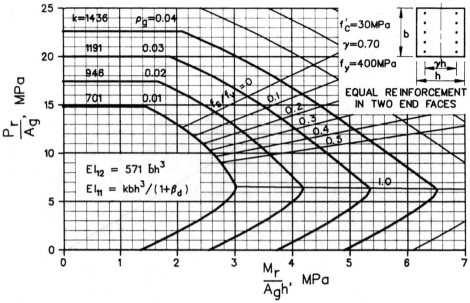

Reproduced, with permission, from Ref. 15.6

Fig. 15.20 *Interaction diagrams for axial load and moment resistance for circular tied or spiral columns*

Reproduced, with permission, from Ref. 15.6

FIG. 15.21 *Interaction diagrams for axial load and moment resistance for circular tied or spiral columns*

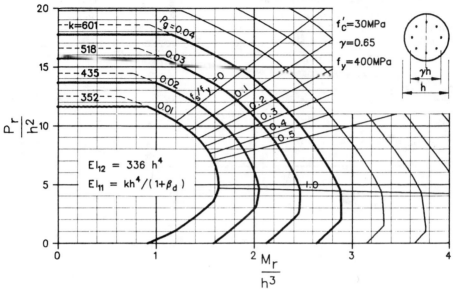

Reproduced, with permission, from Ref. 15.6

Fig. 15.22 *Interaction diagrams for axial load and moment resistance for circular tied or spiral columns*

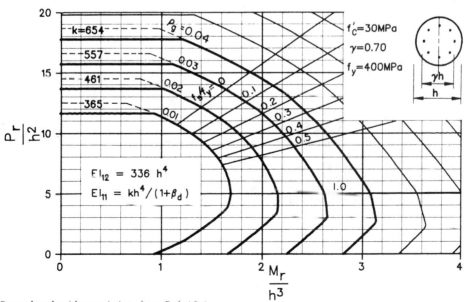

Reproduced, with permission, from Ref. 15.6

EXAMPLE 15.4

A column in a braced frame has an effective length of 3.5 m and the cross-sectional dimensions shown in Fig. 15.23. The service load effects are: dead load = 720 kN, live load = 1040 kN, dead load moment = 120 kN·m, and live load moment = 195 kN·m. All moments are about the major axis. Check the adequacy of the column for this loading condition. $f_c' = 30$ MPa and $f_y = 400$ MPa.

Fig. 15.23 *Example 15.4*

#10 ties at 400

A_s: 6 – #45 bars
$f_c' = 30$MPa
$f_y = 400$MPa

400

450

600

SOLUTION

1. Factored loads and moments

    $$\text{Factored load} = 1.25 \times 720 + 1.5 \times 1040 = 2460 \text{ kN}$$
    $$\text{Factored moment} = 1.25 \times 120 + 1.5 \times 195 = 443 \text{ kN·m}$$
    $$e = M/P = 443/2460 = 0.180 \text{ m}$$

2. Check whether long or short column
    Assuming conservatively that the column bends in symmetrical single curvature bending (that is, $M_1/M_2 = 1$), the limiting slenderness ratio for a short column is:

    $$\frac{k\ell_u}{r} = 34 - 12 \, M_1/M_2 = 22$$

    The radius of gyration, r, may be taken as 0.3 h for a rectangular section (Code (10.11.3)):

$$r = 0.3 \times 600 = 180 \text{ mm}$$
$$\text{Limiting } k\ell_u = 22 \times 180 = 3960 \text{ mm} > 3500 \text{ mm}$$

The slenderness effects can be neglected and the column treated as a short column.

3. Check reinforcement ratio

$$\rho = \frac{6 \times 1000}{400 \times 600} = 0.025 = 2.5 \text{ percent}$$

and is within the limits of 1–8 percent.

4. Compute factored resistance
   For the section, $\gamma = (600 - 150)/600 = 0.75$.
   For the material properties and reinforcement configuration, interaction diagrams in Figs. 15.17–15.19 are applicable. Since interaction diagram for $\gamma = 0.75$ is not included here, the diagram in Fig. 15.19 for $\gamma = 0.7$ will be used. (Note that increasing γ will slightly increase the strength of a section. Hence, using a graph for a γ value lower than the actual γ gives slightly conservative results).

   Here, $e/h = 180/600 = 0.3$. In the graph in Fig. 15.19, drawing a radial line corresponding to $e/h = 0.3$, and interpolating between the curves for $\rho = 2$ percent and $\rho = 3$ percent gives:
   For $\gamma = 0.7$, $e/h = 0.3$, and $\rho = 2.5$ percent

   $\text{slope of line} = \dfrac{\gamma e}{h}$ $\quad P_r/A_g = 12.2 \text{ MPa}$

   $$P_r = 12.2 A_g = 12.2 \times 400 \times 600 \times 10^{-3}$$
   $$= 2928 \text{ kN} > 2460 \text{ kN} \qquad\qquad \text{OK}$$

   and $\qquad\qquad M_r = P_r e = 2928 \times 0.180$
   $$= 527 \text{ kN·m} > 443 \text{ kN·m} \qquad\qquad \text{OK}$$

   The strength is adequate.

5. Check ties (Code (7.6.5)):
   Minimum required tie size for No. 35 bars is No. 10 bars $\qquad\qquad$ OK
   For maximum spacing limits:

   $$16 \times \text{bar diameter} = 16 \times 35.7 = 571 \text{ mm}$$
   $$48 \times \text{tie diameter} = 48 \times 11.3 = 542 \text{ mm}$$
   $$\text{least section dimension} = 400 \text{ mm}$$

   The 400 mm spacing is OK.
   Clear distance between the middle longitudinal bar and the laterally supported bars on either side

   $$= \frac{400 - 2 \times 75}{2} - 35.7 = 89.3 \text{ mm} < 150 \text{ mm}$$

Hence the middle bars need not be enclosed by corners of additional ties. Use of No. 10 ties at 400 mm, as provided, is satisfactory.

15.8 DESIGN OF SECTIONS USING INTERACTION DIAGRAMS

Design of columns is usually done with the aid of interaction diagrams. It is possible to design sections using equilibrium equations following a procedure similar to that used in the analysis problems illustrated in Examples 15.1 to 15.3. However, because of the several factors involved, such a procedure requires many initial assumptions and a process of trial and adjustment. Finally, it is always possible to assume a trial section and then check whether it has the required strength by using the analysis procedures discussed before.

The design procedure using interaction diagrams is illustrated with Examples 15.5 and 15.6.

EXAMPLE 15.5

The analysis of a multi-storey braced building frame was done by considering one floor at a time, as suggested in CSA A23.3-M84(8.9). From the analysis under specified loads, the loads and moments acting on an exterior column at the ground floor level is obtained as follows (see also Fig. 15.24a):

1. Column load and moment due to the loads on a single adjacent span,
    (a) Dead load: axial load on column = 144 kN
                                       moment on column = 96 kN·m
    (b) Live load: axial load on column = 78 kN
                                       moment on column = 52 kN·m

2. Column axial load accumulated from all remaining floors above:
    due to dead load = 1180 kN
    due to live load (allowing for the reduction for large tributary areas) = 370 kN
    The unsupported length of the column is 3.5 m. Design a tied column, taking $f'_c = 30$ MPa, and $f_y = 400$ MPa.

SOLUTION

1. Factored loads
    The weight of the column itself, assumed to act at its top, is calculated for an estimated column size of 500 mm square as:

    $$0.5 \times 0.5 \times 3.5 \times 2.4 \times 9.81 = 20 \text{ kN}$$

    a. For loading condition giving maximum axial load and moment,

    $$\text{axial load, } P_f = 1.25 (144 + 1180 + 20) + 1.5 (78 + 370)$$
    $$= 2352 \text{ kN}$$
    $$\text{moment, } M_f = 1.25 \times 96 + 1.5 \times 52 = 198 \text{ kN·m}$$
    $$e = M_f/P_f = 84.2 \text{ mm}$$

Fig. 15.24 *Example 15.5*

(a)

$A_s = 6 - No. 30$

No.10 at 400

(c)

(b)

b. For the loading condition giving the maximum ratio of bending moment to axial load (that is, live load only on the adjacent span),

$$\text{axial load} = 1.25 (144 + 1180 + 20) + 1.5 \times 78 = 1797 \text{ kN}$$
$$\text{bending moment} = 198 \text{ kN·m}$$

2. **Approximate column size**

   A column with reinforcement placed on two opposite sides is selected, for which design charts in Figs. 15.17–15.19 are applicable. Since the column is in the ground floor, a relatively high reinforcement ratio of $\rho_g = 4$ percent may be selected, so that the same column size with lesser amounts of reinforcement may be used for the columns at the next few levels.

   Figure 15.19 indicates that for $\rho_g = 0.04$, the maximum value of $P_r/A_g = 22.6$ MPa for a column loaded at low eccentricity. This gives:

$$\text{Minimum } A_g = P_f/22.6$$
$$= 2352 \times 10^3/22.6 = 104 \times 10^3 \text{ mm}^2$$

   or, for a square column, $b_{min} = \sqrt{A_g}$
$$= 323 \text{ mm}$$

   Since the column is loaded with $e = 84.4$ mm, which is greater than $0.1\ h$, a somewhat larger size is needed, and a size 400 mm × 400 mm is selected as a trial section.

3. **Check whether slenderness effects can be neglected**

   For braced frames, limiting $k\ell_u/r$ is:

$$\left(\frac{k\ell_u}{r}\right)_{lim} = 34 - 12\ M_1/M_2$$

   Because of some fixity at the foundation, a column at the ground floor level in a braced frame will bend in reverse curvature. However, conservatively assuming $M_1/M_2 = 0$,

$$\left(\frac{k\ell_u}{r}\right)_{lim} = 34$$

$$(k\ell_u)_{lim} = 34 \times 0.3 \times 400 = 4080 \text{ mm}$$

   The actual unsupported length is 3.5 m, and since for braced frames $k \leqslant 1$, the actual $k\ell_u < (k\ell_u)_{lim}$, the slenderness effects can be neglected. (Note that CSA A23.3-M84 (10.11.2.1) stipulates that for braced columns, k be taken as 1.0 unless analysis shows that a lower value may be used.)

4. **Select reinforcement**

   For 400 mm × 400 mm column, assuming 40 mm cover, No. 10 ties, and No. 35 longitudinal bars, $\gamma = 0.65$.
   For loading condition (a):

$$P_f/A_g = 2352 \times 10^3/(400)^2 = 14.7 \text{ MPa}$$
$$M_f/A_g h = 198 \times 10^6/(400)^3 = 3.09 \text{ MPa}$$

For these values, Table 15.18 gives

$$\text{for } \gamma = 0.65 \qquad \rho_g = 2.65 \text{ percent}$$

For loading condition (b):

$$P_f/A_g = 1797 \times 10^3/400^2 = 11.23 \text{ MPa}$$
$$M_f/A_g h = 3.09 \text{ MPa}$$

The ρ required for this condition is less than that for loading case (a).

$$A_s = \rho_g A_g$$
$$0.0265 \times 400^2 = 4240 \text{ mm}^2$$

Six No. 30 bars give very close to the required area. This will be provided and the actual strength for this area verified.

5. Check strength
   The strength of the section designed may now be checked by statics. Selecting the six No. 30 bar arrangement shown in Fig. 15.24b, from the interaction diagram, it is apparent that the failure is in compression. The strain profile is shown in Fig. 15.24b. The forces are (assuming compression reinforcement to yield),

$$C_c = 0.85 \times 0.6 \times 30 \times (0.85x) \times 400 \times 10^{-3} = 5.20x \text{ kN}$$

$$C_s' = 3 \times 700 \,(0.85 \times 400 - 0.85 \times 0.6 \times 30) \times 10^{-3} = 682 \text{ kN}$$

$$T = 3 \times 700 \times 0.85 \times 200\,000 \times \frac{0.003\,(330 - x)}{x} \times 10^{-3}$$

$$= 1071\,(330 - x)/x \text{ kN}$$

$$e = M/P = 198 \times 10^6/2352 \times 10^3 = 84.2 \text{ mm}$$

Taking moments about P_f,

$$682 \times (115.8 - 70) - (5.20x) \times (0.85x/2 - 115.8)$$
$$+ \frac{1071\,(330 - x)}{x}\,(284.2 - 70) = 0$$

Solving,

$$x = 323 \text{ mm}$$

Check,

$$\epsilon_s' = \frac{0.003\,(323 - 70)}{323} = 0.00235 > \epsilon_y = 0.002$$

and compression steel has yielded.

The internal forces are:

$C_c = 5.20 \times 323 = 1680 \text{ kN}$
$T = 1071 \times (330 - 323)/323 = 23 \text{ kN}$
$P_r = 1680 + 682 - 23 = 2340 \text{ kN}$
$M_r = [1680(200 - 0.85 \times 323/2) + 682(200 - 70) + 23(200 - 70)] \times 10^{-3}$
$\quad = 197 \text{ kN·m}$

Check $M_r/P_r = 197/2340 = 0.0842 \text{ m}$ OK

Factored resistances are:

$$P_r = 2340 \text{ kN}$$
and $\qquad\qquad\qquad M_r = 197 \text{ kN·m}$

The actual self weight of the 400 mm square column is 13 kN against the estimated weight of 20 kN. With this correction, the actual factored load is 2340 kN. Hence the factored resistances are adequate.

6. Design of ties
    No. 10 bar size is satisfactory.
    Maximum spacing limits are:

$$16 \times \text{bar diameter} = 16 \times 29.9 = 478 \text{ mm}$$
$$48 \times \text{tie diameter} = 48 \times 11.3 = 542 \text{ mm}$$
$$\text{lateral dimension} = 400 \text{ mm}$$

Provide No. 10 ties at 400 mm spacing. The arrangement shown in Fig. 15.24c may be used.

15.9 COLUMNS OF CIRCULAR SECTION

In previous Sections, the computation of the strength of columns was illustrated with reference to rectangular sections. The procedure for calculating the strength of columns of circular cross section (or of columns of other shapes, bent in a plane containing an axis of symmetry) is identical in principle to that for a rectangular column. However, a circular column with spiral reinforcement has far greater ductility and toughness than a tied column, as shown in Fig. 15.4. Because of this, the maximum limiting axial load resistance of a spiral column is taken slightly larger than that of a similar tied column.

The cross section of a circular column is shown in Fig. 15.25a. The failure of the column may be tensile or compressive, depending on the relative eccentricity of the applied load. In either case, at failure conditions, the maximum compressive strain in the concrete is 0.003. The variation of strain in the cross section is as shown in Fig. 15.25b, where x is the depth to neutral axis. From this diagram, the strains, and hence the stresses, in the various groups of reinforcement can be determined,

Fig. 15.25 *Circular section*

(a) Section

(b) Strain

$$f_s = E_s \epsilon_s \leq f_y$$

0.003

(c) Forces

$\beta_1 x$

$0.85\,\phi_c f_c'$

$C_s = A_s \phi_s f_s$

in terms of x. The internal factored forces in the reinforcement are given by $A_{s1}\phi_s f_{s1}$, $A_{s2}\phi_s f_{s2}$, etc., as shown in Fig. 15.25c.

The concrete area in compression has the shape of the segment of a circle. However, the assumption of an average factored stress of $0.85\phi_c f_c'$ acting over a reduced area of depth $a = \beta_1 x$ is still applicable. On this basis, the magnitude and location of the resultant compressive force in the concrete, C_c, can be determined in terms of x. The internal forces must be in equilibrium with the factored resistance P_r applied at the eccentricity e. Hence the value of x is such that the resultant of all internal forces acts along P_r. This condition determines x. The factored resistance, P_r, is then obtained as the resultant of all internal factored forces in the cross section. These computational procedures were illustrated earlier in Examples 15.1 to 15.3, with reference to a rectangular section. However, in the case of circular sections the distribution of the longitudinal reinforcement bars around the perimeter at varying distances from the extreme fibre in compression and the shape of the area of concrete in compression make the expressions somewhat more complex.

As for rectangular columns, in practice, analysis and design of columns of circular sections are done with the aid of interaction diagrams. The procedure is illustrated with Example 15.6.

EXAMPLE 15.6

The column at the lowest level of a braced frame has an effective length $k\ell_u$ = 2900 mm. The design loads (factored) on the column are: axial load P_f = 3500 kN, and moment M_f = 250 kN·m. Design a circular, spirally reinforced column, f'_c = 30 MPa, and f_y = 400 MPa.

SOLUTION

1. Eccentricity $e = M_f/P_f$
    $$= 250 \times 10^6/(3500 \times 10^3) = 71.4 \text{ mm}$$

2. Approximate column size
    Since the column is at the lowest level, a large reinforcement ratio of ρ = 4 percent may be selected so that the same column size with less reinforcement may be used for upper levels.
    From the chart in Fig. 15.20, for f'_c = 30 MPa, f_y = 400 MPa, γ = 0.6, and ρ = 0.04, the maximum value of P_f/h^2 is 18.6 MPa.

    $$\text{minimum required } h^2 = P_f/18.6$$
    Hence,
    $$= 3500 \times 10^3/18.6 = 188 \times 10^3 \text{ mm}^2$$
    $$h_{\min} = 434 \text{ mm}$$

    The actual e = 71.4 mm is greater than 0.05 h. It is apparent that the actual e/h is in the neighbourhood of 0.16. Hence, the P_f/h^2 for the actual section will be somewhat less than the maximum, and a larger size column is necessary. A column 500 mm in diameter will be tried.

3. Check whether slenderness effects can be neglected.

    $$\left(\frac{k\ell_u}{r}\right)_{\lim} = 34 - 12\, M_1/M_2$$

    Taking, conservatively, M_1/M_2 = 0

    $$(k\ell_u)_{\lim} = 34 \times r$$
    $$= 34 \times 0.25\, h$$
    $$= 34 \times 0.25 \times 500 = 4250 \text{ mm}$$

    Actual $k\ell_u$ = 2900 mm < $(k\ell_u)_{\lim}$, and slenderness effects may be neglected.

4. Reinforcement requirement
   Assuming spirals with No. 10 bars, 40 mm clear cover, and No. 35 longitudinal bars, (Fig. 15.25):

$$\gamma = \frac{500 - 2 \times 40 - 2 \times 11.3 - 35.7}{500} = 0.72$$

The chart in Fig. 15.22 for $\gamma = 0.7$ can be used.

$$P_f/h^2 = 3500 \times 10^3/500^2 = 14.0 \text{ MPa}$$
$$M_f/h^3 = 250 \times 10^6/500^3 = 2 \text{ MPa}$$

For these coordinate values, Fig. 15.22 gives:

$$\text{for } \gamma = 0.7 \qquad \rho = 3.8 \text{ percent}$$

This reinforcement ratio is close to the desired value of 4 percent.

$$\text{Total area, } A_s = 0.038 \times \frac{\pi \times 500^2}{4} = 7460 \text{ mm}^2$$

Select eight No. 35 bars giving an area of 8000 mm², which is only slightly in excess of the requirement. For the reinforcement selected, it is now possible to carry out an exact analysis and determine the actual strengths, as was done for the rectangular section in Example 15.5. However, because of the lengthy computations involved, it is not included here.

5. Design of spirals
   The minimum ratio of spiral reinforcement is (Eq. 15.8):

$$\rho_{s, \text{ min}} = 0.45 \left(\frac{A_g}{A_c} - 1\right) \frac{f_c'}{f_y}$$

The core diameter, D_c, measured to the outside of the spiral:

$$D_c = 500 - 2 \times 40 = 420 \text{ mm}$$

$$\rho_{s, \text{ min}} = 0.45 \left(\frac{\pi \times 500^2/4}{\pi \times 420^2/4} - 1\right) \frac{30}{400} = 0.0141$$

ρ_s is the ratio of volume of spiral reinforcement to the total volume of the core. Considering a length of column equal to the pitch, s, of the spiral (Fig. 15.26):

$$\rho_s = \frac{\text{Volume of spiral in one loop}}{\text{Volume of core in length } s}$$

$$= \frac{a_s \pi (D_c - d_s)}{(\pi D_c^2/4) s} \geq \rho_{s, \text{ min}}$$

Fig. 15.26 *Example 15.6*

where a_s = area of spiral reinforcement
              d_s = diameter of spiral reinforcement
             D_c = diameter of core
and s = pitch of spiral

Rearranging this relation, in order to get the required spacing:

$$s \leqslant \frac{4\,a_s(D_c - d_s)}{D_c^2 \times \rho_{s,\,min}}$$

Substituting $D_c = 420$ mm, and $\rho_{s,\,min} = 0.0141$,

$$s \leqslant \frac{a_s\,(420 - d_s)}{622}$$

To select a proper spiral reinforcement, the spacings with three alternative choices for spiral bar size are tabulated below:

Bar/wire size	d_s mm	a_s mm^2	s_{max} mm	clear spacing mm
No. 10	11.3	100	65.7	54.4
9 mm	9	63.6	42.0	33.0
8 mm	8	50.3	33.3	25.3

CSA A23.3-M84(7.6.4.3) limits the pitch to one-sixth of core diameter = $420/6 = 70$ mm; the clear spacing to not more than 75 mm nor less than 25 mm; and the bar size to not less than 6 mm. All three alternatives above meet these requirements; however, select No. 10 bar spiral at a pitch of 50 mm.

15.10 COMBINED AXIAL COMPRESSION AND BIAXIAL BENDING

15.10.1 GENERAL

In all the discussions so far, columns subjected to bending moment about only one principal axis (uniaxial bending) were considered. Frequently, columns such as those at corners of buildings are subjected to bending moments about both principal axes (biaxial bending). This creates no difficulty for circular columns, as the section can be designed as a column subject to uniaxial bending with the factored moment, M_f, taken as the resultant moment, $M_f = (M_{fx}^2 + M_{fy}^2)^{1/2}$, where M_{fx} and M_{fy} are the factored moments acting about the x and y axes (principal axes), respectively.

In the case of rectangular columns, biaxial bending results in a neutral axis inclined to the principal axes. Further, the neutral axis is not necessarily at right angles to the plane of the resultant applied moment. For a known neutral axis location, the strain distribution causing failure strain (0.003) at the extreme compression fibre can be drawn (see Fig. 15.27). From this the factored stresses and internal forces in the steel and concrete can be computed using the procedures already used in Examples 15.1 and 15.2. The factored resistances, P_r, M_{rx} and M_{ry}, corresponding to these internal forces can readily be determined from statics. In most practical problems, the factored loads and moments are the quantities known, and it is required to design an adequate cross section to carry these loads. However, the complexities of the computation preclude a direct design procedure. Even with a trial section selected, the determination of the location of the actual neutral axis is a lengthy process of trial and error, and hence an analytical computation of its factored resistance is very difficult.

Several simplified and approximate procedures have been proposed (Refs. 15.9, 15.10) for the design of biaxially loaded columns, mostly based on the shape of the interaction (failure) surface; one of these is described in Section 15.10.2. Combinations of the factored axial load, P_r, and biaxial moments M_{rx} and M_{ry} (about x and y axes, respectively) corresponding to failure (that is, $\epsilon_{c,\,max} = 0.003$) of a biaxially loaded column can be represented by a failure surface in a three-dimensional plot of $P_r = M_{rx} = M_{ry}$, as shown in Fig. 15.28. One set of values of P_r, M_{rx}, and M_{ry}, such as those corresponding to the neutral axis location shown in Fig. 15.27c,

Fig. 15.27 *Biaxial bending and compression*

(a)   (b)

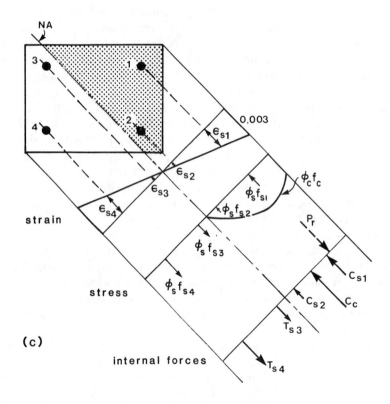

(c)

represents one point on the failure surface. The traces of the failure surface on the x–z and y–z planes correspond to the interaction curves for uniaxial bending about the x and y axes, respectively, and have the shape of the curve in Fig. 15.14.

Fig. 15.28 *Failure surface for biaxially loaded column*

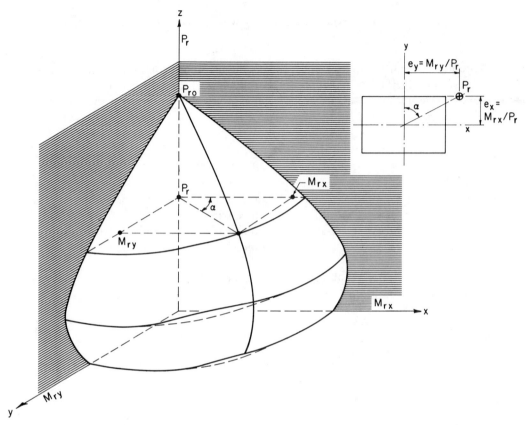

15.10.2 RECIPROCAL LOAD METHOD FOR BIAXIALLY BENT COLUMNS

This approximate method to compute the axial load capacity of a section when the load is applied at eccentricities e_x and e_y was developed by Bresler (Ref. 15.9). It has been found to give satisfactory results. The method is based on an approximate representation of a point on the failure surface, plotted in terms of variables $1/P_n$, e_x, and e_y. Bresler has shown that the nominal load capacity, P_n, at eccentricities of e_x and e_y, can be approximately represented by the equation:

$$\frac{1}{P_n} = \frac{1}{P_{nx}} + \frac{1}{P_{ny}} - \frac{1}{P_{no}} \qquad (15.15)$$

where P_n = nominal axial load capacity at eccentricities of e_x and e_y
       P_{nx} = nominal axial load capacity for load applied at eccentricity e_x only, with $e_y = 0$

P_{ny} = nominal axial load capacity for load applied at eccentricity e_y, with $e_x = 0$

P_{no} = nominal axial load capacity with concentric loading, that is, with $e_x = e_y = 0$

Note that "nominal capacity" here means the resistance computed with the material resistance factors, ϕ_s and ϕ_c, taken as unity. Introducing the material resistance factors, for design purposes, Eq. 15.15 can be rewritten, in terms of the factored axial load resistances, as:

$$\frac{1}{P_r} = \frac{1}{P_{rx}} + \frac{1}{P_{ry}} - \frac{1}{P_{ro}} \tag{15.16}$$

where P_r = factored axial load resistance with eccentricities e_x and e_y
P_{rx} = factored axial load resistance with e_x only and $e_y = 0$
P_{ry} = factored axial load resistance with e_y only and $e_x = 0$
P_{ro} = factored axial load resistance for concentric loading

To facilitate the use of the design charts for uniaxial bending (Figs. 15.16–15.19), Eq. 15.16 may be multiplied by the gross area of section, A_g, to give Eq. 15.17.

Fig. 15.29 *Components of Eq. 15.15*

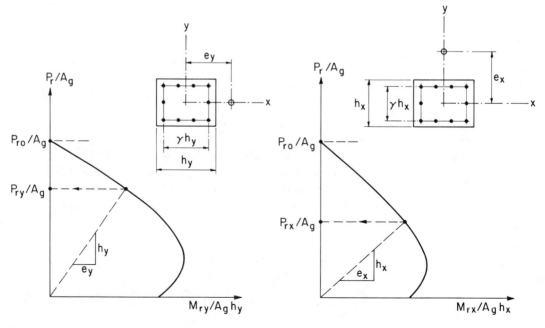

(a) Uniaxial bending about y-axis   (b) Uniaxial bending about x-axis

$$\frac{A_g}{P_r} = \frac{A_g}{P_{rx}} + \frac{A_g}{P_{ry}} - \frac{A_g}{P_{ro}} \qquad (15.17)$$

Equation 15.17 can be used to compute the resistance, P_r, of a given section, loaded with given eccentricities e_x and e_y. The three terms on the right-hand side of Eq. 15.17 can be obtained from the interaction diagrams for uniaxial bending, as indicated in Fig. 15.29. This gives the value of P_r/A_g, and hence P_r, the strength of the section with biaxial eccentricity.

For the design of a biaxially loaded column, a trial section is first selected and its adequacy checked using Eq. 15.17. One approach to select the trial section is to design it as though it is bent uniaxially with a moment slightly greater than the resultant of the applied moments (that is, $M_f = \sqrt{M_{fx}^2 + M_{fy}^2}$) acting about the principal axis about which the larger of the two components acts. The accuracy of the method decreases when the axial force is smaller than the lesser of P_{rb} or $0.1\,\phi_c f_c' A_g$. However, in this range, it is sufficiently accurate and usually conservative to ignore the axial force and design the section for biaxial bending only.

The design of a biaxially loaded column using the method described above is illustrated in Example 15.7. An alternative design procedure, based on an approximate relationship between M_{nx} and M_{ny} for a constant load, P_n, (which defines a load contour on the failure surface) is described in Refs. 15.7 and 15.10.

EXAMPLE 15.7

A corner column in the second floor of a braced frame is subjected to factored loads of $P_f = 1320$ kN, $M_{fx} = 112$ kN·m, and $M_{fy} = 80$ kN·m. The effective length of the column is $k\ell_u = 3000$ mm. Design the column, using a relatively high reinforcement ratio.

$$f_c' = 30 \text{ MPa and } f_y = 400 \text{ MPa}$$

SOLUTION

1. The eccentricities are:

$$e_x = M_{fx}/P_f = 112 \times 10^6 / 1320 \times 10^3 = 84.8 \text{ mm}$$
$$e_y = 80 \times 10^6 / 1320 \times 10^3 = 60.6 \text{ mm}$$

Resultant moment is:

$$M_{fr} = (112^2 + 80^2)^{1/2} = 138 \text{ kN·m}$$

2. Estimate column size
   In practice, the approximate column size may be known from the preliminary designs, and this step may not be necessary.
   A reinforcement ratio of about 4 percent is selected. For 4 percent steel, the

maximum value of P_r/A_g for minimum eccentricity is 22.6 MPa as seen from Fig. 15.17. Hence the minimum area required to carry P_f 1320 kN, with no moment, is:

$$1320 \times 10^3 / 22.6 = 58\ 400\ mm^2$$

A column section having this area will have the eccentricities about both axes greater than 0.1 h. Hence a much larger column area will be required. For trial a 350 mm square section is selected.

3. Check slenderness effects
   The loading causing maximum moment in exterior columns of braced frames bends the column in reverse curvature (see Fig. 10.4). In computing the limiting slenderness ratio for such columns, the value of M_1/M_2 in Eq. 15.1 may be taken conservatively as -0.5. Hence,

$$(k\ell_u/r)_{lim} = 34 - 12\ (-0.5) = 40$$
$$(k\ell_u)_{lim} = 40 \times 0.3 \times 350 = 4200\ mm$$

The column is not slender.

4. Estimate reinforcement required
   Reinforcement may be selected assuming the column to be bent uniaxially with a moment equal to the resultant moment, increased by 10–20 percent. On this basis, taking a 15 percent increase in moment:

$$P_f/A_g = 1320 \times 10^3/350^2 = 10.8\ MPa$$
$$M_f/A_g h = 1.15 \times 138 \times 10^6/350^3 = 3.70\ MPa$$

For these values, Fig. 15.17 gives:

$$\rho_g = 2.7\ percent$$
$$A_s = 0.027 \times 350^2 = 3307\ mm^2$$

Select four No. 35 bars, giving:

$$A_s = 4000\ mm^2,\ which\ gives:$$
$$\rho_g = 4000/350^2 = 0.033$$

5. Check the adequacy of section for biaxial bending
   With No. 10 ties and 40 mm cover:

$$\gamma h = 350 - 2x\ (40 + 11.3) - 35.7 = 212\ mm$$
$$\gamma = 212/350 = 0.60$$

Figure 15.17 is applicable:

$$e_x/h = 84.8/350 = 0.242$$
$$e_y/h = 60.6/350 = 0.173$$

Biaxial bending strength is represented by the Eq. 15.17, as:

$$\frac{A_g}{P_r} = \frac{A_g}{P_{rx}} + \frac{A_g}{P_{ry}} - \frac{A_g}{P_{ro}} \tag{15.17}$$

The reciprocal of the first term on the right-hand side, giving the strength under uniaxial bending about the x-axis with $e_y = 0$, is read from Fig. 15.17 off the radial line $e/h = e_x/h = 0.242$ at $\rho_g = 3.3$ percent as:

$$P_{rx}/A_g = 14.6 \text{ MPa}$$

Similarly, for $e/h = e_y/h = 0.173$ and $\rho_g = 3.3$ percent

$$P_{ry}/A_g = 17.5 \text{ MPa}$$

The reciprocal of the last term, P_{ro}/A_g, which is the intercept of curve for $\rho_g = 3.3$ percent with the vertical axis in Fig. 15.17, may be calculated from Eq. 15.2 as:

$$\frac{P_{ro}}{A_g} = 0.85 \ \phi_c f_c'(1 - \rho_g) + \phi_s f_y \rho_g$$

$$= 0.85 \times 0.6 \times 30 \ (1 - 0.033) + 0.85 \times 400 \times 0.033 = 26.0 \text{ MPa}$$

Substituting in Eq. 15.17,

$$\frac{P_r}{A_g} = \frac{1}{\dfrac{A_g}{P_{rx}} + \dfrac{A_g}{P_{ry}} - \dfrac{A_g}{P_{ro}}}$$

$$= \frac{1}{\dfrac{1}{14.6} + \dfrac{1}{17.5} - \dfrac{1}{26.0}} = 11.47 \text{ MPa}$$

$$P_r = 11.47 \times 350^2 \times 10^{-3} = 1400 \text{ kN}$$

This being greater than biaxially eccentric load applied of 1320 kN, the design is adequate.

PROBLEMS
---

1. A column in a braced frame, with an effective length of 3 m, has the cross section shown in Fig. 15.23. Compute the factored axial load resistance of the column, the load being applied at an eccentricity of: (a) 150 mm about the major axis, and (b) 700 mm about the major axis. The material strengths are $f_c' = 40$ MPa and $f_y = 300$ MPa.
2. Figure 15.30 shows a circular spirally reinforced column at the foundation level in a braced frame. The effective length of the column is 2.7 m. Compute the factored axial load resistance of the column if the load is applied at an eccentricity of 120 mm. Take $f_c' = 30$ MPa, $f_y = 400$ MPa. Also, design the spiral reinforcement.

Fig. 15.30 *Problem No. 2*

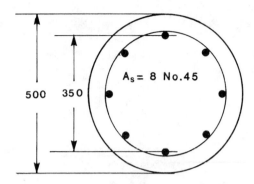

3. Design the cross section for a short rectangular column to carry a factored axial load of 2000 kN and a factored moment of 150 kN·m. Use a reinforcement ratio of 2 percent. Take f'_c = 30 MPa, f_y = 400 MPa.
4. A rectangular column in a braced frame has an effective length of 3 m. The column has to carry an unfactored dead load of 800 kN at an eccentricity of 180 mm and a specified live load of 1200 kN at an eccentricity of 250 mm, both eccentricities being about the major axis. Design a tied column. Take f'_c = 30 MPa and f_y = 400 MPa. Use a reinforcement ratio of about 4 percent.
5. A rectangular corner column at the ground floor level in a braced frame is subjected to a factored axial load of 1600 kN, and factored moments about the two principal axes of 140 kN·m and 190 kN·m, respectively. The slenderness effects can be neglected. Design a tied column. f'_c = 30 MPa and f_y = 400 MPa.
6. For the column in Problem 4, design a circular spirally reinforced column.
7. The biaxially loaded column, designed in Example 15.7, is now subjected to a factored axial load of P_f = 1400 kN and biaxial moments of $M_{fx} = M_{fy}$ = 80 kN·m. Check the adequacy of the design.

REFERENCES

15.1 ACI Committee 315, *Proposed ACI Standard: Details and Detailing of Concrete Reinforcement*, Concrete International, American Concrete Institute, Detroit, Vol. 2, No. 2, Feb. 1980, pp. 73–108.

15.2 *Reinforcing Steel — Manual of Standard Practice*, Reinforcing Steel Institute of Ontario, Willowdale, 1983.

15.3 MacGregor, J.G., Breen, J.E., and Pfrang, E.O., *Design of Slender Concrete Columns*, ACI Journal, Proc., Vol. 67, No. 1, Jan. 1970, pp. 6–28.

15.4 *Commentary on Building Code Requirements for Reinforced Concrete* (ACI 318–83), ACI Committee Report, American Concrete Institute, Detroit, 1983, 155 pp.

15.5   *Reinforced Concrete Column Investigation — Tentative Final Report of Committee 105*, ACI Journal, Proc. Vol. 29, No. 5, Feb. 1933, pp. 275–282.

15.6   *Concrete Design Handbook*, Canadian Portland Cement Association, Ottawa, 1985.

15.7   *Design Handbook, Vol. 2 — Columns*, ACI Special Publication SP-17 A (78), American Concrete Institute, Detroit, 1978.

15.8   *CRSI Handbook*, Concrete Reinforcing Steel Institute, Chicago, 1980.

15.9   Bresler, B., *Design Criteria for Reinforced Concrete Columns Under Axial Load and Biaxial Bending*, ACI Journal, Vol. 57, 1960, pp. 481–490.

15.10   Gouwens, A.J., *Biaxial Bending Simplified*, Reinforced Concrete Columns, ACI Special Publication SP-50, American Concrete Institute, Detroit, 1975, pp. 223–261.

CHAPTER 16

SLENDER COLUMNS

16.1 BEHAVIOUR OF SLENDER COLUMNS

Slender columns carrying axial compression and bending may have their strength reduced because of the increased bending moment caused by the transverse deflection. Figure 16.1 shows a column loaded at an eccentricity, e, at both ends, causing single curvature bending. The applied end moments, $M_o = Pe$ (which may be termed the primary moment), cause bending of the column, which increases the effective eccentricity of the load at sections away from the ends of the column. If Δ is the deflection of the column at its mid-height, the maximum bending moment in the column at this section (Fig. 16.1b) is:

$$M_{\max} = P\,(e + \Delta) \tag{16.1a}$$

$$= M_n + P\,\Delta \tag{16.1b}$$

The additional moment due to the deflection of the column is $P\,\Delta$, and may be termed the "$P - \Delta$ moment." It should be noted that in Eq. 16.1, the deflection, Δ, itself is dependent on the moment. Hence the variation of $M_{\max}$ with P is nonlinear, with $M_{\max}$ increasing at a faster rate as the load, P, increases.

Figure 16.1c shows the maximum axial load-moment interaction diagram for the column cross section. If the column is very short (to be exact, of zero length), its lateral deflection, Δ, is negligibly small, and the maximum moment, $M_{\max} = Pe$, increases linearly with P. In this case, the P–M path, superimposed on the interaction diagram, will be along line OA in Fig. 16.1c. When the load and moment combination reaches $P = P_s$ and $M_{\max} = P_s e$ corresponding to the point A on the interaction diagram, the column fails by crushing of the concrete at the section of maximum moment. If, on the other hand, the column is longer, with increasing load P, the deflection, Δ, is no longer negligible, and the moment $M_{\max} = P_1\,(e + \Delta_1)$ will vary along the curve OB in Fig. 16.1c. Failure occurs at a load P_1, the moment $P_1\,(e + \Delta_1)$, represented by point B. Thus, although the column section and applied load eccentricity, e, are the same, the slenderness has caused a decrease in the load-carrying capacity of the column from P_s for the short column to P_1 for the longer one. In both cases, failure of the column occurs due to the maximum compressive strain in the concrete reaching the limiting value (this may or may not be preceded

558

Fig. 16.1 *Behaviour of slender columns*

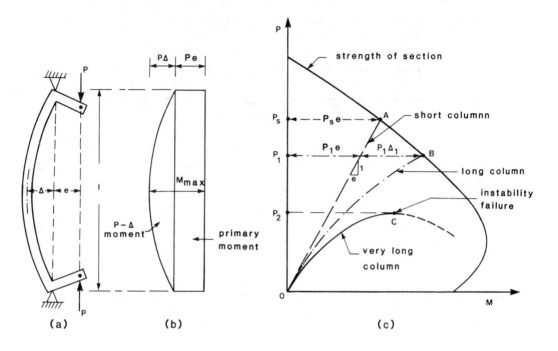

(a)   (b)   (c)

by tensile yielding of the reinforcement on the tension side, depending on which part of the interaction diagram the points *A* and *B* fall (see Fig. 15.6)). Most columns in practical building frames will have this type of failure.

If the column in Fig. 16.1 is very long, the increase in deflection may be such that the variation of maximum moment with increasing *P* follows the path *OC*, with dP/dM reaching zero at point *C*. In this case the column fails by instability and the capacity of the column is given by the peak load attained, equal to P_2. This type of failure may occur in very slender columns in unbraced frames.

The moment magnification due to the $P-\Delta$ effect is also dependent on the ratio of applied end moments, M_1/M_2, (or e_1/e_2). where M_1 and M_2 are the smaller and larger applied end moments, respectively (Fig. 16.2); and the ratio M_1/M_2 is considered positive if the column is bent in single curvature and negative if bent in double curvature. In braced columns bent in symmetrical single curvature (that is, $M_1/M_2 = +1.0$), as shown in Fig. 16.1a, the moment is always magnified by the deflection. In columns braced against sidesway, when the moments are unequal $(M_1/M_2 \neq 1.0)$, and particularly when the column is bent in double curvature, the $P-\Delta$ effect due to column deflection may not always increase the moment above the maximum moment in the undeflected column. This is shown in Fig. 16.2. For the columns in Fig. 16.2, the maximum applied moment (primary moment) is at the ends of the columns, whereas the lateral deflections and related increase in moments occur away from the ends. If the deflection is relatively small, so that the additional moment is $Py = Py_1$ as indicated by curve 1, the maximum moment is not affected by the $P - \Delta$ effect. If, however, the column is very slender, the deflections may be much larger, causing an additional moment $Py = Py_2$ as indicated by curve 2 in Fig. 16.2. In this case the maximum moment is influenced by $P - \Delta$ effects.

Column deflection is increased by creep under sustained loads, and hence the latter may also cause an increase in the $P - \Delta$ moment and thereby a decrease in the column capacity. Another factor that affects the column moment is the changes that may occur in the stiffnesses of either the column or the restraining beams framing into the column. Changes in stiffness may result from axial load, cracking, yielding, and creep.

For a column bent in symmetrical single curvature, with applied end moments of $M_o = Pe$, as in Fig. 16.1a, it can be shown (Ref. 16.1) that under elastic behaviour, the maximum moment is very nearly equal to:

$$M_{max} = M_o + P\Delta$$

$$\simeq M_o \times \frac{1 \times \psi P/P_{cr}}{1 - P/P_{cr}} \qquad (16.2)$$

Fig. 16.2 *Columns braced against sidesway, unsymmetrical end moments—member stability effect*

(a) Single curvature   **(b) Double curvature**

where ψ = a coefficient, = 0.23 for the loadings in Fig. 16.1a (ψ has slightly different, but smaller, values when the applied column moment is due to transverse loading on the column)

P_{cr} = Euler critical load for the column

The values of both ψ and P/P_{or} are very small in the practical range of columns, and hence their product is negligible compared to 1.0. With this approximation, Eq. 16.2 can be simplified to:

$$M_{max} = M_o \times \frac{1}{1 - P/P_{cr}} \tag{16.3}$$

Equations 16.2 and 16.3 are applicable for symmetrical single curvature bending only. In the derivation of Eq. 16.2, it is assumed that the maximum primary moment is at the mid-height of the column where the maximum $P - \Delta$ moment also occurs. The magnification of moment was shown to depend also on the ratio M_1/M_2 (see Fig. 16.2). With unequal end moments, in columns braced against sidesway, the maximum values of the primary moment and of the $P - \Delta$ moment occur at different sections and are not directly additive. For such columns with unequal end moments M_1 and M_2, Eq. 16.3 can be used by replacing M_o with an equivalent uniform moment equal to $C_m M_2$ (Ref. 16.1). The factor, C_m, is derived such that equal end moments of $C_m M_2$ causing symmetrical single curvature bending of the column would give the same maximum moment, M_{max}, as occurs under the actual moments, M_1 and M_2. Thus replacing the actual moment by the equivalent uniform moment, for columns braced against sidesway.

$$M_{max} = M_2 \times \frac{C_m}{1 - P/P_{cr}} \geq M_2 \tag{16.4}$$

where

$$C_m = 0.6 + 0.4 \, M_1/M_2 \geq 0.4 \tag{16.5}$$

Note that for actual symmetrical angle curvature bending, $M_1/M_2 = 1.0$, and Eq. 16.4 reduces to Eq. 16.3.

The foregoing discussion, and the column examples in Figs. 16.1 and 16.2, dealt with braced columns where sidesway is prevented. In this case, the amplification of moment by the factor $C_m/(1 - P/P_{cr})$ is due to the axial load, P, acting on the lateral deflection due to the member curvature relative to the chord line joining the ends of the column. This is termed the *member stability effect*.

A frame which is not braced against sidesway may displace laterally resulting in a relative lateral displacement (lateral drift or sway) between the ends of the column leading to magnification of column end moments, which is discussed in the next paragraph. A frame may be considered as braced against sidesway if the lateral loads in the direction under consideration are resisted by lateral load carrying member or members (such as shear walls, elevator shafts, stairwells, shear trusses, or other types of lateral bracing) extending from the foundation to the upper end of the frame. In such a frame, the lateral drift in the column is negligibly small. A frame is unbraced if the lateral load effects are resisted by the frame itself. Real structures are seldom fully braced or completely unbraced. For the purpose of evaluating lateral drift effects, Ref. 16.5 suggests that when the stability index, $(\Sigma P_f \Delta_o)/(H_f h_s)$, for a storey is not greater than 0.04, the lateral drift effect will not exceed 5 percent of the first order moments and the structure can be considered to be braced. Here H_f is the total factored lateral force acting within the storey, Δ_o is

Fig. 16.3 *Column subjected to sidesway — lateral drift effect*

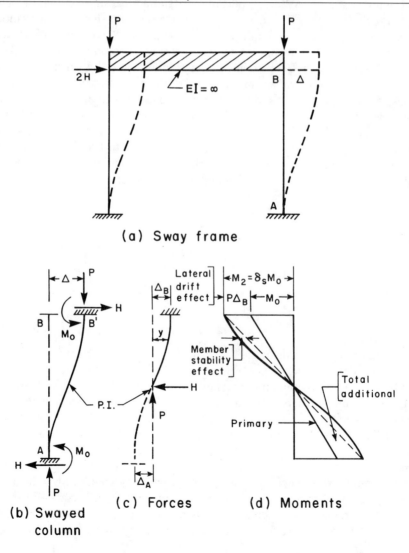

(a) Sway frame

(b) Swayed column

(c) Forces

(d) Moments

the elastically computed first order storey drift due to H_f, h_s is the storey height and the summation is over the columns of the storey.

An unbraced frame may sway laterally due to the effects of lateral loading, and of gravity loading if the loading and/or the frame is unsymmetric. The lateral sway or drift of a column is the relative transverse displacement, Δ, between the ends of the column (Fig. 16.3a). The additional moment resulting from the vertical loads acting on the laterally displaced structure, $P\Delta_A$ and $P\Delta_B$ at column ends in Fig. 16.3d, is termed the *lateral drift effect*. In columns in unbraced frames, the magnifications of the moment may be more significant. Here, the maximum values of the primary moment, M_o, and the maximum values of the lateral drift effect both occur at the column ends and are additive, as shown in Fig. 16.3. Thus the primary moment, amplified by the lateral drift effect, gives the end moment M_1 and M_2 in the column. In addition there is a curvature in the member and resulting transverse displacements at points along the length of the column relative to the chord joining the column ends. This gives rise to a further amplification of the moment over the length of the column due to the "member stability effect" described earlier (Fig. 16.2). However, usually for columns in unbraced frames the lateral drift effect is the more dominant one and the amplification factor due to the member stability effect (Eq. 16.4) may be unity. Figure 16.3 shows a column completely restrained against rotation at the ends and subjected to sidesway. The final moments, shown in (d), are obtained by combining the primary moments with the lateral drift and member stability effects. It can be seen that the maximum applied (primary) moment is always increased by the lateral drift effect, while the increase due to member stability effect may not be significant. Practical frames will have beams of finite flexural stiffness and therefore, in general, column ends are only partially restrained against rotation (Fig. 16.4). In such cases, the sidesway displacement and the additional $P - \Delta$ moment will be even greater. Further, as already explained in Section 15.1.3, columns in unbraced frames have larger slenderness ratios, $k\ell/r$ ($k = 1$ to ∞), than identical columns in braced frames. Hence columns in unbraced frames are weaker than similar columns in braced frames.

Briefly, the behaviour of slender columns can be summarized as follows:

1. The lateral deflection of the column results in significant additional moments which, in general, reduce the load-carrying capacity of the column.
2. The moment magnification factor due to member stability effect depends mainly on the ratio of applied axial load to the critical load of the column and the end moment ratio M_1/M_2 as is evident from the second term on the right hand side of Eq. 16.4.
3. Columns free to sway are weaker than braced columns, and have the primary moment increased by both lateral drift effect and member stability effect. While lateral drift effects apply to columns in a frame with appreciable sway, member stability effects apply to all columns.

What is referred to in this section as "applied moment" or "primary moment" is the moment without inclusion of the $P - \Delta$ effects, and in the case of a column in a frame is the moment determined by a conventional (first order) frame analysis, where the influence of the change in geometry of the frame due to deflections is

Fig. 16.4 *Column sidesway in unbraced building frames*

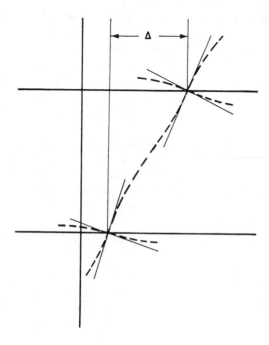

neglected. If a second order analysis is made, including the effects of change in geometry of the frame due to deflections as well as the effects of axial loads, variable moment of inertia and duration of loads on member stiffnesses, a better estimate of the real moments including slenderness effects is obtained. In lieu of such a detailed second order analysis, slenderness effects in compression members are accounted for approximately by magnifying the primary moment for lateral drift effects and member stability effects. The fundamental equations for moment magnifiers are based on behaviour of columns with hinged ends and these must be modified to account for the effect of end restraints. This is done by using an "effective column length," $k\ell_u$, in the computation of slenderness effects. The effective length factor, k, was briefly discussed in Section 15.1.3b. A method for determining k for columns in frames is described in the next section. The radius of gyration $r = (I/A)^{1/2}$, where I is the moment of inertia and A the area of the column section. The value of I is influenced by the extent of cracking and creep of concrete. These effects are considered in Section 16.3.

16.2 EFFECTIVE LENGTH OF COLUMNS IN FRAMES

In Section 15.1.3 it was explained how the k-factor of a column depends on the flexural stiffnesses of the beams framing into it, relative to the stiffness of the column itself. Julian and Lawrence developed a procedure (presented in detail in Ref. 16.2) for determining k for braced and unbraced frames, and also prepared the alignment charts shown in Figs. 16.5 and 16.6 for the graphical determination of k. These charts are in terms of two parameters, ψ_A, and ψ_B, representing the degree of restraint at ends A and B, respectively, of the column under consideration. The charts are based on a simplified elastic analysis of a partial frame, consisting of the column being considered plus all the members framing into the column ends and lying in the plane of bending (Fig. 16.7). Details of the analysis are also presented in Ref. 16.3.

Fig. 16.5 *Effective length factors for columns in braced frames*

$$\psi_i = \frac{(\Sigma EI/\ell_c)_{\text{columns}}}{(\Sigma EI/\ell)_{\text{beams}}} \quad \text{at end i of column}$$

Reproduced, with permission, from Ref. 15.6

Fig. 16.6 *Effective length factors for columns in unbraced frames*

$$\psi_i = \frac{(\Sigma EI/\ell_c)_{\text{columns}}}{(\Sigma EI/\ell)_{\text{beams}}} \quad \text{at end i of column}$$

Reproduced, with permission, from Ref. 15.6

The parameter ψ is given by

$$\psi = \frac{\Sigma \left(\dfrac{EI}{\ell} \text{ of columns} \right)}{\Sigma \left(\dfrac{EI}{\ell} \text{ of beams} \right)} \tag{16.6}$$

where Σ indicates summation for all members rigidly connected to that joint and lying in the plane of buckling of the column. Note that for members braced against

Fig. 16.7 *Members considered in computing k of column*

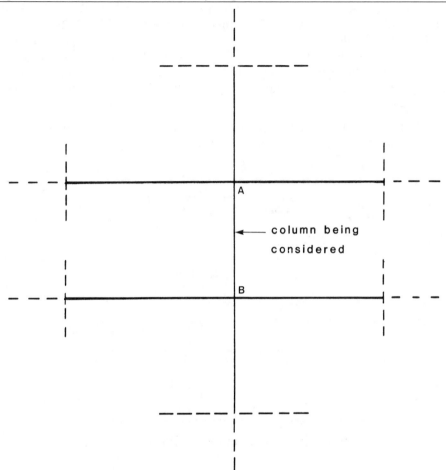

sideway, k may be taken as unity, since the actual value lies in the range of 0.5 to 1.0. For unbraced columns, k will be greater than 1.0.

The EI values used in Eq. 16.6 should reflect the effects of widely varying beam and column reinforcement percentages and of beam cracking (CSA A23.3-M84 (10.11.2.2)). Refs. 16.4 and 16.5 recommend the use of the moment of inertia of cracked-transformed sections for flexural members (beams) and Eq. 16.7 for the EI of the column.

$$EI = 0.2\,E_c I_g + E_s I_{se} \qquad (16.7)$$

where I_g = moment of inertia of the gross concrete section
and I_{se} = moment of inertia of the reinforcement about the centroid of the member cross section

An alternative procedure suggested in Ref. 16.5 is to use $0.5I_g$ for beams (0.5 to account for the effect of more extensive cracking in flexural members) and I_g for columns, with $k\ell/r < 60$. This combination is also suggested in Ref. 16.6. Appendix C to CSA A23.3-M84 also recommends the use of the effective length factors, k, given in Table 16.1 for braced compression members, as an alternative to the use of the monogram in Fig. 16.5. For braced columns and for computing member stability effects, the Code also suggests that k may be taken, conservatively, as unity.

Table 16.1 Effective Length Factors for Columns in Braced Frames (CSA A23-3-M84)

			k		
TOP	Hinged	0.81	0.91	0.95	1.00
	Elastic	0.77	0.86	0.90	0.95
	Elastic	0.74	0.83	0.86	0.91
	Stiff	0.67	0.74	0.77	0.81
MEMBER STABILITY EFFECTS		Stiff	Elastic	Elastic	Hinged
			BOTTOM		

Effective Length Factors for Clause 10.11.2.1

Reproduced, with permission, from Ref. 15.6

16.3 CODE PROCEDURES FOR DESIGN OF SLENDER COLUMNS

16.3.1 GENERAL

A detailed description of the background to the provisions in the Code for the design of slender columns is contained in Ref. 16.7. The Code requires consideration of slenderness effects for all compression members not braced against lateral deflection, and for braced compression members when the ratio $k\ell_u/r$ of the member exceeds the limit set by Eq. 15.1. Analysis has shown that for columns with slenderness ratios below this limit, the decrease in strength due to slenderness effects will be only of the order of 5 percent of the cross-sectional strength.

The design of a slender column involves two parts:

1. The analysis of the structure to determine the forces and moments on the column, allowing for all the slenderness effects discussed in the last two sections.
2. The proportioning of the column cross section to resist the forces and moments determined in part 1 above.

The second part involves the design of the cross-sectional shape, size, and reinforcement details, necessary to resist a given axial load and moment. The procedure for this is the same as for the design of short columns, described in Chapter 15, and hence is not repeated here.

For the first part, CSA A23.3-M84 presents two alternative approaches, namely: (1) a detailed second order structural analysis; and (2) an approximate evaluation of slenderness effects using moment magnifiers. Procedure (1) is preferable in general, and for columns in unbraced frames in particular. For all compression members with $k\ell_u/r > 100$, procedure (1) is mandatory. The moment magnifiers used in procedure (2) are to account for both lateral drift effects, if applicable, and the member stability effects. Comparisons with analytical and test results have shown that procedure (2) gives satisfactory results.

16.3.2 DETAILED ANALYSIS

In this approach (CSA A23.3-M84 (10.10.1)), the forces and moments in the compression members are determined from a second order structural analysis which takes into account all slenderness effects, such as the influence of column and frame deflections on moments, effects of axial loads and variations in moment of inertia on member stiffnesses, and effects of sustained loads. Reference 16.5 lists the minimum required considerations for a satisfactory analysis procedure as follows:

1. use of realistic moment-curvature relationships for members,
2. effect of foundation rotations on displacements,
3. effect of axial loads on the stiffness and carry-over factors for very slender columns ($\ell_u/r > 45$),
4. effects of creep when sustained lateral loads are present,
5. effects of the lateral deflections of the frame, and of the column itself.

To satisfy item 1, it is acceptable to use a linear moment-curvature relation, with the flexural stiffness, EI, computed as (Ref. 16.5):

For columns,

$$(EI)_{col} = E_c I_g (0.2 + 1.2\rho_t E_s/E_c) \tag{16.8}$$

For beams,

$$(EI)_{beam} = 0.5 E_c I_g \tag{16.9}$$

where ρ_t = ratio of total area of reinforcement to area of column.

Procedures for second order analysis are presented in Refs. 16.8–16.10.

16.3.3 APPROXIMATE EVALUATION OF SLENDERNESS EFFECTS

In this procedure (CSA A23.3-M84 (10.11)), the factored axial loads and moments acting on the column are determined from a conventional elastic frame analysis. To account for the slenderness effects, the moment so obtained is magnified by factors δ_b and δ_s, given below, for the member stability effects and lateral drift effects, respectively. The column section is then designed for the factored axial load, P_f, obtained from the analysis, and the magnified factored moment, M_c.

(a) Member Stability Effect

1. Moment magnifier, δ_b.

The magnified factored moment, M_c, allowing for member stability effect, is given by:

$$M_c = \delta_b M_2 \tag{16.10}$$

$$\delta_b = \frac{C_m}{1 - \dfrac{P_f}{\phi_m P_c}} \geqslant 1.0 \tag{16.11}$$

$$P_c = \frac{\pi^2 EI}{(k\ell_u)^2} \tag{16.12}$$

where
$\quad M_c \;=\;$ magnified factored moment, used in the design of section
$\quad M_2 \;=\;$ larger factored end moment on the column, considered positive
$\quad \delta_b \;=\;$ moment magnification factor, to reflect the effects of member curvature between ends of the column
$\quad P_f \;=\;$ factored axial load
$\quad P_c \;=\;$ critical load
$\quad k \;=\;$ effective length factor
$\quad \ell_u \;=\;$ unsupported length of column
$\quad C_m \;=\;$ factor relating actual moment diagram to an equivalent uniform moment diagram
$\quad \phi_m \;=\;$ member resistance factor, $= 0.65$

For columns with no transverse loads between supports, C_m is given by Eq. 16.13. For all other cases, C_m must be taken as unity.

$$C_m = 0.6 + 0.4\, M_1/M_2 \geqslant 0.4 \tag{16.13}$$

where M_1 is the smaller factored end moment on the column associated with the same loading case as M_2, considered positive if the member is bent in single curvature, and negative if bent in double curvature. Equation 16.10 is identical to Eq. 16.4 described earlier, except for the introduction of the member resistance factor in order to express the load ratio, P/P_{cr}, in terms of factored loads.

2. Calculation of end moments M_1 and M_2.

The magnification of column moment by the factor δ_b to account for the member stability effect must be applied for columns in both braced and unbraced frames. For this, the applicable end moments in the column, M_1 and M_2, are required.

For columns in braced frames, the drift of the frame is small even under lateral loads and therefore the magnification of primary end moments due to the lateral drift effect (see Fig. 16.3) can be neglected. Hence in braced frames, the column end moments, M_1 and M_2, to be used in Eq. 16.10 for computing M_c are the smaller and larger end moments calculated by a conventional elastic frame analysis.

For columns in unbraced frames, there are two conditions to be considered. For combinations of vertical loads which do not produce "appreciable lateral deflections," again the lateral drift effect can be neglected. Therefore the moments M_1 and M_2 are taken as those calculated by a conventional elastic frame analysis under these non-sway loads. The Code denotes these moments due to non-sway loads by M_{ns}, so that M_2 is the larger M_{ns} in this case. For this classification, the loads can be assumed to cause no "appreciable lateral deflections" if the resulting lateral deflection, Δ, does not exceed $\ell_u/1500$. Thus typically in a symmetric building frame the load combination $1.25D + 1.5L$ applied uniformly may not cause appreciable lateral deflection, and the end moments obtained from a conventional elastic analysis can be directly used in Eq. 16.10 to compute the associated magnified factored moment for design. This case is illustrated in Fig. 16.8a.

The second case to be considered for columns in unbraced frames is the loading conditions which result in appreciable lateral deflections. This may occur under combinations of vertical loads (unsymmetrical pattern of loading, and/or unsymmetrical frame) or combinations of vertical and lateral loads. With lateral deflections, the column end moments are magnified due to the lateral drift effect (Fig. 16.3). In this case the Code prescribes that the moment, M_s, due to the loads which result in appreciable lateral deflection (calculated by conventional elastic analysis) be first magnified by the factor δ_s to allow for the lateral drift effect (see Section 16.3.3(b)) and then combined with the part of moment, M_{ns}, which does not result in sway, and the so combined moments taken as M_1 and M_2 for computing M_c. This case is illustrated in Fig. 16.8b for the load combination $\alpha_D D + \psi(\alpha_L L + \alpha_Q Q)$. Thus, in this case, the final magnified factored moment to be used in design is:

$$M_c = \delta_b M_2$$

where M_2 = larger $(M_{ns} + \delta_s M_s)$
δ_s = moment magnifier for lateral drift effect (see Section 16.3.3(b) below)

In columns subjected to transverse loading between supports, the maximum primary moment may occur at a section away from the supports. In such cases, the value of the largest calculated moment occurring anywhere along the member must be used as M_2 in Eq. 16.10 and C_m taken as 1.0 in Eq. 16.11.

If calculations show that the factored moments at both ends of a column are zero or so small that the corresponding end eccentricities, $e = M/P$, are less than $(15 + 0.03h)$ mm, then the slender column must be designed for a moment M_2 in Eq. 16.10 corresponding to a minimum eccentricity of load of $(15 + 0.03h)$ mm, about each principal axis applied separately. (This minimum eccentricity is not intended to be applied about both axes simultaneously.) The ratio M_1/M_2 in Eq. 16.13 for

Fig. 16.8 *Calculation of column end moments, M_1 and M_2—columns in unbraced frames*

(i) Loading   (ii) Moment, M_{ns}

(a) Loadings which do not produce appreciable sway

(b) Load combinations which result in appreciable sway

this case is taken as the actual ratio of the calculated end moments if they are not zero, and as 1.0 when calculated end moments are zero.

3. *EI* values and P_c.

In calculating the critical load using the Euler buckling formula in Eq. 16.12, the choice of an *EI* value must reflect the effects of cracking, creep, and nonlinearity in the concrete stress-strain relation. Equations 16.14 and 16.15, specified in the Code in lieu of a more accurate calculation represent lower bound values obtained from an evaluation of theoretical and experimental results (Refs. 16.5, 16.7). Equation 16.14 generally gives more accurate values, while the simpler Eq. 16.15 gives more conservative results. The accuracy of Eq. 16.15 improves for lightly reinforced columns. More accurate values of *EI* can be obtained from moment-curvature relationships for the section.

$$EI = \frac{0.2\, E_c I_g + E_s I_{se}}{1 + \beta_d} = EI_{11}$$ (16.14)

or more conservatively,

$$EI = 0.25 E_c I_g = EI_{12}$$ (16.15)

where notations EI_{11} and EI_{12} for the two values, with subscripts 11 and 12, identifying the respective equation numbers in Section 10 of CSA A23.3-M84, are introduced here for convenience in calculations

I_{se} = moment of inertia of reinforcement about the centroidal axis of the member cross section,

β_d = ratio of maximum design dead load moment to maximum design total load moment, always positive.

The factor β_d approximately accounts for the reduction of EI due to creep under sustained loads. Equation 16.15 is simpler to evaluate. The second term in Eq. 16.14 depends on the amount and disposition of reinforcement. For design computations, it is convenient to express the more accurate value EI_{11} as $kbh^3/(1 + \beta_d)$, where k is a function of concrete strength, reinforcement location and steel ratio. Values of k for the various percentages and arrangements of reinforcement are presented in the interaction diagrams in Ref. 15.6 (see also Figs. 15.16 to 15.22). In terms of k,

$$EI_{11} = kbh^3/(1 + \beta_d) \qquad (16.16)$$

The effective length factor, k, in Eq. 16.12 may be determined using the alignment charts given in Section 16.2, or using Table 16.1. Alternatively, k may be taken as 1.0 in the computation of member stability effects.

(b) Lateral Drift Effect

Frames not braced against lateral deflection may undergo appreciable drift which causes magnification of column end moment (Figs. 16.3, 16.8b(v)). The magnified end moments, $\delta_s M_s$, in such a frame may be calculated by a second order analysis as explained in Section 16.3.2. In lieu of such an analysis, for columns with $k\ell_u/r \leqslant 100$, CSA A23.3-M84 (10.11.7) permits the approximate evaluation of lateral drift effect by multiplying the factored end moments, M_s, due to loads resulting in appreciable drift and calculated by conventional elastic frame analysis by a factor, δ_s, given by Eq. 16.17.

$$\delta_s = \cfrac{1}{1 - \cfrac{\Sigma P_f}{\phi_m \Sigma P_c}} \geqslant 1.0 \qquad (16.17)$$

where $\phi_m = 0.65$ and the summation applies to all columns in the storey that are rigidly attached to beams or footings and offers lateral restraint. Note that in a multistorey building structure, the floor system acts as a stiff diaphragm, equalizing the sway deflections of all the columns in a storey. Hence the moment magnification due to lateral drift effect for all the columns in a storey are related. δ_s given by Eq. 16.17 is applicable to all columns in the storey.

In the calculation of $P_c = \pi^2 EI/(k\ell_u)^2$ in Eq. 16.17, the factor k must be computed as for an unbraced frame (Fig. 16.6) and must be greater than 1.0. The column end moments magnified for the lateral drift effect, $\delta_s M_s$, must be used in Eq. 16.10 to apply the member stability magnifier, δ_b, and the final magnified factored moment, M_c, used for the design of the column (see Section 16.3.3(a)2 and Fig. 16.8b).

Figures 16.3 and 16.8b(v) show that, in frames not braced against sidesway, the beams at a joint are subjected to the amplified column moment. In order to avoid the formation of plastic hinges in the restraining beams, and to maintain the flexural restraint provided by the beams. A23.3-M84 (10.11.8) requires flexural members to be designed for the total magnified end moments, $M_{ns} + \delta_s M_s$, from all of the columns at the joint.

(c) Design Procedure

The column section must be designed to resist the factored axial load, P_f, and the magnified factored moment, M_c. However, the computation of M_c using Eqs. 16.10–16.17 requires a knowledge of the cross-sectional dimensions, as both EI and k depend on them. (It may be recalled that member properties, at least in relative terms, are needed even for the initial structural analysis.) Thus an iterative procedure is inevitable in the design process.

For the design of slender column sections using the moment magnifier method, two approaches may be used:

1. Start with a trial section, compute δ_s, δ_b, and M_c using the properties of the trial section, design the section for P_f and M_c so computed, compare with trial section and revise if necessary.
2. Assume initial δ values, design the section for P_f and M_c, compute actual δ_b and δ_s for the section designed, compare with initial values of δ and revise if necessary.

The design of slender columns is illustrated in Examples 16.1 and 16.2.

EXAMPLE 16.1

A critical loading pattern in a braced building frame produces the following forces in an interior column: $P_f = 3000$ kN; and, at both ends, $M_f = 290$ kN·m. The moments are all due to live load, and cause symmetrical single curvature bending. The unsupported length of the column is 6 m. Design the column. Use $f_c' = 30$ MPa and $f_y = 400$ MPa.

SOLUTION

A tied rectangular section column will be used, so that the interaction diagrams in Figs. 15.17–15.19 can be used.

1. Trial section
    Assume about 3 percent reinforcement. For this, Fig. 15.17 gives the maximum value of $P_f/A_g = 20$ MPa, for a column with minimum eccentricity.
    Minimum $A_g = 3000 \times 10^3/20 = 150 \times 10^3$ mm²
    The column, being relatively long, and subjected to sizable moment, may need a somewhat larger section. Select a section 400 mm wide and 500 mm deep, overall.

Fig. 16.9 *Example 16.1*

4 – No. 35

4 – No. 35
(A_s = 4000mm)

400

70   70

500

2. Check slenderness

$$\text{Limiting } k\ell_u/r = 34 - 12 \, M_1/M_2 \tag{15.1}$$
$$= 34 - 12 \times 1 = 22$$

Since the frame is braced, $0.5 < k < 1.0$. The properties of the beams framing into the columns are not given so that the alignment chart (Fig. 16.5) cannot be used. Hence, conservatively, k may be taken as unity (Code 10.11.2.1)).

$$\text{Actual } k\ell_u/r = 1 \times 6000/(0.3 \times 500)$$
$$= 40 > 22$$

The column is *long* and slenderness effects must be considered.

3. Magnified moment
Using the simpler Eq. 16.15:

$$EI_{12} = 0.25 \, E_c I_g$$
$$= 0.25 \times 5000 \sqrt{30} \times \frac{bh^3}{12}$$
$$= 571 \, bh^3 \text{ (see Fig. 15.17)}$$

If the more detailed Eq. 16.14 is to be used:

$$EI_{11} = \frac{kbh^3}{1 + \beta_d} \tag{16.16}$$

From Fig. 15.17, for 3 percent steel,

$$k = 996$$

Since
$$\beta_d = M_D/M_{(D+L)} = 0,$$

$$EI_{11} \quad 996 \times 400 \times 500^3$$
$$= 4.98 \times 10^{13} \text{ N} \cdot \text{mm}^2$$

$$P_c = \frac{\pi^2 EI}{(k\ell_u)^2} = \frac{\pi^2 \times 4.98 \times 10^{13}}{(1 \times 6000)^2} \times 10^{-3}$$
$$= 13.7 \times 10^3 \text{kN}$$

Magnification factor is (Eq. 16.11):

$$\delta_b = \frac{C_m}{1 - \dfrac{P_f}{\phi_m P_c}}$$

For symmetrical single curvature bending, $(M_1/M_2 = 1)$ and Eq. 16.13 gives $C_m = 1.0$

$$\delta_b = \frac{1}{1 - \dfrac{3000}{0.65 \times 13\ 700}} = 1.51$$

$$M_c = \delta_b M = 1.51 \times 290 = 438 \text{ kN} \cdot \text{m}$$

4. Compute area of steel

$$P_f/A_g = 3000 \times 10^3/(400 \times 500) = 15 \text{ MPa}$$

$$\frac{M_c}{A_g h} = \frac{438 \times 10^6}{400 \times 500^2} = 4.38 \text{ MPa}$$

With 40 mm cover, No. 10 ties and No. 35 bars,

$$\gamma = \frac{500 - (2 \times 40 + 2 \times 11.3 + 35.7)}{500} = 0.72$$

For $P_r/A_g = 15$ MPa, and $M_r/A_g h = 4.38$ MPa, Fig. 15.19 gives:

$$\gamma = 0.7, \quad \rho_g = 3.9 \text{ percent}$$
$$A_s = 0.039 \times 400 \times 500 = 7800 \text{ mm}^2$$

Select eight No. 35 bars, giving area $A_s = 8000$ mm², as shown in Fig. 16.9.

$$\text{Actual } \rho_g = \frac{8000 \times 100}{400 \times 500} = 4 \text{ percent}$$

The section initially selected is adequate and the reinforcement is within normal limits, and hence no revision is needed. The design is slightly on the conservative side because:

(i) $k = 1$ has been used while actual k should be less than 1.0.
(ii) for $\rho_g = 4$ percent and $\gamma = 0.70$, Fig. 15.19 shows $k = 1436$ compared to the 996 used in the *EI* computation,
(iii) actual $\gamma = 0.72$ while curves for $\gamma = 0.70$ are used.

Items (i) and (ii) have the effect of increasing the critical load P_c, and hence decreasing δ_b. Use of the lower γ value also gives conservative results. However the difference is not significant.

--------- **EXAMPLE 16.2** ---------

The frame dimensions, and column sizes based on preliminary calculations for an unbraced frame are shown in Fig. 16.10. The column forces and moments obtained from a conventional frame analysis for separate loadings of 1.25D, 1.5L, and 1.5Q (Q = wind load) are given in Table 16.2 below.

Fig. 16.10 *Example 16.2*

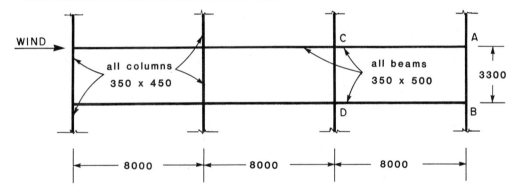

TABLE 16.2 Example 16.2

Column	AB			CD		
Loading	1.25D	1.5L	1.5Q	1.25D	1.5L	1.5Q
Mom. at top, kN·m	−56	−59	−51	−8	54*	−72
Mom. at bottom, kN·m	59	61	75	9	56*	93
Axial load, kN	590	612	67	1100	1130	8

*single curvature case considered

Note that for interior columns, live load is placed on all floors above, but in the immediate neighbouring floors, the loading causing single curvature bending is considered here. Design the columns. Use $f'_c = 30$ MPa and $f_y = 400$ MPa.

SOLUTION

1. Effective length factors

Gross I_g of columns and 0.5 I_g of beams will be used in the calculation of k factors using the alignment charts.

For beams,

$$\frac{0.5\, E_c I_g}{\ell_u} = \frac{1}{2} \times 5000\,\sqrt{30} \times \frac{350 \times 500^3}{12 \times (8000 - 450)}$$

$$= 6.61 \times 10^9\ \text{N} \cdot \text{mm}$$

For columns:

$$\frac{E_c I_g}{\ell_u} = 5000 \times \sqrt{30} \times \frac{350 \times 450^3}{12 \times (3300 - 500)}$$

$$= 26 \times 10^9\ \text{N} \cdot \text{mm}$$

$$\psi = \frac{\Sigma\,(EI/\ell)_{\text{col}}}{\Sigma\,(EI/\ell)_{\text{beam}}}$$

For exterior columns:

$$\psi_A = \Psi_B = \frac{2 \times 26 \times 10^9}{6.61 \times 10^9} = 7.87$$

From Fig. 16.6, $k = 2.68$

For interior columns:

$$\Psi_A = \Psi_B = \frac{2 \times 26 \times 10^9}{2 \times 6.61 \times 10^9} = 3.93$$

From Fig. 16.6, $k = 2.02$

The slenderness ratio is:

$$\frac{k\ell_u}{r} = \frac{2.02 \times 2800}{0.3 \times 450} = 41.9$$

This is less than 100 and therefore slenderness effects may be evaluated by the approximate method.

2. Loading cases and corresponding moments

The load combinations to be considered for applying slenderness effects are:

(i) $1.25D + 1.5L$
(ii) $1.25D + 1.5Q$
(iii) $1.25D + 0.7\,(1.5L + 1.5Q)$

For each of the above cases of loading, the parts of moment due to loads causing appreciable sway, M_s, must be magnified by the sway magnification factor, δ_s, and combined with the non-sway part of the moment, M_{ns}, before applying the magnifier

Table 16.3 Example 16.2

	Column AB (i) 1.25D + 1.5L M_{ns}	Column AB (ii) 1.25D + 1.5Q M_{ns} (1.25D)	Column AB (ii) M_s (1.5Q)	Column AB (iii) 1.25D + 0.7(1.5L + 1.5Q) M_{ns} 1.25D + 1.05L	Column AB (iii) M_s 1.05Q	Column CD (i) 1.25D + 1.5L M_{ns}	Column CD (ii) 1.25D + 1.5Q M_{ns} (1.25D)	Column CD (ii) M_s (1.5Q)	Column CD (iii) 1.25D + 0.7(1.5L + 1.5Q) M_{ns} 1.25D + 1.05L	Column CD (iii) M_s 1.05Q
1 Mom. at top	−115	−56	−51	−97.3	−35.7	46	−8	−72	29.8	−50.4
2 Mom. at bottom	120	59	75	102	52.5	65	9	93	48.2	65.1
3 Axial load, P_f	1202	590	67	1018	47	2230	1100	8	1891	6
4 Eff. length factor, k	2.68					2.02				
5 Ratio, β_d	0.49	0.52		0.42		0.17	0.10		0.39	
6 $P_c = 38\,000/[k^2(1 + \beta_d)]$	3550	3480		3730		7960	8470		6700	
7 δ_s		1.28		1.75			1.28		1.75	
8 $M_{ns} + \delta_s M_s$ at top, $= M_1$	−115	−121		−160		46	−100		−58.4	
9 $M_{ns} + \delta_s M_s$ at bottom, $= M_2$	120	155		194		65	128		162	
10 $C_m = 0.6 + 0.4\, M_1/M_2 \geq 0.4$	0.4	0.4		0.4		0.88	0.4		0.46	
11 For δ_b, $P_c = 38\,000/(1 + \beta_d)$	25 500	25 000		26 760		32 480	34 550		27 340	
12 $\delta_b \geq 1.0$	1.0	1.0		1.0		1.0	1.0		1.0	
13 $M_c = \delta_b M_2$, (kN·m)	120	155		194		65	128		162	
14 P_f, (kN)	1202	657		1065		2230	1108		1897	
15 $M_c/(A_g h)$, MPa	1.69	2.19		2.74		0.92	1.81		2.29	
16 P_f/A_g, MPa	7.63	4.17		6.76		14.2	7.03		12.0	

for the member stability effect, δ_b. For the three-load combinations, the moments M_s, M_{ns} and axial load are tabulated separately for exterior and interior columns in lines 1–3 in Table 16.3.

3. Sway magnification factor, δ_s
   Assuming 40 mm clear cover, No. 10 ties and No. 35 bars,

$$\gamma = \frac{450 - (2 \times 40 + 2 \times 11.3 + 35.7)}{450} = 0.7$$

The value of *EI* will be computed using Eq. 16.14. Assuming $\rho_g = 0.02$, for $\gamma = 0.7$, Table 15.19 gives $k = 946$, so that

$$EI_{11} = \frac{kbh^3}{1 + \beta_d} = \frac{946 \times 350 \times 450^3}{1 + \beta_d}$$

$$= \frac{3.02 \times 10^{13}}{1 + \beta_d} \, \text{N} \cdot \text{mm}^2$$

The Euler load, P_c, is

$$P_c = \frac{\pi^2 EI}{(k\ell_u)^2} = \pi^2 \times \frac{3.02 \times 10^{13}}{1 + \beta_d} \times \frac{1}{(k \times 2800)^2} \times 10^{-6}$$

$$= \frac{38\,000}{k^2\,(1 + \beta_d)} \, \text{kN} \tag{16.18}$$

Values of the effective length factor *k* already determined, factor β_d giving the ratio of maximum factored dead load moment to the maximum factored total load moment, and the Euler load P_c given by Eq. 16.18 are tabulated in Table 16.3 at lines 4–6.

Sway magnification factor, δ_s, is applicable for loading cases (ii) and (iii). Using Eq. 16.17,

$$\delta_s = \frac{1}{1 - \dfrac{\Sigma P_f}{\phi_m \Sigma P_c}} \geq 1.0$$

For loading case (ii) (note that for wind load alone, ΣP_f for a storey $= 0$),

$$\delta_s = \frac{1}{1 - \dfrac{2 \times 590 + 2 \times 1100}{0.65(2 \times 3480 + 2 \times 8470)}} = 1.28$$

For loading case (iii),

$$\delta_s = \frac{1}{1 - \dfrac{2 \times 1018 + 2 \times 1891}{0.65(2 \times 3730 + 2 \times 6700)}} = 1.75$$

The column end moments for computation of member stability effects are $M_{ns} + \delta_s M_s$, and these are tabulated in lines 8 and 9 of Table 16.3. Of these, the numerically larger end moment is taken as M_2 and the smaller as M_1. All columns are bent in double curvature except column CD under loading case (i).

4. Moment magnification for member stability effect
   The modifier, δ_b, for member stability effect is, by Eq. 16.11,

$$\delta_b = \frac{C_m}{1 - \dfrac{P_f}{\phi_m P_c}} \geq 1.0 \qquad \text{(16.19a)}$$

where
$$C_m = 0.6 + 0.4\, M_1/M_2 \geq 0.4 \qquad \text{(16.19b)}$$

In computation of δ_b, the critical load, P_c, may be computed conservatively with the effective length factor, k, taken as 1.0, so that

$$P_c = \pi^2 \frac{EI}{\ell_u^2} = \pi^2 \times \frac{3.02 \times 10^{13}}{1 + \beta_d} \times \frac{1}{2800^2} \times 10^{-3}$$

$$= \frac{38\,000}{1 + \beta_d}\, \text{kN} \qquad \text{(16.19c)}$$

Values C_m, P_c and δ_b calculated using Eqs. 16.19b, a, c are tabulated at lines 10–12 in Table 16.3. In all cases $\delta_b = 1$, indicating that the maximum magnified factored moment to be used in design, $M_c = \delta_b M_2 = M_2$, occurs at the column ends. The values of the magnified factored moment, M_c, and the factored axial load, P_f, to be used for design are given in lines 13 and 14 of Table 16.3.

5. Design of columns
   The design of the column for the moment, M_c, and load, P_f, is made as for short columns, illustrated in Chapter 15. Here use will be made of the interaction diagrams in Figs. 15.17–15.19.
   The stress parameters $M_c/(A_g h)$ and P_f/A_g for the columns are given at lines 15 and 16 in Table 16.3. From the combinations of these parameters for various loading cases and entering chart in Fig. 15.19 for $\gamma = 0.7$, the required reinforcement ratios are:

   for column, AB, $\rho_g = 0.01$ (for loading case (iii)), and

   for column CD, $\rho_g = 0.012$ (for loading case (iii)).

$$A_s = 0.012 \times 350 \times 450 = 1890\ \text{mm}^2$$

Four No. 25 bars give an area of 2000 mm² and $\rho_g = 0.0127$, and may be used uniformly for all columns. Note that EI value was calculated using Eq. 16.14 with the factor k taken as 946 for an assumed steel ratio of 0.02. For $\rho_g = 0.0127$, k and EI will be slightly lower and the computation may be revised for this.

--------------------------------- **PROBLEMS** ---------------------------------

1. For the problem in Example 16.1, check whether a column 350 mm × 500 mm with four No. 35 bars is adequate.
2. Design the columns in the problem in Example 16.2 (Fig. 16.10) if the storey height is 3.5 m, assuming the moments to be the same as given in Table 16.2

REFERENCES

16.1   Johnston, B.G., (ed) *SSRC Guide to Stability Design Criteria for Metal Structures*, 3rd ed., John Wiley and Sons, Inc., New York, 1976.

16.2   Kavanagh, T.C., *Effective Length of Framed Columns*, Transactions, ASCE, Vol. 127, Part II, 1962, pp. 81–101.

16.3   Salmon, C.G., and Johnson, J.E., *Steel Structures — Design and Behaviour*, 2nd ed., Harper and Row, New York, 1980.

16.4   Breen J.E., MacGregor, J.G., and Pfrang, E.O., *Determination of Effective Length Factors for Slender Concrete Columns*, ACI Journal, Proc. Vol. 69, No. 11, Nov. 1972, pp. 669–672.

16.5   *Commentary on Building Code Requirements for Reinforced Concrete ACI 318–83*, American Concrete Institute, Detroit, 1983.

16.6   Collins, M.P., Mitchell, D. and MacGregor, J.G., *Explanatory Notes on CSA Standard A23.3-M84*, Canadian Portland Cement Association, Ottawa, Ontario, Canada, 1985.

16.7   MacGregor, J.G., Breen, J.E., and Pfrang, E.O., *Design of Slender Concrete Columns*, ACI Journal, Vol. 67, No. 1, Jan. 1970, pp. 6–28.

16.8   Wood, B.R., Beaulieu, D., and Adams, P.F., *Column Design by P-Delta Method*, Proc., ASCE, Vol. 102, No. ST2, Feb. 1976, pp. 411–427.

16.9   Wood, B.R., Beaulieu, D., and Adams, P.F., *Further Aspects of Design by P-Delta Model*, Proc., ASCE, Vol. 102, No. ST3, Mar. 1976, pp. 487–500.

16.10 MacGregor, J.G., and Hage, S.E., *Stability Analysis and Design of Concrete*, Proc., ASCE, Vol. 103, No. ST10, Oct. 1977, pp. 19–53.

CHAPTER 17

FOOTINGS

17.1 GENERAL

The loads acting on all structures erected on the ground have to be transmitted to the supporting soil. The part of the structure, usually built below the ground surface, which interacts with the soil base and effects the transfer of loads from the parts above (superstructure) to the soil is identified as the foundation structure or substructure. In reinforced concrete structures, the loads from the superstructure are brought to the foundation level mostly through compression members such as columns, walls and piers. The average specified (service) load stresses in these compression members are usually in the range of 7–20 MPa, whereas the allowable bearing pressures on the soil base may range from 200 to 500 kPa and may be even less. Thus the safe transfer of the high intensity loads to the relatively weaker soil is the function of the foundation structure. When soil of sufficient strength is available within a short vertical distance below the ground surface, the load transfer can be effected by spreading the load over a sufficiently large area around the local point. This type of foundation structure is generally classified as a spread foundation, and consists of different types of footings (Fig. 17.1) and mats. Mat (also called raft) foundations are used when the bearing capacity of the soil is low, or when the columns loads are so large that spread footings, if used, will cover most of the area of the building. They consist of solid two-way slabs (flat plates or flat slabs, upside down), inverted slab-beam-girder systems, or other variations of these, that usually extend over the entire foundation area (Fig. 17.2). Because of their high rigidity, mat foundations reduce both total and differential settlement. In the case of very weak surface and near-surface soils, resort is made to deep foundations, such as piles and caissons.

Footings are the most widely used type of foundation in usual situations and are relatively cheap to build. This chapter is primarily concerned with the design of simple types of footings. The design of more complex forms of foundations is beyond the scope of this text, and for these reference may be made to books on foundation engineering (Refs 17.1, 17.2)

17.2 TYPES OF FOOTINGS

Different types of common footings are shown in Fig. 17.1. A footing supporting a single column is called an isolated footing or an individual column footing. Such

Fig. 17.1 Types of footings

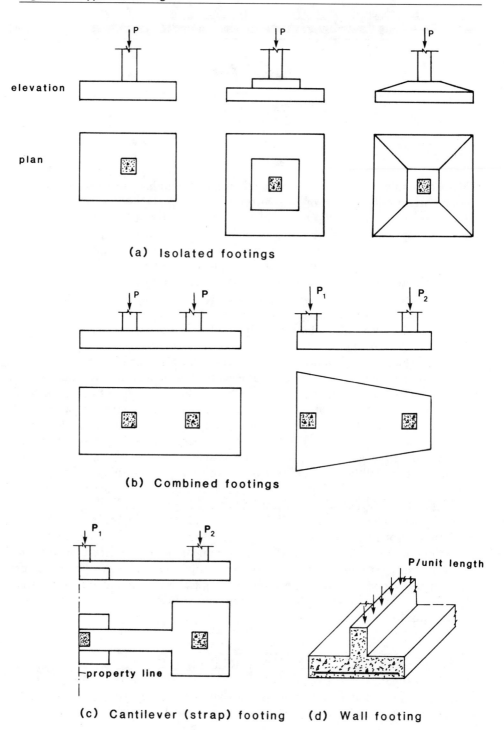

(a) Isolated footings

(b) Combined footings

(c) Cantilever (strap) footing   (d) Wall footing

Fig. 17.2 *Two types of mat foundation (flat plate, flat slab)*

footings may be of uniform thickness, or they may be stepped or sloped, varying in thickness to suit the bending moment variation. In plan, the shape is usually square or rectangular, although other shapes may be used in special situations. Footings supporting two or more columns are termed combined footings (Section 17.6, Fig. 17.1). A variation of this type is the cantilever (or strap) footing.

The soil reaction from below causes a column footing to bend into a saucer-like shape, and hence the footing is generally reinforced in two directions perpendicular to each other and parallel to the edges, similar to a flat slab supported on a column. Wall footings, similar to column footings, distribute the load from the wall to a wider area and are continuous throughout the length of the wall. However, they bend essentially in one direction only, transverse to the length of the wall, and hence are reinforced mainly in the transverse direction.

17.3 ALLOWABLE SOIL PRESSURE

The base area of a footing is selected so as to limit the maximum soil pressure below it, due to the applied loads, to within safe limits. The safe soil pressure that can be allowed at the foundation level is determined using the principles of soil mechanics. The primary considerations in the determination of the allowable soil pressure are: (1) that the soil does not fail under the applied loads; and (2) that the settlements, both overall and differential, are within the limits permissible for the structure. In general, structural damage is caused more by differential settlement than by a uniform vertical settlement or a slight rigid body rotation. Determination of allowable soil pressure is described in detail in Refs. 17.1 and 17.2.

The allowable soil pressure, q_a, given to the structural designer includes a suitable factor of safety. The safety factor ranges from 2 to 6, depending on the type of soil, and related uncertainties and approximations. Since the factor of safety is included in q_a, it is applicable to specified (service) load conditions. Hence the calculation for the required area of footing must be made with q_a and applied specified loads, or alternatively with an increased soil pressure and the factored loads.

For computing required areas of footings under combinations of specified loads including dead plus live plus wind or earthquake, etc., the load combination factor, ψ, given in CSA A23.3-M84 (9.2.4) may be taken into account in computing load effects.

The allowable pressure for a given depth, D, of foundation may be expressed as the **net** or the **gross** value. The net pressure is the value which can be safely carried at the foundation depth, in excess of the existing overburden pressure. The gross pressure is the total pressure that can be carried, including the pressure due to existing overburden. In the design examples given in this chapter, it is assumed that the allowable bearing pressure, q_a, indicated is the gross value. With the gross soil pressure known, the required area of footing must be determined from the total load (including the weight of the footing itself and of the backfill, if any) which is applied at the base of the footing.

If, instead of an allowable soil pressure, q_a, applicable for service load conditions, a factored soil bearing capacity, q_r, corresponding to ultimate limit state is given, the area of footing can be computed for the ultimate limit state under factored loads. In the examples in this chapter it will be assumed that the bearing capacity given corresponds to service load conditions.

17.4 DISTRIBUTION OF SOIL REACTION UNDER FOOTING

The distribution of the soil reaction acting at the base of the footing is somewhat non-uniform, and is dependent on the rigidity of the footing itself and the properties of the soil. However, it is usual practice in structural design, and generally satisfactory, to assume a linear distribution of pressure at the base of the footing.

In a symmetrically loaded footing, where the resultant vertical load passes through the centroid of the footing, as shown in Fig. 17.3a, the distribution of the soil pressure is assumed uniform, and its magnitude, q, is given by

$$q = \frac{P}{A} \leqslant q_a$$

where P is the total vertical load above the foundation level under combinations of specified loads; and A is the base area of the footing. Limiting q to the allowable soil pressure, q_a, will give the minimum required area of footing:

$$\text{required } A \geqslant \frac{P}{q_a} \tag{17.1}$$

The load combination for computing P under service load conditions is (CSA A23.3-M84 (9.2.2)):

Fig. 17.3 *Assumed distribution of soil reaction*

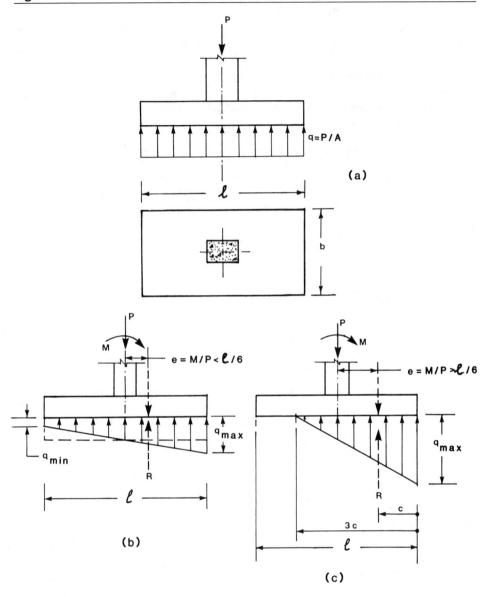

$$P = D + \psi (L + Q + T)$$

where the load combination factor, ψ, has the values given in Section 3.6.

When the resultant vertical load is eccentric with respect to the centroid of the footing, the soil pressure is non-uniform. The eccentricity may result from one or more of the following effects: (1) the column transmitting a moment in addition to the vertical load (Fig. 17.4a); (2) the column carrying a vertical load being offset from the centre (Fig. 17.4b); and (3) the column transmitting a lateral force located above the foundation level, in addition to the vertical load (Fig. 17.4c). In cases (1) and (3) above, if the moment and shear are fixed in direction and proportional to the load in magnitude, it is possible to offset the footing so that the resultant line of thrust passes through the centroid of the footing, and the pressure distribution is uniform (Fig. 17.4d). However, in practice, the moment (and lateral force) may vary in magnitude and alternate in direction (for example, the moments due to wind load), in which case offsetting the footing to obtain a uniform pressure distribution is not possible.

The general case of an eccentric loading, where the resultant vertical load, P, has an eccentricity, e, about one of the principal centroidal axes of the contact area between the footing and the soil below, is shown in Fig. 17.3b. Assuming the footing

Fig. 17.4 *Eccentric loading on footing*

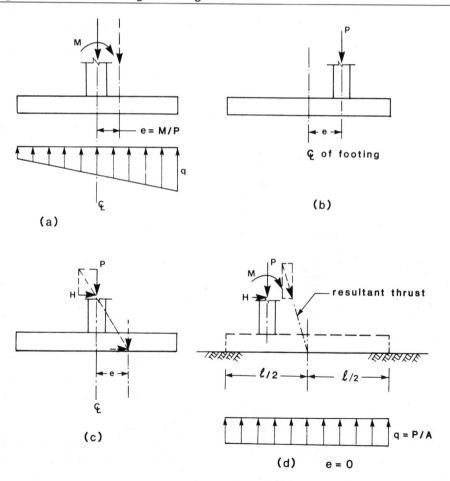

to be rigid, and a linear distribution of soil pressure, the latter may be computed by combining the pressures due to the vertical load and due to the moment. Thus the extreme values of the pressure, q_{min} and q_{max}, are given by

$$q_{min, \, max} = \frac{P}{A}\left(1 \pm \frac{6e}{\ell}\right) \qquad (17.2)$$

where A is the area of the footing ($b \times \ell$ in Fig. 17.3a), and ℓ is the length of the footing in the direction of bending. Equation 17.2 is valid so long as q_{min} is not tensile, that is $e \leq \ell/6$, or the resultant load acts within the middle third of the length, ℓ.

When $e > \ell/6$, Eq. 17.2 will indicate that q_{min} is tensile. Since little tension can be developed between the footing and the soil below, it is common practice to assume no tensile force at the interface, and then the pressure distribution is triangular as shown in Fig. 17.3c. The resultant R of the soil pressure must coincide with the resultant vertical load of eccentricity e. This condition gives the effective width of contact with the soil base as $3c$ as shown in Fig. 17.3c. The maximum soil pressure is twice the average value and is readily computed as:

$$q_{max} = \frac{2P}{3bc} \qquad (17.3)$$

In all cases considered above, the dimensions of the footing must be selected so that q_{max} does not exceed the allowable soil pressure, q_a. It should be realized that differential soil pressure at the base of the footing, such as shown in Fig. 17.3c, may lead to tilting of the footing, which in turn will reduce the degree of fixity at the column base. In general, the footing geometry must be selected to minimize differential soil pressure across the base, and particularly so if the soil is compressible.

17.5 GENERAL DESIGN CONSIDERATIONS FOR FOOTINGS AND CODE REQUIREMENTS

17.5.1 GENERAL

The base area required for a footing is determined from the forces and moments due to unfactored loads, in whatever combinations they may occur, and the allowable soil pressure, q_a (Eq. 17.1). Once the area has been determined, the subsequent structural design of the footing is done for the factored loads, following the Strength Design procedure of CSA A23.3 M84. In order to compute the factored moments, shears, etc., acting at critical sections of the footing, a fictitious factored soil pressure, q_f, corresponding to the factored loads, is calculated using the factored loads.

The major considerations in the structural design of the footing are flexure, shear (both one-way and two-way action), bearing, and development of reinforcement. In these aspects, the design procedures are similar to those for beams and two-way slabs supported on columns. Additional considerations for footings involve the transfer of force from the column to the footing, and in some cases where horizontal forces are involved, safety against sliding and overturning.

CSA A23.3-M84 (15.7) specifies a minimum depth of the footing above bottom reinforcement of 150 mm for footings on soil and 300 mm for footings on piles. For round and regular polygon shaped columns, the critical sections for moment, shear, and development of reinforcement in the footing may be located by replacing the actual column with an equivalent square column having the same area as the actual one.

17.5.2 BENDING MOMENT

The factored moment at any section, such as *AB* in Fig. 17.5a, is determined from the net upward soil pressure, q_f, acting on the part *ABCD* of the footing. Based on results of extensive tests, the section of maximum design moment is taken as: (1) at the face of the column, pedestal, or wall for footings supporting a concrete column, pedestal, or wall (Fig. 17.5a); and (2) halfway between the face and centreline of the wall for footings supporting masonry walls (Fig. 17.5b).

In one-way reinforced footings (for example, wall footings) and in two-way reinforced square footings, the flexural reinforcement is placed at a uniform spacing across the full width of the footing. In two-way reinforced rectangular footings, again the reinforcement in the long direction is uniformly spaced across the full width of the footing. However, in the short direction, the Code requires a larger concentration of reinforcement to be provided within a central band of width, *b*, centred on the column centreline. The amount of reinforcement so required in this band is given by:

$$\text{Reinforcement in band width } b = A_{s,\,\text{short}} \times 2/(\beta + 1) \qquad (17.4)$$

where *b* is the length of the short side of the footing, or the length of the supported wall or column, whichever is greater; $A_{s,\,\text{short}}$ is the total reinforcement in the short direction; and β is the ratio of the long side to the short side of the footing. The above portion of the reinforcement must be uniformly distributed within the width, *b*, and the remainder distributed uniformly in the outer portions of the footing. This is to approximately account for the non-uniform variation of the transverse bending moment along the length of the footing. In reinforced concrete footings, at all sections where flexural tension reinforcement is required, the amount of reinforcement provided must not be less than $1.4/f_y$.

17.5.3 SHEAR AND DEVELOPMENT LENGTH

Shear in footings, in general, involves both one-way and two-way shear. Wall footings are loaded along their full length, and under the soil pressure bend only one way (into a cylindrical surface) like an inverted double cantilever slab. Hence these need be designed only for one-way or beam shear. The critical section for one-way shear is taken, as for beams, at a distance, *d*, from the face of the wall or column

Fig. 17.5 *Critical sections for moment and shear*

(a)

(b)

(Fig. 17.5). Column footings may fail in one-way shear or two-way (punching) shear, and have to be designed for both. The behaviour of footings in two-way shear is identical to that of a two-way slab supported on columns, discussed in Chapter 14. The critical section for two-way shear is taken at a distance $d/2$ from the periphery of the column (Fig. 17.5, see also Fig. 14.16). The design procedures for one- and two-way shear are identical to those discussed in Chapters 6 and 14, respectively. However, shear reinforcing is generally avoided in footings, and the factored shear force, V_f, is kept below the factored shear resistance provided by concrete, V_c, by providing the necessary depth. An exception to this is the situation where the thickness of footing that can be provided is restricted. The factored shear resistance of concrete, V_c, as given in Eq. 6.11 and Eq. 14.20 is taken as:

For one-way (beam) shear,

$$V_c = 0.2\lambda\phi_c\sqrt{f_c'}\, b_w d$$

For two-way shear,

$$V_c = (1 + 2/\beta_c)0.2\lambda\phi_c\sqrt{f_c'}\, b_o d \leq 0.4\lambda\phi_c\sqrt{f_c'}b_o d$$

where, β_c is the ratio of the longer to the shorter dimension of the loaded area. Because of this need to limit the shear stress, the depth of footing is often dictated by shear considerations, and the design for shear usually precedes the flexural design.

The critical sections for development of flexural reinforcement are the sections of maximum moment, and all other sections where there is a change in section or in reinforcement. Further, where the transfer of force at the base of the column, from the column to the footing, is accomplished by reinforcement, such reinforcement must also have adequate development length on each side (see Section 17.5.4).

17.5.4 TRANSFER OF FORCES AT BASE OF COLUMN

All forces (axial, shear and moment) applied at the base of the column (or pedestal) must be transferred to the footing by compression in concrete or by reinforcement, dowels, or mechanical connectors between column and footing. The compressive force in the contact surface between the column and the footing at the top of the footing is limited to the factored bearing resistance, F_{br}, for either surface, given by Eq. 17.5. If the actual compressive force exceeds F_{br}, the excess is carried by reinforcement, dowels, or mechanical connectors. In the case of tensile forces in the column, the entire force must be transferred to the footing by reinforcement, dowels, or mechanical connectors. For transferring the column moment at its base to the footing, it may be necessary to use the same amount of reinforcement in the footing as in the column. All reinforcement provided across the interface to facilitate transfer of column loads must have adequate development length (in compression) on each side of the interface. This reinforcement may consist of the longitudinal reinforcement in the column extended into the footing or may be separate dowel bars (Fig. 17.6a). The diameter of dowels must not exceed that of the column bars by more than one bar size. The reinforcement across the interface must have an area not less than 0.5 percent of the area of the supported column (or pedestal).

The factored bearing resistance, F_{br}, is specified in CSA A23.3-M84 (10.15) as:

$$F_{br} = 0.85\phi_c f_c' A_1\sqrt{A_2/A_1} \tag{17.5}$$
$$\leq 1.7\phi_c f_c' A_1$$

where A_1 is the loaded area; A_2 is the maximum area of the portion of the supporting surface that is geometrically similar to and concentric with the loaded area; and $\sqrt{A_2/A_1}$ must not be taken larger than 2. The definition of area A_2 is illustrated in Fig. 17.6b. The factor $\sqrt{A_2/A_1}$ allows for the increase in strength of the concrete in the bearing area resulting from the confinement offered by the surrounding con-

Fig. 17.6 *Transfer of stresses at column base*

(a) Reinforcement across column-footing interface

(b) Definition of area A_2,
CSA A23·3-M84(10·15)

(c) Shear key

crete. In bearing zones, the high axial compressive stresses give rise to transverse tensile stresses which may lead to lateral splitting, bursting, or spalling of concrete. Where this possibility exists, transverse and confinement reinforcement must be provided to resist these effects.

Capacity to transfer horizontal shear forces between column and footing may be evaluated using the shear–friction method (see Chapter 6). Additional capacity, if required, must be provided by shear keys (Fig. 17.6c) or other mechanical means.

Overturning and sliding of foundations may be significant considerations only when the horizontal loads are very high and the gravity loads relatively low as in the case of foundations of retaining walls. This is seldom the case with footings of columns in usual building structures. However, where these effects are significant, the critical loading condition should be the one maximizing the horizontal loads which tend to destabilize the foundation, and minimizing the gravity loads which stabilize it. Thus all live gravity loads (except those which contribute to the sliding or overturning) are excluded and the dead loads which resist overturning are multiplied by a load factor of 0.85.

Design of wall footings and isolated column footings are illustrated in Examples 17.1–17.3. Combined footings are described in Section 17.6.

EXAMPLE 17.1

Design of Wall Footing

A concrete wall, 325 mm thick, carries unfactored loads per running metre of 80 kN dead load and 220 kN live load. Allowable soil pressure at a depth of 1.25 m below grade is 210 kN/m². Design the footing. Take $f_c' = 20$ MPa, $f_y = 400$ MPa.

SOLUTION

The design is made for a typical 1 m length of wall (see Fig. 17.7).

1. Width of footing
    It is assumed that the allowable soil pressure is the gross pressure at a depth of 1.25 m. Estimating the thickness of footing to be 0.3 m, and assuming the unit weight of concrete and soil as 24 kN/m³ and 16 kN/m³, respectively (it is assumed that the soil will be backfilled to grade), the net soil pressure allowable at a depth 1.25 m is:

    $$q_a = 210 - 0.3 \times 24 - (1.25 - 0.3) \times 16 = 188 \text{ kN/m}^2$$

    Required width of footing is:

    $$b = \frac{80 + 220}{188} = 1.60 \text{ m}$$

2. Factored pressure
    The weight of footing and weight of overburden are uniformly distributed loads and are balanced by the uniform soil pressure, and hence do not cause any bending moment or shear force in the footing. The soil pressure for structural design of the footing is that due to the load transmitted through the wall. (The footing, by itself, buried in the soil is not subjected to moments and shears.)

    $$\text{Factored load} = 1.25 \times 80 + 1.5 \times 220 = 430 \text{ kN}$$

    Soil pressure corresponding to this load is:

Fig. 17.7 *Example 17.1*

concrete wall

GL

325

1250

1000

75

637.5

75

300

1600

$$q_f = \frac{430}{1.6 \times 1} = 269 \text{ kN/m}^2 \text{ (kPa)}$$

For design computations, it may be convenient to take $q_f = 0.269$ MPa (N/mm^2) and to use all dimensions in mm. This will be followed in the examples here.

3. Depth for shear
    Critical section for shear is at a distance d (mm) from the wall face. The factored shear is:

$$V_f = 0.269 \times 1000 \times \left(\frac{1600 - 325}{2} - d \right)$$

$$= 269 \, (638 - d) \text{ N}$$

$$V_c = 0.2\lambda\phi_c\sqrt{f_c'}b_wd = 0.2 \times 1.0 \times 0.6 \times \sqrt{20} \times 1000d = 537d \text{ N}$$

The minimum depth required to limit $V_f \leqslant V_c$ is given by:

$$269(638 - d) \leqslant 537d$$

Solving, $d = 213$ mm

An overall thickness of 300 mm, with a cover of 75 mm and No. 15 bars will give $d = 300 - 75 \quad 16/2 = 217$ mm, which is adequate.

4. Design for moment
   Maximum moment at wall face is:

$$M_f = 0.269 \times 1000 \times \left(\frac{1600 - 325}{2}\right)^2 \times \frac{1}{2} = 54.7 \times 10^6 \, \text{N} \cdot \text{mm}$$

$$K_r = \frac{M_f}{bd^2} = \frac{54.7 \times 10^6}{1000 \times 217^2} = 1.16 \, \text{MPa}$$

From Table 5.4,

$$\rho = 0.00364 \qquad\qquad \rho_{\text{max}} = 0.0153$$
$$\rho_{\text{min}} = 1.4/400 = 0.0035$$
$$A_s = 0.00364 \times 1000 \times 217 = 789 \, \text{mm}^2$$

No. 15 bars at 250 mm gives $A_s = 800$ mm².

Required development length for No. 15 bars (Table 9.1) is 371 mm. Actual length available, assuming an end cover of 75 mm is:

$$\frac{1}{2}\left(1600 - 325\right) - 75 = 563 \, \text{mm} > 371 \, \text{mm} \qquad\qquad \text{OK}$$

The footing dimensions are given in Fig. 17.7.

The Code does not specify any temperature and shrinkage reinforcement for footings. Since it is covered by earth, the temperature stresses may not be significant in a footing. In a wall footing which is in the form of a long narrow strip, some secondary bending moment may develop as a result of slight differential settlements along the length of the wall, and it will be good practice to provide some secondary reinforcement in the longitudinal directions.

For the footing in this example, four No. 15 bars may be provided. (This gives a steel ratio of 0.0017.)

Factored bearing resistance in interface is (without factor $\sqrt{A_2/A_1}$),

$$\begin{aligned}
F_{br} &= 0.85 \, \phi_c f'_c A_1 \\
&= 0.85 \times 0.6 \times 20 \times 325 \times 1000 \times 10^{-3} = 3315 \, \text{kN} \\
&> P_f = 430 \, \text{kN} \qquad\qquad \text{OK}
\end{aligned}$$

EXAMPLE 17.2

Design of Square Isolated Footing

A tied column, 450 mm square, and reinforced with eight No. 35 bars carries unfactored dead load of 1300 kN and live load of 1000 kN. Suitable soil with an allowable soil pressure of 300 kN/m² is available at a depth of 1.5 m. Design a square footing. For the column concrete f'_c is 30 MPa and for the footing is 20 MPa. $f_y = 400$ MPa.

SOLUTION

1. Size of footing
   Assuming a depth of 750 mm for the footing, and unit weights of 24 kN/m³ and 16 kN/m³ for concrete and soil, respectively, the net allowable soil pressure at depth 1.5 m is:

   $$q_a = 300 - 0.75 \times 24 - 0.75 \times 16 = 270 \text{ kN/m}^2$$

   $$\text{Required area of footing} = \frac{1300 + 1000}{270} = 8.52 \text{ m}^2$$

   $$\text{Minimum size} = \sqrt{8.52} = 2.92 \text{ m}$$

   Select a 3 m × 3 m footing.

2. Design for shear
   The factored load $= 1.25 \times 1300 + 1.5 \times 1000 = 3125$ kN
   Factored soil pressure $q_f = 3125/(3 \times 3) = 347$ kN/m²
   $= 0.347$ N/mm²

   (a) One-way or beam shear
   The critical section is at a distance d (in mm) from the column face. Referring to Fig. 17.8b:

   $$V_f = 0.347 \times 3000 \left(\frac{3000 - 450}{2} - d \right) \text{N}$$

   $$= 1040 \, (1275 - d) \text{ N}$$

   Factored shear resistance of concrete, $V_c = 0.2 \times 1 \times 0.6 \times \sqrt{20} \times 3000d$

   $$= 1610d \text{ N}$$

   To limit shear force, V_f, to less than V_c,

   $$1040(1275 - d) \leqslant 1610d$$

   Solving, $\qquad\qquad\qquad\qquad d \geqslant 500 \text{ mm}$

   (b) Two-way shear
   Critical section is at $d/2$ (in mm) from the periphery of the column. Factored shear force is:

   $$V_f = 0.347 \, [3000^2 - (450 + d)^2] \text{ N}$$
   $$V_c = (1 + 2/\beta_c)0.2\lambda\phi_c\sqrt{f'_c}b_od \leqslant 0.4\lambda\phi_c\sqrt{f'_c}b_od$$
   $$= 0.4\lambda\phi_c\sqrt{f'_c}b_od, \text{ since } \beta_c = 1 \text{ (square column)}$$
   $$= 0.4 \times 1 \times 0.6 \times \sqrt{20} \times 4(450 + d)d$$
   $$= 4.29 \, (450 + d)d \text{ N}$$

   Substituting in the equation:

   $$V_f \leqslant V_c$$
   $$0.347[3000^2 - (450 + d)^2] \leqslant 4.29(450 + d)d$$

Fig. 17.8 *Example 17.2*

(a)

(b)

Solving, $d \geq 605$ mm

(Note that it is usually satisfactory, and always conservative, to neglect the upward pressure in the punching failure block, that is, the area $(450 + d) \times (450 + d)$, and assume $V_f = P_f = 3125$ kN in this case. Also, instead of getting an exact solution as above, a depth may be assumed and then the computed factored shear resistance, V_c, compared with V_f. This will greatly simplify the calculations.)

The depth for two-way shear controls in this example. Taking the average d as 605 mm, and assuming a cover of 75 mm and No. 20 bars, the thickness of footing required is:

$$h = 605 + 75 + 20 = 700 \text{ mm}$$

Select $h = 700$ mm. The depth to the centre of reinforcement will differ by one bar diameter for the reinforcement along the two directions. For relatively deep square footings, it is adequate to use an average depth, measured to the interface between the two layers, for the reinforcement computations, and provide the reinforcement as computed in both directions (see Fig. 17.8).

$$d_{av} = 700 - 75 - 20 = 605 \text{ mm}$$

3. Design for flexure
   Factored moment, M_f, at column face is:

$$M_f = 0.347 \times 3000 \, (1275)^2/2 = 8.46 \times 10^6 \text{ N} \cdot \text{mm}$$

$$K_r = \frac{M_f}{bd^2} = \frac{8.46 \times 10^8}{3000 \times 605^2} = 0.771 \text{ MPa}$$

$$\rho = 0.00236 < \rho_{max} = 0.0153$$

$$\rho_{min} = 1.4/400 = 0.0035 > \rho$$

$$\rho = \rho_{min} = 0.0035$$

$$A_s = 0.0035 \times 3000 \times 605 = 6350 \text{ mm}^2$$

Provide twenty-two No. 20 bars at uniform spacing in each direction giving $A_s = 6600$ mm^2 (spacing $= 136$ mm), or alternatively, thirteen No. 25 bars may be used.

Required development length for No. 20 bars (Table 9.1) $= 510$ mm, and for No. 25 bars $= 850$ mm.

Length available $= 1275 - 75 = 1200$ mm, which is adequate with either bar size.

4. Transfer of force at column base
   Actual factored compressive force is:

$$P_f = 3125 \text{ kN}$$

The factored bearing resistance is given by

$$F_{br} = 0.85 \, \phi \, f_c' \, A_1 \, \sqrt{A_2/A_1}$$

For column face, $F_{br} = 0.85 \times 0.6 \times 30 \times 450^2 \times 1 \times 10^{-3} = 3098$ kN
< 3125 kN

For footing, face, $\sqrt{A_2/A_1} = \sqrt{3000^2/450^2} = 6.67$, limited to 2
$F_{br} = 0.85 \times 0.6 \times 20 \times 450^2 \times 2 \times 10^{-3} = 4131$ kN > 3125 kN   OK

The compressive force exceeds the factored bearing resistance of concrete at column face by 27 kN and this must be transferred by reinforcement, dowels, or mechanical connectors. In this case, the minimum reinforcement provided across the interface (see below) is more than adequate to transfer this excess force of 27 kN.

The minimum reinforcement required to be provided across the interface, at 0.5 percent of column area, is:

$$A_s = 0.005 \times 450^2 = 1013 \text{ mm}^2$$

Provide four No. 20 bars giving $A_s = 1200 \text{ mm}^2$.
Development length required for these bars in compression is:

$$\ell_d = \frac{0.24 \, f_y d_b}{\sqrt{f'_c}} = \frac{0.24 \times 400 \times 19.5}{\sqrt{20}} = 419 \text{ mm}$$

or,
$$\not< 0.044 \, f_y d_b = 0.044 \times 400 \times 19.5 = 343 \text{ mm}$$

Available vertical embedment length within the footing is

$$= d - d_b = 605 - 20 = 585 \text{ mm} > 419 \text{ mm} \qquad \text{OK}$$

These bars must also be extended into the column for a length equal to their development length, $\ell_d = 419$ mm. The details of the footing are given in Fig. 17.8.

EXAMPLE 17.3

Design of a Rectangular Footing

Redesign the footing for the column in Example 17.2, if one of the dimensions of the footing is limited to a maximum of 2.5 m due to space restrictions. Use $f_y = 400$ MPa and $f'_c = 25$ MPa.

SOLUTION

1. Size of footing
   From step 1 of Example 17.2, the required area of footing = 8.52 m². With a width of 2.5 m, required length $\ell = 8.52/2.5 = 3.41$ m. Select a rectangular footing, 3.5 m × 2.5 m, as shown in Fig. 17.9.

2. Depth required for shear
   An exact solution for required depth for shear may be obtained using the condition $V_f \leq V_c$, as was done in the previous two examples. However, in this example a trial and error procedure will be used.

   Estimate required footing thickness as 700 mm. Assume cover of 75 mm and No. 25 bars as flexural reinforcement. The bottom layer of reinforcement will be placed in the long direction in which the moment is greater. The effective depths are:
   In long direction, $d = 700 - 75 - 25.2/2 = 612$ mm.
   In short direction, $d = 700 - 75 - 1.5 \times 25.2 = 587$ mm.
   Average d for shear calculation = 600 mm.

Fig. 17.9 *Example 17.3*

(a)

11 No. 25
at uniform
spacing

12 No. 25 bars in 2.5m

(b)

Factored soil pressure $q_f = 3125/(3.5 \times 2.5) = 357$ kN/m^2 = 0.357 MPa.

Critical section for two-way shear is at $d/2 = 300$ mm from column perimeter, as shown in Fig. 17.9b. Factored shear force, V_f, is:

$$V_f = 0.357 [3500 \times 2500 - 1050^2]$$
$$= 2.73 \times 10^6 \text{ N}$$
$$V_c = 0.4 \times 1.0 \times 0.6 \times \sqrt{25} \times 4 \times 1050 \times 600$$
$$= 3.02 \times 10^6 \text{ N}$$
$$V_c > V_f \qquad\qquad\qquad \text{OK}$$

Critical section for one-way shear is at $d = 600$ mm from column face, as shown in Fig. 17.9b.

$$V_f = 0.357 \times 2500 \times 925$$
$$= 826 \times 10^3 \text{ N}$$
$$V_c = 0.2\lambda\phi_c\sqrt{f_c'}b_wd$$
$$= 0.2 \times 1.0 \times 0.6 \times \sqrt{25} \times 2500 \times 612$$
$$= 918 \times 10^3 \text{ N}$$
$$V_c > V_f \qquad\qquad\qquad\qquad\qquad\qquad\qquad \text{OK}$$

The depth is adequate for shear.

3. Design for flexure
   In the long direction,

$$M_f = 0.357 \times 2500 \times 1525^2/2 = 1.04 \times 10^9 \text{ N·mm}$$

$$K_r = \frac{1.04 \times 10^9}{2500 \times 612^2} = 1.11 \text{ MPa}$$

$$\rho = 0.00342 < \rho_{max} = 0.0191$$

$$< \rho_{min} = 0.0035$$

Take $\rho = \rho_{min} = 0.0035$

$$A_{s,\text{ long}} = 0.0035 \times 2500 \times 612 = 5360 \text{ mm}^2$$

Provide eleven No. 25 bars at uniform spacing. Spacing is within permissible limits.
In the short direction,

$$M_f = 0.357 \times 3500 \times 1025^2/2 = 656 \times 10^6 \text{ N·mm}$$

$$K_r = \frac{656 \times 10^6}{3500 \times 587^2} = 0.544 \text{ MPa}$$

Reinforcement required for this is less than the minimum of $\rho = 0.0035$, and
the latter controls.

$$A_s = 0.0035 \times 3500 \times 587 = 7190 \text{ mm}^2$$

Fifteen No. 25 bars give an area of 7500 mm².
   The area of reinforcement to be provided within a central band of width $b = 2.5$ m is:

$$A_s \times \frac{2}{\beta + 1} = 7190 \times \frac{2}{(3.5/2.5) + 1} = 5990 \text{ mm}^2$$

that is, twelve No. 25 bars.
   In this case, provide twelve bars within the middle band of width 2.5 m, and
two bars each on the two outer segments making up a total of sixteen bars (see
Fig. 17.9). The spacings are within limits ($< 3h$ and 500 mm).
   Required development length for No. 25 bars is $0.019\, A_bf_y/\sqrt{f_c'} = 850$ mm
(Table 9.1). Available least length (in the short direction) is $1025 - 75$ (cover) $=$
950 mm which is greater than required 850 mm.

4. Transfer of force at base of column
    Actual factored compressive force at interface = 3125 kN. For footing,
    $\sqrt{A_2/A_1} = \sqrt{2500^2/450^2} = 5.5$, limited to 2.

    $F_{br} = 0.85 \times 0.6 \times 25 \times 450 \times 450 \times 2 \times 10^{-3} = 5164$ kN > 3125 kN OK

    As in Example 17.2, the minimum requirement of four No. 20 bars across the
    interface will suffice, and this will have adequate length of embedment for development.

17.6 COMBINED FOOTINGS

A footing supporting more than a single column or wall is called a combined footing
(Eq. 17.1). In soils having very low allowable soil pressures, various forms of mat
foundations supporting several columns are used (Refs. 17.1–17.3). However, even
in soils having sufficient bearing capacity for the use of individual footings, some
combined footings may become necessary under the following situations:
1. The columns are so closely spaced that individual footings, if used, will overlap
    in areas.
2. An exterior column along the periphery of the building is so close to the property
    line that an individual footing, symmetrically placed, will project beyond the
    boundary.

   The soil pressure under a combined footing may be determined assuming a straight
line distribution, or by analysis as a beam on an elastic foundation. The former
method is approximately true for relatively rigid footings, while for flexible footings
the latter method is more appropriate (Refs. 17.1, 17.4). Two-column footings can
be considered rigid and the soil pressure distribution determined assuming a straight
line variation.

   Figure 17.10 shows examples of two-column footings. It was previously mentioned
that non-uniform soil pressure may lead to the tilting of the footing. In order to
avoid such a situation, the footing geometry is so selected that the centroid of the
footing area coincides with the resultant of the column loads (including consideration
of the moments and lateral forces, if any). With this arrangement, the soil pressure
under the footing may be assumed to be uniform. The combined two-column footing
may be rectangular (or trapezoidal with the wider end near the heavier load), when
the exterior column, which has the space limitation for an independent footing,
carries a lesser load than the interior column (that is, $\bar{x} > s/2$, as shown in Fig.
17.10a,b,c). When the exterior column carries the heavier load, $\bar{x} < s/2$, a trapezoidal
footing, as shown in Fig. 17.10e, is needed to align the load resultant, R, and the
centroid, c, of the footing.

17.7 DESIGN OF TWO-COLUMN COMBINED FOOTING

Referring to Fig. 17.10, in usual situations, the edge distance, a, and the distance,
$\bar{x}$, to the resultant column load are known. The plan dimensions of the footing can
be selected to satisfy the two requirements:

Fig. 17.10 *Two-column combined footing*

1. Area of footing, $A = $ Total load$/q_a$.
2. Centroid of footing area must coincide with resultant of column loads.

When a rectangular footing is used, as in Fig. 17.10a, the second requirement gives $\ell = 2(\bar{x} + a)$, and then $b = A/\ell$. In a trapezoidal footing (Fig. 17.10c,e) usually the length ℓ is selected first and the dimensions b_1 and b_2 computed using the two requirements above.

Once the dimensions of the footing area are determined, the uniform factored soil pressure may be computed as $q_f = $ *factored load/area*. A simplified and usually conservative load transfer arrangement, useful in the structural design of the footing, is shown in Fig. 17.11. The footing is considered as a uniformly loaded wide longitudinal beam supported by transverse beam segments at the column line. The longitudinal distribution of bending moments and shear forces may be determined

Fig. 17.11 *Assumed load transfer in two-column footing*

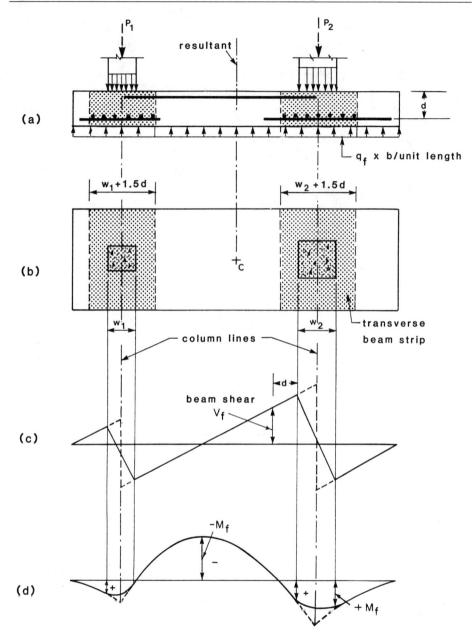

from statics, as shown in Fig. 17.11c,d. The critical sections for one-way or beam shear are at a distance *d* from the column face, and for two-way shear at *d*/2 from the column perimeter. The footing depth, *d*, is determined from shear strength considerations. The flexural reinforcement in the longitudinal direction is designed for the positive moment at the face of the column (reinforcement placed at the bottom) and for the maximum negative moment (top reinforcement) between the columns. Since the design sections for both shear and moment are outside the column width, the shear and moment diagrams may be obtained assuming the column loads as concentrated at their centrelines, as shown by the dashed lines in Fig. 17.11c,d. The values of the design shear and moments are unaffected by this simplification.

In order for the footing to act as a wide beam supported at the column lines, two narrow transverse bands of the footing below the columns act as transverse beams (see Fig. 17.11b). These two bands, acting as transverse beams cantilevered from the columns and supporting the longitudinal wide beam, must be designed for the associated bending moment and shear force. The effective width of footing acting as a transverse beam may be taken approximately as the width of column plus $0.75d$ on either side of the column (Ref. 17.1).

The design of a rectangular combined footing is illustrated in Example 17.4. The procedure just described is the conventional approach assuming the footing as rigid. Slightly more economical designs will result by the use of the beam-on-elastic foundation method (Refs. 17.1, 17.4). More detailed discussions of various types of spread foundations are contained in Refs. 17.1 and 17.2.

--- **EXAMPLE 17.4** ---

The layout and relevant data for an exterior and adjacent interior column are given in Fig. 17.12. The allowable soil pressure at the base of the footing, 1.5 m below grade, is 250 kN/m². Concrete for the footing has $f'_c = 25$ MPa, and $f_y = 400$ MPa. Design a combined footing.

SOLUTION

1. Footing dimensions

   Assuming a depth of footing of 750 mm, and unit weights of concrete at 24 kN/m³ and of soil at 16 kN/m³, net allowable soil pressure is:

   $$q_a = 250 - 0.75 \times 24 - 0.75 \times 16 = 220 \text{ kN/m}^2$$

   Area of footing required is given by:

   $$A = (600 + 400 + 1000 + 800)/220 = 12.7 \text{ m}^2$$

   In arranging footing dimensions to obtain a uniform soil pressure distribution, q_f, at factored loads, the distance $\bar{x}$ (see Fig. 17.10b) to be used must be the value corresponding to factored loads (rather than the value of $\bar{x}$ at service loads. $\bar{x}$ at service loads and at factored loads may be somewhat different if the dead load to live load ratios are different for the two columns).

   The factored loads are:

   For exterior column, $P_{f1} = 1.25 \times 600 + 1.5 \times 400 = 1350$ kN
   For interior column, $P_{f2} = 1.25 \times 1000 + 1.5 \times 800 = 2450$ kN

Fig. 17.12 Example 17.4

Column	1(Ext.)	2(Int.)
Size	400 mm²	500 mm²
Reinforcement	4–No. 35	4–No. 35 plus
		4–No. 30
Dead Load	600 kN	1000 kN
Live Load	400 kN	800 kN
f'_c	30 MPa	30 MPa
f'_y	400 MPa	400 MPa

$$\bar{x} = \frac{2450 \times 4.5}{(1350 + 2450)} = 2.90 \text{ m}$$

Since $\bar{x} > s/2 = 2.25$ m, a rectangular combined footing can be used. To locate the centroid of the rectangular footing at a distance $\bar{x}$ from the centre of the exterior column, the footing length must be:

$$\ell = 2 \times (2.90 + 0.20) = 6.2 \text{ m}$$

Selecting a length of 6.2 m, the required width of footing, b, for providing an area of 12.7 m² is:

$$b = \frac{12.7}{6.2} = 2.05 \text{ m}$$

Select a footing size 6.2 m × 2.10 m, as shown in Fig. 17.13a. The actual factored soil pressure at factored load is:

$$q_f = \frac{1350 + 2450}{6.2 \times 2.1} = 292 \text{ kN/m}^2$$

$$= 0.292 \text{ MPa}$$

2. Shear force and bending moment in the longitudinal direction

Fig. 17.13 *Design example 17.4*

(a) Footing dimensions

(b) Loading

(c) Shear in longitudinal beam (kN)

(d) Moments, kN.m

(e) Transverse beams

Considering the footing in the longitudinal direction as a wide beam, supported on the columns, the distributed upward load per unit length of this beam is (Fig. 17.13b):

$$= 292 \times 2.1 = 613 \text{ kN/m}$$

The distribution of shear and moment are shown in Fig. 17.13c, d. The section of zero shear, a distance x from the exterior edge is obtained as

$$613 \, x = 1350$$
$$x = 2.20 \text{ m}$$

The maximum moment at this section is:

$$M = \frac{613 \times 2.2^2}{2} - 1350 \times (2.2 - 0.2) = -1217 \text{ kN·m}$$

3. Design depth for shear
   (a) Wide-beam action
       The maximum shear is at the inside face of an interior column, and equal to 1377 kN. The shear at the critical section a distance d (mm) from the face of the column is (Fig. 17.13c):

$$V_f = 1377 - 613 \times \frac{d}{1000} \text{ kN} = (1377 - 0.613d) \text{ kN}$$

$$V_c = 0.2\lambda\phi_c\sqrt{f'_c}b_w d = 0.2 \times 1.0 \times 0.6 \times \sqrt{25} \times 2100d \times 10^{-3}$$
$$= 1.26d \text{ kN}$$

Substituting in $\qquad\qquad V_f \leqslant V_c,$

$$1377 - 0.613d \leqslant 1.26d$$

Solving, $\qquad\qquad d \geqslant 735 \text{ mm}$

   (b) Two-way shear at exterior column
       An exact value for the required d can be obtained by setting up the equation $V_f \leqslant V_c$ and solving for d, as was illustrated in Example 17.2. Here, the stress will be checked using the value $d = 735$ mm determined for one-way shear.
       The critical section is at $d/2 = 368$ mm from the column perimeter (Fig. 17.13a). The shear is:

$$V_f = 1350 - 292 \times (0.4 + 0.368)(0.4 + 0.735)$$
$$= 1095 \text{ kN}$$

For square column, $\beta_c = 1$, and

$$V_c = 0.4\lambda\phi_c\sqrt{f'_c}b_o d$$
$$= 0.4 \times 1.0 \times 0.6 \times \sqrt{25} \times (768 \times 2 + 1135) \times 735 \times 10^{-3}$$
$$= 2356 \text{ kN}$$
$$V_f < V_c \qquad\qquad\qquad\qquad\qquad\qquad\qquad\qquad \text{OK}$$

(c) Two-way shear at interior column
Using $d = 735$ mm at the critical section (Fig. 17.13a):

$$V_f = 2450 - 292 (0.5 + 0.735)^2 = 2004 \text{ kN}$$

V_f is obviously $< V_c$ at interior column since even for the exterior column $V_c = 2356$ kN.

It is seen that one-way shear controls in this example, and the effective depth required is 735 mm. Select an overall thickness of 850 mm. With a clear cover of 75 mm and assuming No. 25 bars, the depth provided is:

$$d = 850 - 75 - 25/2 = 762 \text{ mm}$$

4. Design of longitudinal flexural steel
Maximum negative moment is:

$$M_f = 1217 \text{ kN} \cdot \text{m}$$

$$K_r = \frac{M_f}{bd^2} = \frac{1217 \times 10^6}{2100 \times 762^2} = 1.00 \text{ MPa}$$

$$\rho = 0.00307 < \rho_{max} \text{ (Table 5.4)}$$

and $$< \rho_{min} = 0.0035$$

ρ_{min} controls, and $A_s = 0.0035 \times 2100 \times 762 = 5600$ mm². Provide twelve No. 25 bars divine $A_s = 6000$ mm², and a spacing of 175 mm which is satisfactory.

Development length for No. 25 top bars is (Table 9.1) $1.4 \times 850 = 1190$ mm. The length available beyond the peak moment section (see Fig. 17.13b, d) is in excess of this on both sides, and hence OK.

The positive moment in the zone near the interior column is far less than the peak negative moment and, therefore, here also the minimum reinforcement requirement governs. Hence, provide twelve No. 25 bars here also. Required development length for bottom bars is 850 mm and the length provided is in excess of this (Fig. 17.13b).

5. Design of transverse beams (Fig. 17.13e)
(a) Transverse beam under exterior column
The total load on the beam is equal to the load transferred to the column, 1350 kN. Load per unit length of beam is:

$$= \frac{1350}{2.1} = 643 \text{ kN/m}$$

Projection of beam beyond column face:

$$= \frac{2.1 - 0.4}{2} = 0.85 \text{ m}$$

Moment in transverse beam at column face is:

$$M_f = 643 \times 0.85^2/2 = 232 \text{ kN·m}$$

Effective depth for transverse beam (reinforcement placed above longitudinal reinforcement) $= 850 - 75 - 1.5 \times 25.2 = 737$ mm.
Width of beam $= w + 0.75d = 400 + 0.75 \times 737 = 953$ mm.

$$K_r = \frac{232 \times 10^6}{953 \times 737^2} = 0.449 \text{ MPa}$$

For this, Table 5.4 shows that minimum reinforcement controls.

$$A_s = \rho_{min} \times bd$$
$$= 0.0035 \times 953 \times 737 = 2458 \text{ mm}$$

Provide eight No. 20 bars giving $A_s = 2400$ mm^2.
Required development length is 510 mm.

$$\text{Available length} = 850 - \text{cover} = 775 \text{ mm} \qquad \text{OK}$$

The beam shear in the transverse beam must be checked at a distance $d = 737$ mm from the column face. This is $850 - 737 = 113$ mm from the edge of the beam. Obviously, the shear at this section is very small and hence shear stresses will be well within limits.

(b) Transverse beam under interior column

$$\text{Load per unit length} = \frac{2450}{2.1} = 1167 \text{ kN/m}$$

Beam projection from column face $= 0.8$ m

$$\text{Moment at column face, } M_f = 1167 \times \frac{0.8^2}{2} = 373 \text{ kN·m}$$

Width of beam $= 500 + 2 \times 0.75 \times 737 = 1606$ mm

$$K_r = \frac{373 \times 10^6}{1606 \times 737^2} = 0.428 \text{ MPa}$$

Again the minimum reinforcement requirement controls.

$$A_s = 0.0035 \times 1606 \times 737 = 4142 \text{ mm}^2$$

Provide fourteen No. 20 bars.

Development length available $= 800 - 75 = 725$ mm > 510 mm   OK

Once again, the critical section for beam-shear is at $d = 737$ mm from column face, which is 63 mm from the edge of the footing where shear is negligibly small.

6. Transfer of force at base of column
For exterior column, factored bearing resistance of footing is:

$$F_{br} = 0.85 \times 0.6 \times 25 \times 400^2 \times 10^{-3} = 2040 \text{ kN} > 1350 \text{ kN} \quad \text{OK}$$

At base of interior column, factored bearing resistance (even without the factor $\sqrt{A_2/A_1}$) is:

$$F_{br} = 0.85 \times 0.6 \times 25 \times 500^2 \times 10^{-3} = 3188 \text{ kN}$$

Factored compressive force $= 2450 \text{ kN} < F_{br}$ \hfill OK

In both cases, provide the minimum reinforcement as follows:

Exterior column, $0.005 \times 400^2 = 800 \text{ mm}^2$;
provide four No. 25 bars.
Interior column, $0.005 \times 500^2 = 1250 \text{ mm}^2$;
provide four No. 25 bars.

Vertical development length required for the No. 25 bars, in compression, is:

$$0.24 f_y d_b / \sqrt{f_c'} \geqslant 0.044 f_y d_b$$

The depth available in the footing is greater than this, and hence adequate.

7. Details of reinforcement
Figure 17.14 gives a sketch of the reinforcement details. Some of the longitudinal reinforcement is (arbitrarily) extended to the full length of the footing in order to: (1) provide support for the transverse reinforcement under the exterior column; and (2) to provide some reinforcement in the large (otherwise unreinforced) area of concrete at the top outside of the interior column.

PROBLEMS

---

1. For two columns in Example 17.4, assuming that there is no property line limitation, design and detail independent square footings.
2. A circular column, 650 mm in diameter, carries unfactored loads of 100 kN dead load and 800 kN live load. The column is reinforced with eight No. 30 bars, and has $f_c' = 30$ MPa and $f_y = 300$ MPa. The column centreline is 1.2 m from the property line. Allowable soil pressure at a depth of 1.2 m below grade is 240 kN/m². Design a rectangular footing for the column. For footing, assume $f_c' = 20$ MPa and $f_y = 300$ MPa.
3. The layout of an exterior and adjacent interior column in a building is shown in Fig. 17.15. Design a combined footing for the columns. Allowable soil pressure at base of footing, 2 m below grade, is 300 kN/m². Use $f_c' = 25$ MPa, $f_y = 400$ MPa.

Fig. 17.14 *Details of reinforcement, Example 17.4*

Fig. 17.15 *Problem 3*

	Ext. Col.	Int.Col.
Size	450 mm sq	600 × 450 mm
D.L	750 kN	1200 kN
L.L	600 kN	1000 kN

REFERENCES

17.1 Bowles, J.E., *Foundation Analysis and Design*, 3rd ed., McGraw-Hill Book Co., New York, 1982.

17.2 Peck, R.R., Hanson, W.E., and Thornburn, T.H., *Foundation Engineering*, 2nd ed., John Wiley & Sons, Inc., 1974.

17.3 ACI Committee 336, *Suggested Design Procedures for Combined Footings and Mats*, ACI Journal, Proc. Vol. 63, No. 10, Oct. 1966, pp. 1041–1057.

17.4 Kramisch, F., and Rogers, P., *Simplified Design of Combined Footings*, ASCE Journal, Soil Mech. Div., Proc. Vol. 87, No. SM5, Oct. 1961, pp. 19–44.

CHAPTER 18

SPECIAL PROVISIONS FOR SEISMIC DESIGN

18.1 GENERAL

The National Building Code of Canada (Refs. 18.1, 18.2) presents the minimum design requirements for earthquake-resistant design in order to ensure an acceptable level of safety. These requirements take into consideration the characteristics and probability of occurrence of earthquakes, characteristics of the structure and the foundation, and the amount of damage that is considered tolerable.

The minimum lateral seismic force at the base of a structure specified in NBC 1985 is given by Eq. 2.3. In this equation, the numerical coefficient K reflects the material and type of construction, damping, ductility and/or energy absorptive capacity of the structure. Types of construction that have ductility and damping characteristics and have been shown to perform satisfactorily in earthquakes are assigned lower values of K. Values for K are given in NBC 1985 (4.1.9.1.(7)), and range from 0.7 to 3.0. By providing ductility and energy-absorption capability in the structure, the seismic forces can be minimized. For cast-in-place, continuously reinforced concrete structures, the value of the coefficient K ranges from 0.7 to 1.30 depending on the ductility and damping that are built-in. To meet the ductility requirements associated with these reduced seismic forces, the structure must be designed and detailed to meet the special provisions specified in CSA A23.3-M84 (Section 21). For reinforced concrete structures without special provisions for ductility in the lateral load resisting system (for example, a conventional reinforced concrete building not more than 60 m in height and conforming to the requirements in the Code Sections 1–20 and 21.9 and having nominal ductility), the value for K is 1.3. The lateral force resisting systems specifically considered in the Code are ductile moment-resisting space frames and/or ductile flexural walls.

Ductility may be defined as the ability of a structure or member to undergo inelastic deformations beyond the initial yield deformation with no decrease in the load resistance. Yielding, in this context, is flexural. A *ductile moment-resisting frame* is a frame of continuous construction consisting of flexural members and columns designed and detailed to accommodate reversible lateral displacements after the formation of plastic hinges without decrease in strength. A *flexural wall* is a reinforced concrete structural wall cantilevering from the foundation and designed and detailed to be ductile and to resist seismic forces and dissipate energy through flexural yielding at one or more plastic hinges.

18.2 ROLE OF DUCTILITY IN SEISMIC DESIGN

Although a general qualitative definition of ductility was given in the preceding Section, it is difficult in practice to quantify the degree of ductility of a structure. For an under-reinforced beam section in flexure, a typical moment-curvature $(M - \psi)$ relation is shown in Fig. 18.1a. Based on the idealized $M - \psi$ relationship, a *ductility factor*, in terms of curvature, may be expressed as the ratio ψ_u/ψ_y, where ψ_y is the curvature at first yield (idealized) and ψ_u is the maximum curvature at the section. The value of ψ_u/ψ_y is a section property and can be computed relatively easily using the principles described in Chapter 4. In the case of a single beam member as shown in Fig. 18.1b, the definition of a ductility factor is more difficult, as it could be in terms of the curvature at a section, the rotation (θ) at a joint, or the displacement (Δ) (Fig. 18.1b) at a selected point. The ductility factor obtained by the three methods will differ, and furthermore, the rotations and displacements will depend on several factors such as the span dimensions, shape of moment diagram, support restraint, etc. The problem is more complex when it comes to the ductility factor of an entire structure. However, qualitatively, a ductile structure will have a load-displacement response as shown in Fig. 18.1c. This ductility is achieved by ensuring member section responses similar to that in Fig. 18.1a so that an adequate number of plastic hinges (see Section 10.7.1) would develop under extreme lateral seismic forces.

Ductility helps reduce the forces in members and dissipate energy due to seismic effects. This is illustrated schematically in Fig. 18.2, for the simple case of one cycle of static loading and unloading. For the same energy input, the maximum force, P_u, induced in a system responding elastically is greater than that in a system responding in an elastic-plastic manner. Further, the latter system dissipates a substantial part of the energy by inelastic deformation. However, the maximum displacement, δ_u, of the elastic-plastic (ductile) system is more than that in the elastic (linear) system. Thus, while ductility helps in reducing induced forces and in dissipating energy, it also demands that the deformations that the system can accommodate without damage be larger.

Forces and moments in the elements of the lateral force resisting system due to earthquake effects are usually determined by a linear elastic analysis of the system under static loads equivalent to the earthquake. The equivalent static loads are determined as prescribed in NBC 1985 (see Section 2.2 and Eq. 2.3) and the magnitude of these forces is directly proportional to the numerical coefficient K. Thus, for the same ground motion, a ductile moment-resisting space frame $(K = 0.7)$ will have induced lateral forces which are only 35 percent of those if the structure has little ductility and damping $(K = 2.0)$. However, the ductile frame has to sustain inelastic deformations which are several times the elastic deformations calculated for these reduced lateral forces. Since reinforced concrete is relatively less ductile in compression and shear, dissipation of seismic energy is best achieved by flexural yielding. Hence weakness in compression and shear in relation to flexure is avoided.

In a structure composed of a ductile frame and/or a ductile flexural wall conforming to CSA A23.3-M84 (Section 21), the inelastic response is developed by the formation of plastic hinges (flexural yielding) in the members as shown in Fig. 18.3.

Fig. 18.1 *Concept of ductility*

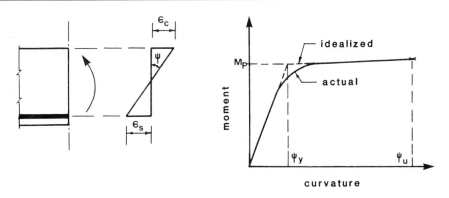

(a) Moment–curvature relation for under–reinforced section

(b) Member behaviour (schematic)

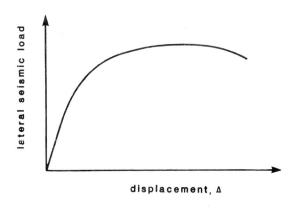

(c) Structure behaviour

Fig. 18.2 *Influence of ductile structural response on member forces and energy dissipation*

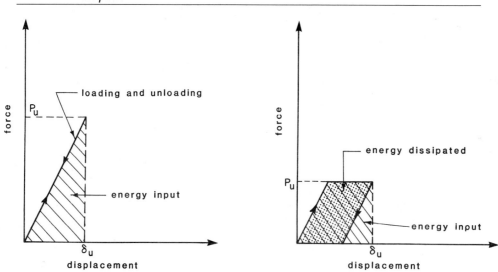

(a) Elastic response

(b) Elastic–plastic response

For ductile frames, hinges may form in the beams or in the columns as shown in Fig. 18.3a. Plastic hinges in beams have larger rotation capacities than column hinges and mechanisms involving beam hinges have larger energy-absorptive capacity on account of the larger number of beam hinges. Furthermore, in the event of residual deformation and damage, a column is more difficult to straighten and repair than a beam. Hence it is preferable to design the frame so as to form the inelastic deformations in the beams rather than the columns.

18.3 MAJOR DESIGN CONSIDERATIONS

18.3.1 GENERAL

The objective of the special seismic design provision in CSA A23.3-M84 (Section 21) is to accomplish a satisfactory inelastic structural response similar to that shown in Fig. 18.1c, by ensuring adequate ductile flexural behaviour (with ability to undergo large inelastic reversible deformations) for individual members such as beams, flexural walls and columns, and by preventing other non-ductile types of failure. This leads, in relative terms, to combinations of strong columns and weak beams, members strong in shear and weak in flexure, and stronger foundations and weaker superstructure. It may be recalled from earlier chapters (in particular, Chapters 4, 6 and 15) that the major factors which contribute to member (flexural) ductility are:

Fig. 18.3 *Formation of inelastic deformations (plastic hinges) in ductile structure*

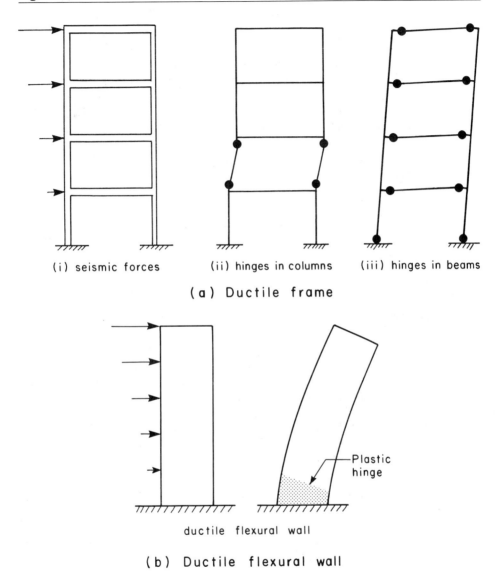

(i) seismic forces   (ii) hinges in columns   (iii) hinges in beams

(a) Ductile frame

—Plastic hinge

ductile flexural wall

(b) Ductile flexural wall

(1) a low tensile reinforcement ratio and/or the use of compression reinforcement; (2) adequate shear reinforcement to provide a shear strength in excess of the flexural strength; (3) confinement of concrete and compression reinforcement by closely spaced hoops or spirals; and (4) proper detailing with regard to anchorage of reinforcement, splicing, minimum reinforcement ratios, etc. Similarly, continuity assists in development of more inelastic response and thereby moment redistribution and energy dissipation at several hinges.

18.3.2 CODE REQUIREMENTS

The design and detailing requirements for beams, columns and their joints, forming part of the lateral (seismic) force resisting system in ductile moment-resisting frames are given in clauses 21.3, 21.4, 21.6 and 21.7 of CSA A23.3-M84, and for frame members not considered as part of the lateral force resisting system in clause 21.8. Requirements for ductile flexural walls are given in clauses 21.5 and 21.7. Clause 21.9 gives the requirements for building members with nominal ductility, so that they can be classified as cast-in-place continuously reinforced concrete construction with a *K*-value of 1.3. These Code requirements are presently considered to be the minimum requirements necessary to enable the structure to sustain a series of oscillations into the inelastic range of response without critical deterioration in strength. Some of the pertinent considerations and the application of these requirements are briefly presented below (see also Ref. 18.3).

18.3.3 STRUCTURAL FRAMING, ANALYSIS AND PROPORTIONING

In a structure designed to resist earthquake forces in a ductile manner, it is expected that a severe earthquake will induce lateral deformations and oscillations sufficient to develop reversible plastic hinges at suitable locations in the ductile frame and the ductile flexural walls. The structural scheme should be such that these plastic hinge formations in members, or failure of individual elements, will not lead to instability or progressive collapse. Building redundancy into the structural system is desirable as it assists in the development of alternative load paths thereby helping redistribution of forces, avoidance of progressive collapse and dissipation of energy.

In addition to adequate strength to resist the earthquake forces, the structure must have sufficient stiffness to limit the lateral deflection or drift. The distribution of forces to the elements of the lateral force resisting system is usually based on a linearly elastic model of the system acted on by the equivalent factored static loads. Allowing for some plastic deformations in the structural system, Ref. 18.1 recommends that the anticipated drift due to earthquake forces be taken as three times the deflection obtained from the elastic analysis under the equivalent static loads. Assuming at least all the horizontal structural members to be fully cracked is likely to result in better estimates of the possible drift than using the uncracked stiffness for all members (Ref. 18.3).

The limiting value of drift depends on the acceptable damage to the nonstructural components such as plaster, glass panels, etc., and a recommended maximum for interstorey drift is 0.005 times the storey height under the specified earthquake loads (Ref. 18.2). The effect of drift on the vertical load-carrying capacity of the lateral force resisting system must also be taken into account. The analysis must also account for the effects of members not forming part of the lateral force resisting system and of nonstructural members on the response of the structure to earthquake motion.

18.3.4 MATERIALS

The fact that yielding is permitted to occur at certain locations in the structure also necessitates the avoidance of too much over-strength at such locations. Over-strength will result in the member not yielding at the expected lateral load level as intended, and this may force adjoining elements and/or foundations to receive higher seismic load and suffer consequent damage. The plastic moment capacity of under-reinforced sections is primarily dependent on the yield stress of steel. Further, yield strength far in excess of that specified may lead to excessive shear and bond stresses as the plastic moment is developed. Hence, the Code prescribes the use of reinforcing steel conforming to CSA G30.16 for resisting earthquake-induced flexural and axial forces. This steel has more closely controlled chemical composition and mechanical properties, better elongation and a specified upper limit on the yield stress (see Table 1.3). It also has an ultimate tensile strength, f_u, which is at least 1.25 times the tensile yield strength, f_y. The length of the hinging region along the member axis and hence the inelastic rotation capacity of hinging regions is found to increase with the ratio f_u/f_y. CSA G30.12 Grade 300 and 400 reinforcement is also permitted, but only if it satisfies similar stress-strain characteristics.

High strength, low density concrete has a relatively lower ultimate compressive strain than normal density concrete, which may adversely affect the rotation capacity of flexural members. In the absence of experimental and field data on behaviour of members with structural low density concrete subjected to displacement reversals in the nonlinear range, the Code limits the maximum strength of structural low density concrete for use in seismic resistant design to 30 MPa.

18.3.5 FOUNDATION

As plastic deformations are permitted to occur at suitable locations in the structure under severe earthquakes, the maximum earthquake loads transmitted to the foundation are not governed by the factored loads, but by the loads at which actual yielding takes place in the structural elements transferring the lateral loads to the foundation, such as frame members and walls. Hence, the Code (cl. 21.2.2.3) requires that the *factored* resistance of the foundation system be sufficient to develop the nominal moment capacity (that is, the resistance computed using ϕ_c and $\phi_s = 1.0$) and the corresponding shears of the frames and walls. Under factored load conditions this means a relatively stronger foundation and weaker superstructure combination, ensuring ductile behaviour of the superstructure without serious damage to the foundation.

18.3.6 FLEXURAL MEMBERS IN DUCTILE FRAMES

As shown in Section 4.4 and Fig. 4.7, the ductility of flexural members decreases with increasing steel ratios. In order to ensure significant ductile behaviour even under reversals of displacements into the inelastic range, to avoid congestion of steel and to limit the shear stresses in beams of usual proportions, the Code limits the tensile reinforcement ratio to a maximum of 0.025. To avoid sudden brittle failure of a beam as the cracking moment of the section is reached, a minimum reinforcement ratio of $\rho = 1.4/f_y$ must be provided at both the top and bottom for the entire length of the member, with at least two bars being placed on each side. The reversible plastic hinges in flexural members usually develop near continuous supports such

Fig. 18.4 *Required factored shear resistance for beams*

(a) Loading

$1.25D + 0.7(1.5L + 1.5Q)$

(b) Plastic hinges due to seismic forces

(c) Loading on member

$$V_r, \text{required} = \frac{w_f \ell_n}{2} + \frac{M_{pr,L} + M_{pr,R}}{\ell_n}$$

(d) Shear force

$V_r, \text{required} = V_{max}$

as at beam-column connections (Fig. 18.3a(iii)). A minimum positive moment capacity must be provided at such supports. Hinging regions are subjected to very large and reversible strains and the concrete shell outside the hoop reinforcements may spall off. Splices of flexural reinforcement are not permitted in and near possible plastic hinge locations. Lap splices, when used elsewhere, must be enclosed within hoop or spiral reinforcement. Transverse reinforcement for confining concrete and to support longitudinal bars must also be provided in hinging regions, as explained below.

The bar extensions must provide for possible unpredictable shifts in the inflection points which may occur under combined loadings including seismic loads. Tests have shown that at interior beam-column connections, the continuation of top and bottom beam reinforcement through the column will develop the theoretical flexural strength of the beams. Hence such bars must be made continuous wherever possible. In other cases both top and bottom bars must be extended to the far face of the confined column core, and anchored in accordance with Code cl. 21.6.1.3. The confined core is the portion of the member cross section confined by the perimeter of the transverse reinforcement conforming to the requirement in Section 21 of the Code, measured from out-to-out of the transverse reinforcement.

Flexural members of lateral force resisting ductile frames are assumed to yield at the design earthquake. Hence the required factored shear resistance of flexural members is related to the flexural strength of the members rather than to the factored shear force obtained from the lateral-load analysis. The members must be capable of developing reversible plastic hinges without premature non-ductile failure in shear. Hence flexural members must have a factored shear resistance exceeding the maximum probable shear force in the member as plastic hinges develop at both ends of the beam as in Fig. 18.3a(iii). As these beams are expected to develop large inelastic rotations at these hinges, the strains in the flexural reinforcement at the face of the joint may be well in excess of the yield strain and may go into the strain-hardening range. To allow for this possibility of strain-hardening, and to get a conservative estimate of the factored shear resistance required, the Code specifies that the moment at the plastic hinges be taken as the "probable moment resistance," M_{pr}, which is the moment resistance of the section calculated using $1.25f_y$ as the stress in the tension reinforcement, and f'_c as the strength of the concrete, and with the material resistance factors, ϕ, taken as unity. In addition to these moments, M_{pr}, at each end of the beam, the factored gravity loads acting on the beam also cause beam shear. Thus the required factored shear resistance of the beam under the load combination $\alpha_D D + \psi(\alpha_L L + \alpha_Q Q)$ is as shown in Fig. 18.4. This over-estimation of the shear strength requirement of the beam in relation to flexural strength ensures that the member develops plastic hinges before shear failure.

Because of the alternating nature of the shear force due to seismic effects, the direction of the associated diagonal tensile stress also alternates, as shown in Fig. 18.5. Hence inclined web reinforcement, which is effective for shear in one direction,

Fig. 18.5 *Type of web reinforcement for reversed shear*

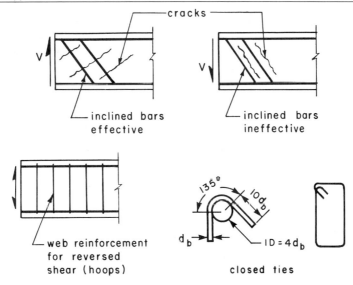

will not be effective when the shear is reversed. Web reinforcement for seismic design must be placed perpendicular to the longitudinal reinforcement, and must be provided throughout the length of the member. Since plastic hinges are likely to form at the ends of flexural members near the beam-column connections, in order to confine the concrete and to prevent buckling of the longitudinal bars, such regions must be provided with closely spaced hoops or spirals enclosing the main flexural reinforcement. At sections where inelastic rotations occur alternately in opposite directions, the reinforcement on the opposite faces also alternates in tension and compression. Furthermore, after a few cycles of inelastic rotations, the concrete cover may spall off. Hence the web reinforcement in such regions must be in the form of closed hoops with their free ends bent at 135° and adequately anchored into the core concrete (Fig. 18.5). Further, in the shear design of flexural members of lateral force resisting frames by the Simplified Method (Chapter 6), the shear resistance of concrete, V_c, is ignored.

18.3.7 COLUMNS

The Code requires that at beam-column joints, the sum of the *factored* moment resistance (with $\phi_c = 0.6$ and $\phi_s = 0.85$) of the columns framing into the joint be at least 1.1 times the sum of the *nominal* moment resistance (that is, with $\phi_c = \phi_s = 1.0$) of the beams and girders framing into the joint. The nominal resistance of the beams must be computed including the effects of the slab reinforcement within a distance of three times slab thickness on either side of the beam. This requirement, shown in Fig. 18.6a, must be satisfied for beam moments acting in either direction. The requirement is intended to ensure the formation of plastic hinges in the beams rather than in the columns (strong column–weak beam combination). Note that the column resistances are computed conservatively (with inclusion of ϕ-factors) compared to the beam resistances. Again, lap splices are not permitted near the ends of the column where spalling of the concrete shell is likely to occur.

The factored shear resistance of columns must be greater than the larger of (1) the shear force due to the factored loads, and (2) the shear force in the column corresponding to the development of the *probable* moment resistances in the beams framing into it. The latter condition and corresponding column shears are shown in Fig. 18.6b. This allows for possible over-strength of the beams and thereby ensures avoidance of shear failure of columns.

Apart from the normal transverse reinforcement required for columns, special transverse confining reinforcement must be provided near the joints (high moment regions) and on both sides of any section where flexural yielding may occur due to inelastic lateral displacement of the frame. In columns which may develop plastic hinges, such as at the base of the building, and columns supporting discontinued stiff members like walls or trusses, this special transverse confining reinforcement is required for the full height of the column.

18.3.8 JOINTS IN DUCTILE FRAMES

The design requirements for beam-column joints in ductile frames given in the Code (cl. 21.6) are based on Ref. 18.4. The joints are zones of heavy concentration of

Fig. 18.6 Column resistance requirements

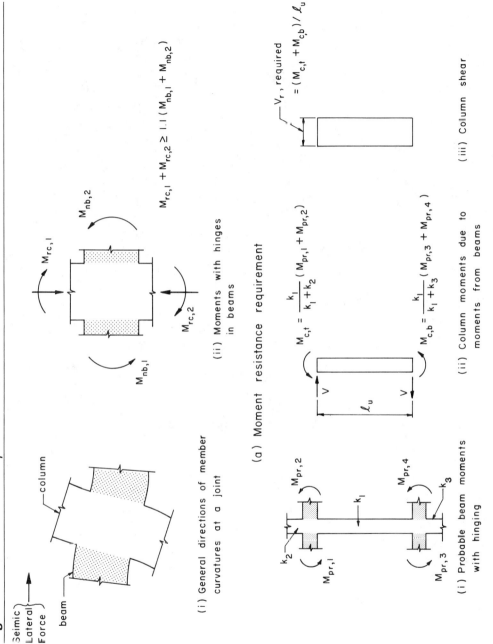

(i) General directions of member curvatures at a joint

(ii) Moments with hinges in beams

$$M_{rc,1} + M_{rc,2} \geq 1.1 (M_{nb,1} + M_{nb,2})$$

(a) Moment resistance requirement

$$M_{c,t} = \frac{k_1}{k_1 + k_2} (M_{pr,1} + M_{pr,2})$$

$$M_{c,b} = \frac{k_1}{k_1 + k_3} (M_{pr,3} + M_{pr,4})$$

(i) Probable beam moments with hinging

(ii) Column moments due to moments from beams

$$V_r, \text{ required} = (M_{c,t} + M_{c,b}) / \ell_u$$

(iii) Column shear

(b) Shear resistance requirement

stresses. Joints must have adequate strength to facilitate the development of large inelastic reversible rotations in the beams at the face of the joints. The flexural stresses transferred to the faces of the joint by the beams generate high shear forces within the joint on horizontal shear planes as shown in Fig. 18.7. As already mentioned in Section 18.3.6, to allow for the very large strains in and the possibility of strain-hardening of flexural reinforcement in the beams at the faces of the joint, the Code specifies that the factored shear force in the joint be determined assuming a tensile stress of $1.25 f_y$ in the beam reinforcement (Fig. 18.7).

Tests have indicated that the shear strength of joints is not sensitive to the amount of shear reinforcement and may be taken as a function of only the compressive strength of concrete (Ref. 18.3). On this basis, the Code recommends that the factored shear resistance of the joint be taken as:

1) for externally confined joints,

$$V_r = 2.4 \, \lambda \phi_c \, \sqrt{f_c'} \, A_j$$

Fig. 18.7 *Horizontal factored shearing forces on an interior joint*

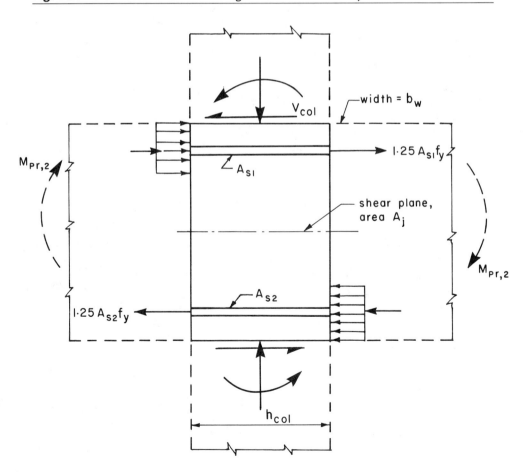

and 2) for others,

$$V_r = 1.8 \, \lambda \phi_c \, \sqrt{f_c'} \, A_j$$

where A_j is the minimum cross-sectional area within a joint in a plane parallel to the axis of the reinforcement generating the shear in the joint (equal to the lesser of A_g of the column or $2b_w h_{col}$); and a joint is considered to be externally confined, if members frame into all vertical faces of the joint and if at least three-fourths of each face of the joint is covered by the framing member.

Irrespective of the magnitude of the shear force in the joint, transverse confining (hoop) reinforcement must be provided through the joint around the column reinforcement. Where the joint is externally confined, the amount of transverse hoop reinforcement may be reduced in that portion of the joint. At a column where the width of beam exceeds the width of column, the longitudinal beam reinforcements located outside the column core must also be provided with lateral confinement, either by beams framing into the joint laterally or by transverse hoop reinforcement through the joint. Since flexural reinforcement is expected to develop stresses of the order of $1.25f_y$ at the face of the joint and is furthermore subjected to stress reversals, the development length requirements within the joint are also very critical.

The connection zone is an area of high concentration of reinforcement, consisting of beam bars, column bars and connection hoops. Many failures in structures under seismic loading can be traced to poor detailing, particularly of beam-column joints. Extreme care is needed in detailing the reinforcement at the connection in order to provide for proper stress transfer and to avoid congestion and placing difficulties for both reinforcement and concrete.

18.3.9 DUCTILE FLEXURAL WALLS

Ductile flexural walls are vertical members cantilevering from the foundation and subjected to bending moment, horizontal shear and axial load. In ductile walls, inelastic deformation usually involves plastic hinges at the base, as shown in Fig. 18.3b and/or at locations of abrupt changes in strength and stiffness of the lateral load resisting system. In the absence of such abrupt changes, plastic hinges will form within the lower half of the wall height. Unlike a beam, a flexural wall is relatively thin and deep, and is subjected to substantial axial forces. These must be designed as axially loaded beams capable of forming reversible plastic hinges with sufficient rotation capacity. The wall being thin, the regions of high compressive strain must be checked for instability (buckling). Stability of the compressed zone can be improved by thickening this zone or by providing lateral support such as by a flange or cross wall. Flanged walls, having I or channel sections, have higher bending resistance and ductility, and lesser likelihood of lateral buckling at the compressed flange, and may be used in elements such as elevator shafts.

CSA A23.3-M84 (21.5) gives minimum design and detailing requirements for ductile flexural walls. The wall is reinforced with distributed reinforcement provided in both vertical and horizontal directions, and with vertical reinforcement concentrated at each end of the wall. The distributed reinforcement provides the shear resistance, controls the cracking, inhibits local breakdown in the event of severe cracking during an earthquake, and also resists temperature and shrinkage stresses. The concentrated and the distributed vertical reinforcement provides the flexural and axial load resistance.

The concentrated vertical flexural reinforcement near the ends must be tied together by transverse ties, as in a column, in order to provide confinement of the concrete and to ensure the yielding, without buckling, of the flexural bars on the compression side, when a plastic hinge is formed. A minimum ductility is ensured by placing an upper limit on the depth to the neutral axis.

Because of minimum reinforcement requirements, and the possibility of strain-hardening in the steel, etc., the actual moment capacity of a flexural wall may be greater than the design moment, M_f. To prevent a premature brittle shear failure of the wall before the development of its full plastic resistance in bending, the Code requires the wall to be designed to resist a shear force greater than the shear that is present when the wall develops a plastic hinge with the probable moment resistance, M_{pr}. M_{pr} is computed using $1.25f_y$ as the stress in the tension reinforcement and f_c' as the strength of the concrete and without the application of any resistance factors. Because of possible severe shear cracking under cyclic loads, the shear carried by the concrete (cl. 11.3) is neglected in the region of the plastic hinge.

18.4 RECENT ADVANCES

The purpose of this chapter is to explain the background to the seismic design provisions of the Code. A detailed discussion of seismic design of reinforced concrete structures is beyond the scope of this book. Rapid advances are being made in this area and recent publications on this topic (for example, Refs. 18.5–18.10) may be consulted for more details.

REFERENCES

18.1   *National Building Code of Canada 1985*, Part 4: Structural Design, National Research Council of Canada, Ottawa, 1985, pp. 151–180.

18.2   *The Supplement to the National Building Code of Canada — 1985*, Ch. 4: Commentaries on Part 4 of NBC 1985, Commentary J, National Research Council of Canada, Ottawa, 1985, pp. 221–241.

18.3   ACI Committee 318, *Commentary on Building Code Requirements for Reinforced Concrete* ACI 318–83, American Concrete Institute, Detroit, 3rd printing, 1983.

18.4   ACI-ASCE Committee 352, *Recommendations for Design of Beam-Column Joints in Monolithic Reinforced Concrete Structures*, (ACI 352 R-76, Reaffirmed 1981), American Concrete Institute, Detroit, 1976, 19 pp.

18.5   Arnold, C. and Reitherman, R., *Building Configuration and Seismic Design*, John Wiley & Sons, Inc., New York, N.Y., 1982, 296 pp.

18.6   Applied Technology Council, *Tentative Provisions for the Development of Seismic Regulations for Buildings*, ATC 3-06, National Bureau of Standards, special publication 510, U.S. Government Printing Office, Washington, D.C., 1978, 514 pp.

18.7   Dowrick, D.J., *Earthquake Resistant Design*, John Wiley & Sons, Chichester, U.K., 1977.

18.8   *Reinforced Concrete Structures in Seismic Zones*, ACI Publication SP-53, American Concrete Institute, Detroit, 1977.

18.9   *Reinforced Concrete Structures Subjected to Wind and Earthquake Forces*, ACI Publication SP-63, American Concrete Institute, Detroit, 1980.

18.10   *Earthquake Effects on Reinforced Concrete Structure*, ACI Publication SP-84, American Concrete Institute, Detroit, 1985.

APPENDIX A

BOOKKEEPING

Bookkeeping refers to the art or practice of keeping a systematic record of design correspondence, data, and calculations. Because the design of a project may span several months or even years, and may be worked on by many designers, and because calculations may cover thousands of pages, a good system of bookkeeping is mandatory if designers are to work efficiently. Benefits derived from the use of a sound, uniform bookkeeping system are as follows:

1. Assists the designers to proceed with the design in a systematic and previously proven manner.
2. Ensures the organization and effective indexing of design calculations.
3. Facilitates the checking of calculations.
4. Facilitates the transfer of information to drawings.
5. Minimizes repetition of calculations and wasted design effort.

Some basic guidelines or bookkeeping hints which should be followed by designers in designing a structure such as a multistorey building are the following:

1. Do all calculations on special paper provided for the purpose.
2. Ensure that every page is indexed and that all data such as project number, date, designer, etc., is recorded. Each major division should have a "General" subdivision which summarizes codes, design and analysis methods, loads, material properties and other data required for the design of the components or elements in that division.
3. Keep calculations in three ring binders. Binders ensure that pages do not get lost and facilitate amending of design files.
4. Where possible, calculations should be arranged in tabular format.
5. Reference all calculations to the applicable code provisions, design aids, or technical publications.
6. For each structural system or component being designed, clearly define the geometry by referring to sketches or drawings or provide a sketch showing span, cross section, tributary width, stiffness, restraint conditions, openings or variations in bent widths.
7. For each structural system or component being designed, identify the source of all loading data and illustrate the loads by sketches if appropriate.
8. The derivation of all forces should be shown on a sketch, if necessary.
9. Cross reference calculations if previous results are used.
10. Identify each member being designed by grid references or member marks.

FIG. A-1 *Format A for recording vertical information on the "Model Building"*

Floor Level	Fl-Fl. Height (M)	Elev. (M)	Ht.Ab. Ground (M)	Ht.Ab. Bsmt (M)
PARA'PT ROOF	2.40	174.39	97.78	99.56
25	5.66	168.73	92.12	93.90
24	3.60	165.13	88.52	90.30
23	3.60	161.53	84.92	86.70
22	3.60	157.93	81.32	83.10
21	3.60	154.33	77.72	79.50
20	3.60	150.73	74.12	75.90
19	3.60	147.13	70.52	72.30
18	3.60	143.53	66.92	68.70
17	3.60	139.93	63.32	65.10
16	3.60	136.33	59.72	61.50
15	3.60	132.73	56.12	57.90
14	3.60	129.13	52.52	54.30
13	3.60	125.53	48.92	50.70
12	3.60	121.93	45.32	47.10
11	3.60	118.33	41.72	43.50
10	3.60	114.73	38.12	39.90
9	3.60	111.13	34.52	36.30
8	3.60	107.53	30.92	32.70
7	3.60	103.93	27.32	29.10
6	3.60	100.33	23.72	25.50
5	4.50	95.83	19.22	21.00
4	4.50	91.33	14.72	16.50
3	4.50	86.83	10.22	12.00
2	4.50	82.33	5.72	7.50
1	4.50	77.83	1.22	3.00
BSMT. CAISSON	3.00	74.83		

11. Separate by underlining or highlighting final component dimensions, reinforcement, etc., and transfer applicable results (walls, beams, column, footing designs), directly to schedules.

12. Utilize reusable formats to present dates, identify members, or display results. A vertical format, (Fig. A-1), which represents, to scale, the number of floors, floor to floor heights, elevations, and distances above the basement floor is convenient for:

    a. Column rundowns.
    b. Column designs.
    c. Lateral force calculations.
    d. Manual lateral load analyses.

e.   Displaying, graphically, column loads.

f.   Displaying, graphically, lateral forces.

A horizontal format, (Fig. A.2), which consists of a floor plan containing only grid lines and basic geometry is convenient for showing:

a.   Floor loads on loading drawings.

b.   Slab and beam identification marks.

c.   Column identifications groupings, and tributary areas.

d.   Wall and core tributary areas.

e.   Slab reinforcement results.

f.   Openings, framing details.

g.   Foundation layouts.

Both these formats are used repeatedly throughout the various stages of design.

Fig. A.2   *Format B for recording horizontal information on the "Model Building"*

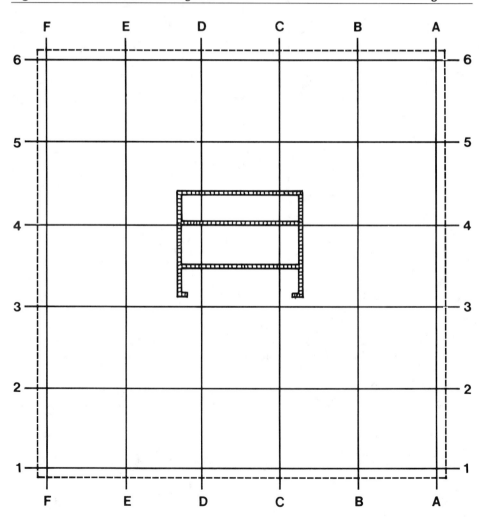

APPENDIX B

DESIGN AIDS

INDEX